X-Ray Diffraction Methods
in Polymer Science

X-Ray Diffraction Methods
in Polymer Science

LEROY E. ALEXANDER

Professor of Chemistry and Senior Fellow
Mellon Institute
Carnegie-Mellon University

ROBERT E. KRIEGER PUBLISHING COMPANY
HUNTINGTON, NEW YORK
1979

Original Edition 1969
Reprint Edition 1979 with corrections

Printed and Published by
ROBERT E. KRIEGER PUBLISHING COMPANY, INC.
645 NEW YORK AVENUE
HUNTINGTON, NEW YORK 11743

Copyright © 1969 by
JOHN WILEY & SONS INC.
Reprinted by arrangement

Printed in the United States of America

Library of Congress Cataloging in Publication Data

Alexander, Leroy Elbert, 1910-
    X-ray diffraction methods in polymer science.

    Reprint of the ed. published by Wiley-Interscience, New York
in the Wiley series on the science and technology of materials.
    Includes bibliographies.
    1. X-ray crystallography.   2. Polymers and polymeriza-
tion.   3. X-rays—Diffraction.   I. Title.
[QD945.A365 1979]   547'.84   78-23488
ISBN 0-88275-801-2

To Harold Klug

Preface

This monograph is directed rather specifically to the needs of scientists engaged in the study of "linear" synthetic macromolecules, including many of some technological interest. Only secondary attention is given to polymers of biological significance — first, because of obvious spatial limitations and, second, in recognition of the availability of a number of excellent reviews treating applications of x-rays to this important class of polymers.

On confronting the manifold and somewhat bewildering array of x-ray diffraction effects generated by polymers, the neophyte is likely to experience an initial frustration, which, if he is persistent, is shortly displaced by the pleasant discovery that much can be learned about this intriguing class of compounds with only a modest grasp of diffraction principles complemented by his own enlarging experience. On the other hand, the more quantitative procedures require certainly not less than the theoretical background presented in this monograph. The amazing and truly infinite variety exhibited by the diffraction patterns of polymers is, of course, to a large degree a direct consequence of the numerous types of lattice disorders and distortions to which they are prone and which they manifest with a ubiquity and intensity unique among chemical compounds.

As has been true of so many scientific books, a prime motive for writing the present monograph has been to assemble in one place much of the relevant information that was previously distributed widely in the literature. As the writing progressed, the increasingly rapid proliferation of the literature made it all too evident that even though the topics treated were selected on the basis of their apparent current importance to polymer analysis anything like a complete or comprehensive treatise was out of the question. I am all too aware of numerous deserving subjects that have had to be excluded.

The introduction to x-ray diffraction by polymers, given in Chapter 1, is not intended to be self-sufficient. Instead, its purpose is to map out for the reader those theoretical areas of which an understanding in some depth is prerequisite to a full exploitation of the power of diffraction methods in analytical and structural investigations. To assist the serious reader in acquiring this background, numerous references are cited. At the same time the present text emphasizes the theoretical expressions that have practical application but, for the most part, omits the mathematical derivations of these expressions. Considerable effort has been devoted to making clear the precise significance of these mathematical formulas, in particular by means of numerical examples. Thus the dominant emphasis in this monograph is rather more on quantitative than qualitative interpretations of experimental data. It is hoped that this aspect will prove not unwelcome in consideration of the more qualitative emphasis that has characterized most of the recent review papers dealing with the structural and morphological interpretation of polymer diffraction effects.

This acknowledged stress on quantitative modes of interpretation calls for a note of caution. Analytical and structural properties that are expressible as numerical parameters necessarily have correspondences to preconceived hypothetical models, and accordingly such numerical characterizations of polymers reveal the true states of the specimens in question only insofar as the models are faithful images of the actual systems (which are, correctly speaking, unknown). The uncertainties attending such interpretations are compounded just at present by the extremely rapid evolution in our concepts of the physical constitution of polymers. Hence the structural and morphological interpretations that the reader will find distributed through the text represent only an imperfect and momentary glimpse of a rapidly changing picture. Put another way, the structural and morphological concepts contained in the text represent some of the facets of current thinking in the field but possess a very uncertain degree of "permanent" value. On the more positive side it may reasonably be hoped that the diffraction methods presented will prove more durable than the interpretations that can be derived from them.

As another result of the considerable attention paid to quantitative interpretations, greater than expected difficulties have been encountered with nomenclature, particularly in relation to the mathematical statements of the theory of small-angle x-ray scattering (Chapter 5). Although a strenuous effort has been made to adhere to the preferred usage of the related literature, numerous departures therefrom have been required to maintain a reasonably systematic and self-consistent scheme of symbolism. In this connection the alphabetical listing of symbols employed in Chapter 5, which appears in Table 5-7, should prove helpful to the reader both within that chapter and elsewhere in the book.

Compared with *X-Ray Diffraction Procedures* (John Wiley and Sons, New York, 1954, coauthored with Harold P. Klug), the present monograph pays less attention to details of experimental technique. An attempt has been made to direct the reader to the necessary information of this kind by means of numerous references throughout the body of the text to the "Procedures" as well as to other appropriate works.

In the course of the preparation of the manuscript I have been extraordinarily fortunate in obtaining the cooperation in various capacities of many prominent polymer scientists as well as experts in x-ray diffraction and other special fields. In fact, the contributions of a number of them have been so important that to a very real degree the finished monograph embodies something of the characteristic stamp of each. For critically reading portions of the text it is a pleasure to thank S. Castellano, E. S. Clark, C. R. Desper, P. J. Flory, T. G. Fox, P. H. Hermans, H. P. Klug, O. Kratky, J. H. Magill, M. E. Milberg, F. A. Miller, R. L. Miller, S. S. Pollack, W. Ruland, W. O. Statton, R. S. Stein, R. F. Stewart, A. Turner-Jones, and Z. W. Wilchinsky. For supplying unpublished illustrative figures or data I am greatly indebted to E. S. Clark, O. Kratky, J. H. Magill, S. S. Pollack, A. Turner-Jones, J. W. Visser, J. H. Wakelin, Jr., Z. W. Wilchinsky, and a number of commercial suppliers of instrumentation, all of whose contributions are acknowledged in the appropriate figure or table captions. I also gratefully acknowledge the permission of R. L. Miller to include a revised edition of his well-known crystallographic tables as Appendix 3, and I wish to thank R. Bonart for help with the theory of diffraction by paracrystalline structures.

Deserving of special mention are the sustained interest, unfailing encouragement, and general moral support supplied by so many individuals, of whom I wish to name in particular E. S. Clark, H. P. Klug, O. Kratky, J. H. Magill, and W. O. Statton. I am grateful to Dr. Paul C. Cross, Vice-President for Research of Carnegie-Mellon University, for his support of the project and for providing the excellent facilities of Mellon Institute for the preparation of the manuscript. I am indebted also to Professor P. M. de Wolff of the Technological University of Delft, who was my host during a sabbatical year in which the initial outlines of the project took shape. I am appreciative of the expert photographic assistance of E. G. Beggs and Miss Carol Dovedot and acknowledge with thanks the assistance of Mrs. Maureen Buccigrossi with the figure copy and the invaluable help of my wife Eleanor with the figure copy and proofing and in numerous other capacities. In conclusion I acknowledge my special indebtedness and express my deep appreciation to Miss Mary L. Condy not only for her outstanding services in the typing and production of the manuscript but also for her unfailing willingness to go beyond the call of duty to achieve excellence, an increasingly scarce commodity in today's world.

Pittsburgh, Pennsylvania   LEROY E. ALEXANDER
May 1969

Contents

1   Introduction to X-Ray Diffraction by Polymers   1

  1-1   Ideal Crystals   5
        1-1.1   Crystal Systems and Space Lattices   5
        1-1.2   Crystallographic Planes   9

  1-2   Crystallization in Polymers   12
        1-2.1   Importance of Stereoregularity   12
        1-2.2   Tacticity   15
        1-2.3   Polymer Single Crystals   16
        1-2.4   Spherulitic Crystallization   17
        1-2.5   Intermediate Crystalline Structures   20
        1-2.6   Texture in Polymers   21

  1-3   X-Rays and Matter   26
        1-3.1   Properties of X-rays   26
        1-3.2   Scattering of X-rays by Matter   29

  1-4   Wide-Angle Diffraction by Polymers   33
        1-4.1   The Reciprocal Lattice   33
        1-4.2   The Geometry of Diffraction   35
        1-4.3   The Intensity of Diffraction   39
        1-4.4   Amorphous Specimens   43
        1-4.5   Randomly Oriented Microcrystalline
                Specimens   45
        1-4.6   Preferentially Oriented Microcrystalline
                Specimens   50

  1-5   Small-Angle Scattering by Polymers   62

  General References   64

  Specific References   64

2   Instrumentation   **66**

2-1   Principal Diffraction Techniques   66
      2-1.1   Photographic versus Counter Technique   66

2-2   Preparation and Mounting of the Specimen   68
      2-2.1   Geometry of Transmission Techniques   68
      2-2.2   Unoriented Specimens   72
      2-2.3   Oriented Specimens   73
      2-2.4   Mounting Fibers   76
      2-2.5   Specimen Arrangements for Diffractometry   77

2-3   Photographic Techniques   82
      2-3.1   Choice of Photographic Film   82
      2-3.2   Microdensitometry of X-Ray Photographs   82
      2-3.3   Wide-Angle Cameras   85
      2-3.4   Crystal-Monochromatized Diffraction
              Photographs   97
      2-3.5   Small-Angle Scattering   102

2-4   Diffractometry   113
      2-4.1   General Features of Counter Diffractometers   113
      2-4.2   Counters for X-Ray Measurements   115
      2-4.3   Monochromatization in Diffractometry   119

2-5   The Optical Diffractometer   126
      2-5.1   Principles   127
      2-5.2   The Design of Taylor and Lipson   128
      2-5.3   Preparation of the Masks   130

General References   133

Specific References   133

3   Degree of Crystallinity in Polymers   **137**

3-1   The Concept of Crystallinity   137

3-2   Basic Principles of the X-Ray Method   138

3-3   Experimental Procedures   143
      3-3.1   Crystallinity from the Integrated Crystalline
              and Amorphous Scatter with Allowance for
              Lattice Imperfections (Method of Ruland)   143

3-3.2 Crystallinity from the Intensity of Amorphous Scatter, Lattice Imperfections Not Considered 151

3-3.3 Crystallinity from the Intensities of Both the Crystalline and Amorphous Scatter, Regression Curve, Lattice Distortions Not Considered 165

3-3.4 Crystallinity from Differential Intensity Measurements, Crystallinity Index 176

3-4 Crystallinity from Other Physical Measurements 189
3-4.1 Density 189
3-4.2 Infrared Absorption 192
3-4.3 Nuclear Magnetic Resonance (NMR) 194

General References 196

Specific References 197

4 Preferred Orientation in Polymers 200

4-1 Diffraction Patterns of Oriented Polymers 199

4-2 Representation of Preferred Orientation 202
4-2.1 Classification of Orientation Modes 206

4-3 Preparation of Pole Figures 209
4-3.1 Transmission Technique, Sheet Specimen 215
4-3.2 Reflection Technique, Sheet Specimen 223
4-3.3 Other Techniques 227
4-3.4 Special Instrumentation 229
4-3.5 Illustrative Preparation of a Pole Figure 234

4-4 Analytical Description of Preferred Orientation 241
4-4.1 Axial Orientation 241
4-4.2 Higher Orientation Modes 252
4-4.3 Other Methods of X-Ray Analysis 259

4-5 Optical Birefringence 268
4-5.1 Axial Orientation 269
4-5.2 Biaxial Orientation 270
4-5.3 Experimental Measurement of Birefringence 271
4-5.4 Illustrative Applications 272

General References 277

Specific References 277

5 Macrostructure From Small-Angle Scattering   **280**

5-1   Preliminary Considerations   283
      5-1.1   Angular Nomenclature   283
      5-1.2   The Intensity of Small-Angle Scattering   283
      5-1.3   Absolute Intensity Measurements   285
      5-1.4   Slit versus Pinhole Collimation   286
      5-1.5   The Correction of Collimation Errors   287

5-2   Generalized Systems   290
      5-2.1   Scattering Power   290
      5-2.2   The Invariant, Q or $\widetilde{Q}$ 291

5-3   Systems of Two Phases, Each of Uniform Electron
      Density   293
      5-3.1   Scattering Power   293
      5-3.2   Rule of Constancy of $s^4 I(s)$ or $s^3 \widetilde{I}(s)$ 294
      5-3.3   Specific Inner Surface, O_s 294
      5-3.4   Correlation Function   296

5-4   Dilute Particulate Systems — Two Phases, Each of
      Uniform Electron Density   297
      5-4.1   General Principles   297
      5-4.2   Globular Particles   300
      5-4.3   Rodlike Particles   304
      5-4.4   Lamellar Particles   306
      5-4.5   Linear Polymers in Solution   307
      5-4.6   Polydispersity   313
      5-4.7   Globular Particles: Example of Lysozyme   314
      5-4.8   Rodlike Particles: Example of Air-Swollen
              Cellulose   319

5-5   Dense Systems   327
      5-5.1   Diffuse Scattering and Voids   327
      5-5.2   Discrete Interferences   332

General References   353

Specific References   353

6 Microstructure From Wide-Angle Diffraction   **357**

6-1   Molecular Structure   357
      6-1.1   Configuration and Tacticity   358
      6-1.2   Conformational Nomenclature   360
      6-1.3   Factors Determining Chain Conformation   366
      6-1.4   Randomly Oriented Systems   368

6-1.5   Cylindrically Symmetrical Systems   376
6-1.6   Helical Molecules   388

6-2   Crystal Structure   405
6-2.1   Crystal Structure of Poly(ethylene adipate)   408
6-2.2   Further Illustrative Structure   419

General References   419

Specific References   420

7   Lattice Distortions and Crystallite Size   423

7-1   The Scherrer Equation   423

7-2   The Treatment of Hosemann   424
7-2.1   Concept of the Paracrystal   424
7-2.2   Distortions of the First and Second Kinds 425
7-2.3   Separation of Size and Distortion Parameters   427
7-2.4   Applications to Linear Polyethylene   429

7-3   The Analysis of Buchanan and Miller   437
7-3.1   Method of Fourier Transforms   437
7-3.2   Method of Integral Breadths   438
7-3.3   Application to Isotactic Polystyrene   440
7-3.4   General Conclusions   451

General References   452

Specific References   452

Appendix 1   Fourier and Fourier-Bessel Transforms   455
Appendix 2   Fourier Analysis of Reflection Profiles by the
              Method of Stokes   467
Appendix 3   Crystallographic Data for Various Polymers   473
Appendix 4   Mass Absorption Coefficients   524
Appendix 5   X-Ray Wavelengths   529
Appendix 6   Atomic Scattering Factors   533
Appendix 7   Filters for X-Rays   540
Appendix 8   Formulas for Calculating Interplanar Spacings,
              d_{hkl} 542
Appendix 9   Condensed Tables of Lorentz and Polarization
              Factors   544
Appendix 10   Atomic Weights   548
Appendix 11   Miscellaneous   Physical   and   Numerical
              Constants   552
Name Index   553
Subject Index   557

X-Ray Diffraction Methods
in Polymer Science

1

Introduction to X-Ray Diffraction by Polymers

The special usefulness of x-ray diffraction in the study of solid substances lies in its ability to distinguish ordered from disordered states. Thus it is common knowledge that liquids or glasses (so-called amorphous materials) produce x-ray patterns of a diffuse nature consisting of one or more halos, whereas well-crystallized substances yield patterns of numerous sharp circles or spots. (See Figure 1-1.) This capacity to reveal the degree of ordering in solid substances makes x-ray diffraction well suited to the investigation of polymers, which, compared with simple chemical compounds, either crystallize more poorly or can be crystallized only under carefully controlled experimental conditions.

The ions or molecules of simple inorganic and organic compounds, when condensed from the melt or solution, tend to arrange themselves in a regular manner in three dimensions, forming crystals. Thus consider for simplicity the regular arrangement of molecules in two dimensions represented in Figure 1-2. We see that this arrangement can be naturally related to a plane lattice of points defined by two intersecting sets of parallel lines. The entire hypothetical two-dimensional crystal is composed of a very large number of identical parallelograms with edges a and b, which may be spoken of as unit cells of the structure. Each cell contains one complete sample of the entire structure and gives all the information needed to correctly visualize the overall arrangement. The extension of these ideas to three-dimensional structures is not difficult.

Compared with the customary ease with which simple ionic and molecular compounds can be crystallized, the great length and often

1

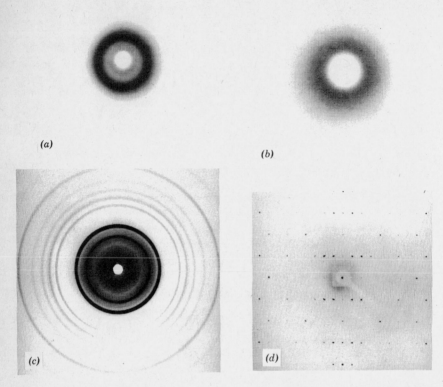

Figure 1-1 X-Ray diffraction patterns of crystalline and noncrystalline substances: (a) liquid (benzene); (b) glass (microscope slide); (c) polycrystalline powder (quartz); (d) single crystal (zinc dimethyldithiocarbamate).

appreciable irregularity of polymer molecules present formidable hindrances to the attainment of the three-dimensional regularity that is required for the growth of a crystal. Thus it may be supposed that a great deal of "thermal wriggling" of the long molecules is required to disentangle regions of the disordered liquid phase and permit ordering to occur. Some polymers cannot be crystallized; thus they consist solely of molecules in a disordered array and their x-ray patterns resemble those of glasses and liquids. (See Figure 1-1.) Other polymers are partially crystalline; their x-ray patterns contain relatively sharp crystalline spots as well as one or more amorphous halos. A mixed pattern of this type is also generated by certain elastomers that undergo partial crystallization on being stretched. (See Figure 1-3.)

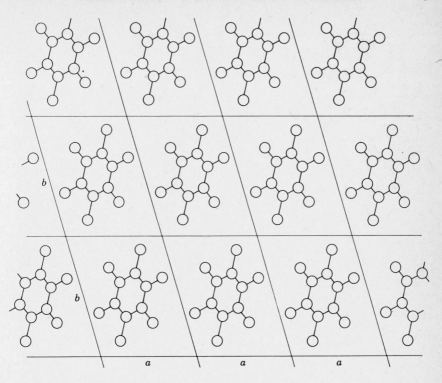

Figure 1-2 Plane molecules arranged on a two-dimensional lattice.

Until rather recently such a mixed pattern was commonly regarded as evidence that the specimen consists of a three-dimensional mosaic of crystallites and intervening amorphous regions (*the two-phase concept*). Although this picture is undoubtedly basically correct for many partially crystalline polymers, especially when the degree of crystallinity is low, it is now recognized that similar sharp and diffuse diffraction effects can be generated simultaneously by a crystal lattice containing certain kinds of defects[1–3] (*the crystal-defect concept*). In fact the occurrence in polymer crystals of slip and lattice dislocations that are similar to the same phenomena in metals has been conclusively demonstrated[4,5].

Whether we favor the *two-phase* concept, the theory of the *crystal defect* state, or some combination of both views—or whether we adopt some other perspective—the accumulated evidence makes it clear that crystallization in polymers is less perfect by an order of magnitude or so than crystallization in simple ionic and molecular

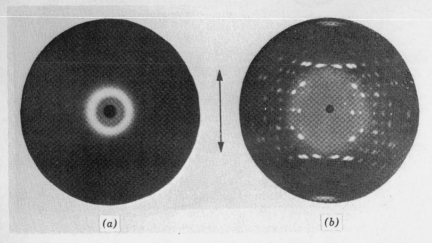

Figure 1-3 X-Ray diffraction photographs of (a) unstretched and (b) stretched polyiso-butene. Fiber axis vertical. [Fuller, Frosch, and Pape, *J. Am. Chem. Soc.*, 62, 1905 (1940). Copyright 1940 by the American Chemical Society. Reprinted by permission of the copyright owner.]

compounds. Even in such relatively ideal crystalline substances imperfections of several kinds are almost invariably present, whereas in the crystalline lattices of polymers disorders and distortions on a much larger scale tend to prevail, and this, combined with the unusual morphological features of polymer crystals, requires that a wide variety of x-ray diffraction procedures be employed. In any given situation it may be difficult to select the optimal technique in advance since the choice rests on so many factors; for example, the stereo-chemistry of the polymer molecule, the thermal and mechanical pre-treatment, the amount of material available, the method of preparing the specimen for x-ray analysis, and the presence or absence of preferred orientation. Thus the most fruitful application of x-ray crystallographic techniques to polymers demands more than ordinary versatility and ingenuity.

It is expedient to preface the subject of x-ray diffraction by poly-mers with preliminary descriptions of (a) ideal crystals, (b) the nature of crystallization in polymers, (c) polymer morphology, and (d) physical aspects of x-rays and their interaction with matter. Some familiarity with these essentials is most certainly prerequisite to an understanding of x-ray diffraction by polymers. Although most readers possess a prior knowledge of some of these topics, it cannot be assumed that they are versed in them all. Therefore we shall first sketch this essential background, at the same time suggesting some general references at the end of this chapter for those who may wish to improve their grasp of particular topics. In the last portion of the chapter we shall then present a general introduction to diffraction by polymers.

1-1 IDEAL CRYSTALS

1-1.1 Crystal Systems and Space Lattices

Depending on the configuration of the molecular chains and their arrangement with respect to one another, any given structure can be assigned to one of six crystal systems characterized by the ratios of the lengths of its unit-cell edges—a, b, and c—and the three angles defined by those edges—$\alpha = \angle b,c$, $\beta = \angle a,c$, $\gamma = a,b$. The characteristics of the six crystal systems are given in Table 1-1.† In addition the given structure can be shown to possess one of the 14 Bravais space lattices, of which one is triclinic, two are monoclinic, four orthorhombic, two tetragonal, two hexagonal, and three cubic. The unit cells of the 14 space lattices are shown in Figure 1-4. It will be noticed that cells numbered 1, 2, 4, 8, 9, 10, and 12 are simple (primitive) cells, whereas all the others contain one or three additional lattice points, which are situated either at the centers of one or all of the faces (end- or face-centered lattices) or at the geometrical center of the cell (body-centered lattice).

Table 1-1   The Six Crystal Systems

System	Axial Translations (unit-cell constants)	Angles between Crystal Axes (degrees)
Triclinic	$a \neq b \neq c$	$\alpha \neq \beta \neq \gamma \neq 90$
Monoclinic	$a \neq b \neq c$	$\beta \neq 90$
		$\alpha = \gamma = 90$
Orthorhombic	$a \neq b \neq c$	$\alpha = \beta = \gamma = 90$
Tetragonal	$a = b \neq c$	$\alpha = \beta = \gamma = 90$
Hexagonal:		
Hexagonal division	$a = b \neq c$	$\alpha = \beta = 90$
		$\gamma = 120$
Rhombohedral		
division	$a = b = c$	$\alpha = \beta = \gamma \neq 90$
Cubic	$a = b = c$	$\alpha = \beta = \gamma = 90$

†For the sake of simplicity we have chosen to disregard the internal symmetry of the structure, the ultimate criterion on which the choice of the crystal system depends. The edges and angles of the unit cell suffice for the present treatment.

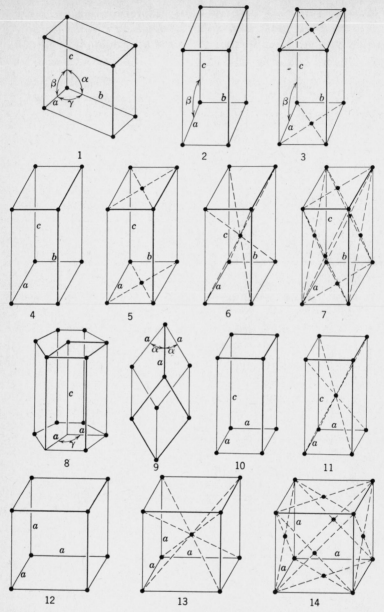

Figure 1-4 Unit cells of the fourteen space lattices: 1—triclinic; 2 and 3—monoclinic; 4, 5, 6, and 7—orthorhombic; 8—hexagonal; 9—rhombohedral; 10 and 11—tetragonal; 12, 13, and 14—cubic. (Klug and Alexander, *X-Ray Diffraction Procedures*, Wiley, New York, 1954.)

As an example consider the arrangement of molecular chains in an ideal crystal of nylon 66, poly(hexamethylene adipamide), as shown in Figure 1-5 [6]. Because of the very great, and indefinite, length of any single molecule, it is manifestly not practicable to select a unit cell in a fashion analogous to that of Figure 1-2 with one, two, or more discrete molecules. However, if we neglect the terminal features of the long molecular chains, it is possible to define a smaller unit cell that contains a discrete number of monomer units. Figure 1-5 shows how we can select a triclinic unit cell that contains the substance of one hexamethylene-adipamide unit, $-\mathrm{HN(CH_2)_6NHCO(CH_2)_4CO-}$, as was first shown by Bunn and Garner [6]. We may note that, although one monomer unit lies parallel to each c-edge of the unit cell, any given monomer is shared by the four unit cells that come in contact at this edge, with the result that each cell actually contains the substance of only one monomer unit. We should also observe that one monomer unit may be regarded as associated with each lattice point. Thus one

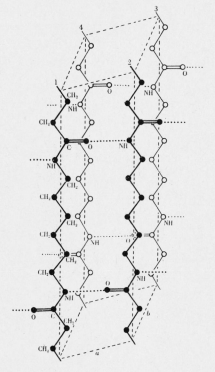

Figure 1-5 Arrangement of molecular chains in nylon 66, poly(hexamethylene adipamide). (Bunn and Garner [6].)

monomer unit is the characteristic *motif* of the nylon 66 structure; the entire ideal crystal consists of a three-dimensional assemblage of a very large number of duplicates of this motif. If the lattice translations (also edge lengths of the unit cell) are a, b, and c, and if we choose to place the first motif at the origin of the triclinic coordinate system, another motif will be found at any point ja, mb, nc, where j, m, n are integers ranging from zero to very large positive or negative values.

Figure 1-6 shows how the molecular arrangement in *cis* 1,4-polybutadiene[7] is based on a c-centered monoclinic unit cell (No. 3 of Figure 1-4) with $a = 4.60$, $b = 9.50$, $c = 8.60$ Å, and the angle $\beta = 109°$. Molecular segments consisting of two monomer residues,

$$-CH_2CH=CHCH_2CH_2CH=CHCH_2-$$

traverse the cell in the c-direction, one along each c-edge and one at the center of the c-face. When the structure is viewed in its entirety, each cell is seen to contain the substance of two chain segments, or four monomer units. In the *cis* 1,4-polybutadiene structure the

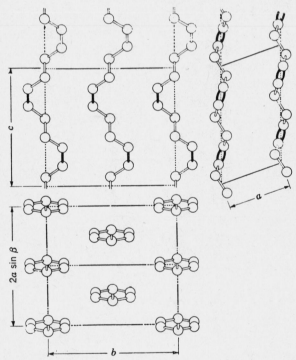

Figure 1-6   Structure of *cis* 1,4-polybutadiene as seen in three projections. Monoclinic unit cell with $a = 4.60$, $b = 9.50$, $c = 8.60$ Å; $\beta = 109°$. (Natta and Corradini[7].)

characteristic motif is a two-monomer segment. If once again we place the first motif at the origin of the crystallographic coordinate system (this time monoclinic), the c-centered lattice will duplicate the motif not only at all points ja, mb, nc as before but also at all points $(j+\frac{1}{2})a$, $(m+\frac{1}{2})b$, nc.

Before concluding these remarks on the relationship between actual structures and their space lattices we must emphasize that the choice of origin in the structure is perfectly arbitrary; that is, this choice may be made so that the lattice points either do or do not coincide with any component atom of the motif. Nevertheless, in the solution of crystal-structure problems one choice of origin may be most advantageous mathematically because of the particular symmetry possessed by the motif.

1-1.2 Crystallographic Planes

Consider again briefly a two-dimensional, or plane, lattice such as that illustrated in Figure 1-2. It is shown in Figure 1-7 how it is possible to draw straight lines that pass through the lattice points in various directions, such as A, B, C, and D. Although rows of points may be discerned in many directions, the density of points in some lines is greater than in others. As shown below, lattice lines may be designated by their intercepts on the X- and Y-axes.

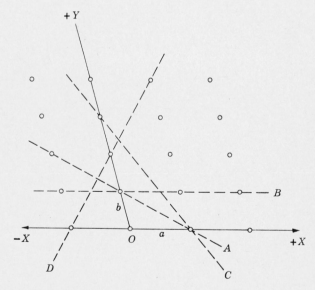

Figure 1-7 Various linear rows of points in a two-dimensional lattice.

Lattice Line	Intercepts	Designation
A	$1a$ and $1b$	1, 1
B	∞ and $1b$	∞, 1
C	$1a$ and $3b$	1, 3
D	$-1a$ and $2b$	$-1, 2$

Similarly it is possible in the three-dimensional lattice of an actual crystal to lay out crystallographic planes that pass through the lattice points in various orientations. The three planes—ABC, DEF, and $DE\infty$—in Figure 1-8 intersect the axes of an orthorhombic space lattice in different ways. As in the two-dimensional case, we may choose to denote each plane by its intercepts on the coordinate axes. Thus plane ABC with intercepts $1a$, $2b$, and $1c$ may be designated 121, DEF may be designated 243, and $DE\infty$ may be designated 24∞. For mathematical operations, however, it is preferable to adopt the Miller crystallographic indices, which are obtained from the unit intercepts by taking their reciprocals and clearing of fractions. As outlined in Table 1-2, the planes ABC, DEF, and $DE\infty$ can then be assigned the Miller indices (212), (634), and (210). The plane faces of crystals are crystallographic planes of relatively simple indices; the simpler their indices, the more densely populated with lattice points are the planes—and the more likely are these planes to be observed as actual crystal faces; for example, crystal faces of indices (100), (110), (111), and (102) are much more likely to be developed than those of indices (135), (029), or (347).

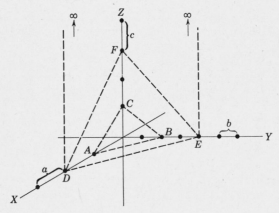

Figure 1-8 Three planes intersecting the axes of an orthorhombic space lattice in different ways. (Klug and Alexander, *X-Ray Diffraction Procedures*, Wiley, New York, 1954.)

**Table 1-2   Derivation of Miller Indices from
the Plane Intercepts[a]**

Plane	Intercepts	Reciprocals of Intercept Multiples	Cleared of Fractions	Miller Indices (hkl)
ABC	$1a, 2b, 1c$	$\frac{1}{1}\frac{1}{2}\frac{1}{1}$	212	(212)
DEF	$2a, 4b, 3c$	$\frac{1}{2}\frac{1}{4}\frac{1}{3}$	634	(634)
DE∞	$2a, 4b, \infty c$	$\frac{1}{2}\frac{1}{4}\frac{1}{\infty}$	210	(210)

[a]Data from Figure 1-8.

Another important feature of crystallographic planes is illustrated by means of the two-dimensional lattice net in Figure 1-9. This net may be regarded as one ab section of an orthorhombic space lattice. From the diagram it is clear that any given set of Miller indices hkl refers not to just one plane but rather to a family of parallel planes that permeates the entire crystal and is characterized by a particular interplanar spacing $d(hkl)$. It can also be seen from Figure 1-9 how planes of simple indices, such as (100) and (010), are more

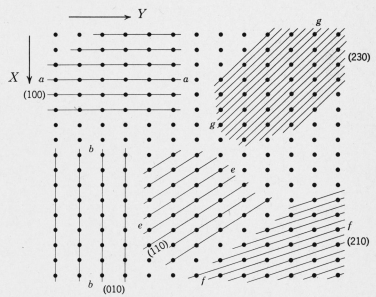

Figure 1-9   Plane traces on an ab net of an orthorhombic space lattice. (Klug and Alexander, *X-Ray Diffraction Procedures*, Wiley, New York, 1954.)

densely populated with lattice points than planes of more complicated indices—for example, (210) and (230).

For an orthorhombic crystal with axial translations (unit-cell dimensions) a, b, c the interplanar spacing for planes (hkl) is given by

$$d(hkl) = \left[\left(\frac{h}{a}\right)^2 + \left(\frac{k}{b}\right)^2 + \left(\frac{l}{c}\right)^2\right]^{-1/2}. \tag{1-1}$$

For a cubic crystal $a = b = c$, and the expression simplifies to

$$d(hkl) = \frac{a}{\sqrt{h^2 + k^2 + l^2}}. \tag{1-2}$$

The formulas for the interplanar spacing appropriate to the several crystal systems are given in Appendix 8.

It is important to be able to sketch readily a plane of any given Miller indices on a set of coordinate axes; for example, planes (110) have intercepts $1a$, $1b$, ∞ on the X, Y, Z axes, respectively, as depicted in Figure 1-10. The figure also shows planes (220) with intercepts $\frac{1}{2}a$, $\frac{1}{2}b$, ∞; planes (020) with intercepts ∞, $\frac{1}{2}b$, ∞; as well as planes (321), ($\overline{1}12$), and ($\overline{1}\overline{1}1$). The interplanar spacing $d(hkl)$ for planes of given indices is, then, simply the perpendicular distance between the plane as drawn and the origin of coordinates.

Note that in Figure 1-10 planes (110) and (220) are parallel and therefore have identical slopes with respect to the coordinate axes; however, the interplanar spacing of the family (220) is one-half that of planes (110). Hence, although either set of indices is an equally valid description of a visible crystal face, the two sets of planes clearly bear different relationships to the internal atomic structure of the crystal. Similarly planes (330) are parallel to planes (110), but $d(330) = \frac{1}{3}d(110)$; planes (440) are parallel to planes (110), but $d(440) = \frac{1}{4}d(110)$; and so on. Similar considerations hold for planes of other indices.

1-2 CRYSTALLIZATION IN POLYMERS

1-2.1 Importance of Stereoregularity

In the light of the foregoing discussion of crystal systems and space lattices we may rightly deduce that stereoregularity of a polymer chain and its substituent groups is a prerequisite for crystallization; for example, with reference to the poly(vinyl chloride) chain pictured in Figure 1-11, if successive chlorine atoms were randomly placed in left- and right-handed positions, it would be impossible for neighboring molecules to fit together in a regular manner. However, in

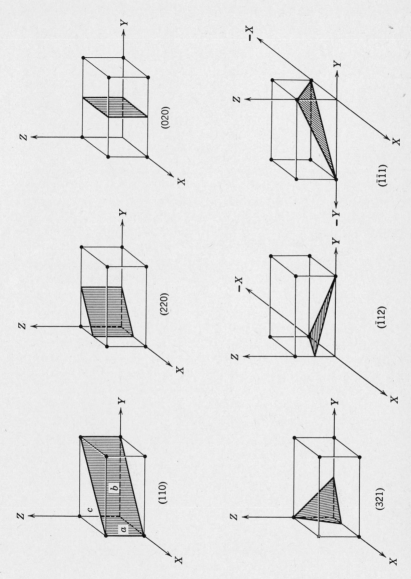

Figure 1-10 Diagrams of planes having various Miller indices in a unit cell with edges a, b, c.

13

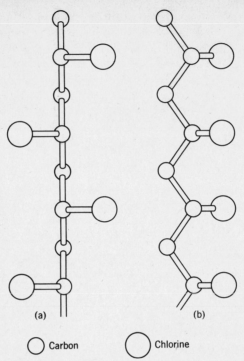

(a)   (b)

⬤ Carbon   ⬤ Chlorine

Figure 1-11 Molecular configuration of poly(vinyl chloride) as projected (*a*) parallel and (*b*) perpendicular to the plane of the zigzag carbon chain. For simplicity the hydrogen atoms are not shown.

poly(vinyl chloride) polymerized with the aid of stereospecific catalysts left and right placements alternate with sufficient regularity to permit a considerable portion of the polymer to crystallize.

Prior to the development of stereospecific catalysis [8] many synthetic polymers could not be crystallized at all or only very imperfectly; for example, poly(vinyl chloride), polystyrene, polypropene, poly-1-butene, and poly(methyl methacrylate). Nevertheless, without the benefit of stereospecific polymerization techniques a number of natural and synthetic linear polymers did crystallize; thus the following may be cited: gutta percha, natural rubber, polyisobutene, polyethylene, poly(vinyl alcohol), and poly(vinylidene chloride). These polymers crystallize because identical or spatially equivalent side groups are attached in a highly regular manner to carbon atoms of the main chain. In both gutta percha and natural rubber every fourth carbon along the chain bears a methyl side group and is doubly bonded to the carbon atom following it in the chain; however, the configuration across the double bond is *trans* in gutta percha but *cis* in natural rubber. In polyethylene each carbon atom has two attached hydrogen atoms and is equivalent to every other carbon atom. In polyisobutene every second carbon atom has two attached methyl groups, whereas in poly(vinylidene chloride) every second carbon atom has two chlorine atoms. In poly(vinyl alcohol) every second chain carbon atom bears a hydrogen atom and a hydroxyl group, but crystallization is nevertheless possible because the sizes of these two substituents are both small and not too different in relation to the stereochemical features of the main carbon chain.

Crystallization of linear polymers is hindered not only by insufficient stereoregularity along the molecular chains but also by the occurrence of branching or cross-linking, which, if very extensive, gives rise to three-dimensionally bonded, glasslike systems in which independently mobile molecules are not present. Examples of such three-dimensional structures are phenol-formaldehyde resins.

1-2.2 Tacticity

The widespread application of stereospecific catalysis subsequent to Natta's announcement[8] in 1955 has resulted in a multiplicity of new crystalline polymers, including in particular stereoregular polymers of numerous olefins and diolefins. Natta originated a system of nomenclature for the characterization of ideally stereoregular polymers, which may be briefly explained at this point. Strictly speaking, a linear polymer may be termed *tactic* only when all the monomer units that comprise the main chain have the same configuration or when opposing configurations follow each other in an orderly way. Furthermore, we are principally concerned with polymers that have repeating monomer units solely in head-to-tail arrangements, such as

$$-CH_2CHCH_2CHCH_2CHCH_2CHCH_2CH-$$
$$\begin{array}{ccccc} | & | & | & | & | \\ R & R & R & R & R \end{array}$$

rather than, for example,

$$-CH_2CHCHCH_2CH_2CHCH_2CHCH_2CH_2CH-.$$
$$\begin{array}{cccccc} | & | & | & | & | \\ R & R & R & R & R \end{array}$$

Tactic polymers are in turn divided into *isotactic* and *syndiotactic* classes. In *isotactic* polymers the substituent groups are all in right-handed or all in left-handed positions, whereas in *syndiotactic* polymers the R groups alternate regularly in right- and left-handed positions. These arrangements may be represented as follows:

$$
\begin{array}{cccc}
\underset{C}{H_2} & \underset{C}{H_2} & \underset{C}{H_2} & \underset{C}{H_2}
\end{array}
$$

isotactic

$$
\begin{array}{cccc}
C & C & C & C \\
R\;H & R\;H & R\;H & R\;H
\end{array}
$$

$$
\begin{array}{cccc}
\underset{C}{H_2} & \underset{C}{H_2} & \underset{C}{H_2} & \underset{C}{H_2}
\end{array}
$$

syndiotactic

$$
\begin{array}{cccc}
C & C & C & C \\
R\;H & H\;R & R\;H & H\;R
\end{array}
$$

All polymers that possess a considerable degree of randomness in right- and left-handed placements are referred to as *atactic*. The subject of tacticity is treated in more detail in Chapter 6.

1-2.3 Polymer Single Crystals

Although the now-celebrated announcements in 1957 of the growth of single crystals of polyethylene[9–11] from dilute solutions were generally regarded as an epochal discovery, earlier reports of equivalent and closely related phenomena went largely unnoticed; for example, we may cite the conclusion of Storks[12,13] that chain folding must occur on the basis of electron-microscopic examination of oriented crystallites in thin polymer films, Schlesinger and Leeper's[14] growth of large single crystals of α-gutta from dilute solution, and Jaccodine's[15] discovery of spiral growth steps in polyethylene obtained from hot xylene solution.

The characteristics of polyethylene single crystals shown in the electron micrograph of Figure 1-12 are also typical of single crystals of other synthetic polymers obtained subsequently by various investigators. The crystals are thin and lamellar in habit and nearly always too small to be visible in the light-optical microscope. Furthermore, the thickness of a lamella is surprisingly uniform from one polymer to another, being of the order of 100 Å, whereas the dimensions in the plane of the lamella are very much larger — typically of the order of several microns[9–11]. Electron-diffraction studies have proved that the molecular chains are normal to the lamellae, which has the consequence that the molecular chains must be folded back and forth between the two principal faces of each lamella[12, 13].

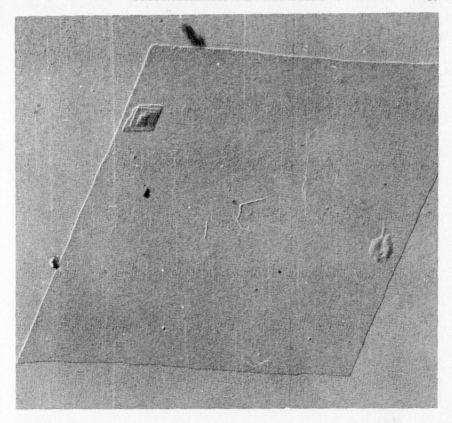

Figure 1-12   Electron micrograph of a polyethylene single crystal. Magnification 20,000×. (Courtesy of E. I. du Pont de Nemours and Company, Inc; micrograph by P. H. Geil.)

1-2.4   Spherulitic Crystallization

Sufficiently rapid growth from the melt or, in many cases, from relatively concentrated solutions results in the formation of dendritic crystalline structures termed *spherulites.* In thick specimens they tend to assume a spherical shape, whereas in thin films they appear circular. Optical and x-ray studies have shown that (a) the entire structure of a given spherulite originates from a single primary nucleus and (b) crystallization proceeds continuously from the nucleus outward. Formation of spherulites is now recognized to be the most characteristic mode of crystallization from the melt.

Although the crystallization within a spherulite is a continuous phenomenon, uniformity of crystalline orientation is not preserved. Rather, the original lamella that grew from the primary nucleus branches at points of imperfection (dislocations) in the crystal lattice to yield secondary lamellae, which bend and twist as they grow and in turn branch to produce tertiary lamellae, and so on. The overall result is the production of a multitude of fibrous lamellae that develop radially outward in all directions from the nucleus until stopped by the impingement of neighboring spherulites on one another. This phase of growth is often referred to as *primary crystallization*. Next further, or *secondary*, *crystallization* proceeds within each spherulite, transforming a portion of the remaining amorphous material. X-ray studies demonstrate that a particular crystalline direction— for example, the *a*- or *b*-axis, coincides with the axis of each of the primary radial lamellae of the spherulite. Furthermore, x-ray and optical studies lead to the general conclusion that the molecular chains tend to orient themselves tangentially in spherulites. By convention the *c*-axis is most commonly chosen as the direction parallel to the molecular chains.

Spherulites in thin films can be advantageously viewed in the light-optical microscope with polarized light. Figure 1-13 shows the typical appearance of spherulites in a thin polymer film as observed

Figure 1-13   Spherulites in a thin film of nylon 77 observed between crossed Polaroids. Magnification 150×. (Courtesy of J. H. Magill.)

between crossed Polaroids (equivalent to Nicol prisms). The simple black interference Maltese cross is characteristic of a radial structure in which the radiating crystalline lamellae *do not* twist. When these crystalline elements *do* undergo a rather regular twisting in the process of crystallizing outward from the original nucleus, concentric extinction bands are superposed on the pattern of the Maltese cross, as illustrated in Figure 1-14. When primary crystallization within a polymer specimen has become complete, the spherulites contact one another along interfaces resembling domain boundaries in metals, as may be seen in Figure 1-15, and the entire assemblage resembles a jigsaw puzzle composed of polygons of various shapes.

The primary nuclei from which polymer spherulites grow exhibit statistically random orientations, and this diversity of orientations is further compounded by the typical radial morphology with which growth of the primary lamellar fibers proceeds. Additional multiplicity of orientations is introduced by the phenomena of branching, bending, and twisting during the growth of crystallites. The reader will appreciate therefore that a spherulitically crystallized polymer specimen tends to display in toto statistically random crystalline orientations.

Figure 1-14 Spherulite of nylon 77 showing banded interference pattern observed between crossed Polaroids. Magnification 650×. (Courtesy of J. H. Magill.)

Figure 1-15 Typical appearance of adjoining sphèrulites when primary crystallization is complete. Low-molecular-weight L-poly(propylene oxide) crystallized at 30°C after fusion at 100°C. (Courtesy of J. H. Magill.)

1-2.5 Intermediate Crystalline Structures

Under appropriate conditions it is possible to obtain a wide range of crystalline structures that are intermediate in complexity between single crystals and spherulites. Slow crystallization from more concentrated solutions, especially at higher temperatures, leads to the formation of crystalline objects termed *axialites*. Their appearance varies greatly with the angle of view[16, 17]. From some directions axialites appear to be sheaflike splaying aggregates of crystal lamellae—as illustrated in Figures 1-16 and 1-17*a*, *c*, and *h*— whereas from other viewing angles they exhibit polygonal forms and other characteristics of polymer single crystals, although in a thick, compact version, as shown in Figure 1-17*g*, *i*, *a'*, *e'*.

When spherulitic growth is greatly restricted in one dimension, as when crystallization takes place in a very thin molten film, platelike structures of various types may appear[18–20]. A thick aggregate of single-crystal lamellae having a common nucleus and orientation, and presenting a polygonal appearance when viewed from at least one direction, is referred to as a *hedrite*[18]. Figure 1-18 shows a square hedrite of poly(ethylene oxide) growing in a cooling molten film. Compared with polymer single crystals hedrites are very thick, consisting of 100 or more lamellae as a rule. In thin films of nylon

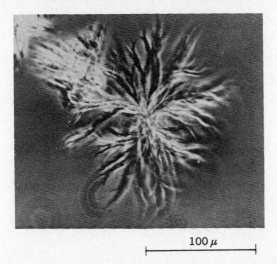

100 μ

Figure 1-16 View of axialite showing sheaflike splaying aggregation of crystal lamellae. (Bassett, Keller, and Mitsuhashi[16].)

56 spherulitic arrays of wedge-shaped platelets have been observed (see Figure 1-19)[20]. Like a single crystal, each individual platelet displays extinction at intervals of 90° on being rotated between crossed polarizers. Such an array of platelets constitutes a striking example of a crystal habit that is intermediate between a single crystal and a spherulite.

Evidence from electron-optical and x-ray studies increasingly supports the conclusion that a lamellar habit of crystallization is very general among polymers, and not alone in special structures such as hedrites and axialites but in bulk and molded materials and drawn fibers as well. This in turn has the important implication that chain folding is a prevalent, if not invariable, feature of polymer crystallization. This subject is discussed further in Chapter 5.

1-2.6 Texture in Polymers

An important aspect of the microstructure of polymers concerns their *texture*; that is, the arrangement of the crystalline regions, or crystallites, with respect to each other and to possible intervening amorphous regions or voids[21]. Texture is most significant in polymers wherein the molecular chains and crystallites have a preferred orientation as the result of mechanical treatment such as drawing or rolling—or sometimes as the result of highly directional temperature gradients during crystallization.

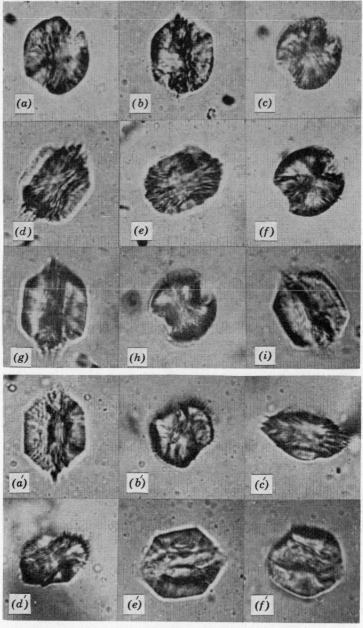

Figure 1-17 View of axialites from two tilting series: $(a) - (i)$, one axialite viewed from nine different directions; $(a') - (f')$, another axialite viewed from six different directions. Magnification 750×. (Bassett, Keller, and Mitsuhashi[16].)

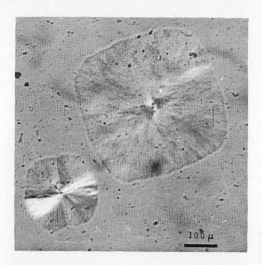

Figure 1-18 Square hedrite of poly(ethylene oxide) growing in a thin molten film as it cooled to room temperature. The smaller structure is a birefringent sheaf with non-birefringent growth occurring normal to the axis of the sheaf. Crossed Polaroids, first-order red plate. (Geil[18].)

Figure 1-19 Spherulitic array of wedge-shaped platelets in thin film of nylon 56. (Magill[20].)

Most highly drawn, crystalline polymer fibers show characteristic x-ray interferences parallel to the fiber axis at very small scattering angles, as illustrated in Figure 1-20a. An explanation for this phenomenon, advanced some time ago by Hess and Kiessig[22], is that the molecular chains are grouped in bundles, or fibrils, oriented approximately parallel to the fiber axis and that crystalline and amorphous

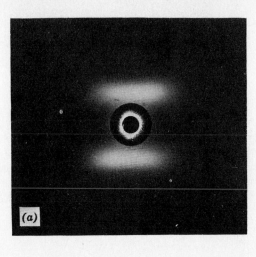

Figure 1-20 (a) Small-angle-scattering diagram of hot-drawn linear polyethylene (110°C). Nickel-filtered CuKα radiation; specimen-to-film distance 400 mm. [Bonart and Hosemann, *Kolloid-Z., Z. Polymere*, **186**, 16 (1962). Reprinted by permission of Dr. Dietrich Steinkopff, Darmstadt, Germany, *Kolloid-Z., Z. Polymere*.] (b) Small-angle quadrant diagram of a poly(ethylene terephthalate) fiber. Photograph by Schmidt. [Stuart, *Ann. N.Y. Acad. Sci.*, **83**, 3 (1959). Copyright, The New York Academy of Sciences, 1959; reprinted by permission.]

regions alternate in a rather regular manner along the fibril axis. (See Figure 1-21*a*.) Sometimes the small-angle reflections are observed to divide into two diagonal (quadrant) reflections (Figure 1-20*b*), which can be explained on the basis of the Hess and Kiessig model if adjoining fibrils pack together so that the amorphous regions in a given fibril are adjacent to the crystalline regions in its neighboring fibrils, as sketched in Figure 1-21*b*. A three-dimensional model that incorporates these features is shown in Figure 1-22.

The mounting evidence for the widespread occurrence of lamellar crystallization and chain folding in polymers has tended to cast doubt on the validity of the Hess and Kiessig model. In particular it is hard to reconcile these new concepts with the propagation of highly extended molecular chains through successive crystalline and amorphous regions, an essential feature of the Hess-Kiessig model. Nevertheless, for drawn but unannealed fibers there is some correlation between the intensity of the meridional reflection at small angles and the degree of crystallinity as determined with x-rays, in apparent support of this model. On the other hand, subsequent annealing produces an anomalously large increase in the intensity of the meridional reflection from the standpoint of the model of Hess and Kiessig. This phenomenon is discussed in Section 5-5.2.

In fiber textures, such as described above, the crystallographic axis coincident with the molecular chains, usually designated *c*, tends to orient parallel to the fiber axis, while *a* and *b* distribute themselves randomly about this direction. More complex textures characterize

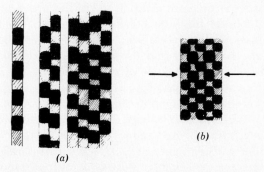

Figure 1-21 (*a*) Polymer fibrils composed of alternating crystalline and amorphous regions. (*b*) Proposed arrangement of crystalline and amorphous regions in neighboring fibrils to explain quadrant diffraction diagrams. [Bonart and Hosemann, *Kolloid-Z.*, *Z. Polymere*, **186**, 16 (1962). Reprinted by permission of Dr. Dietrich Steinkopff, Darmstadt, Germany, *Kolloid-Z., Z. Polymere.*]

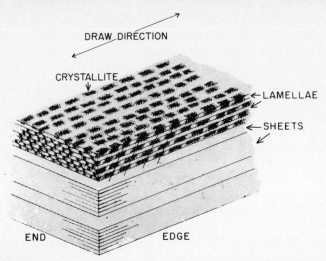

Figure 1-22 Sketch of possible polymer texture in oriented poly(ethylene tereph-thalate). [Statton and Godard, *J. Appl. Phys.*, **28**, 1111 (1957).]

the higher degrees of preferred orientation that may be induced by extrusion of films or by a combination of drawing and rolling. This subject is explored in some detail in Chapter 4.

1-3   X-RAYS AND MATTER

1-3.1   Properties of X-Rays

The x-rays of importance in diffraction studies are generated when high-energy electrons impinge on a metal target, the more useful being iron, copper, and molybdenum. At a sufficiently high x-ray-tube voltage the x-ray beam is found to possess a spectrum of the kind shown in Figure 1-23. It consists of two parts, a broad band of continuously varying wavelengths (the *continuous*, or *white*, radiation) with a broad energy maximum in the neighborhood of 0.5 Å and a *characteristic* spectrum superposed on it consisting of two lines, $K\alpha$ and $K\beta$, the $K\alpha$ line being the more intense. Actually the $K\alpha$ line is itself a doublet with components $K\alpha_1$ and $K\alpha_2$ separated by such a small wavelength interval that in most x-ray diffraction patterns they are resolved only at large values of 2θ. In studies of polymers the separation of the $K\alpha$ doublet is seldom seen.

On traversing matter direct or diffracted x-rays are absorbed to an extent that is determined by the absorption coefficient of the sub-

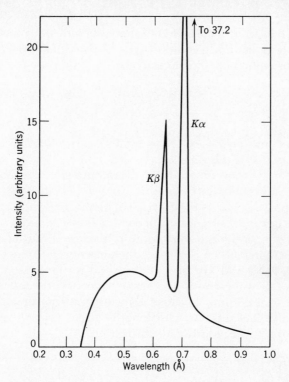

Figure 1-23 Intensity curve for x-rays from a molybdenum target operated at 35 kVp. [After Ulrey, *Phys. Rev.*, **11**, 401 (1918).]

stance concerned for x-rays of the given wavelength. The linear absorption coefficient μ is the natural logarithm of the ratio of the x-ray intensity incident on a specimen to the intensity of the ray after it has penetrated 1 cm of the material. Thus if I_0 and I are the incident and transmitted intensities, respectively, and t is the thickness of the specimen in centimeters,

$$\frac{I}{I_0} = e^{-\mu t}, \tag{1-3}$$

and

$$\frac{1}{t} \log_e \frac{I_0}{I} = \mu \qquad \text{(in cm}^{-1} \text{ units)}. \tag{1-4}$$

The mass absorption coefficient is obtained by dividing the linear coefficient μ by the density of the material, ρ. The quantity μ/ρ is independent of the physical and chemical state of the substance, which is not true of μ. Appendix 4 lists the numerical values of μ/ρ in units of cm²/g for most of the elements and the commonly used x-ray wavelengths. For a compound or mixture of known composition the mass absorption coefficient can be calculated from the expression

$$\frac{\mu}{\rho} = \sum_i w_i \left(\frac{\mu}{\rho}\right)_i, \tag{1-5}$$

in which w_i and $(\mu/\rho)_i$ are the weight fraction and mass absorption coefficient, respectively, of component i.

Except for certain pronounced discontinuities, the absorption of any given substance for x-rays increases rapidly with increasing wavelength, as shown in Figure 1-24. The discontinuity shown in the absorption curve of zirconium occurs at 0.689 Å and is referred to as the K-absorption edge of zirconium because x-rays of wavelength just smaller than 0.689 Å possess sufficient energy to eject an electron from the K-shell and are highly absorbed in the process. Contrariwise, x-rays of wavelength just larger than 0.689 Å are only weakly absorbed by zirconium because they are not energetic enough to eject K-shell electrons. At wavelengths larger than those portrayed in the figure three additional absorption edges—L_I, L_{II}, and L_{III}—

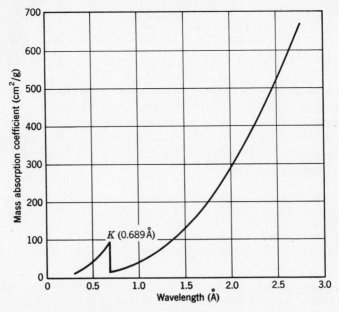

Figure 1-24   Absorption coefficient of zirconium for x-rays as a function of wavelength.

occur, corresponding to the critical energy levels associated with the ejection of electrons from the three L subshells. For other elements similar absorption spectra are found, except that the positions of the absorption edges shift toward lower wavelengths with increasing atomic number.

Returning now to the emission spectrum, described briefly in the first paragraph of this section, we see that it is not possible to obtain a beam of pure $K\alpha$ x-rays directly from the x-ray tube. However, the direct x-ray beam can be passed through an appropriate absorbing filter that will reduce the intensity of the $K\beta$ line to a negligible level while depressing the $K\alpha$ line only moderately. For a particular x-ray target element the proper β-filter will evidently be the element whose K-absorption edge falls between the wavelengths of the $K\alpha$ and $K\beta$ lines of the target. Thus zirconium, with its K-absorption edge at 0.689 Å, is an appropriate filter material for MoK radiation with $K\alpha = 0.71069$ and $K\beta_1 = 0.63225$ Å. Niobium, with its K-absorption edge at 0.65291 Å, can also be used. Nickel-filtered CuKα radiation (1.54178 Å) is the most generally useful radiation for polymer studies. When a β-filter is employed, the sharp diffraction effects are produced only by the $K\alpha$ x-rays, while the residual continuous spectrum has the effect of generating a diffuse background. Appendix 5 presents a tabulation of the wavelengths of the K-emission spectra and absorption edges of the elements. Appendix 7 lists the proper choice of filter material and the thickness required for effective $K\beta$ filtration of the x-radiations commonly used in diffraction studies.

Radiation that is much more nearly monochromatic can be obtained by inserting a crystal monochromator in the direct x-ray beam and reflecting the $K\alpha$ line from a set of planes of the monochromator on the specimen to be studied. All wavelengths other than $K\alpha$ are rigorously eliminated except for relatively weak harmonics; that is, wavelengths in the continuous spectrum equal to submultiples of the $K\alpha$ wavelength. For example, CuKα radiation, with $\lambda = 1.54178$ Å, might be accompanied by the harmonics

$$\frac{\lambda}{2} = 0.77089 \text{ Å} \quad \text{and} \quad \frac{\lambda}{3} = 0.51393 \text{ Å}.$$

1-3.2 Scattering of X-Rays by Matter

Coherent and Incoherent Scattering. When x-rays of a given wavelength impinge on the electrons in an atom, the electrons become secondary emitters of x-rays and we speak of the atoms as "scattering" the incident x-rays. Two kinds of scattering take place —

coherent, or *unmodified*, and *incoherent*, or *modified*. The major part of the energy of scattering goes into *coherent* scattering, which can be utilized for structural studies because it occurs without change of wavelength and, moreover, without loss of phase relationship between the incident and scattered rays. Because the electrons occupy an appreciable volume about the atomic nucleus, instead of being concentrated at a single point, the x-rays scattered coherently by the various electrons within one atom are more or less out of phase with each other to an extent that depends on the size of the scattering angle 2θ. Thus the overall coherent-scattering power of the atom diminishes with 2θ; the shape of the coherent-scattering curve as a function of angle is called the *scattering factor*, or *form factor*, of the atom (represented by f).

At zero scattering angle the x-rays that are scattered coherently by the electrons within one atom are all in phase, with the result that the scattering factor at $2\theta = 0°$ is proportional to the number of electrons and therefore to the atomic number of the atom concerned. The convention has been adopted of setting the atomic scattering factor at zero angle, f_0, equal to the number of electrons present. It can be realized that the x-ray scattering power of light elements— such as hydrogen, carbon, nitrogen, and oxygen—is much smaller than that of heavy elements. This has important consequences in the study of polymers, which are composed chiefly of light elements.

We now shall discuss briefly the *incoherent* scattering because it has an important bearing on the analysis of the diffuse x-ray diffraction effects produced by polymers. This kind of scattering has no consistent phase relationship to the incident x-rays, and, furthermore, it consists of a band of wavelengths centered at a wavelength somewhat longer than that of the direct beam. The incoherent scattering is also frequently termed *Compton* scattering after A. H. Compton, who first explained the shift in wavelength as resulting from the encounter of an x-ray photon with a loosely bound or free electron[23].

The incoherent scattering contributes to the diffuse background in an x-ray diffraction pattern. Therefore in quantitative studies of amorphous or highly disordered polymers, which produce more or less diffuse patterns, it is necessary to effect a separation of the coherent and incoherent scattering since the structural features are related to the coherent, but not to the incoherent, scattering. It is normally not feasible to accomplish this separation experimentally; however, the magnitude of the modified scattering can be calculated with considerable accuracy and subtracted from the experimental diffraction pattern.

The intensity of the incoherent (modified) scattering of an atom in electron units can be expressed as follows [24]:

$$I_{inc} = R \left(Z - \sum_n \left| f_{nn} \right|^2 - \sum_{m \neq n} \sum \left| f_{mn} \right|^2 \right), \tag{1-6}$$

where

$$f_{mn} = \int \Psi_m^* \exp{(i\boldsymbol{\rho} \cdot \mathbf{r})} \Psi_n \, dv. \tag{1-7}$$

In these expressions Z is the atomic number, f_{nn} is the scattering factor of the nth electron, f_{mn} is an exchange term due to the interaction of the mth and nth electrons, $\boldsymbol{\rho}$ is the reciprocal-lattice vector [see (1-11)], $\mathbf{r}$ is the position vector within the atom, Ψ_m^* and Ψ_n are electronic wavefunctions, and R is the Breit-Dirac electron recoil factor, which has a value close to unity except for elements of low atomic number. Calculations based on accurate Hartree-Fock wavefunctions for a large number of atoms and ions have been performed with the aid of (1-6), and the numerical values of I_{inc}/R versus $(\sin \theta)/\lambda$ are available in tabular form [25]. These include all the atoms of low atomic number that are important in polymer studies. In all precise diffraction studies of light elements (including polymers) that involve an evaluation of I_{inc}, the factor R should be calculated.

$$R = \left(\frac{\nu'}{\nu} \right)^3 \tag{1-8}$$

$$R = \left(1 + \frac{2h}{mc\lambda} \sin^2 \theta \right)^{-3}, \tag{1-9}$$

where h is Planck's constant, m is the mass of the electron, c is the velocity of light, λ is the wavelength of the incident x-rays, and ν and ν' are the frequencies of the x-rays before and after scattering, respectively. The importance of including the recoil factor R in calculations of the intensity of incoherent scattering by light elements has been emphasized by Ergun [26] and by Ruland (see Section 3-3.1). Ergun points out that R should be set equal to $(\nu'/\nu)^2$ rather than $(\nu'/\nu)^3$ when the number of photons per unit area per unit time is measured, as when using proportional counters and scalers.

Figure 1-25 shows the experimental scattering curve (A) of a noncrystalline polyisoprene specimen normalized to electron units and compares it with (B) the calculated total independent scattering, (C) the calculated incoherent scattering, and (D) the theoretical independent coherent scattering. The scattering curves are plotted as a function of $S = 4\pi(\sin \theta)/\lambda$. It will be noticed that the incoherent scattering is zero at $S = 0$ and increases with scattering angle until

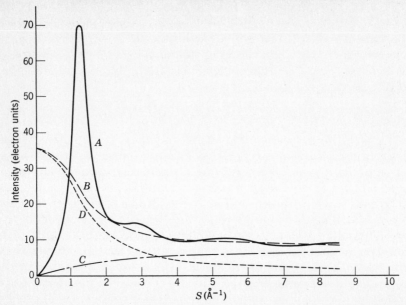

Figure 1-25 Experimental and independent scattering curves for a synthetic poly-isoprene sample: A — experimental scattering; B — total independent scattering; C — incoherent scattering; and D — independent coherent scattering. (Klug and Alexander, *X-Ray Diffraction Procedures*, Wiley, New York, 1954.)

at large values of S it is much larger than the independent coherent scattering. This serves to emphasize that an accurate knowledge of the theoretical incoherent-scattering curve is indispensable for the proper correction of experimental scattering curves prior to the analysis of the coherent scattering and making structural deductions therefrom.

Fluorescent X-Rays. It has been already explained that x-rays of wavelengths just shorter than the K-absorption edge of a given material are strongly absorbed by it, with ejection of K-electrons. A second consequence of this absorption of x-rays is the emission by the absorber of x-rays of its own K-spectrum, which bear no phase relationship to the exciting (incident) x-rays and thus can contribute only a general diffuse scattering to the diffraction pattern. Thus iron-containing compounds (K-edge of $Fe = 1.743$ Å) fluoresce strongly in $CuK\alpha$ x-rays (1.542 Å), and chromium-containing compounds (K-edge of $Cr = 2.070$ Å) fluoresce strongly when irradiated with $FeK\alpha$ x-rays (1.937 Å), resulting in diffraction patterns with very dense backgrounds. Such "fluorescence blackening" of diffraction diagrams can be avoided by selecting an x-ray target with a $K\alpha$ wavelength that is either longer or *considerably* shorter than the wavelengths of the K-absorption edges of the elements that constitute the specimen. Fortunately the problem of x-ray fluorescence is not encountered in preparing the diffraction patterns of polymers consisting exclusively of light elements.

1-4 WIDE-ANGLE DIFFRACTION BY POLYMERS

For both experimental and theoretical reasons it is useful to separate x-ray diffraction effects into small- and wide-angle regions according to the size of the angle of deviation from the direct beam, customarily designated 2θ. Small-angle diffraction, or scatter, refers to effects observed at angles smaller than about 2 or 3°. Wide-angle diffraction encompasses effects that are observed at all larger angles, theoretically extending up to 180°. For polymers most, if not all, of the diffraction effects are observed at angles well below 90°.

1-4.1 The Reciprocal Lattice

To understand the geometry of x-ray diffraction by a single crystal it is of great help to employ the reciprocal-lattice concept[27]. In relation to any crystal lattice in real space we visualize a reciprocal lattice in reciprocal space, as illustrated in two dimensions in Figure 1-26 by means of the ac net of a monoclinic space lattice. One unit-cell face (heavy black lines) of the direct lattice has edges a and c and the interaxial angle β. At right angles to the edge c there are rows of reciprocal-lattice points with translations of a^*, and at right angles to the edge a there are rows with translations of c^*. Thus a reciprocal-lattice net is built up with translations (reciprocal-cell edge lengths) of a^* and c^* and the interaxial angle $\beta^* = 180° - \beta$. It should be noted that the vector $\mathbf{a}^*$ is oriented normal to the planes (100) in real space and that the magnitude of $\mathbf{a}^*$ is $1/d_{100}$.

In three dimensions the relationship between the direct and reciprocal lattices is also relatively simple except for the triclinic lattice, wherein all three interaxial angles differ from 90°. For orthogonal crystal systems it is easy to show that

$$a^* = \frac{1}{d_{100}} = \frac{1}{a},$$

$$b^* = \frac{1}{d_{010}} = \frac{1}{b},$$

$$c^* = \frac{1}{d_{001}} = \frac{1}{c}.$$

The quantities a^*, b^*, and c^* may be expressed in reciprocal angstrom units, Å^{-1}, whereas a, b, c, and the corresponding d quantities are expressed in angstroms. For any crystal system the volumes of its direct and reciprocal unit cells are reciprocally related, $V = 1/V^*$. The volume V may be expressed in Å^3 and V^* in units of Å^{-3}. General equations relating the dimensions of the direct and reciprocal unit cells for any crystal system are [28]:

$$a^* = \frac{bc \sin \alpha}{V},$$
$$b^* = \frac{ca \sin \beta}{V}, \qquad (1\text{-}10)$$
$$c^* = \frac{ab \sin \gamma}{V}.$$

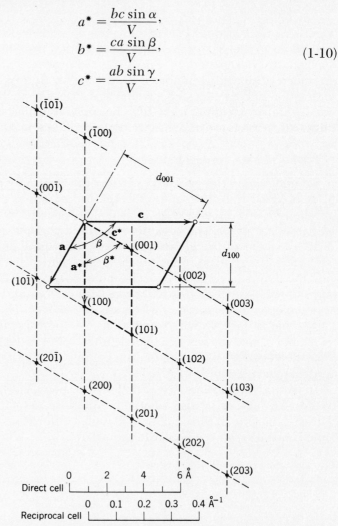

Figure 1-26 Two-dimensional diagram of the relationship between the direct (solid lines) and reciprocal (broken lines) lattices.

It is important to note that a given set of planes (hkl) in the direct lattice is represented in the reciprocal lattice by a lattice *point* (hkl). Moreover, the reciprocal-lattice vector $\boldsymbol{\rho}_{hkl}$ that joins that point to the origin is always perpendicular to the planes (hkl) in the direct cell, and its magnitude is equal to the reciprocal of the interplanar spacing:

$$| \boldsymbol{\rho}_{hkl} | = \frac{1}{d_{hkl}}. \tag{1-11}$$

From these considerations it can be appreciated that the reciprocal lattice indeed bears a reciprocal relationship to the direct crystal lattice.

1-4.2 The Geometry of Diffraction

We now proceed to discuss the diffraction of x-rays by an assemblage of atoms as if they were point scatterers with 2θ-dependent scattering factors. If a parallel beam of x-rays of wavelength λ were to fall on a linear row of regularly spaced atoms at an angle Δ, as shown in Figure 1-27, it can be seen that the secondary waves generated by the individual atoms would be in phase only when the paths DE and FG differed by some whole number of wavelengths, which is to say that

$$DE - FG = m\lambda, \tag{1-12}$$

where m is an integer. For a lattice spacing a

$$DE = a \cos \epsilon \quad \text{and} \quad FG = a \cos \Delta,$$

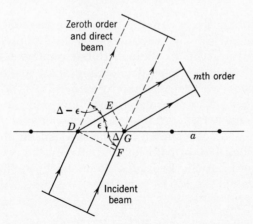

Figure 1-27 Conditions for diffraction from a row of atoms. (Klug and Alexander, *X-Ray Diffraction Procedures*, Wiley, New York, 1954.)

with the result that (1-12) becomes

$$a(\cos \epsilon - \cos \Delta) = m\lambda, \qquad\qquad (1\text{-}12')$$

which is the geometrical condition for constructive interference of the scattered x-rays.

For a three-dimensional lattice with axial translations a, b, and c it can be shown that three equations of the form of $(1\text{-}12')$ must be simultaneously satisfied in order for constructive interference to occur and a diffracted ray to be observed:

$$a(\cos \epsilon_1 - \cos \Delta_1) = m\lambda,$$
$$b(\cos \epsilon_2 - \cos \Delta_2) = p\lambda, \qquad\qquad (1\text{-}13)$$
$$c(\cos \epsilon_3 - \cos \Delta_3) = q\lambda.$$

From the form of these equations it can be realized that only selected orientations of the crystal with respect to the incident x-ray beam of wavelength λ will yield diffracted rays, because the three equations must for any given orientation yield constructive interference in one and the same direction. The three relationships of (1-13) are known as the Laue equations after the discoverer of x-ray diffraction, Max von Laue, who enunciated them in 1912 [29].

It is possible to solve (1-13) simultaneously to give a very useful equation relating the x-ray wavelength λ, the interplanar spacing d, and one-half the angle of deviation of the diffracted rays from the incident x-rays, θ:

$$n\lambda = 2d \sin \theta. \qquad\qquad (1\text{-}14)$$

This is the familiar Bragg equation, or Bragg law [30]. By means of the simple geometrical construction shown in Figure 1-28 the validity of this equation can be demonstrated on the assumption that x-rays are "reflected" by a set of crystallographic planes in much the same way as light rays are reflected by a mirror (*specular* reflection). The figure shows x-rays of wavelength λ impinging at the angle θ on two adjacent planes of a set (hkl) separated by an interplanar distance d. "Reflected" x-rays are shown making the same angle θ with these planes. The difference in path length between the rays reflected by planes No. 1 and 2 is $2b$. From the geometry of the figure it can be seen that the reflected waves from the two planes will be in phase when

$$n\lambda = 2b = 2d \sin \theta,$$

where n is an integer. Figure 1-28 illustrates the case $n = 1$, which is

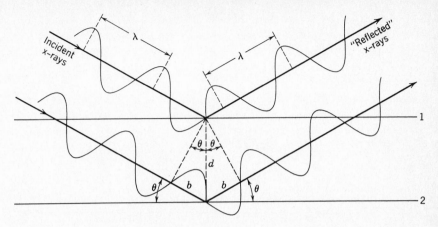

Figure 1-28 Geometry of the Bragg reflection analogy.

referred to as reflection of first order from the given planes. At particular larger values of θ n will equal 2, 3, 4, $\cdots$, giving rise to reflections of second, third, fourth, and higher orders from the same set of planes. For all values of θ that do not satisfy (1-14) the reflected (diffracted) rays will be out of phase with each other and no actual reflected x-rays will be observed.*

It is useful to picture the geometrical conditions for diffraction from a set of planes $(h'k'l')$ with the aid of the reciprocal-lattice concept presented in the preceding section[31]. Imagine the center of the crystal and the origin of the reciprocal lattice to lie at the point O in Figure 1-29. We also visualize a sphere of radius $1/\lambda$ with origin at B as shown, which will be termed the *sphere of reflection*. The unit vector s_0 designates the direction of the incident x-ray beam. The crystal and its reciprocal lattice are now imagined to be tilted so that the reciprocal-lattice point $P_{h'k'l'}$ intersects the sphere of reflection as shown. Under these conditions diffraction can occur from planes $(h'k'l')$ and a diffracted ray is emitted in the direction s, so that

$$\boldsymbol{\rho}_{h'k'l'} = \frac{s - s_0}{\lambda} . \tag{1-15}$$

Also, in the right triangle AOP

$$\frac{|\boldsymbol{\rho}_{h'k'l'}|}{2/\lambda} = \frac{\lambda}{2d_{h'k'l'}} = \sin\theta \tag{1-16}$$

if we bear in mind that $|\boldsymbol{\rho}| = 1/d$ [see (1-11)].

*Actually the diffracted rays suffer a phase retardation of $\pi/2$ relative to the incident rays, but this does not affect the geometry of diffraction and so has been omitted from Fig. 1-28 for simplicity.

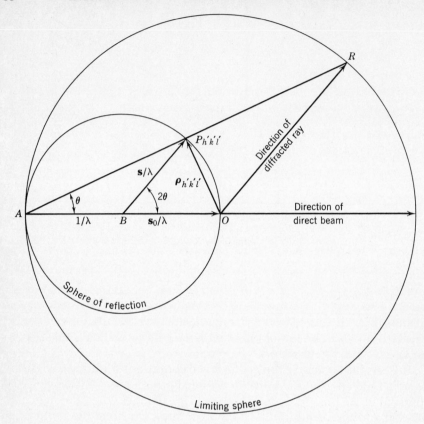

Figure 1-29   Geometrical conditions for diffraction from planes $(h'k'l')$ defined with the aid of the reciprocal-lattice concept.

Now in general the vector $\boldsymbol{\rho}$ is the sum of three primary reciprocal-lattice translations $h'\mathbf{a}^*$, $k'\mathbf{b}^*$, and $l'\mathbf{c}^*$:

$$\boldsymbol{\rho}_{h'k'l'} = h'\mathbf{a}^* + k'\mathbf{b}^* + l'\mathbf{c}^*.$$

If h', k', and l' have a common factor n such that $h' = nh$, $k' = nk$, and $l' = nl$,

$$\boldsymbol{\rho}_{h'k'l'} = n(h\mathbf{a}^* + k\mathbf{b}^* + l\mathbf{c}^*) = n\boldsymbol{\rho}_{hkl}.$$

Equation 1-16 may then be written

$$\frac{n|\boldsymbol{\rho}_{hkl}|}{2/\lambda} = \frac{n\lambda}{2d_{hkl}} = \sin\theta,$$

or

$$n\lambda = 2d \sin \theta, \tag{1-14}$$

which is the Bragg equation again.

From the foregoing discussion it might be concluded that, since the radius of the sphere of reflection is precisely $1/\lambda$ and since the reciprocal-lattice point P_{hkl} is infinitely sharp, the conditions for diffraction from any given set of planes are exceedingly critical. This would indeed be the case for truly monochromatic x-rays and an ideal crystal. However, in actual practice the wavelength employed is not really discrete but is replaced by a very small range of wavelengths, and an actual crystal has a finite size and exhibits certain kinds of imperfections. Therefore the surface of the sphere of reflection has a finite thickness, and the reciprocal-lattice nodes are small domains with some characteristic shape and volume. The latter observation is particularly true of crystalline polymers because of the characteristically small crystallite size and the relatively large magnitude of lattice imperfections.

1-4.3 The Intensity of Diffraction

The preceding section was concerned with the geometrical conditions that govern the directions in which diffracted rays are emitted by a single crystal. The present section is devoted to the factors that govern the intensities of these diffracted rays. At the outset it should be realized that in an actual structure there is no necessity for any atoms to lie precisely on a given crystallographic plane. In fact whether or not this occurs depends on the choice of the origin of coordinates that define the unit cell, and often this choice is a perfectly arbitrary one. In terms of some fixed origin in the structure, the positions of the atoms relative to a given set of planes is the most significant factor determining the intensity of the ray "reflected" by those planes. This important functional relationship is embodied in the structure factor $F(hkl)$.

Structure Factor F(hkl). Letting x, y, z be fractional coordinates of the unit-cell edges a, b, c, respectively, we consider a unit cell containing N atoms in which the nth atom is located at x_n, y_n, z_n. The structure factor can then be defined by [see Appendix 1, (T-10)]

$$F(hkl) = \sum_{n=1}^{N} f_n \exp\left[2\pi i(hx_n + ky_n + lz_n)\right] \tag{1-17}$$

$$
\begin{aligned}
= \sum_{n=1}^{N} f_n [&\cos 2\pi(hx_n + ky_n + lz_n) \\
&+ i \sin 2\pi(hx_n + ky_n + lz_n)],
\end{aligned}
\tag{1-18}
$$

where f_n is the atomic scattering factor of the nth atom and the summation encompasses all the atoms in one unit cell.

If the real and imaginary terms of (1-18) are symbolized by A and B, respectively, the magnitude (modulus) of F is

$$|F(hkl)| = (A^2 + B^2)^{1/2}, \qquad (1-19)$$

and $|F(hkl)|^2 = A^2 + B^2$ may be regarded as the ideal intensity of the reflection (hkl). Now $|F|^2$ can be determined directly from the experimental intensity of the reflection concerned, but the phase relationship of $|F|^2$, given by $\phi = \tan^{-1}(B/A)$, as shown in Figure 1-30, cannot be directly determined in an analogous way.

A number of experimental factors modify the ideal intensity $|F|^2$ to yield the intensity that is actually observed, $I(hkl)$:

$$I(hkl) = P \cdot L \cdot j \cdot A \cdot |F|^2. \quad (1-20)$$

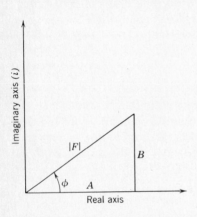

Figure 1-30 Real and imaginary components of the structure factor F with the phase $\phi = \tan^{-1}(B/A)$.

Polarization Factor **P.** The x-rays emitted by an x-ray tube target are unpolarized, but after being scattered by the electrons of an atom they are plane polarized to a degree that is dependent on the Bragg angle θ:

$$P = \tfrac{1}{2}(1 + \cos^2 2\theta). \qquad (1-21)$$

If crystal-monochromatized radiation is employed and θ' is the Bragg angle for the reflecting planes of the monochromator, the polarization factor becomes

$$P = \frac{1 + \cos^2 2\theta' \cos^2 2\theta}{1 + \cos^2 2\theta'}. \qquad (1-22)$$

Lorentz Factor **L.** This factor is basically dependent on the *reflecting time*; that is, the time during which a given family of planes reflects x-rays under a certain set of experimental conditions. The reflecting time arises in part from the lack of true parallelism and true monochromaticity of the incident x-ray beam. Furthermore, the reflecting time is a function of the inclination of the reflecting planes to the axis of rotation of the crystal. For a single crystal rotating in a monochromatic beam the Lorentz factor takes the form

$$L = \frac{1}{\sin \theta \cos \theta}, \qquad (1\text{-}23)$$

whereas for randomly oriented crystalline powders irradiated with monochromatic x-rays it becomes

$$L = \frac{1}{\sin 2\theta \sin \theta}. \qquad (1\text{-}24)$$

Numerical values of the Lorentz and polarization factors calculated with (1-21), (1-23), and (1-24) are tabulated in Appendix 9.

In structural studies and other quantitative investigations based on the intensities of fiber patterns (see Section 1-4.6) it has been common practice to employ the Lorentz factor in the form given by (1-23). However, de Wolff[32] has shown that the following expressions are more satisfactory:

Equatorial (paratropic) reflections:

$$L(hk0) = (\sin^2\theta \cos \theta)^{-1}. \qquad (1\text{-}25)$$

General reflections:

$$L(hkl) = (\sin^2\theta \cos \theta \sin \phi_{hkl,z})^{-1}. \qquad (1\text{-}26)$$

Meridional (diatropic) reflections:

$$L(00l) = (\sin^2\theta \cdot \cos \theta \cdot t)^{-1}. \qquad (1\text{-}27)$$

In these equations $\phi_{hkl,z}$ is the angle between the normal to the reflecting planes and the fiber axis, and t is given to a reasonable approximation by

$$t = 0.815 \, \beta_{1/2} \text{ radians,}$$

where $\beta_{1/2}$ is the azimuth at which the intensity of the reflection falls to one-half its peak value (at $\beta = 0°$). The reader is referred to Chapter 4 for a precise exposition of the significance of the angles $\phi_{hkl,z}$ and β in describing preferred orientation.

Multiplicity Factor j. This is the number of different sets of planes [different (hkl)'s] that contribute to a single observed reflection. For moving-film methods such as the Weissenberg and precession techniques applied to single crystals and for single-crystal diffractometry $j = 1$ because the intensity is measured at but one reciprocal-lattice node (hkl) at a time. However, when polymer fibers or films are examined or when crystals are rotated through

360° during preparation of the diffraction photograph, certain reflections will be superimposed, with resulting values of j that are larger than unity. In the diffraction patterns of randomly oriented crystalline powders, crystalline polymers included, all reflections from sets of planes with the same d value will coincide at one diffraction angle 2θ, as can be realized by an inspection of the Bragg equation (1-14). For a crystalline powder belonging to the cubic system

$$d(hkl) = \frac{a}{\sqrt{h^2 + k^2 + l^2}},$$

from which it is evident that all combinations of h, k, and l characterized by the same quadratic sum, $h^2 + k^2 + l^2$, will contribute to a composite diffraction halo at one and the same angle 2θ; for example, the following 12 sets of planes will contribute to a single powder reflection, which therefore has a multiplicity of $j = 12$:

$$
\begin{array}{cccc}
110 & \bar{1}10 & 1\bar{1}0 & \bar{1}\bar{1}0 \\
101 & \bar{1}01 & 10\bar{1} & \bar{1}0\bar{1} \\
011 & 0\bar{1}1 & 01\bar{1} & 0\bar{1}\bar{1}
\end{array}
$$

Absorption Factor A.† The absorption factor for x-rays of a particular crystal or other specimen depends not only on its elemental composition and the wavelength of the x-rays (see Section 1-3.1) but also on the size and shape of the specimen. Theoretical values of A for spheres and cylinders have been calculated, and a method has been outlined for computing A for a crystal of any shape[33]. Of special importance in polymer investigations are the absorption factors appropriate for flat sheets and for fibers, which tend to approximate cylinders in form.

Temperature Factor. Because of thermal vibrations of the atoms in a crystal, the ideal intensity is reduced by a factor of the form $\exp(-2M)$, where M is related to the mean square displacement, $\bar{\mu}_s^2$, of the atoms from their rest positions, in a direction normal to the reflecting planes, by the expression

$$M = 8\pi^2 \bar{\mu}_s^2 \frac{\sin^2 \theta}{\lambda^2}, \tag{1-28}$$

$$M = B \frac{\sin^2 \theta}{\lambda^2}. \tag{1-29}$$

†This factor is perhaps more justifiably termed the *transmission factor*, as it is referred to in [33].

If the amplitudes of the atomic vibrations are independent of direction in the crystal, the temperature factor is said to be *isotropic*. Moreover, if the vibrational amplitudes of all the constituent atoms are the same, the temperature factor of the crystal is defined by a single numerical value of the coefficient B. Under these conditions the temperature factor exp $(-2M)$ may be included with the factors of (1-20), and $F(hkl)$ possesses precisely the significance given by (1-17) and (1-18). However, the accuracy of present-day intensity measurements is often good enough to justify the assignment of different isotropic B values to some or all of the atoms in the unit cell. When this is true, the temperature factor cannot be expressed separately in (1-20) but must be incorporated into the structure factor itself, so that (1-17) becomes

$$F(hkl) = \sum_{n=1}^{N} f_n \exp\left(-B_n \frac{\sin^2 \theta}{\lambda^2}\right) \exp\left[2\pi i(hx_n + ky_n + lz_n)\right]. \quad (1\text{-}30)$$

Very precise single-crystal structural investigations frequently lead to the conclusion that the thermal motions of some or all of the atoms are *anisotropic*. This can be allowed for by replacing the isotropic coefficients B_n in (1-30) by a more complex expression, but this degree of refinement is seldom if ever justified in the analysis of polymer structures.

1-4.4   Amorphous Specimens

At the beginning of this chapter it was noted that noncrystalline, or so-called amorphous, materials produce diffuse x-ray patterns that consist principally of a few halos. (See again Figure 1-1.) Although there exists in such a specimen no short- or long-range order of a truly crystalline nature, there is still present a minimal kind of short-range order consisting of most probable distances between neighboring atoms. Thus the distances between bonded atoms and second nearest neighbors within a polymer chain are subject to little variation, and furthermore the distance of closest approach of neighboring chains is a vector distance with a high probability of occurrence. These modes of quasi-short-range order are responsible for the interference halos observed in the diffraction patterns of noncrystalline polymers.

Debye [34] has shown that for any system of atoms, regardless of its state of order or disorder, that assumes all orientations in space with equal probability, the x-ray intensity in electron units scattered at the angle 2θ, equivalent to $S = (4\pi \sin \theta)/\lambda$, is given by [see also

Appendix 1, (T-14)]†

$$I = \sum_m \sum_n f_m f_n \frac{\sin Sr_{mn}}{Sr_{mn}}. \qquad (1\text{-}31)$$

In this expression f_m and f_n are the atomic scattering factors of the mth and nth atoms, respectively, and r_{mn} is the vector distance separating atoms m and n. The double summation encompasses all the atoms in the assemblage. Equation 1-31 makes it possible in principle to calculate the scattering curve corresponding to any desired atomic arrangement. However, in practice this procedure is obviously limited to rather small atomic aggregates, even when electronic computers are used.

Two means of interpreting experimental data in the light of (1-31) are available. Formerly the usual practice was to compare the experimental scattering curve with the scattering curves of various structure models, whereas as a general rule most recent investigators have chosen to invert the experimental intensity curve with the aid of the Fourier integral theorem to yield the radial-distribution function (refer again to Appendix 1),

$$4\pi r^2 \rho_r = 4\pi r^2 \rho_0 + \left(\frac{2r}{\pi}\right) \int_0^\infty S\, i(S)\, \sin rS\, dS, \qquad (1\text{-}32)$$

where ρ_r is the number of atoms per unit volume at a distance r from every other atom taken successively as the reference atom, ρ_0 is the average atomic density in the sample, and

$$i(S) = \frac{I}{Nf^2} - 1. \qquad (1\text{-}33)$$

Expressions 1-32 and 1-33 in the forms given are applicable only to systems that are composed of one kind of atom, but they can be generalized so as to apply to specimens of more complex constitution, as explained in Appendix 1. The radial-distribution method is greatly limited because it employs one-dimensional intensity data (I as a function of S) and consequently can yield only one-dimensional structural information (ρ as a function of r). Nevertheless, for the investigation of randomly oriented (spherically symmetrical) polymer specimens it is the only x-ray method available, and in favorable

---

†With reference to (1-15) and Figure 1-29, it should be noted that S is a scalar quantity equal to 2π times the magnitude of the reciprocal-lattice vector $|\boldsymbol{\rho}| = |\mathbf{s} - \mathbf{s}_0|/\lambda$. This convention is adhered to because of its current widespread usage in spite of its possible confusion with the diffracted-beam vector $\mathbf{s}$.

situations it can provide useful structural information. This topic is developed at greater length in Section 6-1.4.

1-4.5 Randomly Oriented Microcrystalline Specimens

In bulk and molded polymer specimens the orientations of the crystalline regions, which may be designated *crystallites*, tend to be statistically random. As has been already explained, this is a natural consequence of the mechanisms of nucleation and crystal growth in the melt or concentrated solutions, as presently understood. Because of the minuteness of the crystallites, with dimensions typically less than 0.1 micron, even a sample amounting to a few milligrams contains a vast number of crystallites, which results in all orientations being represented in equal proportions. Under these circumstances a monochromatic x-ray beam will generate a typical powder-diffraction pattern, such as that shown in Figure 1-31 of a partially crystalline polystyrene specimen recorded on a flat film. The diffraction effects consist of a number of relatively sharp concentric circles superposed on a background of diffuse x-ray scatter and at least one noticeable amorphous halo.

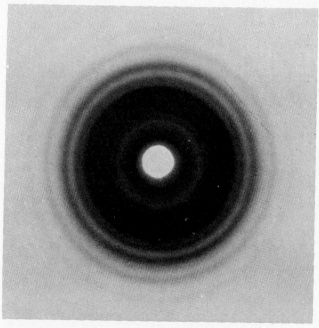

Figure 1-31 Powder diffraction pattern of partially crystalline, unoriented polystyrene recorded on a flat film. (Courtesy of S. S. Pollack.)

The reciprocal-lattice concept can help us to understand how such a diffraction diagram originates. First, referring briefly again to Figure 1-29, we recall that a discrete diffracted ray is emitted in the direction **BP** (equivalent to s or **OR**) when a single crystal is tilted so that a reciprocal-lattice node $P_{h'k'l'}$ falls on the sphere of reflection. In the case of a microcrystalline specimen characterized by random crystallite orientations, however, the reciprocal-lattice vector $\boldsymbol{\rho}$, or **OP**, corresponding to the set of planes $(h'k'l')$ assumes all orientations with equal probability, with the consequence that the point P is dispersed over all positions on the surface of a sphere of radius

$$|\boldsymbol{\rho}_{h'k'l'}| = \frac{1}{d_{h'k'l'}},$$

as indicated in Figure 1-32. This sphere intersects the sphere of reflection in the circular zone shown in perspective as a dotted ellipse. Reflection from planes $(h'k'l')$ will therefore take place in all directions defined by this zone in reciprocal space, and it can be seen that the diffracted rays, thought of as originating at the point B, lie on a cone of semiapex angle 2θ concentric with the undeviated beam. In the same way another reciprocal-lattice node with a different index

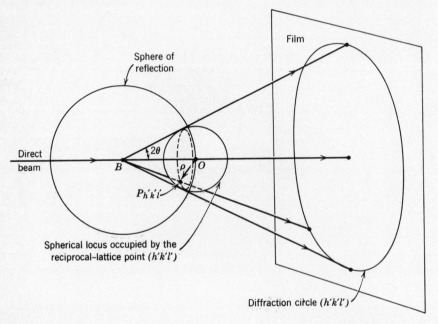

Figure 1-32   Reciprocal-lattice model of the origin of a powder reflection.

triplet (hkl) will generate a diffraction cone of different angle, and so on.

Figure 1-33 depicts the sum total of diffraction from a crystalline powder or polymer specimen as an assemblage of concentric cones. In this illustration diffraction cones are portrayed in both the forward-reflection $(0° < 2\theta < 90°)$ and back-reflection $(90° < 2\theta < 180°)$ regions, which is characteristic of well-crystallized ionic and molecular compounds. However, the extreme minuteness of polymer crystallites, together with their relatively imperfect character, tends to eliminate distinct diffraction effects in the back-reflection region. There are two principal photographic methods of registering powder patterns. In the flat-film method a plane film is placed a few centimeters from the specimen and perpendicular to the undeviated beam, whereupon most of the cones in the forward-reflection region are intercepted, giving a pattern of concentric circles. (See Figure 1-31.)

A second way of recording the diffraction pattern, which intercepts both the forward- and back-reflection cones, is shown in Figure 1-34. A narrow beam of monochromatic x-rays irradiates the small cylindrical sample X, and short sections of the diffraction cones are intercepted by the strip of film P, which is cylindrically arranged about X. This method, referred to as the Debye-Scherrer technique, is of special utility for the qualitative analysis of crystalline mixtures by means of their powder-diffraction patterns, which requires the precise measurement and comparison of interplanar spacings (d values). In this connection it should be emphasized that the d values and

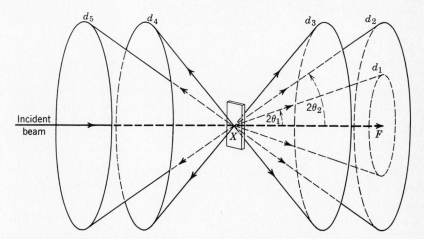

Figure 1-33 Diffraction of x-rays by a randomly oriented polycrystalline specimen. (Klug and Alexander, *X-Ray Diffraction Procedures*, Wiley, New York, 1954.)

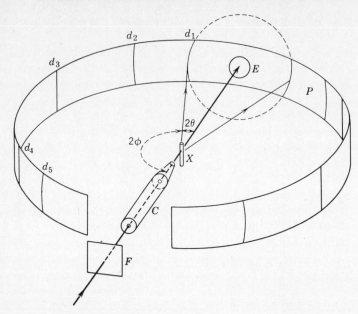

Figure 1-34 Geometrical features of the Debye-Scherrer technique. (Klug and Alexander, *X-Ray Diffraction Procedures*, Wiley, New York, 1954.)

relative intensities of the lines comprising the powder pattern of a given substance are both perfectly reproducible and characteristic of it, so that a powder pattern may be likened to a fingerprint. From measurements of the line positions on properly calibrated powder films the d spacings can be calculated with the aid of the Bragg equation:

$$d = \frac{\lambda}{2} \cdot \frac{1}{\sin \theta}.$$

It is more convenient, although less accurate, to measure the d values by superimposing on the pattern a scale of d and reading off the numerical values directly, as shown in Figure 1-35. Transparent plastic scales for various x-ray wavelengths and camera diameters are available commercially.[†]

The American Society for Testing and Materials (ASTM)[‡] in 1942 published a first edition of x-ray powder-diffraction data cards for about 1300 inorganic and organic compounds and minerals. Now the file is enlarged annually with successive supplements, and more than

[†] Obtainable from N. P. Nies, 969 Skyline Drive, Laguna Beach, California 92651.
[‡] 1916 Race Street, Philadelphia, Pennsylvania 19103.

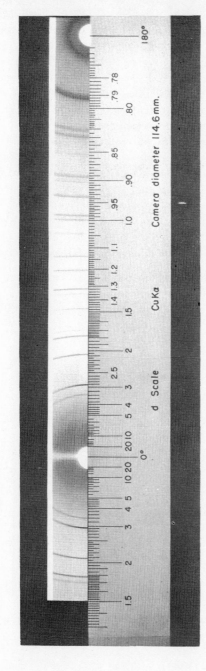

Figure 1-35 Scale of *d* values for measuring Debye-Scherrer patterns. (Klug and Alexander, *X-Ray Diffraction Procedures*, Wiley, New York, 1954.)

49

15,500 compounds were included by 1968. A systematic procedure based on the d values of the three strongest lines is followed to identify the crystalline constituents of mixed powders or phases. Because of the difficulty of preparing polymers of perfectly reproducible crystalline constitution, the number of polymers that are included in the ASTM powder-diffraction file is as yet very limited. Nevertheless, on the basis of its wider applications, the file can be of significant value in most polymer research laboratories. Figure 1-36 is a reproduction of card No. 13–836 for polystyrene. The d values of the three strongest lines in order of decreasing intensity are listed in the upper left-hand corner. The fourth value given, 10.9 Å, is the line of largest interplanar spacing. The pattern of polystyrene was reported by Okamura and Higashimura† and consists of eight lines, whose d values and relative intensities ($100\,I/I_1$) are tabulated at the right. ‡

1-4.6   Preferentially Oriented Microcrystalline Specimens

We shall describe here the general features of *axial* orientation (also called *uniaxial* or *fiber* orientation) and the diffraction effects to be expected. In Chapter 4 this topic is expanded to include higher types of orientation with some practical examples.

13-836

d	5.41	4.81	10.9	10.9	$(C_8H_8)_N$
I/I_1	100	100	75	75	POLYSTYRENE

Rad. λ Filter Dia.	d Å	I/I_1	hkl	d Å	I/I_1	hkl
Cut off I/I_1	10.9	75				
Ref. OKAMURA AND HIGASHIMURA, J. POLYMER SCI. 46 539–41	6.21	25				
(1960)	5.41	100				
	4.81	100				
Sys. S.G.	4.09	50				
a₀ b₀ c₀ A C	3.60	25				
a β γ Z Dx	3.25	25				
Ref.	2.82	25				
ε a n ω β ε γ Sign						
2V D mp Color						
Ref.						

Figure 1-36 Format of ASTM diffraction data card for polystyrene. (Courtesy of American Society for Testing and Materials.)

†*J. Polymer Sci.*, **46**, 539–541 (1960).
‡The reader's attention is also directed to the useful compendium of diffractometric polymer patterns prepared by J.W. Turley (see General References).

Consider first an ideal single crystal rotating about a principal crystallographic axis that is perpendicular to an incident beam of monochromatic x-rays. To simplify the illustration further, suppose that the unit cell is orthorhombic and rotation is about c. Figure 1-37 shows small sections of the zeroth and lth hk reciprocal-lattice nets and indicates how, during the rotation, the reciprocal-lattice points cut through the sphere of reflection, generating diffracted rays. An inspection of Figure 1-38 shows that all diffracted rays with $l = 0$ lie on the surface of a flat cone (the zero, or equatorial, level), whereas rays with $l \neq 0$ lie on a cone with semiapex angle

$$\epsilon_l = \cos^{-1} \frac{l\lambda}{c} \quad (l\text{th level}), \tag{1-34}$$

in which $l = 1$ for the first level, 2 for the second level, and so on. It can be seen that (1-34) is equivalent to the Laue equation (1-12') with $\Delta = 90°$. When the diffraction pattern is registered on a cylindrical film coaxial with the crystal-rotation axis, a single-crystal rotation photograph of the kind shown in Figure 1-39 is obtained. If the radius of the film is r_F, the height of the lth layer line above the equator is

$$y_l = r_F \cot \cos^{-1} \frac{l\lambda}{c}, \tag{1-35}$$

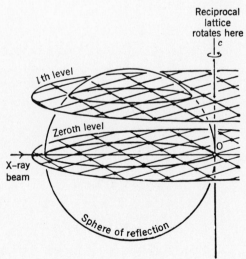

Reciprocal lattice rotates here

c

lth level

Zeroth level

X–ray beam

Sphere of reflection

Figure 1-37 Reciprocal lattice passing through sphere of reflection as it rotates. (Bunn, *Chemical Crystallography*, 2nd ed., Clarendon Press, Oxford, 1961.)

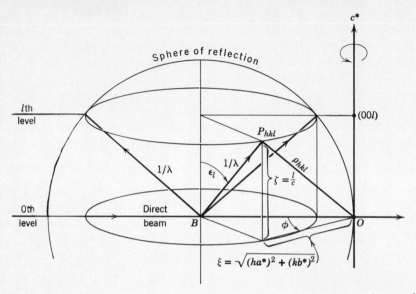

Figure 1-38 Reciprocal-lattice model of diffraction by an orthorhombic single crystal rotating about its c-axis.

from which it is evident that the crystallographic translation c along the crystal-rotation axis can be determined directly from experimental values of y_l, r_F, l, and λ.

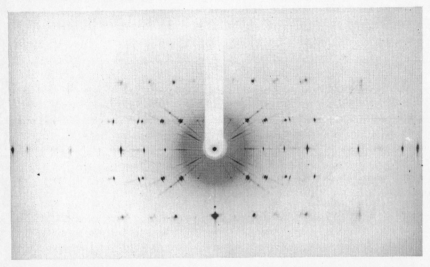

Figure 1-39 Rotation photograph of a single crystal of enstatite ($MgSiO_3$) showing some disordering. Rotation axis vertical. (Courtesy of S. S. Pollack.)

Drawn fibers of synthetic polymers and many natural fibers give fiber patterns that resemble rotation photographs of single crystals; for example, the reader may compare Figure 1-39 with the fiber patterns given in Figures 1-3 and 1-40. In preparing a fiber pattern

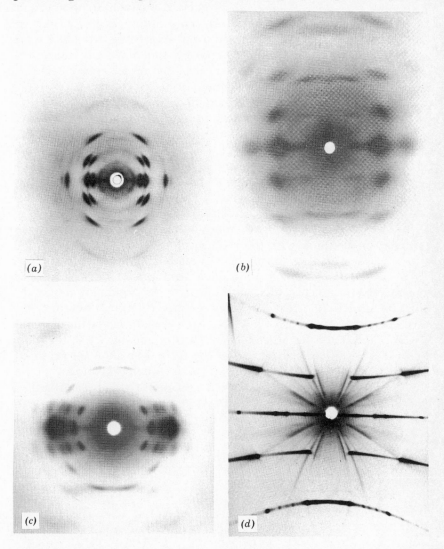

Figure 1-40 Fiber patterns prepared with fiber axis vertical: (a) β-poly-L-alanine [Bamford et al., *Nature*, **173**, 27 (1954)]; (b) natural silk [Astbury, *Endeavor*, **1**, 70 (1942)]; (c) poly(ethylene terephthalate) (Courtesy of I. Matsubara and S. S. Pollack); (d) chrysotile asbestos (Courtesy of S. S. Pollack).

no rotation of the fiber is required since all orientations of the crystallographic axes a and b about the fiber axis (taken to be c) are inherently present in the fiber, which is equivalent to saying that the polymer fiber possesses cylindrical symmetry about the fiber axis. Compared with the reflections in a single-crystal rotation photograph, the reflections of a fiber pattern are more diffuse and often noticeably arclike in shape. These diffraction effects arise from the following properties of the crystallites composing the fiber: (a) their very small dimensions, (b) the presence of lattice imperfections, and (c) departures from ideal fiber orientation.

It is frequently of great value to deduce the dimensions of a unit cell that will be compatible with a given fiber pattern. In order to accomplish this it is necessary to assign indices (hkl) to all the observed reflections in a manner that is consistent with the selected unit cell. By way of introduction we shall consider here a relatively high degree of symmetry for polymer structures—namely orthorhombic. (The reader may find it worthwhile at this point to refer to Table 1-1 and Figure 1-4.) In an orthorhombic crystal the unit-cell edges a, b, and c are mutually perpendicular and unequal; furthermore, they coincide in direction with the reciprocal-lattice translations a^*, b^*, and c^*, respectively, as can be realized from a consideration of reciprocal-lattice geometry. (See again Section 1-4.1.) Let us suppose that the fiber axis coincides with c, or c^*. As explained above, the unit-cell dimension c can be determined directly from a measurement of the height of the lth layer line above the equator by means of (1-35). We also know that for all reflections on the equator $l = 0$, for those on the first level $l = 1$, for those on the second level $l = 2$, and so on. It remains to determine the indices h and k of each reflection.

We commence with the equatorial reflections, which have indices of the type $(hk0)$. By reference to Appendix 8 it will be seen that the interplanar spacing corresponding to a set of planes $(hk0)$ of an orthorhombic crystal is

$$d(hk0) = \left(\frac{h^2}{a^2}+\frac{k^2}{b^2}\right)^{-1/2} = \frac{a}{\left[h^2+\dfrac{k^2}{(b/a)^2}\right]^{1/2}},$$

from which

$$\log d = \log a - \tfrac{1}{2}\log\left[h^2+\frac{k^2}{(b/a)^2}\right]. \qquad (1\text{-}36)$$

A chart for indexing the reflections of a rectangular reciprocal-lattice net can be constructed by plotting for each index pair (h, k) a curve of b/a versus $\log d$. It is convenient to use semilog paper, labeling the ordinate scale b/a and the abscissa d. The term $\log a$ may be arbitrarily assigned a value of unity. The resulting chart, which is shown in Figure 1-41, is a simplified version of the well-known Hull-Davey chart[35, 36] for indexing tetragonal powder patterns. Although it is two- rather than three-dimensional, a tetragonal chart can be used to index an orthorhombic fiber pattern because one layer line (reciprocal-lattice level) is indexed at a time, which is a two-dimensional problem.

To index the equatorial reflections the observed d spacings are plotted along the edge of a strip of paper to the same logarithmic scale as that of the abscissa of the chart. The strip, kept in a horizontal position, is then translated over the chart in both horizontal and vertical directions until a fit is found between all the plotted d spacings and the lines on the chart. Figure 1-42 shows the position of the strip when a fit was achieved for the d spacings of the four equatorial reflections observed by Natta and Corradini[37] in the fiber pattern of poly(vinyl chloride), which is orthorhombic with $a = 10.6$, $b = 5.4$, $c = 5.1$ Å, fiber axis c. Bunn[38] has described the

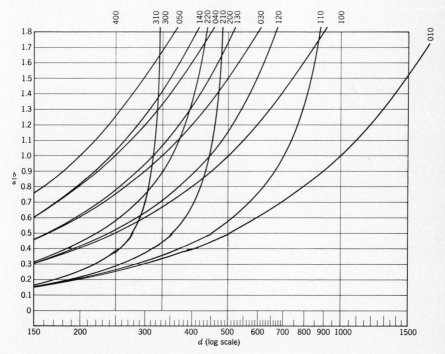

Figure 1-41   Hull-Davey chart for indexing a rectangular reciprocal-lattice net.

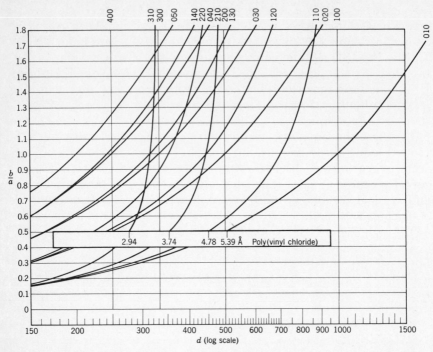

Figure 1-42   Indexing of the four equatorial ($hk0$) reflections of poly(vinyl chloride).

construction and use of another useful chart for indexing rectangular zero-level reciprocal-lattice nets.

Once indices have been selected for the equatorial reflections on the basis of a tentative unit cell, it is necessary to confirm this choice by indexing the higher-level reflections. This can be most satisfactorily accomplished by employing the reciprocal-lattice concept. For the geometry of a rotation or fiber photograph, the position of a given point P in the reciprocal lattice can be expressed by resolving its position vector $\boldsymbol{\rho}_{hkl}$ into components ζ and ξ parallel with and perpendicular to the axis of rotation, respectively, as depicted in Figure 1-38. It is clear from the geometry that

$$| \boldsymbol{\rho}_{hkl} |^2 = \zeta^2 + \xi^2. \tag{1-37}$$

Because of the cylindrical symmetry of rotation, the azimuthal angle ϕ has no pertinence to the present treatment. Figure 1-43 presents the geometrical relationship between the coordinates ζ, ξ of a reciprocal-lattice point P in reflecting position and the corresponding coordinates x, y of the reflection P' on a cylindrical film. Following

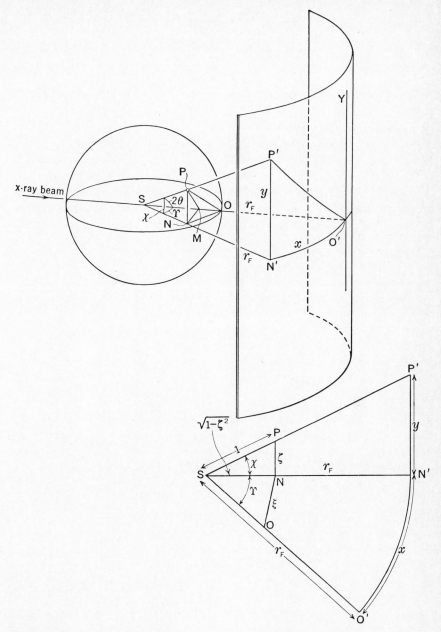

Figure 1-43 Relationship between the reciprocal-lattice coordinates, ζ and ξ, and cylindrical-film coordinates x and y. (Buerger, *X-Ray Crystallography*, Wiley, New York, 1942.)

Buerger's excellent treatment[39] at this point, it can be shown that

$$x = \frac{2\pi r_F}{360°} \cos^{-1}\left(\frac{2 - \zeta^2 - \xi^2}{2\sqrt{1-\zeta^2}}\right), \tag{1-38}$$

$$y = r_F \frac{\zeta}{\sqrt{1-\zeta^2}}, \tag{1-39}$$

where r_F is the film radius.

Since the principal features of the fiber patterns of polymers are fully encompassed in the forward-reflection region, the usual practice is to record such patterns on flat films. The nonequatorial lines then become hyperbolas, as is clearly seen in the fiber diagram of asbestos (Figure 1-40d). The diffraction geometry is now that shown in Figure 1-44. If D is the perpendicular distance from the specimen to the film,

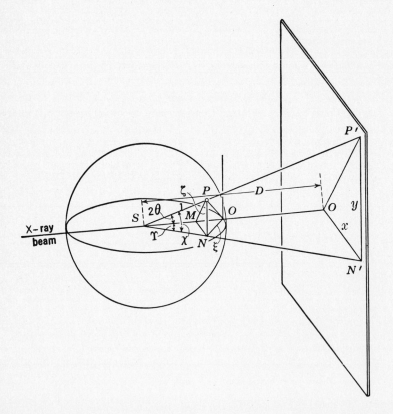

Figure 1-44   Relationship between the reciprocal-lattice coordinates, ζ and ξ, and flat-film coordinates x and y. (Buerger, *X-Ray Crystallography*, Wiley, New York, 1942.)

the film coordinates x and y can be expressed in terms of ζ and ξ as [39]

$$x = D \tan \cos^{-1}\left(\frac{2 - \zeta^2 - \xi^2}{2\sqrt{1 - \zeta^2}}\right),\tag{1-40}$$

$$y = D\frac{\zeta}{\cos 2\theta}$$

$$= D\frac{2\zeta}{2 - \zeta^2 - \xi^2}.\tag{1-41}$$

With the aid of (1-38) through (1-41) Bernal charts [31] can be prepared for any desired values of r_F or D, giving directly the ζ- and ξ-coordinates of any point on a cylindrical or flat fiber diagram. Such a scale printed on transparent plastic can be superposed on the fiber pattern, permitting the ζ and ξ of each reflection to be read off.† Figure 1-45 shows a Bernal chart for a flat film (sample-to-film distance 5.00 cm).

Let us now suppose that the rectangular equatorial a^*b^* reciprocal-lattice net has been indexed on the basis of an orthorhombic cell by using a chart such as shown in Figure 1-41. It is easy to verify these $(hk0)$ indices with the reciprocal-lattice construction shown in Figure 1-46, which is plotted to the scale of poly(vinyl chloride) as an example ($a^* = 1/10.6$, $b^* = 1/5.4$ Å^{-1}). The ξ values of the four equatorial reflections of poly(vinyl chloride) are plotted along some radial line OQ, and arcs about O are then drawn from the four points ξ_1, ξ_2, ξ_3, and ξ_4 through the a^*b^* net. If the previous indexing was correct, each arc should intercept the reciprocal-lattice node that bears the originally selected indices $(hk0)$.

Next the higher layer lines can be indexed in exactly the same way using the same a^*b^* net in conjunction with separate radial plots of the ξ values corresponding to the reflections that comprise each of these layer lines. In some cases, because of systematically absent classes of reflections in the equatorial pattern [for example, $(h00)$ with h odd, $(0k0)$ with k odd, or $(hk0)$ with h or k odd], it may prove necessary to select a new reciprocal translation a^* or b^* (or both) equal to one-half the value deduced from the equatorial data. Conversely, the new direct cell will have one or both of its edges a and b doubled compared with the originally chosen cell. In any given indexing problem a proposed unit cell is acceptable only if it fully

†Transparent-plastic Bernal charts are obtainable from N. P. Nies, 969 Skyline Drive, Laguna Beach, California 92651 and from Polycrystal Book Service, P. O. Box 11567, Pittsburgh, Pennsylvania 15238.

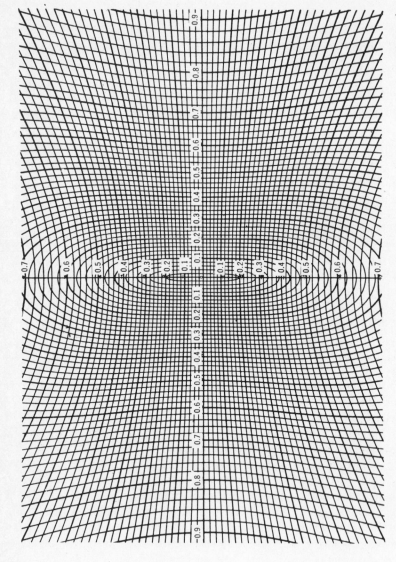

Figure 1-45 Bernal chart for flat film; crystal-to-film distance −5 cm. (Buerger, *X-Ray Crystallography*, Wiley, New York, 1942.)

60

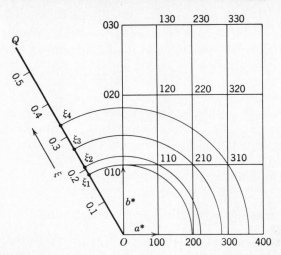

Figure 1-46 Illustration of the use of ξ-coordinates and a scale drawing of a reciprocal-lattice level to index fiber patterns.

satisfies the indexing of both the zero- and higher-level reflections.

A careful consideration of Figures 1-37 and 1-38 will show that for ideally oriented fibers in which c^* coincides with the fiber axis reciprocal-lattice points with indices $(00l)$ cannot intersect the sphere of reflection, so that reflections cannot appear on the meridian of the diffraction pattern unless the fiber is somewhat inclined with respect to the direct beam. The same will be true of reciprocal-lattice points with small h and k indices on the higher levels that describe circles too small to intersect the sphere of reflection. Nevertheless, actual fibers frequently possess a sufficiently wide spread of orientations about the ideal fiber axis to generate meridional reflections when perpendicular to the x-ray beam.

Because certain reflections in a given layer line may possess nearly identical ξ values, it is frequently true of fiber patterns that some pairs of reflections overlap to such an extent that the assignment of unique indices is not possible. Thus in Figure 1-46 the ξ values of (200) and (010) are too similar to permit a choice to be made between them. However, the subsequent analysis of the crystal structure[37] demonstrated that (010) was extinguished for reasons of structural symmetry, thus establishing (200) as the correct indices. The indexing procedures just described become more difficult to apply when the a^*b^* net is not rectangular and especially so when the c^*-axis of the reciprocal lattice is not perpendicular to the a^*b^* net—as is true of triclinic unit cells, and also of monoclinic cells when a or c, rather than the unique axis b, coincides with the fiber axis. The validity of unit cells deduced in such circumstances is frequently subject to some uncertainty. These difficulties can sometimes be lessened by inducing in the polymer specimens higher modes of orientation, with the resultant resolution of certain reflections that are superposed in ordinary fiber patterns.

The subject of preferred orientation in polymers divides itself naturally into two parts: (a) the identification of the mode of orientation present in a given specimen and (b) the measurement of the degree of perfection with which the ideal mode is approached in the specimen. These topics are considered in Chapter 4.

1-5 SMALL-ANGLE SCATTERING BY POLYMERS

We have seen that x-rays are, in effect, scattered by the electrons in matter and that interference, or diffraction, occurs between x-rays scattered by the electron "clouds" that surround the various atomic centers. In a more general sense we may say that the x-ray interference effects result from variations in electron density from one point to another in the material. We have also seen that, if the atoms are regularly arranged according to a space lattice, the scattering angle 2θ is related to the interplanar spacing d by the Bragg equation, $n\lambda = 2d \sin \theta$. This formula indicates the existence of a reciprocal relationship between the interatomic distance and $\sin \theta$. It is also important to note in this relationship that λ and d are of the same order of size, which results in the appearance of diffraction effects over a wide range of angles (wide-angle diffraction).

Because of the reciprocity between interatomic distance and $\sin \theta$, inhomogeneities of colloidal dimensions generate x-ray scattering and interference effects at very small angles, typically less than $2°$ with the wavelength of $CuK\alpha$, 1.542 Å. It should be emphasized that this small-angle scattering has no dependence on the inhomogeneities of atomic dimensions that give rise to the wide-angle diffraction. From the standpoint of the small-angle scattering, the concentrations of electrons on the atomic sites, which constitute the basis for the description of crystal structures, might as well be replaced by continuous distributions of electrons within the unit cells. Only the fluctuations in electron density over much larger distances, typically 30 to 1000 Å, determine the nature of the small-angle scattering. Therefore rather perfect single crystals, pure phases in general, and other homogeneous substances do not scatter x-rays at very small angles.

The two kinds of inhomogeneity that are most likely to be responsible for small-angle x-ray scattering from solid polymers are (a)
alternation of crystalline and amorphous regions, which in general
will possess different electronic densities, and (b) the presence of
microvoids dispersed throughout the solid polymer matrix[40–42].
The intensity of small-angle scattering increases with the degree of
contrast between the electron densities of the two or more kinds of
regions that produce the heterogeneity. Thus it is relatively large
for a system of discrete particles separated by interstices or the
inverse, a dispersion of micropores in a solid medium, but relatively
small for a heterogeneous system of crystalline and amorphous
regions that differ but slightly in density. Likewise, colloidal solutions
of globular protein molecules or diluted fibrillar systems, such as
cellulose swollen in water or alkali solutions, scatter x-rays with an
intensity that is dependent on the difference in the electron densities
of the solvent and the "particles."

The experimental requirements for useful small-angle-scattering
experiments are very exacting in comparison with the requirements
for wide-angle measurements. This is due, in the first place, to the
necessity of measuring the scattered radiation at angles that are very
close to the undeviated beam, which in turn requires very sharp
collimation of the incident x-rays. Not only is well-designed and
-engineered apparatus required but the alignment of the components
is critical, and long periods of time are required for recording the
weakly scattered x-rays. In the second place, it is usually necessary to
evacuate the apparatus in order to eliminate air-scattered x-rays,
which at small angles are of sufficient intensity to distort or mask
features of the small-angle scattering emitted by the specimen.
Third, for the analysis of diffuse small-angle scattering it is mandatory
that wavelengths other than the desired characteristic radiation
(for example, CuKα with $\lambda = 1.542$ Å) be removed by means of a
crystal monochromator, although under some circumstances balanced
filters suffice. (See Section 2-4.3.) In recent years the quantitative
analysis of diffuse scattering has been greatly improved by replacing
photographic cameras with counter diffractometers. For the study
of discrete small-angle interferences, such as shown in Figure
1-20, the use of β-filtered radiation is often acceptable unless a
quantitative knowledge of the intensities and reflection profiles is
desired, in which case more strictly monochromatic x-rays are
preferable.

GENERAL REFERENCES

L. V. Azaroff and M. J. Buerger, *The Powder Method in X-Ray Crystallography*, McGraw-Hill, New York, 1958.

J. Bandrup and E. H. Immergut (editors), *Polymer Handbook*, Interscience, New York, 1966.

C. W. Bunn, *Chemical Crystallography*, 2nd ed., Oxford Press, London, 1961.

G. L. Clark (editor), *Encyclopedia of X-Rays and Gamma Rays*, Reinhold, New York, 1963.

B. D. Cullity, *Elements of X-Ray Diffraction*, Addison-Wesley, Reading, Mass., 1956.

P. H. Geil, *Polymer Single Crystals*, Interscience, New York, 1963.

B. Ke (editor), *Newer Methods of Polymer Characterization*, Interscience, New York, 1964. Chapter 4, "Optical Methods," by R. S. Stein and Chapter 6, "Small-Angle X-Ray Studies," by W. O. Statton.

H. P. Klug and L. E. Alexander, *X-Ray Diffraction Procedures*, Wiley, New York, 1954.

H. S. Peiser, H. P. Rooksby, and A. J. C. Wilson (editors), *X-Ray Diffraction by Polycrystalline Materials*, Institute of Physics, London, 1955.

W. O. Statton, "Characterization of Polymers," in *Handbook of X-Rays in Research and Analysis*, E. Kaelble (editor), McGraw-Hill, New York, 1967. Chapter 21.

H. A. Stuart (editor), *Die Physik der Hochpolymeren*, Springer-Verlag, Berlin, 1955. Volume 3, *Ordnungszustände und Umwandlungsercheinungen in festen Hochpolymeren Stoffen.*

J. W. Turley, *X-Ray Diffraction Patterns of Polymers*, The Dow Chemical Company, Midland, Mich., January 1965.

SPECIFIC REFERENCES

[1] R. Bonart, R. Hosemann, F. Motzkus, and H. Ruck, *Norelco Reporter*, 7, 81 (1960).

[2] R. Hosemann, *J. Appl. Phys.*, 34, 25 (1963).

[3] P. H. Lindenmeyer, *J. Polymer Sci.*, Part C, 1, 5 (1963).

[4] D. A. Zaukelies, *J. Appl. Phys.*, 33, 2797 (1962).

[5] V. F. Holland, *J. Appl. Phys.*, 25, 1351, 3241 (1964).

[6] C. W. Bunn and E. V. Garner, *Proc. Roy. Soc. (London)*, A189, 39 (1947).

[7] G. Natta and P. Corradini, *Nuovo Cimento*, 15, Suppl. 1, 111 (1960).

[8] G. Natta, P. Pino, P. Corradini, F. Danusso, E. Mantica, G. Mazzanti, and G. Moraglio, *J. Amer. Chem. Soc.*, 77, 1708 (1955).

[9] P. H. Till, *J. Polymer Sci.*, 24, 301 (1957).

[10] A. Keller, *Phil. Mag.*, 2, 1171 (1957).

[11] E. W. Fischer, *Z. Naturforsch.*, 12a, 753 (1957).

[12] K. H. Storks, *J. Amer. Chem. Soc.*, 60, 1753 (1938).

[13] K. H. Storks, *Bell Laboratories Record*, 21, 390 (1943).

[14] W. Schlesinger and H. M. Leeper, *J. Polymer Sci.*, 11, 203 (1953).

[15] R. Jaccodine, *Nature*, 176, 305 (1955).

[16] D. C. Bassett, A. Keller, and S. Mitsuhashi, *J. Polymer Sci.*, Part A, 1, 763 (1963).

[17] A. Keller, *Polymer*, 3, 393 (1962).

[18] P. H. Geil, *Polymer Single Crystals*, Interscience, New York, 1963.

[19] H. J. Leugering, *Kolloid Z.*, 172, 184 (1960).

[20] J. H. Magill, *J. Polymer Sci.*, Part A, 3, 1195 (1965).

[21] C. W. Bunn and T. C. Alcock, *Trans. Faraday Soc.*, 41, 317 (1945).

[22] K. Hess and H. Kiessig, *Z. physik. Chem.*, 193, 196 (1944).

[23] A. H. Compton, *Phys. Rev.*, 21, 715 (1923); 22, 409 (1923).

[24] *International Tables for X-Ray Crystallography*, Vol. III, Kynoch Press, Birmingham, 1962, p. 248.

[25] Ibid., pp. 250–252.

[26] S. Ergun in *Chemistry and Physics of Carbon*, Vol. 3, P. L. Walker, Jr. (editor), Marcel Dekker, New York, 1968, pp. 211–288.

[27] P. P. Ewald, Z. *Krist.*, **56**, 129 (1921).

[28] *International Tables for X-Ray Crystallography*, Vol. II, Kynoch Press, Birmingham, 1959, p. 106; M. J. Buerger, *X-Ray Crystallography*, Wiley, New York, 1942, p. 360.

[29] M. von Laue, *Sitz. math. phys. Klasse bayer. Akad. Wiss.*, p. 303 (1912); *Ann. Physik*, **41**, 971 (1913).

[30] W. L. Bragg, *Proc. Cambridge Phil. Soc.*, **17**, 43 (1913).

[31] J. D. Bernal, *Proc. Roy. Soc. (London)*, **A113**, 117 (1926).

[32] P. M. de Wolff, *J. Polymer Sci.*, **60**, S34 (1962).

[33] *International Tables for X-Ray Crystallography*, Vol. II, Kynoch Press, Birmingham, 1959, pp. 291–312.

[34] P. Debye, *Ann. Physik*, **46**, 809 (1915).

[35] A. W. Hull and W. P. Davey, *Phys. Rev.*, **17**, 549 (1921).

[36] W. P. Davey, *A Study of Crystal Structure and Its Applications*, McGraw-Hill, New York, 1934, p. 603.

[37] G. Natta and P. Corradini, *J. Polymer Sci.*, **20**, 251 (1956).

[38] C. W. Bunn, *Chemical Crystallography*, 2nd ed., Oxford Press, London, 1961, pp. 153–154, 460.

[39] M. J. Buerger, *X-Ray Crystallography*, Wiley, New York, 1942, pp. 137 ff.

[40] P. H. Hermans, D. Heikens, and A. Weidinger, *J. Polymer Sci.*, **35**, 145 (1959).

[41] W. O. Statton, *J. Polymer Sci.*, **22**, 385 (1956).

[42] W. O. Statton, *J. Polymer Sci.*, **58**, 205 (1962).

2

Instrumentation

2-1 PRINCIPAL POLYMER DIFFRACTION TECHNIQUES

In appropriate circumstances practically all x-ray diffraction equipment useful in the study of solid materials is also applicable to polymers. That a diversity of experimental techniques may be profitably employed is a direct consequence of the wide range of states of order and crystalline texture assumed by polymers. Thus the apparatus that is best suited to obtaining a specific kind of information (for example, presence or absence of crystallinity, size of crystalline regions, degree of crystallinity, mode of preferred orientation, porosity, and crystal structure) may be quite different from that suited to another goal. In fact the skilled worker may sometimes be obliged to desert the classical experimental arrangements and devise novel procedures to seek out information of previously uncharted types. Lest the newcomer be discouraged, however, it should be emphasized that three well-standardized diffraction techniques serve very satisfactorily for the great majority of polymer analyses. These techniques, which are given primary emphasis in this chapter, are the following:

1. Transmission technique with flat film (sometimes referred to as Laue technique).
2. Fiber-diffraction technique with a cylindrical camera.
3. Small-angle-scattering technique with (a) pinhole or (b) slit collimators.

2-1.1 Photographic versus Counter Technique

Diffraction procedures useful in polymer studies may also be classified into two types according to whether the patterns are recorded photographically or by means of a counter diffractometer

66

(so-called direct recording of the diffracted intensities). In photographic recording the entire portion of the pattern that is of interest is registered simultaneously, which offers the following advantages:

1. It is not necessary for the x-ray source to be highly stable.
2. The entire diffraction pattern (or a substantial portion of it) can be inspected at once.

The principal weaknesses of the photographic method are the following:

1. Exposures of several or many hours are required.
2. Intensities cannot be determined with highest quantitative accuracy because of the inherent difficulties that affect the conversion of photographic-film blackening to intensity [1].

The prime advantage of counter registration is the relatively high accuracy with which intensities can be measured. In principle the accuracy of measurement is limited only by the number of counts recorded since the standard deviation (σ) in measuring N counts is $N^{1/2}$ and the relative standard deviation is $N^{-1/2}$. On the other hand, counter measurements suffer from two principal disadvantages: (a) high stability of the x-ray source is mandatory since the various elements of the diffraction pattern are recorded sequentially rather than simultaneously (as in the case of film), with the result that the measurements are likely to extend over a considerable period of time, and (b) as a general rule it is not practicable to survey the diffraction pattern as a whole. With respect to the sensitivity of counter measurements, it is instructive to observe that 5000 $CuK\alpha$ x-ray photons are registered on photographic film as a small spot of minimum detectable intensity [2], whereas 5000 photons can be registered by a counter with an accuracy of 1.4% when the background is negligible.

We may appropriately conclude this introductory section by emphasizing the special importance of the flat-film technique for analyzing polymers. Since nearly complete diffraction patterns of unoriented and axially oriented polymers can be recorded on flat films by means of exposures of several hours, such photographs constitute convenient, compact, and characteristic pictorial records of the various specimens under investigation. Even a brief inspection of a pattern can tell the experienced investigator a great deal about the constitution of a new polymer specimen; for example, it may supply the answers to these basic questions:

1. Is the specimen crystalline or amorphous?
2. Is the ordering, if present, two- or three-dimensional?

3. Approximately what fraction is crystalline?
4. Is preferred orientation of the crystallites present? If so, what is the degree of alignment?
5. Is the crystalline development relatively perfect or imperfect?
6. If the pattern shows a high degree of orientation, what is the translation period along the fiber axis?
7. Is a particular specimen similar or dissimilar to various others of similar chemical constitution that are encompassed by an investigation?

2-2 PREPARATION AND MOUNTING OF THE SPECIMEN

Undoubtedly the preparation and mounting of the specimen constitutes the most important experimental element in the x-ray analysis of polymers. The principles and techniques that are involved are to a large degree independent of the method employed for registering the pattern (photographic or diffractometric) and of the particular photographic camera used. Therefore we shall attempt to present a somewhat unified treatment of specimen handling preliminary to a description of the most useful cameras and diffractometric apparatus.

2-2.1 Geometry of Transmission Techniques

In a number of techniques the pattern is generated by transmission of the direct beam through a flat polymer sheet. When the direct beam is incident perpendicularly on the sheet (*normal-beam-transmission technique*) and when diffraction effects at only rather low angles are to be observed, the optimum thickness of the specimen (corresponding to maximum intensity of the diffracted beam) is given to a good approximation by [3]

$$t_m = \frac{1}{\mu},$$ (2-1)

μ being the linear absorption coefficient of the specimen for the x-ray wavelength employed. (Refer again to Section 1-3.1.) When $CuK\alpha$ radiation is employed, $\mu/\rho = 5$ to 7 cm^2/g for typical polymers composed only of light elements (carbon, oxygen, nitrogen, and hydrogen) from which μ ranges from 5 to 8 cm^{-1} for the usual density range of 1.00 to 1.25. [Table 2-1 presents an illustrative calculation of the linear absorption coefficient of a polymer, poly(hexamethylene adipamide).] Equation 2-1 leads to an optimum specimen thickness of about 0.2 cm for low-angle diffraction. Figure 2-1 shows that for larger diffraction angles (2θ), however, the path of the beam through the specimen, $l_1 + l_2$, is appreciably increased, causing increased absorption of the x-rays as well as geometrical broadening of the reflections. For these reasons the usual practice is to adopt a specimen thickness of 0.5 to 1.0 mm.

Table 2-1 Calculation of the Linear Absorption Coefficient of Poly(hexamethylene adipamide) for CuKα Radiation

Chemical composition: $C_{12}H_{22}N_2O_2$

Molecular weight: $\sum_i n_i A_i = 226.32$

Densities (g/cm³):
  Crystalline, 1.24
  Amorphous, 1.09

Mass absorption coefficient:[a]

$$\frac{\mu}{\rho} = \sum_i w_i \left(\frac{\mu}{\rho}\right)_i$$

Atom	n	A	μ/ρ[b] (cm²/g)	$n_i A_i$	w_i	$w_i(\mu/\rho)_i$
Carbon	12	12.011	4.60	144.132	0.6368	2.929
Hydrogen	22	1.008	0.435	22.176	0.0980	0.043
Nitrogen	2	14.008	7.52	28.016	0.1238	0.931
Oxygen	2	16.000	11.5	32.000	0.1414	1.626
				226.32	1.0000	5.529

Mass absorption coefficient (cm²/g): 5.529
Linear absorption coefficients (cm⁻¹):

  Crystalline, $1.24 \times 5.529 = 6.83$
  Amorphous, $1.09 \times 5.529 = 6.03$

[a]Equation 1-5.
[b]Numerical values from *International Tables for X-Ray Crystallography*, Vol. III, Kynoch Press, Birmingham, 1962, p. 162. See also Appendix 4.

When the normal-beam-transmission technique is used for accurate intensity measurements, it is necessary to apply absorption corrections to the experimental intensities for 2θ values differing much from 0°. With reference to Figure 2-1, let A be the cross-sectional area of the direct beam where it enters the specimen of thickness t, and let I_0 be the intensity diffracted by unit volume of the specimen at the

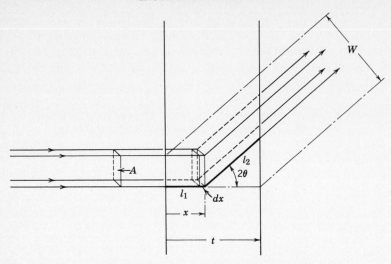

Figure 2-1   Geometry of the normal-beam-transmission technique.

angle 2θ under the hypothetical condition of no absorption. Then the intensity diffracted at the angle 2θ by the volume element $A\,dx$ at depth x is

$$dI = I_0 A \exp\left[-\mu\left(l_1 + l_2\right)\right] dx$$

$$= I_0 A \exp\left[-\mu\left(x + \frac{t-x}{\cos 2\theta}\right)\right] dx. \tag{2-2}$$

The total intensity diffracted at the angle 2θ is obtained by integrating (2-2) between the limits $x = 0$ and t:

$$I_{2\theta} = I_0 A \exp\left(-\mu t \sec 2\theta\right) \int_0^t \exp\left[-\mu x\left(1 - \sec 2\theta\right)\right] dx$$

$$= \frac{I_0 A}{\mu(1 - \sec 2\theta)} \exp\left(-\mu t \sec 2\theta\right) \left\{1 - \exp\left[-\mu t\left(1 - \sec 2\theta\right)\right]\right\}. \tag{2-3}$$

For $2\theta = 0°$ the total diffracted intensity would be

$$I_{0°} = I_0 A t \exp\left(-\mu t\right). \tag{2-4}$$

Therefore the absorption correction to be applied to intensities measured at $2\theta > 0°$ is

$$\frac{I_{0°}}{I_{2\theta}} = \frac{\mu t(1 - \sec 2\theta)}{\exp\left[\mu t(1 - \sec 2\theta)\right] - 1} \qquad \text{(normal-beam transmission). (2-5)}$$

Figure 2-2 shows the geometry of the transmission technique with

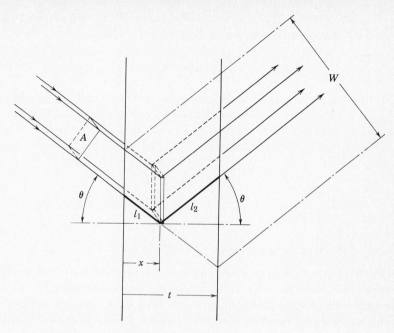

Figure 2-2   Geometry of the symmetrical-transmission technique.

the specimen symmetrically inclined with respect to the incident and diffracted beams (*symmetrical-transmission technique*). This arrangement is mainly suited to diffractometry and has the advantage that the diffraction pattern may be measured at angles as large as $2\theta = 90°$ without the absorption correction becoming excessive because of obliquity of the incident or diffracted beams.

Referring to Figure 2-2, we see that, if the cross-sectional area of the incident beam is A, the infinitesimal volume element irradiated at depth x is $A \sec \theta \, dx$, and the intensity diffracted at the angle 2θ by this element is

$$dI = I_0 A \sec \theta \exp \left[-\mu (l_1 + l_2) \right] dx$$
$$= I_0 A \sec \theta \exp \left(-\mu t \sec \theta \right) dx. \qquad (2\text{-}6)$$

Proceeding as before, we obtain for the total intensity diffracted by the specimen at the angle 2θ

$$I_{2\theta} = I_0 A \sec \theta \int_0^t \exp \left(-\mu t \sec \theta \right) dx$$
$$= I_0 A t \sec \theta \exp \left(-\mu t \sec \theta \right). \qquad (2\text{-}7)$$

Division of (2-4) by (2-7) gives for the absorption correction to be applied to intensities measured at $2\theta > 0°$

$$\frac{I_{0°}}{I_{2\theta}} = \frac{\exp\left[-\mu t(1 - \sec\theta)\right]}{\sec\theta} \quad \text{(symmetrical transmission).} \quad (2\text{-}8)$$

For the symmetrical-transmission geometry the specimen thickness that corresponds to maximum diffracted intensity as a function of θ can be found by differentiating (2-7) with respect to t and equating to 0. The result is

$$t_m = \frac{1}{\mu \sec\theta}, \quad (2\text{-}9)$$

which may be compared with (2-1) for low-angle diffraction with a normally incident beam.

2-2.2   Unoriented Specimens

Specimens that are composed of randomly oriented crystalline and amorphous material (refer again to Sections 1-4.4 and 1-4.5) can be mounted with simple techniques since there is no texture to furnish an axis or plane of reference in the specimen (which would require a particular orientation with respect to the x-ray beam and the photographic film or counter). For Debye-Scherrer cameras the specimens should be, or at least approximate, cylinders of about 1-mm diameter. Cylindrical specimens may be prepared by (a) filling a thin-walled plastic or glass tube with bulk polymer powder, (b) molding bulk polymer into cylindrical shape, (c) molding or rolling bulk powder into cylindrical shape on a supporting wire or glass fiber using a little adhesive, (d) mechanically cutting or turning a molded specimen into cylindrical shape, or (e) building up a sheet of many thin lamellae and then trimming to a roughly cylindrical cross section.

For flat-film photographs cylindrical specimens can also be used, but it is usually more convenient to employ sheet specimens in conjunction with the transmission technique and perpendicularly incident beam, as discussed in the preceding section. Bulk polymer powders can be mounted without introducing undesirable binding media by pressing them firmly in a hole, about 5 mm in diameter, drilled in a metal plate of the desired thickness. The specimen plate may then be affixed directly to the end of the collimator by means of a small bracket or it may be mounted on a specimen stand between the collimator and flat-film cassette. Frequently it is possible to press polymer powders, without adhesives and with only light pressure to avoid deformation of the microstructure, into coherent pellets suitable for x-ray analysis. If an amorphous binder is required—such as gum tragacanth, collodion, or Duco cement—it must be remembered that such materials add a certain amount of diffuse x-ray scatter to the pattern of the polymer itself. Therefore amorphous binders should be employed only when absolutely necessary, especially when poorly crystalline polymers are being investigated[4].

2-2.3   Oriented Specimens

Natural and synthetic fibers exhibit various degrees of *axial orientation*. (See Section 1-4.6.) Small- and wide-angle photographs are most commonly prepared with the fiber axis normal to the incident x-ray beam, the direction on the film parallel to the fiber axis being designated the *meridian* and the direction at right angles the *equator*. A location on the film can be designated by a radial distance r from the undeviated beam and an azimuth β with respect to the meridian. These conventions are illustrated in Figure 2-3.

For studies of polymer sheets and films, and more comprehensive studies of fibers, it is necessary to use a versatile specimen holder such as is shown in Figure 2-4.† The bundle of fibers or sheet specimen cut to the shape of a strip is placed between the clamps (1) and tightened with the screws (2). The specimen is brought to a desired degree of extension by action of the screws (3), and the overall extension is read on the scale (4). The 360° azimuthal circle (5) permits the fiber axis to be tilted at any desired azimuth with the vertical. Alternatively the specimen can be rotated about circle (5) at constant speed by means of the synchronous motor (6), permitting azimuthally averaged photographs to be prepared. This capability is useful when it is necessary to average out the effects of preferred orientation. By employing counter registration and synchronizing a strip-chart potentiometer recorder with rotation of the specimen, a useful chart record of intensity versus azimuth can be obtained. The device depicted in Figure 2-4 is available with or without the horizontal setting circle (7), which varies the inclination of the specimen with respect to the x-ray beam—an adjustment that is important for more detailed measurements of preferred orientation, as discussed in Chapter 4.

Figure 2-5 shows a very useful device for building up and compressing laminated specimens as well as for subsequently supporting

†Rigaku Denki Company, Ltd., 9-8 2-Chome Sotokanda Chiyoda-ku, Tokyo, Japan. Available in the United States from the General Electric Company, Analytical Measurement Business Section, West Lynn, Massachusetts 01905.

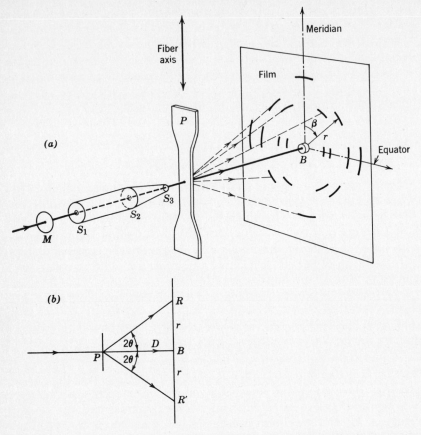

Figure 2-3 Geometry of diffraction by specimens with axial (fiber) orientation: (*a*) schematic perspective; (*b*) dimensional relationships.

them during the diffraction experiments. It is designed primarily for mounting over the inner end of the beam collimator of the general-purpose camera described in Section 2-3.5 and shown in Figure 2-28. However, it can also be employed in other photographic techniques and in the diffractometric measurement of pole-figure data. (See Section 4-3.4.)

For the analysis of specimens with the incident beam parallel to, or making small angles with, the machine (*M*) or transverse (*T*) direction, appropriately cut strips of polymer film or sheet are stacked in the cavity (1) and the slider (2) is moved to the left, causing the tongue (3) to enter the cavity and compress the layers together. The loose ends are then trimmed flush with the surfaces of the plate (4),

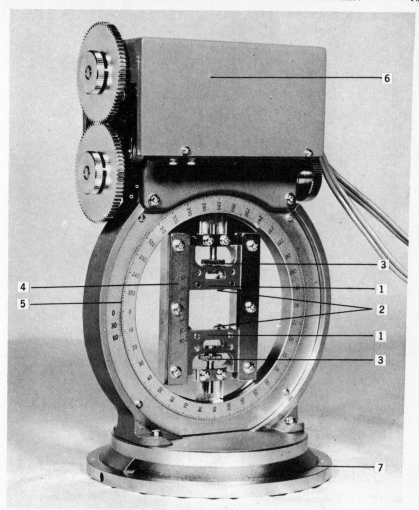

Figure 2-4 Versatile holder for oriented specimens. (Courtesy of General Electric Company, Analytical Measurement Business Section.)

thus providing a specimen whose thickness is the same as that of the plate (4). For intensity measurements with the beam parallel to, or making small angles with, the normal direction (N), a stack of layers of the desired thickness is affixed to the plate (4) so as to cover the open cavity. Typical dimensions of tongue and cavity are 3.4 × 3.4 mm in the plane of the slide and 1.5 mm in thickness.

A simple manually adjustable azimuthal holder constructed from a

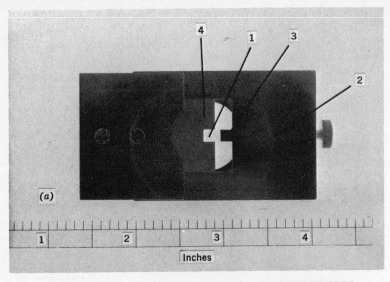

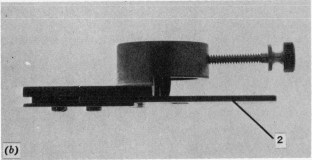

Figure 2-5 Device for compressing and supporting laminated polymer specimens during diffraction experiments: (*a*) bottom view; (*b*) edge view. (Courtesy of E. S. Clark and W. H. Warhus Company.)

machinists's bevel protractor has been employed by the author in diffractometric measurements of elastomers [5]. (See Chapter 3, p. 154.) Other useful azimuthal holders have been described in the literature [6–8]. More sophisticated specimen-orienting apparatus for the diffractometric preparation of complete pole figures is described in Chapter 4.

2-2.4 Mounting Fibers

Because of the small diameter of individual fibers it is necessary for x-ray analysis that a considerable number be assembled into a parallel bundle. Figure 2-6 shows how long fibers may be wound tightly on a small rectangular frame, parallel to one another and several layers deep, to furnish a specimen of sufficient thickness to yield a reasonably intense diffraction pattern [9]. Short-cut (staple) filaments are much more difficult to handle and perhaps may be best mounted individually by gluing the ends, admittedly a tedious process. Statton[4] describes a combination pinhole collimator and fiber holder that facilitates the preparation of fiber bundles for photographic analysis (Figure 2-7). A stainless-steel cylinder that holds the lead pinhole system has, as an integral component, a little shelf, located beside the outlet pinhole, on which the fibers are wound. By varying the number of turns in one layer and the number of layers, the effective thickness and width of the specimen intercepting the primary beam can be regulated as desired.†

2-2.5 Specimen Arrangements for Diffractometry

Polymer patterns can be measured with an x-ray counter diffractometer by means of one of the three geometrical arrangements shown in Figure 2-8. The optimal specimen thickness and correction for absorption of x-rays by the specimen have already been discussed for arrangements *b* and *c*. (See Section 2-2.1 and Figures 2-1 and 2-2.) Method *a* is the conventional geometry for x-ray diffractometry of

†Obtainable from the W. H. Warhus Company, 406 Rowland Park Boulevard, Carrcroft, Wilmington, Delaware 19803.

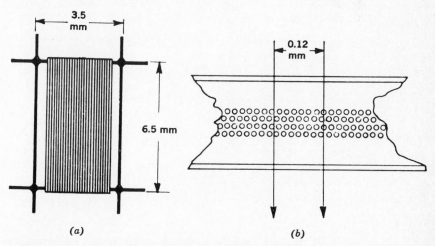

(a)　　　　　　　　　　　　(b)

Figure 2-6 (*a*) Wire frame with fibers wound on it. (*b*) Top view of cross section through cell containing the sample. (Hermans, Heikens, and Weidinger [9].)

Figure 2-7 Combination fiber holder and pinhole collimator. (Statton[4].)

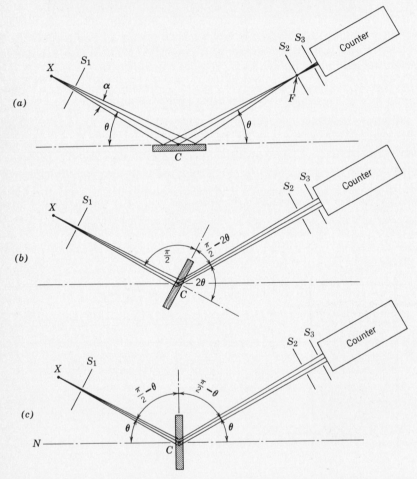

Figure 2-8 Three geometrical arrangements of specimen with respect to x-ray source and counter for diffractometry: (*a*) symmetrical reflection; (*b*) normal-beam transmission; (*c*) symmetrical transmission.

crystalline powders, which with certain limitations can likewise be applied to polymers. It is most directly suited to molded or laminar specimens with flat surfaces. As indicated in Figure 2-8a, the x-ray beam diverges from a sharp line focus X (represented by a point in the diagram) as limited by the divergence slit S_1 to a small angular range α, usually 1 to 4°. If the distances XC and CF are equal and if the surface of the specimen makes equal angles (θ) with the direct and diffracted beams as shown, the rays diffracted from the surface of the flat specimen converge to an approximate line focus at F, pass through the receiving slit S_2 and scatter slit S_3, and enter the counter window.

It is customary to prescribe as the minimum acceptable thickness of the flat specimen a value that is large enough to yield maximum diffracted intensity, which is [10]

$$t \geqslant \frac{3.2}{\mu} \frac{\rho}{\rho'} \sin \theta, \qquad (2\text{-}10)$$

wherein ρ and ρ' are respectively the densities of the solid material constituting the powder and of the powder (including interstices). Since for solid polymers $\rho = \rho'$ and μ for $CuK\alpha$ radiation is about 6 cm^{-1}, it is evident that thicknesses of several millimeters are required to satisfy the foregoing criterion, even at moderately small diffraction angles. In actual practice such a large thickness is seldom acceptable because the theoretically sharp focus of diffracted rays at F is greatly broadened at the smaller Bragg angles, distorting the profiles of crystalline reflections and necessitating the use of very wide receiving and scatter slits to avoid loss of diffracted intensity. An inspection of Figure 2-9 shows that a flat specimen of thickness t,

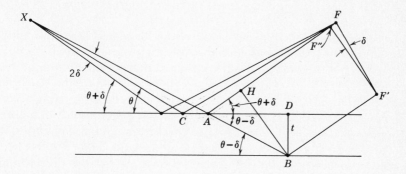

Figure 2-9 Broadening of focus of x-rays diffracted by a low-absorbing specimen in diffractometry.

where t is smaller than specified by (2-10), broadens the theoretical focus F over the range FF'. From the geometry of the figure it can be seen that the linear breadth of the focus is

$$FF' = \frac{F'F''}{\cos \delta}$$

$$= \frac{HB}{\cos \delta}$$

$$= \frac{t \sin 2\theta}{\sin (\theta - \delta) \cos \delta}$$

$$= \frac{2t \sin \theta \cos \theta}{(\sin \theta \cos \delta - \cos \theta \sin \delta) \cos \delta}.$$

To a very good approximation we may let $\sin \delta = \delta$ and $\cos \delta = 1$, so that

$$FF' = \frac{2t \cos \theta}{1 - \delta \cot \theta},$$

and with $XC = CF = d$, the angular displacement of F' from F is

$$\Delta 2\theta = -\frac{57.3}{d} \times FF' \text{ degrees}$$

$$= -\frac{114.6\, t}{d} \frac{\cos \theta}{1 - \delta \cot \theta} \text{ degrees.} \qquad (2\text{-}11)$$

Since the receiving slit S_2 is centered on F, in order to encompass the entire diffracted beam its angular width ω_s must not be less than twice the magnitude of $\Delta 2\theta$ given by (2-11), or approximately

$$\omega_s \geqslant 229.2 \frac{t}{d} \frac{\cos \theta}{1 - \delta \cot \theta} \text{ degrees} \qquad \text{(symmetrical reflection).} \qquad (2\text{-}12)$$

In order to minimize broadening of the focus and distortion of the diffraction-line profiles of low-absorbing specimens, such as polymers, it is best to employ sheet specimens that are much thinner than specified by (2-10) (for example, of 0.5- to 1.0-mm thickness for CuKα radiation) and then to apply a correction factor to compensate for the variation of absorption with 2θ, as has been already explained for transmission techniques. The correction factor for the symmetrical-reflection technique is easily derived from Figure 2-10. Once again letting A be the cross-sectional area of the direct beam at its point of intersection with the specimen, and assuming parallel radiation for simplicity, we can write for the intensity diffracted at

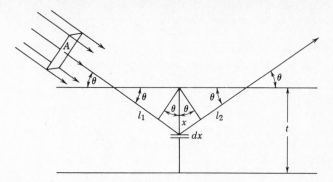

Figure 2-10 Geometry of diffraction of x-rays by the symmetrical-reflection technique.

the Bragg angle θ by an infinitesimal thickness dx at depth x:

$$dI = I_0 A \csc \theta \exp (-2\mu x \csc \theta) \, dx.$$

Integration between the limits $x = 0$ and t gives for the total intensity diffracted at angle θ

$$I_t = \frac{I_0 A}{2\mu} [1 - \exp (-2\mu t \csc \theta)], \tag{2-13}$$

and for a value of t that is large enough to satisfy (2-10), which is effectively ∞,

$$I_\infty = \frac{I_0 A}{2\mu}. \tag{2-14}$$

Therefore the absorption correction to be applied to intensities from specimens that are thinner than prescribed by (2-10) is

$$\frac{I_\infty}{I_t} = [1 - \exp (-2\mu t \csc \theta)]^{-1} \quad \text{(symmetrical reflection).} \tag{2-15}$$

The reader is referred to the literature [11, 12] for a detailed treatment of the correction of line shapes for the effect of low absorption by the specimen.

For practical reasons the symmetrical-reflection technique should be used only at 2θ angles larger than about 30°. At smaller angles not only is the broadening of the focus F greater but the irradiated area of sample surface becomes large, and the requirement that the irradiated surface be truly planar becomes more important and at the same time more difficult to satisfy. In this connection it may be observed also that it is extremely difficult to mount an aggregate of fibers so as to present a large flat surface to the divergent x-ray beam of the technique shown in Figure 2-8a.

It is recommended, then, that the geometrical arrangement b or c of Figure 2-8 be used when $2\theta < 30°$. As explained in a preceding section, in general the symmetrical-transmission geometry c is superior to the normal-beam transmission geometry b; in fact, when the measurements do not extend to angles larger than 90°, it is sometimes quite satisfactory to dispense with the reflection technique a entirely and to record the intensities throughout the angular range of diffraction by the symmetrical-transmission method c. It must be emphasized, however, that the intensities recorded by transmission will be much smaller than those measured at the same Bragg angles by reflection since the parafocusing[13] principle is utilized in the latter but not in the former technique. This intensity disadvantage of the transmission technique is aggravated by the fact that it requires an incident beam of small divergence (usually 1° or less) to preclude serious loss of resolution as a consequence of divergence of the diffracted rays.

2-3 PHOTOGRAPHIC TECHNIQUES

2-3.1 Choice of Photographic Film

The same kinds of film that are employed in a diversified x-ray diffraction laboratory are also useful in the study of polymers. Except when very precise measurements are to be made, high-speed, relatively coarse-grained emulsions should be used in order to keep exposure times as short as possible. Examples of such films are Eastman type KK, Eastman No Screen, and Ilford type G. In addition to their graininess and higher speed, films of this class display a higher level of chemical fog than do slower, fine-grained films such as Eastman type AA and Ilford type B or C. Such fine-grained films are to be preferred when highest resolution of the details of a diffraction pattern is required, as when using microcameras or making certain kinds of small-angle-scattering measurements. For a detailed comparison of various domestic and foreign x-ray films the reader is referred to two articles in the literature[14, 15]. Manufacturers will usually furnish on request the speed, grain, and fog-level characteristics of the films that they supply.

2-3.2 Microdensitometry of X-Ray Photographs

Although counter diffractometers are finding increasing use in precise polymer studies, a well-equipped polymer diffraction laboratory must nevertheless provide facilities for the quantitative measurement of intensities from photographic patterns. To serve this function there should be available a modern recording micro-densitometer (microphotometer) that possesses the following capabilities.

1. Measurement of optical densities as high as 4.

2. Scanning of the pattern radially (direction r in Figure 2-3) in order to measure the intensity distribution as a function of 2θ.

3. Scanning of the pattern circularly about the $0° 2\theta$ point in order to determine the intensity distribution azimuthally (angle β in Figure 2-3).

It must be emphasized that well-engineered and versatile instruments with these basic features are inherently rather expensive. Several firms currently offer units of satisfactory quality.†

Figure 2-11 shows the optical system of a modern double-beam microphotometer based on the Dobson[16] design. Light from the lamp is chopped by the rotating shutter and passes along two identical optical paths to give superposed images on the sensitive surface of the photocell. The left-hand beam passes through a movable optical-density wedge controlled by a servomotor, while the right-hand beam passes through the specimen film F to be measured. The motor that rotates the shutter also drives a small a-c generator whose output bears a fixed phase relationship to the alternating emf produced by the photocell when the two beams are of unequal intensity. When the left-hand beam is the stronger, the output of the photocell is 90° in advance of that from the a-c generator; when the right-hand beam is stronger, the phase of the photocell output is 90° behind. An automatic control system makes use of this change of phase to bring the density wedge into a position of balance. The movement of the wedge is coupled to a potentiometer that records the light transmission by means of a pen-chart recorder. If the absorption gradient of the density wedge is properly chosen, relative optical density, rather than percent transmission, will be displayed directly on the chart recorder. The condensers C_1 and C_2 focus images of the straight lamp filament on both the wedge and the film under examination, and the objectives O_1 and O_2, and eyepieces E_1 and E_2, produce enlarged images of the lamp filament in the planes of the apertures

†Joyce, Loebl and Co., Ltd., Gateshead 11, England (distributed in the United States by National Instrument Laboratories, Inc., 12300 Parklawn Drive, Rockville, Maryland 20852); N. V. Nederlandsche Instrumentfabriek (Nonius), Delft, Netherlands (available in the United States through Enraf-Nonius, Inc., 130 County Courthouse Road, Garden City Park, New York 11040); Crystal Structures, Ltd., 339 Cherry Hinton Road, Cambridge, England.

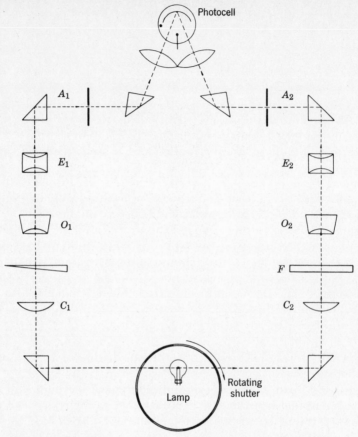

Figure 2-11 Optical system of a modern double-beam microphotometer. (Courtesy of Crystal Structures, Ltd.)

A_1 and A_2. Figure 2-12 shows the appearance of a modern double-beam automatic-recording microdensitometer.

In addition to the three basic capabilities stated above a functional microdensitometer for polymer studies should also provide the following features: (a) convenient manual positioning of the film carriage in X-Y translational coordinates, (b) choice of several aperture shapes and sizes, and (c) means for displaying an enlarged image of the general region of the film being examined in order to permit the diffraction spot in question to be located readily. For a treatment of the principles that govern the photographic recording and measurement of x-ray intensities the reader is referred to other sources [1, 17–19].

Figure 2-12 A modern double-beam automatic-recording microdensitometer. (Courtesy of Joyce, Loebl and Company, Ltd.)

2-3.3 Wide-Angle Cameras

Flat-Film Cameras. These cameras are sometimes referred to as Laue, monochromatic-pinhole, or transmission cameras. For the experimental arrangement the reader is referred once more to Figure 2-3. The direct beam passes through the β-filter M; is collimated by a pinhole system with apertures S_1, S_2, and S_3; penetrates the polymer specimen P; and is finally intercepted by the small, concave lead button B. The figure shows a drawn specimen, with axis of preferred orientation vertical, generating a typical fiber pattern of prominent equatorial reflections and less conspicuous interferences on the first- and second-layer lines. For highly monochromatic photographs the filter M must be replaced by a flat crystal monochromator, a topic that is discussed in some detail in Section 2-3.4. The circular apertures S_1 and S_2 define a beam of the order of 0.5 to 1.0 mm in diameter, and the guard aperture S_3 excludes from the film x-rays scattered by S_2. Even when the incident beam is crystal monochromatized, air-scattered x-rays tend to darken the central region of the film around B. This blackening can be (a) reduced by placing the beam stop as close as possible to the specimen without obscuring any significant details near the center of the diffraction pattern or (b) eliminated by evacuating the camera. The beam stop can be supported on a wire, or crossed wires, flexible enough to allow the stop to be displaced momentarily a few millimeters in order to imprint on the film an image of the direct beam, thus defining the $0°\,2\theta$ reference point. The film is held in a light-tight cassette that is loaded in the darkroom and then replaced in precisely its original position on the supporting bracket. The distance from sample to film (D in Figure 2-3b) may normally be made equal to one-half the width of the diffraction pattern wanted, which permits the registration of diffracted rays up to $2\theta = 45°$. Thus for a diffraction pattern 10×10 cm in size D may be set equal to 5 cm, or it may be made somewhat smaller if a larger range of 2θ is needed.

Commercially produced flat cameras provide for the mounting of collimator, specimen, and film cassette on separate brackets that slide to any desired relative positions along a track, or ways. Such a flat-camera assembly is shown in Figure 2-13. It can be seen that in this unit the film cassette can be mounted in either the forward (transmission) or back-reflection location.

With reference to Figure 2-3b once more, it will be observed that the values of 2θ corresponding to particular diffraction effects on the film can be determined accurately only if the distance D is known precisely, since $\tan 2\theta = r/D$. However, because of the wide physical variabil-

Figure 2-13 Flat camera useful for transmission or back-reflection measurements. (Courtesy of General Electric Company, Analytical Measurement Business Section.)

'ity of polymer specimens, in practice it is difficult to measure D with high accuracy or to keep it reproducible during the preparation of diffraction patterns from successive specimens. For a series of specimens that are of the same thickness and mounted in an identical manner it is possible to maintain a reproducible sample-to-film distance with the aid of a special gauge block[20]; D can be determined with reasonable accuracy by replacing the polymer specimen with a standard crystalline substance, mounted identically, which produces sharp diffraction effects at known angles (for example, quartz or silicon powder).

A Buerger-type precession camera[21, 22] can be used to prepare excellent diffraction patterns of polymer fibers and films[23]. The technique is worthwhile mainly for highly oriented specimens, in which case the use of an appropriate layer-line screen results in a diffraction pattern that is an undistorted plane section of the Fourier transform of the specimen. This method also has the advantage of including the meridional reflections, which in principle are not accessible to the stationary flat-film technique. (See Chapter 1, p. 61.) Its chief disadvantage is that very long exposures are required when the layer-line screen is of sufficiently narrow aperture to provide good resolution. Figure 2-14 shows precession patterns of oriented polyethylene made with the beam incident on the specimen in three mutually perpendicular directions. For a detailed analysis of the precession method as applied to fibers the reader is referred to a paper by M. V. King[24].

Statton has designed an evacuatable box camera with pinhole collimators that, although primarily useful for small-angle-scattering studies, can also be employed effectively for wide-angle diffraction. For the latter purpose a fixed sample-to-film distance of 3 or 5 cm is used in conjunction with a $3\frac{1}{4} \times 4\frac{1}{2}$-in. ($8.25 \times 10.80$-cm) sheet of film. For a fuller description the reader is referred to Section 2-3.5.

Cylindrical Cameras. In Section 1-4.6 it was explained that crystalline polymer fibers with a high degree of axial orientation yield diffraction photographs resembling those obtained by rotating single crystals. Although the principal features of such fiber patterns can be recorded on flat films, a cylindrical camera is required for the registration of the complete patterns. (Refer again to Figures 1-34, 1-35, and 1-39.) A cylindrical camera fitted with a goniometer head for orienting single crystals constitutes a very satisfactory instrument for photographically recording the complete intensity data from fibers for x-ray structural analysis—as is illustrated, for example, by investigations of Holmes, Bunn, and Smith[25] and of Kinoshita[26].

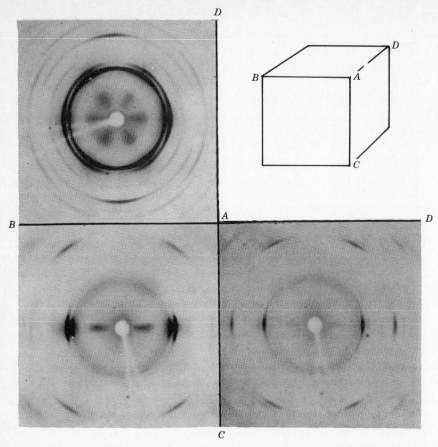

Figure 2-14 Precession photographs of an oriented polyethylene specimen cut in the shape of a small cube. Three perpendicular directions of incident beam as indicated by lettering of cube faces and patterns. Exposure 1 hr with nickel-filtered Cu$K\alpha$ radiation, 45 kVp, 15 mA; precession angle 30°, 20-mm layer-line screen 34.6 mm from specimen. (Courtesy of E. S. Clark and E. I. du Pont de Nemours and Company, Inc.)

Figure 2-15 shows a commercial general-purpose camera that can be easily adapted to fiber studies.† Part *a* is a close-up view of the beam collimator *H* and the goniometer head *G*, which has two arcs and two translating slides for tilting and centering the specimen, respectively. Figure 2-15*b* shows the camera with a telemicroscope *T* in place for viewing the specimen and a flat-film cassette *F* in the back-reflection position. Alternatively the flat cassette can be placed in the forward position. In part *c* the camera is seen with a semicylindrical

† Obtainable from Pye Unicam, Ltd., York Street, Cambridge, England.

cassette S in place for recording diffraction effects below $2\theta = 45°$. Also available, but not shown, is a 360° cylindrical cassette of 30-mm radius for obtaining the complete diffraction pattern.

Strictly speaking, it is not necessary to rotate a polymer fiber during the preparation of the diffraction photograph (see again Section 1-4.6); however, rotation has the effect of ensuring truly cylindrical centering of the specimen in instances when the specimen is not itself cylindrically symmetrical or cannot be accurately centered for some reason. Figure 2-16 shows fiber photographs of polytetrafluoroethylene prepared in the cylindrical camera at 25°C (a) with nickel-filtered $CuK\alpha$ radiation and (b) with crystal-monochromatized $CuK\alpha$ radiation [27]. (See also Section 2-3.4.)

Debye-Scherrer Cameras. In Section 1-4.5 the geometry of Debye-Scherrer cameras was described and their usefulness for the

(a)

Figure 2-15 General-purpose camera useful for fiber studies. (Courtesy of Pye Unicam, Ltd.)

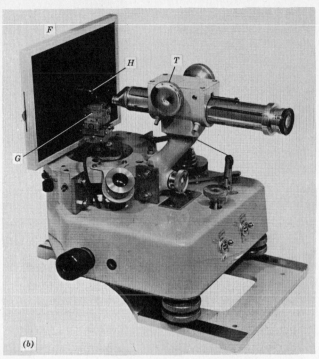

Figure 2-15 (Continued from page 89)

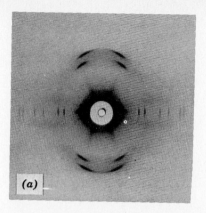 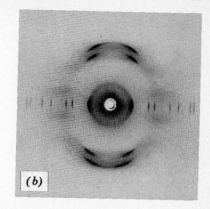

Figure 2-16 Fiber patterns of polytetrafluoroethylene prepared in a cylindrical camera at 25°C with (*a*) nickel-filtered and (*b*) crystal-monochromatized Cu$K\alpha$ radiation. (Courtesy of E. S. Clark.)

qualitative analysis of crystalline materials was pointed out. A reexamination of Figure 1-34 will show that a Debye-Scherrer camera is actually equivalent to a cylindrical camera of very limited height. The special utility of such a camera is to register the patterns of randomly oriented microcrystalline specimens, which of course do not need to be mounted on a goniometer head. Instead, short, rod-shaped specimens can be mounted on small cylindrical brass bases, about 0.125 in. in diameter, that fit directly in the specimen-holding and -centering devices of commercial Debye-Scherrer cameras [28]. In contrast to standard cylindrical cameras, Debye-Scherrer cameras require only a narrow film strip to record representative segments of all the diffraction circles comprising the diffraction pattern.

Figure 2-17 depicts the four important ways of placing the film in a Debye-Scherrer camera: (a) regular, (b) precision, (c) Straumanis asymmetric, and (d) Wilson asymmetric. For the analysis of well-crystallized inorganic and organic powders position *c* is widely preferred because it permits an accurate determination of the film radius from measurements of the diffraction pattern alone. From an inspection of Figure 2-17*c* it can be seen that the angular distance on the film between the centers of the low-angle diffraction arcs *a* and the high-angle arcs *b* is equal to 180°. This reference distance permits the calculation of the effective film radius and therefrom accurate values of 2θ and *d* for every line measured. Wilson [29] modified the Straumanis asymmetric arrangement as portrayed in Figure 2-17*d*,

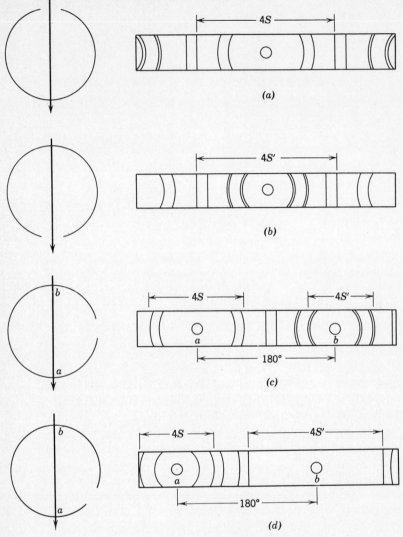

Figure 2-17 The four important Debye-Scherrer film positions: (*a*) regular; (*b*) precision; (*c*) asymmetric (Straumanis); (*d*) asymmetric (Wilson).

which enables the film radius to be determined when no reflections are present at values of 2θ larger than 90°. This geometry can evidently be useful in measuring some crystalline polymer patterns. (See Figure 2-20*c*.)

For the qualitative identification of microcrystalline substances it nearly always suffices to read the interplanar d spacings from the film directly with the aid of an appropriate scale. (See again Figure 1-35.) In fact this procedure is suited to the majority of polymer patterns inasmuch as sharp diffraction effects are usually not present at angles larger than 45°. However, when angular measurements of the utmost accuracy are demanded, the camera can be fitted with a device for applying to the film fiducial marks at accurately known angular positions [30]; alternatively a crystalline calibrating powder can be intimately mixed with the polymer during the preparation of the x-ray specimen or coated on the surface of a previously prepared solid polymer specimen.

Figure 2-18 shows commercial Debye-Scherrer cameras† (based on the design of Buerger [31, 32]) of 57.3- and 114.6-mm diameter that provide patterns particularly suited to angular measurements since

†Philips Electronic Instruments, 750 South Fulton Avenue, Mount Vernon, New York 10550.

Figure 2-18 Small and large Debye-Scherrer powder cameras based on the design of Buerger [31, 32]. (Courtesy of Philips Electronic Instruments.)

1 mm on the film is equal to 2 and 1° 2θ, respectively. In these cameras the film is held snugly against the inner cylindrical surface by a special clamping mechanism. A light-tight cover protects the film from visible radiation during the exposure, and loading and unloading of the film must be performed in the darkroom. Figure 2-19 shows a 114.6-mm Debye-Scherrer camera† that can be evacuated to provide polymer diffraction patterns free of air-scattered x-rays. Figure 2-20 shows Debye-Scherrer patterns of three polymers, the third being that of highly crystalline linear polyethylene prepared with the film in Wilson's asymmetric position.

†Richard Seifert and Company, Hamburg 13, Germany.

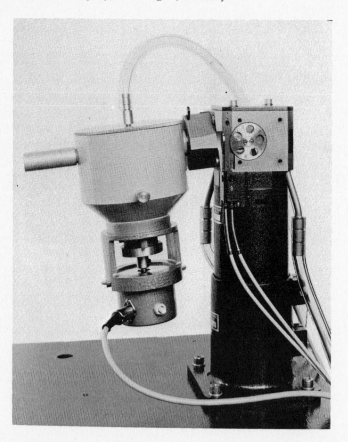

Figure 2-19 Debye-Scherrer vacuum camera. (Courtesy of Richard Seifert and Company.)

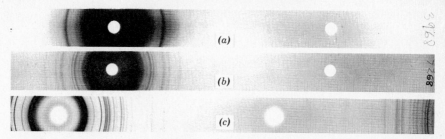

Figure 2-20 Debye-Scherrer patterns of three polymers: (*a*) poly(vinyl alcohol); (*b*) poly(propylene oxide); (*c*) low-molecular-weight linear polyethylene (Marlex 50), with film in Wilson's asymmetric position. (Courtesy of S. S. Pollack.)

Microcameras. By employing x-ray-beam collimators of very fine bore together with small sample-to-film distances, it is possible to examine selectively very small regions in a polymer specimen; for example, we may cite the application of microbeam techniques to the study of spherulites[33] and to tracing variations in crystallinity and crystallite orientation in cold-drawn single fibers[34, 35].

The term "microbeam" is commonly accepted as referring to an x-ray beam collimated by pinholes smaller than 0.1 mm (100 microns) in diameter. Very acceptable collimators are provided by lead-glass capillary tubes of fine bore (for example, marine barometer tubing [36]). Figure 2-21 is a longitudinal section of the collimator system employed in the microcamera built by Chesley[36]. The collimators are ¾-in. lengths of capillary tubing with bores ranging from 25 to 100 microns in diameter. The rubber gasket and small piece of cellophane over the forward (outer) end of the capillary permit the camera to be evacuated. The lead diaphragms with openings slightly larger than the bore of the capillary exclude unwanted x-rays and light from the collimator. In this camera the polymer specimen may be either affixed directly to the back (inner) end of the collimator tube or held in a taut or drawn state by means of a small holder mounted just behind the end of the collimator. A flat piece of film 35 mm square is held in a microcassette at distances of up to 1.5 cm from the specimen. The direct beam, after penetrating the sample, issues through a small hole in the cassette holder and is observed on a fluorescent screen at the back of the camera. Two motions of the specimen holder are provided, making it possible to select the precise region of the specimen to be placed in the x-ray beam.

Figure 2-22 shows the appearance of the assembled Chesley camera mounted on the x-ray-tube housing. The outlet for evacuation of the camera can be seen at the lower right. Figure 2-23 is a

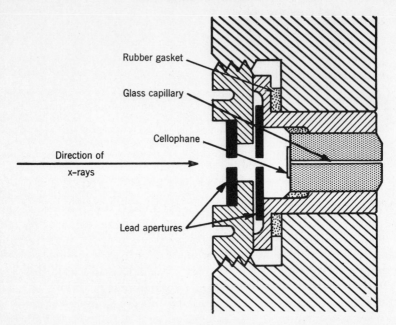

Rubber gasket

Glass capillary

Cellophane

Direction of
x-rays

Lead apertures

Figure 2-21 Longitudinal section of collimator system used in the microcamera designed by Chesley [36]. (Courtesy of Philips Electronic Instruments.)

micro-diffraction pattern, obtained with the Chesley camera, of a nylon 56 fiber displaying preferred orientation. A commercial version of this camera has been on the market for some time.† Also available are microcameras incorporating a semicylindrical film arrangement; these are mounted on brackets that are attached directly to the x-ray-tube housing to improve geometrical stability and retention of alignment.‡

When an ordinary x-ray tube with a focal spot about 1×10 mm in size is used it is not difficult to align a microcamera, but the intensity of the beam will be very low since only a small portion of the focal spot can transmit x-rays through the fine collimator aperture. Under these conditions exposures of many hours will be required even with a specimen of optimal breadth and thickness, and fibers of small diameter will not yield useful diffraction patterns even with very long exposures. However, by mating the microcamera to a microfocus x-ray

†Central Research Laboratories, Inc., Red Wing, Minnesota, 55066; Philips Electronic Instruments, 750 South Fulton Avenue, Mount Vernon, New York 10550.
‡Hilger and Watts, Ltd., Hilger Division, 98 St. Pancras Way, Camden Road, London, N.W. 1, England. Available in the United States from Jarrell-Ash Company, 590 Lincoln Street, Waltham, Massachusetts 02154.

Figure 2-22 Assembled microcamera of Chesley design mounted on x-ray-tube housing. (Courtesy of Philips Electronic Instruments.)

tube it becomes possible to irradiate the specimen with a much more intense beam, thereby providing useful diffraction patterns with practicable exposure times even when the collimator bore and specimen are both small. It must be emphasized that the alignment of a microcollimator and microsource is a critical operation and that the preservation of optical alignment is also more difficult.

2-3.4 Crystal-Monochromatized Diffraction Photographs

Flat-Crystal Monochromators. In polymer studies demanding precise measurements of the intensities or angular positions of x-ray interferences β-filters do not furnish characteristic radiation of the requisite purity. (In this connection see again Section 1-3.1.) A much more monochromatic beam can be obtained by reflecting (actually *diffracting*) the direct x-ray beam from the surface of a flat crystal plate that has been cleaved or cut with its surface parallel to the diffracting planes. Very strong beams can be obtained from the (002) planes of urea nitrate or pentaerythritol, but these crystals deteriorate under x-rays and must be periodically replaced. Durable crystals

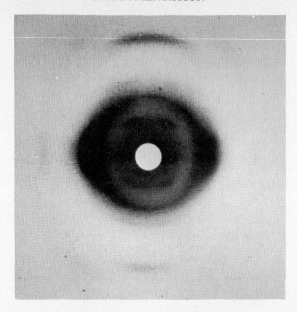

Figure 2-23 Microdiffraction pattern of a nylon 56 fiber. Fiber axis vertical. Magnification 4.4×. (Courtesy of J. H. Magill and S. S. Pollack.)

that produce monochromatic beams of reasonable intensity are rock salt [planes (200) with $d = 2.81$ Å] and fluorite [planes (111) with $d = 3.15$ Å]. Lists of crystals that are suitable for monochromators and a description of their properties may be found in the literature [37–39].

A point of superiority of fluorite as a monochromator is that the harmonic wavelengths (refer to Section 1-3.1) can be eliminated for all practical purposes. The intensity of the first harmonic $\lambda/2$ is inherently exceedingly small because the structure factor of the planes (222) [identical with the second-order reflection from the (111) planes] is nearly zero. Moreover, the higher harmonics, such as $\lambda/3$ and $\lambda/4$, will not be present in the spectrum of the x-ray tube if it is operated at a voltage V less than 24,000 since the lower wavelength limit of the general radiation, given by [40]

$$\lambda_{min} = \frac{12,400}{V}, \tag{2-16}$$

is then larger than $\lambda/3$, which is 0.514 Å for $CuK\alpha$ radiation. Therefore the beam diffracted by the fluorite crystal under these conditions will consist exclusively of $CuK\alpha$ radiation of effective wavelength 1.5418 Å, which for polymer investigations may be regarded as being strictly monochromatic. (In this connection the reader should see Section 1-3.1.)

The reader may refer to Figure 2-16 for an illustration of the improvement in a cylindrical fiber photograph that can be obtained by employing a flat-crystal monochromator. In addition to the monochromatic diffraction effects of part *b* the nickel-filtered pattern *a* contains six radial streaks near the center caused by the general radiation transmitted by the filter as well as a $CuK\beta$ reflection from the powerfully diffracting planes (100) (innermost reflection on the equator). Flat-crystal monochromators can be used to similar advantage in preparing Debye-Scherrer photographs of polymers.

The beam that is produced by a flat monochromator is weaker than the direct beam by a factor of 10 or more for even the most efficient crystals. In principle, therefore, the exposure times will need to be correspondingly increased; however, in actual practice a time factor of from 3 to 6 is generally adequate because of the greatly improved contrast that is realized between the diffraction effects and the diffuse background. Manufacturers of x-ray diffraction equipment can supply suitable holders for mounting and orienting monochromatizing crystals. A relatively simple and inexpensive device is shown in Figure 2-24. It is attached directly to the x-ray-tube housing in order to establish a stable geometrical relationship between the crystal and the focal spot of the x-ray tube.

Curved-Crystal Monochromators. By combining a parafocusing powder camera and a focusing curved-crystal monochromator,

Figure 2-24 Simple holder for mounting a flat monochromator crystal. (Courtesy of Philips Electronic Instruments.)

Guinier[41] designed a camera that is capable of producing exceptionally high-quality diffraction patterns of polycrystalline and semicrystalline substances. In particular these cameras provide the advantages of (a) greatly reduced exposure times compared to cameras incorporating flat monochromators, (b) exceptional resolution and dispersion of the details of the diffraction pattern, and (c) very low backgrounds because of the elimination of the general radiation spectrum and air scattering. Although a camera with these refinements is not required for most polymer studies, it can be of great value when very precise measurements of line profiles and positions are wanted[42–44].

A Guinier camera that has been further developed by P. M. de Wolff[45] has the x-ray optical arrangement shown in Figure 2-25. X-rays from a narrow line source X are (a) diffracted by the ground and elastically bent, thin quartz crystal C (Johansson-type monochromator[46]), (b) transmitted by the sample S, and (c) brought to a line focus at D. The rays diffracted by the sample come to foci at various points on the film F, each corresponding to a particular set of planes (hkl). The diffraction pattern thus formed is a linear function of 2θ and is limited to the angular range 0 to 90°. The specimen, film, and direct-beam focus D all lie on the inner cylindrical circumference of the camera. The Guinier-de Wolff camera has a diameter of 114.6 mm and is quadruple in construction, permitting the diffraction patterns of four specimens to be prepared simultaneously on a single 45×180-mm sheet of film for ease of comparison. The large dispersion of 2 mm per degree of 2θ is twice that of a Debye-Scherrer camera of the same radius. The curvature of the monochromator crystal can be continuously and precisely varied so as to focus x-rays of any wavelength between 0.7 and 2.5 Å, and the entire camera body can be evacuated

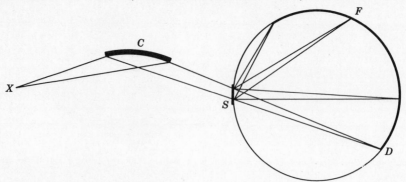

Figure 2-25 Optical arrangement of a Guinier-de Wolff camera.

to eliminate air-scattered radiation. Figure 2-26 shows a well-engineered quadruple Guinier-de Wolff camera that has been made available commercially.† Figure 2-27 shows patterns of four polyallene specimens prepared in this camera. For optimal focusing the specimen should take the form of a lamella, about 0.5 mm thick, with the same curvature as the camera. Actually, however, the sharpness of the pattern is impaired very little by mounting the specimen on a

† N. V. Nederlandsche Instrumentfabriek (Nonius), Delft, Netherlands. Available in the United States through Enraf-Nonius, Inc., 130 County Courthouse Road, Garden City Park, New York 11040.

Figure 2-26 Guinier-de Wolff quadruple focusing camera. Cover removed. [Courtesy of N. V. Nederlandsche Instrumentfabriek (Nonius).]

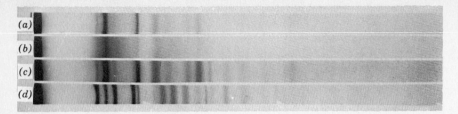

Figure 2-27   Guinier-de Wolff quadruple photograph of four polyallene specimens: (a) and (c) predominantly phase I, (b) largely amorphous, (d) predominantly phase II. Exposure for 45 min at 40 kVp and 21 mA, camera evacuated, specimen stationary. (Courtesy of J. Visser.)

small flat frame tangent to the circumference of the camera, which is a more convenient technique. Polymer films and sheets may be supported directly on the specimen frame, and coherent powders may be pressed into channels provided in the frame. Noncoherent powders may be mixed with a minimal amount of Vaseline or silicone grease to form a thick paste that is then applied as a thin layer on a supporting sheet of cellophane or Mylar, this in turn being supported on the specimen frame. The Vaseline produces weak lines at $d = 4.13$ and 3.72 Å, which may be helpful in calibrating the 2θ scale.

2-3.5   Small-Angle Scattering

Pinhole Collimation.   For the preparation of an undistorted small-angle-scattering pattern it is necessary to collimate the beam with a pair of fine pinholes rather than slits. This method has the disadvantage of requiring very long exposures to compensate for the weakness of the beam that is incident on the specimen. Nevertheless, when the nature of the internal inhomogeneity and texture of the specimen are not known in advance, as is true of most exploratory investigations of polymers, pinhole collimators must be employed. The sharp pinhole collimation required by Bolduan and Bear [47] in their precise studies of collagen (see Figure 5-23a) called for exposures as long as 156 hr. However, cameras suitable for general-purpose work can be designed so as to effect a practical compromise between intensity and resolution.

Figure 2-28 is a cutaway diagram of a very practical general-purpose camera designed by W. O. Statton† to facilitate the preparation

---

†Available from W. H. Warhus Company, 406 Rowland Park Boulevard, Carrcroft, Wilmington, Delaware 19803.

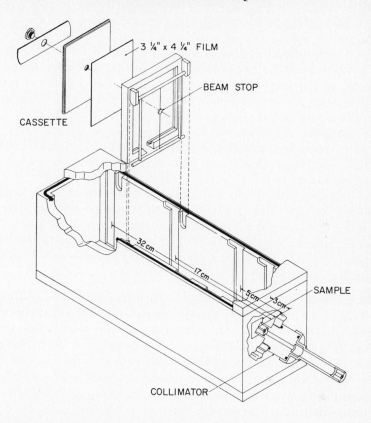

3 ¼" x 4 ¼" FILM

BEAM STOP

CASSETTE

32 cm

17 cm

5 cm   3 cm

SAMPLE

COLLIMATOR

Figure 2-28 Cutaway diagram of small- and medium-angle camera designed by Statton. (Courtesy of W. H. Warhus Company.)

of scattering patterns down to moderately small angles without unduly long exposures. Matched pinhole apertures either 0.38 or 0.51 mm in diameter are spaced 15 cm apart, and discrete sample-to-film distances of 3, 5, 17, and 32 cm are provided. The figure also shows an exploded diagram of the $3\frac{1}{4} \times 4\frac{1}{4}$-in. flat-film cassette and indicates how the undeviated beam is intercepted by means of a small lead plug mounted at the front of the cassette on a fine wire. With its cover in place this camera can be evacuated to about 5 mm Hg, which eliminates perceptible blackening of the film due to air scatter in exposures as long as 100 hr. Accessories available for this camera include a specimen rotator and a heater capable of warming the specimen to 250°C. (Small-angle patterns of polymers prepared in a camera of this type are shown in Figure 5-1.)

H. Kiessig[48] has also developed a general-purpose camera† well suited to the preparation of small-angle-scattering diagrams of polymers. A longitudinal section of this camera is shown in Figure 2-29. The small cylindrical beam-collimating tube at the left is fitted with interchangeable pinhole diaphragms of 0.3- and 1.0-mm diameter. The specimen is mounted on a special holder that fits on the end of the collimator, and its temperature can be elevated to as much as 300°C with the aid of a small heater available as an accessory. The pattern is recorded on a circular flat film that can be placed at fixed distances of 200 and 400 mm from the specimen, as well as the minimum distance of 100 mm, by adding two interchangeable cylindrical sections to lengthen the camera. When the 0.3-mm pinholes are used and the film is placed at a distance of 400 mm from the specimen, Bragg spacings up to 250 Å can be recorded. Provision is made for evacuation of the camera to eliminate air-scattered x-rays. Figure 2-30 shows the Kiessig camera with Polaroid cassette in place before the x-ray tube.

Slit Collimation. When pinholes are replaced by slit collimators the intensity of scattering is greatly increased and the exposure times can be correspondingly reduced. However, the substitution of slits causes a substantial distortion of the theoretical distribution of scattered intensity, which may be dealt with in either of two ways: (a) by mathematical corrections or (b) by employing slits of sufficient length to permit application of the special theory of scattering by infinitely long slits. The second approach is generally the more satisfactory. However, slit collimation must not be applied to polymer specimens indiscriminately. It is appropriate for any specimen that

†Available from Richard Seifert and Company, Hamburg 13, Germany.

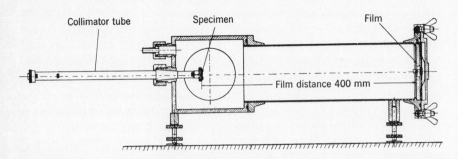

Collimator tube   Specimen   Film

Film distance 400 mm

Figure 2-29 Longitudinal section of Kiessig's small-angle-scattering camera. (Kiessig [48b]; reprinted by permission of Dr. Dietrich Steinkopff, *Kolloid-Z., Z. Polymere.*)

Figure 2-30 Kiessig small-angle-scattering camera with Polaroid cassette in place before the x-ray tube. (Courtesy of Richard Seifert and Company.)

has no preferred orientation (so-called isotropic specimens), but for highly oriented materials it gives useful results only when the slits are either precisely parallel or perpendicular to the orientation axis of the specimen. This topic is discussed in considerable detail in Chapter 5.

Worthington[49] constructed an evacuatable camera with slit collimation for use with a high-intensity, rotating-anode x-ray tube that could be operated at 30 kV (peak) and 80 mA. As a consequence of the high intensity of the incident beam generated by this apparatus he was able to obtain the familiar small-angle pattern of discrete spacings from collagen in only 45 minutes with copper radiation. For this exposure the 1×6-mm rectangular focal spot was "viewed" by the collimator at the small take-off angle of only 2°, thus furnishing a sharp "line" source, which was oriented parallel to the slits of the collimator. In this camera the three collimating apertures, specimen holder, and film cassette are supported by individual brackets that slide on an optical bench. With the camera dimensions listed in Table 2-2 and nickel-filtered $CuK\alpha$ radiation it was possible to resolve a spacing of 1000 Å from the direct beam.

Fankuchen and Jellinek[50] first suggested the use of a double-crystal spectrometer for measuring small-angle scattering down to

Table 2-2   Dimensions Used by Worthington [49] To Resolve a 1000-Å Spacing from the Direct Beam[a]

Slit[b] widths:	
First	0.0034
Second (defining aperture)	0.0022
Third (guard aperture)	0.0057
Distance from:	
First to second slit	9.9
Second to third slit	4.2
Third slit to specimen	0.4
Specimen to film	9.0
Width of lead beam stop	0.026
Height of lead beam stop	1.0

[a]All dimensions are in inches.
[b]The slits consist of pairs of parallel lead straight edges.

extremely small angles. Figure 2-31 shows the geometrical arrangement[51], which is equivalent to a small-angle camera with slit collimation of extremely high resolving power. Two flat crystals of high perfection, usually calcite or quartz, are placed in the so-called parallel position, as shown, and the sample in the form of a transmitting plate is placed between them. The first crystal plays the role of a fixed-beam collimator and "reflects" parallel monochromatic x-rays through the specimen to the second crystal, which acts as an analyzer of the x-rays scattered at small angles by the specimen. The intensity

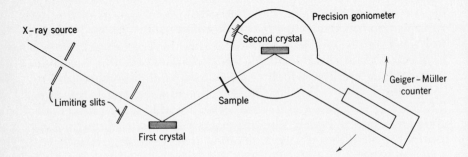

Figure 2-31 Double-crystal spectrometer for small-angle-scattering measurements. (Guinier and Fournet [51].)

scattered by the sample at some angle 2θ is the difference between the intensity measured by the counter with the specimen in place and the intensity with the specimen removed. The angle of measurement is varied by rotating the second crystal. Parasitic scattering from the first crystal is largely excluded from the counter by limiting slits, as shown in Figure 2-31, but it is not possible to suppress the parasitic scattering produced by the second crystal. For this reason the double-crystal spectrometer is primarily suited to locating discrete interferences at very small angles rather than to accurate intensity measurements. In principle, when the crystals are very perfect, the instrument is capable of measuring d spacings as large as 20,000 Å, although practical difficulties reduce this limit considerably. Further advantages of this instrument are (a) relatively wide beams can be used, thus preserving good intensity, and (b) extremely sharp resolution of discrete reflections can be had. In their investigations of biological fibers Kaesberg, Ritland, and Beeman[52] employed a double-crystal spectrometer that utilized high-quality ground and etched calcite crystals. The analyzer crystal and counter could be set to a desired angle with a precision of 0.5 second of arc or better, and discrete reflections could be resolved from the direct beam down to angles as small as 100 seconds of arc.

Kratky[53–58] has constructed a camera of novel design, the U-bar camera, which, when precisely machined and aligned, excludes from the film all parasitic scatter originating from the collimating apertures and at the same time permits intensity measurements to be made to scattering angles as small as 16 seconds of arc (equivalent to $d = 20,000$ Å with $CuK\alpha$ radiation). The unusual collimating arrangement, which is geometrically equivalent to a pair of long slits, is shown in longitudinal section in Figure 2-32 and in schematic perspective in Figure 2-33. The breadth and divergence of the direct beam are limited by the upper face of the metal block B_2, underside of the bridge B_1, and lower face of the block B_3. The beam transmitted by this collimating system is intercepted by the beam stop P immediately in front of the film. It is essential for optimum performance that the beam-defining faces of B_2 and B_3 be perfectly plane and that they coincide precisely with one and the same horizontal plane, indicated in Figure 2-32 by the plane trace H. For observations at the lowest possible scattering angles the upper edge of P should also coincide with H. The breadth of the direct beam Sl (Figure 2-33) can be adjusted by varying the height of the bridge B_1. To accentuate intensity at the expense of resolving power a short front-base length a is used, whereas to improve resolution at the

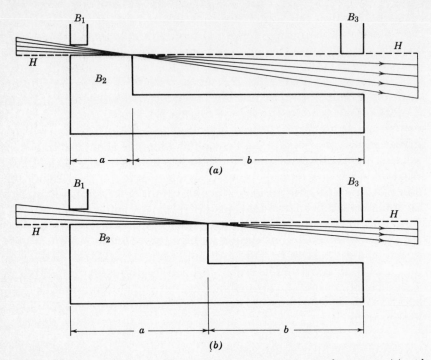

Figure 2-32 Longitudinal sections of collimating system in Kratky camera: (*a*) with short front base for high intensity; (*b*) with long front base for high resolution.

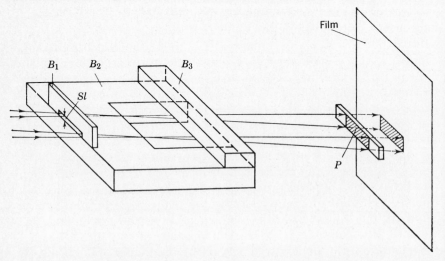

Figure 2-33 Schematic perspective of Kratky camera. (Kratky [58]; reprinted by permission of Pergamon Press.)

expense of intensity a long front base is needed, as depicted in Figure 2-32*a* and *b*, respectively.

If the limiting surfaces of B_2 and B_3 are not highly planar, there will not be a complete absence of diaphragmatic scattering. Steel was chosen as the fabrication material for B_2 and B_3 by Kratky because of its superior machining characteristics in spite of the excitation of secondary iron fluorescence x-rays when the incident x-rays are CuKα. Correct construction and alignment serve to exclude such secondary x-rays, as well as ordinary diaphragmatic scatter, from the film. Figure 2-34 shows one commercial version of the Kratky U-bar camera. Models are currently offered by several firms.† For the measurement of weak interferences and continuous diffuse scattering, the photographic chamber of the camera can be evacuated and a monochromatic incident beam from a Johansson-type curved-crystal monochromator can be used. Up-to-date models of the apparatus are also provided with accessories permitting the intensity pattern to be

†Siemens Aktiengesellschaft, D-75 Karlsruhe 21, Postfach 21 1080, Germany; Siemens America, Inc., 350 Fifth Avenue, New York, New York 10001; Richard Seifert and Company, Hamburg 13, Germany; Anton Paar K. G., Kärntnerstrasse 322, Graz-Strass-gang, Austria; Eastern Scientific Sales Company, 43 East Main Street, Marlton, New Jersey 08053.

Figure 2-34 View of Kratky small-angle camera in place at the x-ray tube. (Courtesy of Siemens and Halske.)

measured with a counter, in which case monochromatization is accomplished with the aid of Ross balanced filters in conjunction with pulse-height discrimination. (See Section 2-4.3.) In view of the exceptional capabilities of the Kratky camera it is hardly surprising that the alignment is very critical.

Another camera that is especially designed for precise intensity measurements at small angles has been developed by Luzzati and Baro[59]. This instrument, shown in Figure 2-35, incorporates a number of features that are of advantage in the measurement of the continuous scattering from dilute dispersions of proteins and other colloids. The incident beam is provided by a Johannson-type quartz monochromator crystal augmented with a pair of collimating slits. By means of highly precise alignment of the apertures, monochromator, and narrow focal spot of a semi-fine-focus x-ray tube the $CuK\alpha_2$ ray can be resolved from the $CuK\alpha_1$ ray and excluded, rendering the incident beam very highly monochromatic. In order to achieve and preserve such critical optical alignment it is essential that a number of experimental precautions be observed, including (a) integral mounting of the monochromator on the x-ray-tube housing, (b) periodic polishing of the target surface of the x-ray tube to maintain planarity, (c) very precise regulation of the voltage and current across the x-ray tube, (d) operation of the tube at a low current level (typically 5 mA), and (e) close control of the ambient temperature. Liquid specimens are held in thin cells with windows of mica, which is highly transparent to x-rays and produces no x-ray scattering at small angles. During the exposures the camera is normally covered with a vacuum hood (not shown in Figure 2-35) to eliminate air-scattered x-rays, and provision is made for controlling the temperature of the specimen at any point within the range −30 to +350°C. The Luzzati-Baro camera is capable of recording scattering intensities to a minimum angle of 0.12°, equivalent to a Bragg spacing of about 800 Å with $CuK\alpha_1$ radiation. The angular resolution is exceptionally high, being about ±0.02°.

Franks Camera. Franks[60, 61] has designed a small-angle camera that produces an intense focused beam by means of total reflection of the incident x-rays from two crossed glass plates. As shown in Figure 2-36, x-rays from a microfocus tube are reflected successively by two mutually perpendicular elastically bent glass plates, transmitted by the guard slit and specimen, and brought to a "point" focus on the photographic film. Thus the geometry is equivalent to pinhole rather than slit collimation. Total reflection

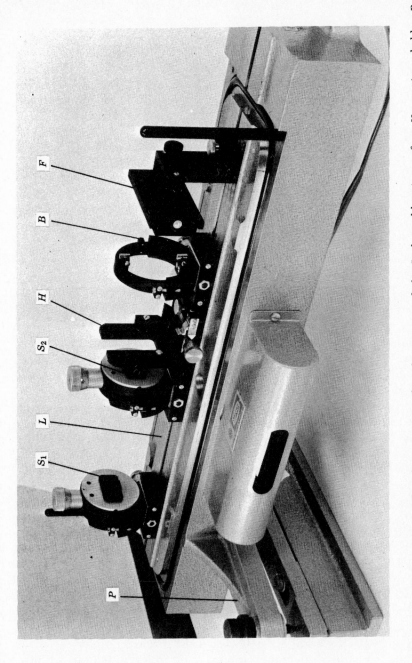

Figure 2-35 Luzzati-Baro small-angle camera. S_1 — first slit; S_2 — second slit; L — lapped bearing surface; H — specimen holder; B — beam-stop support; F — film holder; P — leg of three-point support. (Courtesy of Jarrell-Ash Company.)

111

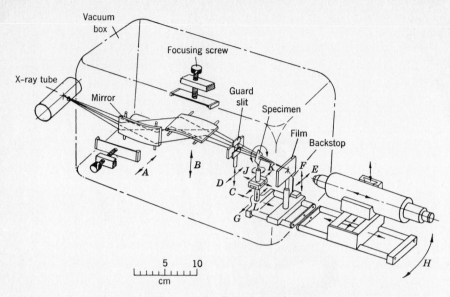

Vacuum box

Focusing screw

X-ray tube

Mirror

Guard slit

Specimen

Film

Backstop

A

B

D

C

J

K

F

E

L

G

H

5   10
cm

Figure 2-36 Schematic perspective diagram of Franks small-angle camera. (Franks [61]; reprinted by permission of The Institute of Physics and The Physical Society.)

of $CuK\alpha$ radiation occurs at a glancing angle of less than 11 minutes of arc, and proper adjustment of the angle also results in a considerable degree of monochromatization. In the Franks camera the convergent beam at its point of intersection with the specimen is only 50 to 100 microns in diameter, and an equivalent Bragg spacing as large as 1000 Å can be resolved from the direct beam.

The Franks camera is well suited to precise studies of polymers by the pinhole-collimation technique. In tests with stained rat-tail tendon [61] a 10-min exposure clearly resolved the first-order 640-Å spacing of collagen as well as a number of higher orders. In the course of their investigations of biological specimens G. F. Elliott and C. R. Worthington [62] improved the design of Franks so as to simplify the alignment procedure and permit extended exposure times when necessary. It should be stated that the performance of the Franks camera is very sensitive to its optical alignment and to the focal-spot size and loading of the x-ray tube.

Two other small-angle cameras that are particularly effective in the study of biological substances have been developed. A. Elliott [63] has modified the Franks design by replacing the total-reflecting glass plates with a gold-plated cast-resin toroidal mirror for more efficient focusing and higher intensity of $CuK\alpha$ radiation. Elliott's camera is capable of resolving spacings of up to about 200 Å from the direct beam.† In the investigation of muscle Huxley and Brown [64] have obtained clean, highly resolved small-angle patterns with greatly reduced exposures by employing a novel optical arrangement consisting of a bent gold-plated Franks mirror close to the x-ray source, followed by a "crossed" bent quartz focusing monochromator to yield a powerful "point" focus. With this camera the collagen pattern could be obtained in as little as 10 sec and living muscle gave useful patterns in 10 min.

2-4 DIFFRACTOMETRY

The inherent advantages of counters for the accurate measurement of x-ray intensities have been pointed out already. For wide-angle measurements of polymers $(5° \leq 2\theta \leq 90°)$ unmodified commercial counter diffractometers can be used with the specimen mounted preferably in one of the two symmetrical fashions shown in Figures 2-8a and c. If the usual divergence and receiving slits are replaced with special collimators that greatly limit the beam divergence, certain kinds of small-angle-scattering measurements can be made. The author and co-workers [65] have used a pair of identical collimators, each having two slits limiting the divergence to 0.05°, with the specimen in the symmetrical-transmission position (Figure 2-8c) and with balanced filters of nickel and cobalt to improve the monochromaticity of the copper x-rays. Although the experimental counting rates were impracticably small for very weakly scattering specimens, the arrangement was adequate for revealing discrete small-angle reflections from polymers down to $0.04°2\theta$ [66].

2-4.1 General Features of Counter Diffractometers

Figure 2-37 shows the appearance of a modern commercial diffractometer, and Figure 2-38 is a "flow" diagram indicating the functional relationship among the components. The output counts can be monitored or registered in several ways. The pulses from the scaler are customarily fed into a pulse-integrating circuit, which actuates a counting-rate meter for visual monitoring. The simplest method of recording the diffraction pattern is to conduct the output current from the pulse-integrating circuit to a potentiometer strip-

†The Elliott toroidal camera is available commercially from Hilger and Watts, Ltd., 98 St. Pancras Way, Camden Road, London, N.W. 1, England, and is marketed in the United States by Engis Equipment Company, 8035 Austin Avenue, Morton Grove, Illinois 60053.

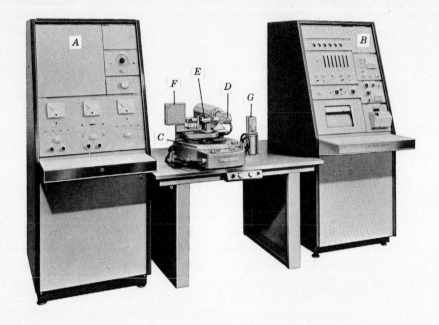

Figure 2-37 Overall view of an x-ray diffractometer. A — high-voltage-generator cabinet; B — detector cabinet; C — 2θ goniometer; D — x-ray tube for diffractometry; E — specimen; F — counter; G — x-ray tube for photographic techniques. (Courtesy of General Electric Company, Analytical Measurement Business Section.)

chart recorder, which is driven at constant speed during the process of scanning the 2θ range of interest. The linearity of response of a good potentiometer recorder is adequate for most quantitative studies of polymer pattens, particularly if a slow 2θ scanning speed is employed in conjunction with a moderately small value of the time constant of the pulse-integrating circuit[67].

For optimal accuracy in recording the diffracted intensities the diffraction pattern must be step scanned; that is, counts must be recorded at successive fixed locations in 2θ separated by a small fixed angular increment such as 0.05 or 0.10°. In the study of oriented specimens similar stepwise measurements may be required as a function of azimuth β, which is accomplished by turning the specimen in its own plane with the counter held stationary. (Refer to Section 2-2.3.) Manual step-scanning procedures are evidently laborious. However, manufacturers now can furnish automatic step-scanning

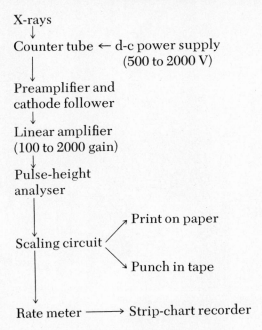

Figure 2-38   Functional relationship among the components of an x-ray diffractometer.

accessories complete with paper-tape printers to record the total counts at each point.

Leading manufacturers have recently made available sophisticated solid-state (transistorized) circuits and associated electromechanical devices for automatically controlling the operation of diffractometers, the angular settings and output counts being registered on punched paper tape or cards. The input control may be applied directly by a small computer or through the medium of prepunched control tape. Such apparatus obviously offers great advantages both from the standpoint of reliability and rapidity of data collection and that of data processing, which can be performed by a computer. Similar automation techniques can be applied in small-angle-scattering studies.

2-4.2   Counters for X-Ray Measurements

We shall summarize rather briefly the characteristics of the three types of counters that are useful in x-ray measurements: Geiger-Müller, proportional, and scintillation. Figure 2-39 shows some representative counter tubes that are designed for x-ray measurements. For detailed information on the application of counters to

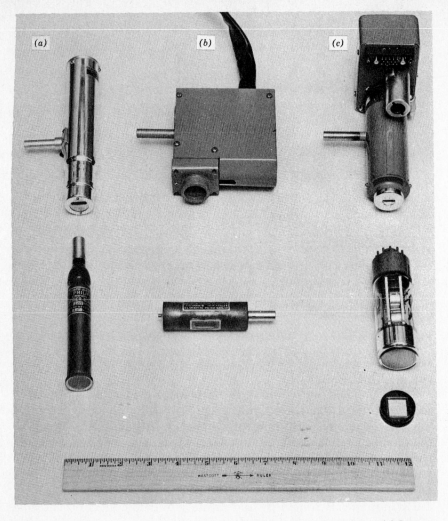

Figure 2-39   Some interchangeable counter tubes for x-ray measurements: (*a*) Geiger-Müller counter; (*b*) xenon-filled proportional counter; (*c*) scintillation counter. The proportional and scintillation counters have cathode-follower circuits, which are mounted directly with the counter on the goniometer arm. The complete detector units are shown in the upper part of the photograph. [Dowling, Hendee, Kohler, and Parrish, *Philips Tech. Rev.*, **18**, 262 (1956–1957).]

x-ray measurements the reader should consult the manufacturers or refer to the suggested reading list at the end of this chapter. Any given counter (the term includes both the tube itself and the related circuitry required in the process of quantum counting) possesses a characteristic *dead time*, or *resolving time*, τ, as a result of which linearity between the counting rate and the intensity of the measured radiation is lost at sufficiently high counting rates. If N_o and N are the actual and ideal counting rates, respectively, in counts per second (cps), for losses up to about 20% N is given to a good approximation by

$$N = \frac{N_o}{1 - N_o\tau}. \tag{2-17}$$

Geiger-Müller Counters [68]. These were the first counters to be developed. They are electrically simple and possess high reliability and stability, but they are unsuitable for the measurement of intense x-ray beams because of their large resolving times, which result in losses of about 2% for a counting rate of 100 cps. Nevertheless, Geiger-Müller counters are highly satisfactory for measuring weak x-ray-scattering effects such as the small-angle scattering from colloidal solutions [69, 70]. The pulses produced by a Geiger-Müller counter are all of approximately the same amplitude regardless of the energies of the incident x-ray photons, which eliminates pulse-amplitude discrimination as a means of monochromatization.

Proportional Counters. These counters have resolving times of the order of 2 to 3 μsec; hence the response is linear to counting rates of at least 5000 cps, and (2-17) can be used to correct for losses at much higher counting rates. Proportional counters derive their name from the property that the pulse amplitudes are proportional to the energies of the exciting x-ray photons (hence to $1/\lambda$), which permits a degree of monochromatization to be achieved by the use of pulse-height discrimination. Compared with the amplitudes of Geiger pulses, those of proportional counters are weaker by a factor of 10^3, which calls for high-gain linear amplification in order to trigger the scaling circuits.

The quantum-counting efficiency (fraction of incident x-ray quanta actually counted) of a Geiger-Müller or proportional counter depends on the x-ray wavelength, dimensions of the counter tube, window composition and thickness, and nature and pressure of the filling gas. Figure 2-40 presents the quantum-counting efficiencies as a function of wavelength for typical Geiger-Müller, proportional, and scintillation counters. It will be noticed that Geiger-Müller and proportional counters have a degree of inherent wavelength selectivity. In particular the counting efficiency of a proportional counter is sharply enhanced over a narrow region immediately below the K-absorption edge of its filling gas. Thus a krypton-filled counter is selectively

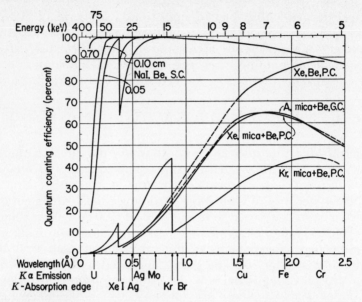

Figure 2-40 Calculated quantum-counting efficiency versus wavelength of typical x-ray counters. G.C.—Geiger-Müller counter, P.C.—proportional counter, NaI—sodium iodide crystal, S.C.—scintillation counter. Window compositions—beryllium (Be) and mica + Be. Filling gases—argon, krypton, and xenon. [After Taylor and Parrish, *Rev. Sci. Instruments*, 26, 367 (1955).]

sensitive to MoKα radiation with λ = 0.711 Å since the krypton K-absorption edge falls at 0.86 Å.

Scintillation Counters. As shown in Figure 2-41, scintillation counters consist of two basic elements—a thin, waferlike, fluorescent single crystal and a photomultiplier tube. The crystal *SC*, which is usually a sodium iodide crystal activated with 1% thallium, is contained in a light-tight jacket and mounted behind a beryllium x-ray

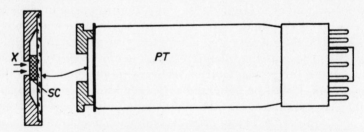

Figure 2-41 Schematic representation of a scintillation counter. *X*—x-rays; *SC*—fluorescent crystal; *PT*—photomultiplier tube. (Dowling et al., op. cit.)

window. When an x-ray photon strikes the crystal, it generates a quantity of fluorescent light that is proportional to the energy of the photon. The fluorescent light is transformed by the photomultiplier tube *PT* into an electric pulse with an amplitude that is also proportional to the energy of the exciting photon.

The resolving time of a scintillation counter (including circuitry) is 2 to 3 μsec, so that the response is linear up to at least 5000 cps, as in the case of a proportional counter. Referring once more to Figure 2-40, we see that the counting efficiency is not only very high but also practically uniform over the entire wavelength region that is useful in x-ray diffraction. For this reason pulse-height discrimination is more essential than when proportional counters are employed, since these are self-monochromatizing to some degree.

2-4.3   Monochromatization in Diffractometry

Pulse-Height Discrimination. When x-rays of a discrete wavelength, or energy, are measured with a proportional or scintillation counter, the amplitudes of the pulses display a bell-shaped frequency distribution about the most probable value A, as illustrated in Figure 2-42. A pulse-height-discriminator circuit accepts only those pulses whose heights exceed some base-line voltage V_B and are less than $V_B + \Delta V$, ΔV being the window width; for example, with reference to the figure, if $V_B = 15$ and $\Delta V = 10$ volts, only pulses with amplitudes between 15 and 25 volts will be passed by the discriminator. Even with a window width of 10 volts, a substantial fraction of the pulses generated by the monochromatic x-ray beam will be lost. An increase in ΔV would increase the fraction of pulses accepted, but it would also diminish the monochromatizing effect of the pulse-height discriminator.

From the above considerations it will be realized that pulse-height discrimination is largely ineffectual in excluding pulses due to wavelengths differing only slightly from the wavelength on which the discriminator window is centered. On the other hand, it is effective in rejecting pulses generated by x-rays of much longer or shorter wavelengths—for example, harmonics. (Refer to Section 1-3.1.) An important side benefit of pulse-height discrimination is its rejection of practically all pulses due to electronic noise, which are characteristically much smaller in amplitude than the desired x-ray pulses. Figure 2-43 shows the striking effectiveness of this method combined with β-filtration in removing the general radiation of short wavelengths from the spectrum of a copper-target x-ray tube.

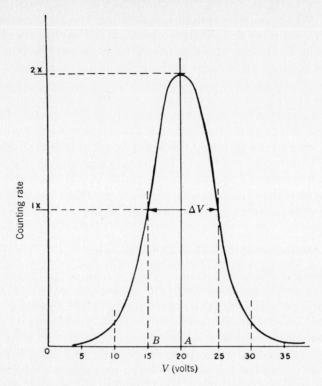

Figure 2-42 Pulse-amplitude distribution generated when a monochromatic x-ray beam is measured by a proportional or scintillation counter. [After Miller, *Norelco Reporter*, 4, (2), 37 (1957).]

Ross Balanced Filters [71, 72].   Prior to studying this section the reader should review Section 1-3.1. The most convenient way of monochromatizing the x-ray beam when employing counters is to use balanced filters. The experimental technique consists in making intensity measurements while alternately interposing in the x-ray beam two filters whose absorption edges are just above and just below the wavelength of the characteristic x-rays. Appendix 7 lists appropriate filter materials and their proper thicknesses for the commonly used x-radiations. By way of illustration we now consider the case of copper radiation and refer to Figure 2-44. The thicknesses of nickel and cobalt, or their oxides in the form of pellets, must be meticulously adjusted until the transmission of the general radiation at a wavelength just shorter than the nickel absorption edge (1.488 Å) is the same (balanced) for both filters. Ideally, at the same time a similar balance should be achieved at a wavelength just longer than the

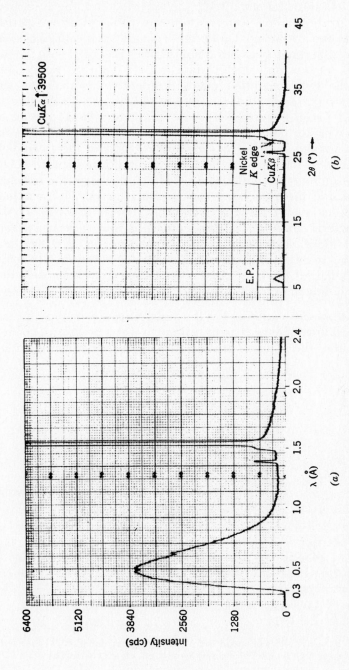

Figure 2-43 Spectra from copper-target x-ray tube obtained with a scintillation counter (*a*) without and (*b*) with pulse-height discrimination. The small peak E.P. is due to the escape peak phenomenon. (After Dowling et al., op. cit.)

121

cobalt edge (1.608 Å). Figure 2-44b shows a plot of μt, the absorption exponent, for both filters as a function of wavelength after a perfect balance has been achieved. Under these somewhat idealized circumstances the difference between the intensities recorded with the two filters will be due only to the narrow *pass band* of wavelengths between the two absorption edges. Evidently this radiation will consist almost entirely of the very intense CuKα characteristic line.

In actual practice perfect balance of the two filters cannot be achieved over the entire general-radiation spectrum of the x-ray tube even though balance may be rather good in the immediate neighborhood of the pass band. However, by utilizing pulse-height discrimination together with balanced filters, the exclusion of wavelengths outside the pass band can be made practically complete. Ruland[73] has analyzed the optimum thickness of filter pairs on the basis of several criteria, with the conclusion that most satisfactory balance of nickel and cobalt occurs at thicknesses of 0.0068 and 0.00755 mm, respectively. Instrumentation for automatically interchanging balanced filters in x-ray diffractometry has been designed [74] and is now commercially available.†

Crystal Monochromators. For the highest degree of monochromatization in diffractometry it is necessary to use a crystal monochromator in either the incident or diffracted beam. For polymer studies

†Supplied by Tem-Pres Research, Inc., 1526 William Street, State College, Pennsylvania 16801.

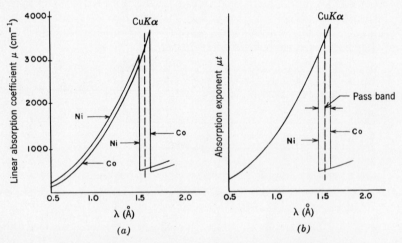

Figure 2-44 Balanced filters of nickel and cobalt for copper radiation: (a) absorption coefficient μ versus wavelength; (b) absorption exponent μt, after balancing.

the most generally satisfactory arrangements are the two shown in Figure 2-45 with the crystal in the diffracted beam. In the terminology suggested by Lang[75] these may be designated the RR (reflection specimen-reflection monochromator) and TR (transmission specimen-reflection monochromator) geometries.

In the RR arrangement parafocused x-rays diffracted by the flat specimen AOB pass through the receiving slit S_1, as in the conventional powder diffractometer, but instead of entering the counter directly they first diverge and are then refocused by the curved monochromator CTD to a focus at the counter slit S_2. The crystal may be both bent and ground (Johansson[46]) or bent only (Johann[76]), which gives a less perfect focus, but in either case the x-rays are focused sufficiently to pass the narrow slit S_2 and enter the counter G. In Figure 2-45a the upper circle is the 2θ circle of the diffractometer, and the lower one represents the focusing circle of the monochromator. It is important to note that, when the monochromator is placed in the diffracted beam, the entire monochromator-counter assembly remains rigid and rotates as a unit about the diffractometer axis O. By means of refined adjustments of the monochromator it is possible with the RR arrangement to separate the $K\alpha_1$-$K\alpha_2$ doublet and permit only the $K\alpha_1$ component to enter the counter.

The TR arrangement (Figure 2-45b) is preferable to the RR geometry for scanning the portion of the diffraction pattern below $2\theta = 30°$, for reasons given on p. 81. An exposition of the several merits of the TR geometry for measurements in the low-2θ region has been given by de Wolff[77]. Not only is it highly recommended for the precise determination of line positions in the patterns of well-crystallized substances but it is equally valuable in the measurement of more diffuse polymer patterns when high monochromaticity is the primary consideration. The TR arrangement is also convenient for the measurement of the beam reflected directly by the monochromator (specimen AOB removed), thus providing the experimental basis of reference for absolute intensity measurements, which have a special value in certain polymer investigations.

In the TR geometry the monochromator and counter assembly also revolves as a rigid unit about the goniometer axis O, while the specimen revolves at one-half this speed, thus preserving the symmetrical-transmission relationship between the incident and diffracted beams (see again Figure 2-8c). For the RR technique Lang[75] used as monochromator a quartz plate $35 \times 10 \times 0.3$ mm cut parallel to the $(10\bar{1}1)$ planes and elastically bent to the proper radius with the aid of a crystal-bending and -supporting device similar to that described

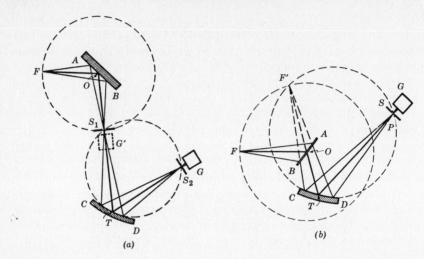

Figure 2-45 Two arrangements for x-ray diffractometry with a curved-crystal monochromator in the diffracted beam: (a) reflection specimen-reflection monochromator (RR); (b) transmission specimen-reflection monochromator (TR). (Lang[75].)

by Wooster et al.[78]. Banerjee[79] constructed an RR monochromator for the General Electric XRD–3 diffractometer incorporating a ground and plastically bent rock-salt crystal to provide a sharp focus. With this apparatus the counting rates were reduced by a factor of only 7 when the monochromator assembly was substituted for the standard counter receiver—a very favorable ratio. Figure 2-46a is a close-up of Banerjee's crystal mount, and 2-46b is a view of the entire goniometer. Ogilvie[80] designed an RR-type monochromator for the Norelco diffractometer that was subsequently made available commercially[81].

Use has also been made of monochromator crystals placed in the direct beam, despite two disadvantages of this arrangement: (a) a very stable crystal is required to resist deterioration under the impact of the direct x-ray beam and (b) if secondary fluorescence x-rays are emitted by the specimen, they are not prevented from entering the receiving counter as is the case with the monochromator in the diffracted beam. With the incident x-ray beam obtained from a flat pentaerythritol monochromator wide-angle studies of cellulose and various synthetic fibers have been carried out at the Textile Research Institute, Princeton, New Jersey[82, 83]. The Princeton instrument incorporates an auxiliary counter that continuously monitors the primary beam in order to compensate for variations with time in the output of the x-ray generator. In this connection it may be pointed out

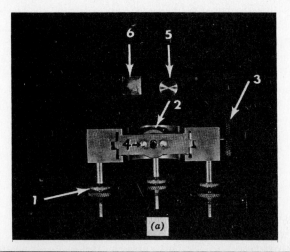

Figure 2-46 Diffracted-beam crystal monochromator for the General Electric XRD–3 diffractometer.

(*a*) Crystal-mounting device: 1 tripods for adjusting tilt of horizontal plane; 2 threaded nut to provide movement normal to the plane of the paper (the nut is held against the front brass panel [with holes marked 4] by means of compression screws and Teflon bushings dropped into a recess directly underneath the holes 4); 3 worm drive for angular adjustment of the crystal in the horizontal plane; 5 flat crystal holder; 6 bent crystal holder.

(*b*) View of entire goniometer. (Banerjee [79].)

that the stability of present-day x-ray generators is so much improved that beam monitors are no longer regarded as essential in precise diffractometric intensity measurements.

Luzzati, Witz, and Baro[70, 84] have adapted a Philips diffractometer to small-angle-scattering measurements, as indicated in Figure 2-47, by placing a bent quartz monochromator crystal M in the direct beam and utilizing primary collimating slits F_1 and F_3 of sufficient length to permit direct application of the theory of scattering by slits of infinite length. The slits F_2 and F_4 minimize the amount of parasitic scatter received by the counter. This diffractometer, like the camera of Luzzati and Baro[59] is tailored to the measurement of the continuous scattering from dilute solutions of proteins, and similarly the minimum angle at which the scattering curve can be resolved from the direct beam is only moderately small, 0.002 radian, which is equivalent to a Bragg spacing of 750 Å with CuKα radiation. However, unlike the camera, the diffractometer has the important feature of providing for absolute intensity measurements; the intensity of the incident beam reflected by the monochromator on the specimen can be measured with the aid of an accurately calibrated set of absorbing filters. The additional refinements of experimental technique observed in the use of the Luzzati-Baro camera, which were described in Section 2-3.5, are also followed in operating the diffractometer.

Ruland[85] has discussed in some detail diffractometric apparatus and techniques for very precise measurements of diffuse x-ray scattering from fibers.

2-5 THE OPTICAL DIFFRACTOMETER

The optical diffractometer is finding increasing use as an auxiliary research tool in the investigation of polymer structure by x-ray diffraction methods. The reader is referred to Appendix 1 for a brief description of the principles underlying the preparation of optical transforms. In essence an optical analog apparatus is employed to simulate the Fourier transform, or diffraction pattern, of a two-dimensional structure model. In polymer studies the almost exclusive method of utilizing optical transforms is to proceed from structure model to diffraction pattern rather than the reverse. For this process a knowledge of the phases corresponding to various points on the transform is normally of no consequence, and structure models can be evaluated simply and directly by comparisons of their optical transforms with the experimental x-ray diffraction pattern.

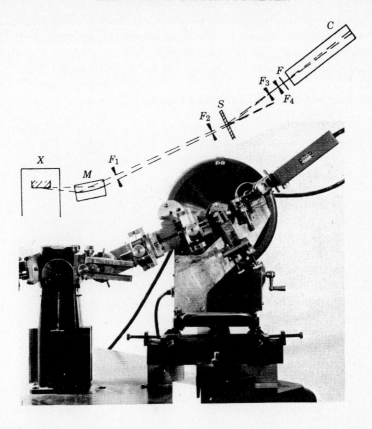

Figure 2-47 Philips diffractometer with curved-crystal monochromator in the direct beam. X x-ray tube; M monochromator; F_1 and F_3 slits defining beam; F_2 and F_4 slits limiting parasitic scatter; S specimen; F calibrated filter; C Geiger-Müller counter. (Luzzati, Witz, and Baro [70].)

2-5.1 Principles

In common practice a broad beam of parallel monochromatic light impinges perpendicularly on an opaque mask punched with an array of small holes that corresponds to the molecular model. The resulting scattering of light by the holes and the mutual interference of the scattered waves (Fraunhofer diffraction) generate the optical transform which can be observed directly with a microscope or recorded on photographic film. We point out here that the intensity in an optical transform is proportional to the square of the amplitude (modulus) of the equivalent Fourier transform; nevertheless, the characteristic (response) curves of most films are such that a photographically registered optical transform tends to resemble more directly the amplitude diagram. The discussion that follows is based principally on publications of Taylor [86] and of Taylor and Lipson [87, 88].

2-5.2 The Design of Taylor and Lipson

Figure 2-48 shows the optical system of the diffractometer that was developed over a considerable period of time in the laboratories of Lipson and Taylor at the College of Science and Technology, Manchester, England. Light from the primary source S_0 is passed through a monochromatizing filter, focused by the condenser lens L_0, and transmitted by the pinhole S_1, which serves as the secondary source. The divergent light from S_1 is collimated by the first main lens L_1, transmitted by the mask, and focused by the second main lens L_2 at the point F. The purpose of the mirror M is to permit the operator to view the transform pattern through the microscope T while manipulating the mask. For direct viewing a microscope with a 1-in. objective and 10× eyepiece is used, and for photographic recording the film is placed at F.

A compact mercury-vapor arc lamp serves as the primary source. A gelatine "mercury yellow" filter is used to isolate a spectral band about 100 Å wide encompassing the mercury yellow lines. The lenses L_1 and L_2 are 12 in. in diameter and have focal lengths of approximately 150 cm. The optical resolution that is attainable photographically is 120 to 140 lines per millimeter, and $\frac{1}{140}$ mm subtends 1 second of arc at a distance of 1.5 m. Figure 2-49 shows the optical diffractometer manufactured by R. B. Pullin and Company, Ltd.,† after the design of Taylor and Lipson.

The overall dimensions of the Taylor-Lipson apparatus are relatively large for the reason that larger masks are much easier to prepare and handle than smaller ones, and the effect of large masks is to confine the diffraction to smaller angles, which in turn calls for large instrumental dimensions. In regard to the standards to be met in the construction and alignment of an optical diffractometer, we may quote from Taylor and Lipson [87], "In the diffractometer . . . the purpose of the optical components is to control the relative phases of the various beams of light involved; in an ideal diffractometer a perfectly parallel beam, of constant amplitude and phase across the whole aperture of the system, should impinge on the mask, and the phase

†Currently marketed by Fleet Electronics, Ltd., 49 Pont Street, London, S.W.1, England.

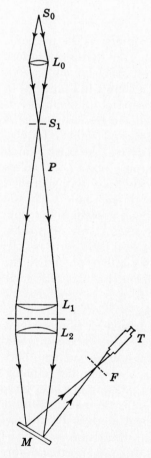

Figure 2-48 Optical system of the Taylor-Lipson (Manchester) optical diffractometer. (Taylor and Lipson [87].)

Figure 2-49 General view of the diffractometer manufactured by R. B. Pullin and Company, Ltd., based on the Taylor-Lipson (Manchester) design. (Taylor and Lipson [87].)

changes introduced by the lower lens between the mask and the focal spot must be identical for all beams parallel to a given direction. These conditions are remarkably difficult to achieve in practice." Four important prerequisites for the production of high-quality diffraction patterns are (a) use of the best quality lenses that one can afford, (b) high coherency of the source, (c) very precise alignment of the optical components, and (d) optimal focusing. The fulfillment of conditions c and d demands a very high degree of rigidity of the entire apparatus.

2-5.3   Preparation of the Masks

A vitally important aspect of optical diffractometry is the preparation of the masks. Because of the difficulties inherent in this operation, it is necessary to accept a compromise between precision and speed. In many ways a pantograph punch, such as that shown in Figure 2-50, is the most generally useful device for the preparation of punched-card structure models. In the instrument pictured the cross-wire A is the tracing point of a pantograph linkage designed so that B follows A with a reduction of 12:1. The framework C carries the mask, which is punched by the punch and die at the stationary point D. Punches and dies can be readily interchanged to provide holes with a graded series of diameters from $\frac{1}{2}$ to $2\frac{1}{2}$ mm. The area of each hole is chosen so as to correspond approximately to the scattering power of the atom that it respresents. For polymers consisting exclusively of carbon, nitrogen, and oxygen (neglecting hydrogen atoms) little error results from making all the holes of one size. In the pantograph shown the maximum area that can be punched is 5×5 cm. Besides stiff cardboard, fully exposed and processed x-ray film is a good opaque medium from which to prepare punched masks.

When very-small-scale masks are required, several photographic and photoetching processes can be of special value [89]. Compared with the card-punching technique the successful use of such methods requires more skill and experience. However, photographic reducing techniques alone are applicable when masks that contain a very large number of scattering points are required, as in the investigation of paracrystalline lattices or very complex helical models. Bonart and Hosemann [90] have made extensive optical-diffraction studies of complex-structure models with the aid of a diffractometer that differs somewhat from that of Taylor and Lipson. (Figures 5-29 and 5-30 show typical masks and the related diffraction patterns from the work of these investigators.)

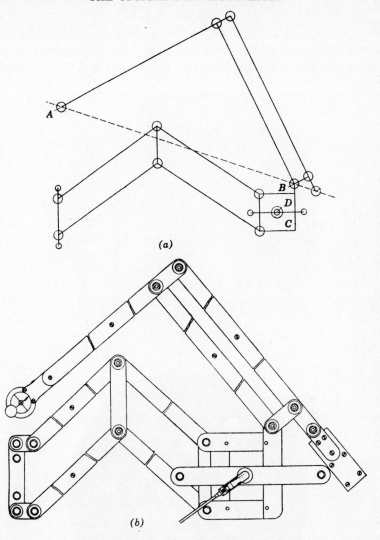

(a)

(b)

Figure 2-50 (a) Schematic and (b) outline drawings of pantograph punch. (Taylor and Lipson [87].)

Taylor and Lipson suggest calibrating each transform pattern with a scale of marks separated by a uniform interval of 1 Å^{-1} by punching along one edge of each mask a straight row of holes spaced 1 Å apart on the same scale that was used in punching the array of holes that represent the structure model (for example, 2 cm = 1 Å). For a mask punched with 1-mm holes the diameter of the useful part of the transform pattern produced by the Taylor-Lipson diffractometer is only about 2 mm, from which it is evident that an enlargement of 50- to 100-fold is mandatory for convenient interpretative work. A microprojector can be used for this purpose. Predecki and Statton

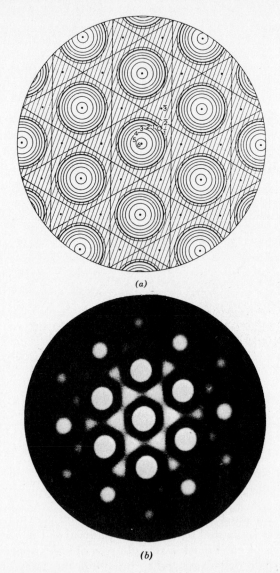

(a)

(b)

Figure 2-51 (a) Numerically calculated and (b) optical transforms of the benzene ring. (Taylor and Lipson [87].)

[91] fitted the R. B. Pullin instrument with a "Polaroid back" so that the diffraction pattern could be first enlarged with a microscope and then photographically recorded. As a simple illustration of the use of the Taylor-Lipson optical diffractometer Figure 2-51 compares the optical transform of the benzene ring with the equivalent numerically computed transform.

In addition to the instrumentation described in this chapter some discussion of more specialized apparatus is given in the succeeding chapters. In particular the reader is referred to Chapter 4 for a treatment of texture, or pole-figure, goniometers.

GENERAL REFERENCES

M. E. Bergman, "Microcameras," *Norelco Reporter*, 6, 96 (1959).

B. D. Cullity, *Elements of X-Ray Diffraction*, Addison-Wesley, Reading, Mass., 1956, Chapter 7.

P. H. Dowling, C. F. Hendee, T. R. Kohler, and W. Parrish, "Counters," *Philips Tech. Rev.*, 18, 262 (1956–1957).

A. Guinier and G. Fournet, *Small-Angle Scattering of X-Rays*, Wiley, New York, 1955, Chapter 3.

International Tables for X-Ray Crystallography, Vol. III, Kynoch Press, Birmingham, 1962, pp. 73–88, 133–156.

H. P. Klug and L. E. Alexander, *X-Ray Diffraction Procedures*, Wiley, New York, 1954, Chapters 4, 5 and 12.

W. Parrish, "Counters," *Philips Tech. Rev.*, 17, 206 (1956).

W. Parrish and T. R. Kohler, "Counters," *Rev. Sci. Instruments*, 27, 795–808 (1956).

H. S. Peiser, H. P. Rooksby, and A. J. C. Wilson (editors), *X-Ray Diffraction by Polycrystalline Materials*, Institute of Physics, London, 1955, Chapters 3, 4, 5, 7, and 11.

W. P. Riley, "Counters and Pulse-Height Discrimination," *Norelco Reporter*, 7, 143 (1960).

W. Ruland, "Precise Diffractometry," *Norelco Reporter*, 14, 12 (1967).

W. O. Statton, "Characterization of Polymers," in *Handbook of X-Rays in Research and Analysis*, E. Kaelble (editor), McGraw-Hill, New York, 1967, Chapter 21.

W. O. Statton, "Small-Angle X-Ray Studies of Polymers," in *Newer Methods of Polymer Characterization*, B. Ke (editor), Interscience, 1964, Chapter 6.

J. Taylor and W. Parrish, "Counters," *Rev. Sci. Instruments*, 26, 367 (1955).

Z. W. Wilchinsky, "Diffractometry by Transmission," in *Advances in X-Ray Analysis*, Vol. 8, Plenum Press, New York, 1965, p. 151.

SPECIFIC REFERENCES

[1] H. P. Klug and L. E. Alexander, *X-Ray Diffraction Procedures*, Wiley, New York, 1954, pp. 108–110, 364–374.

[2] W. A. Wooster, G. N. Ramachandran, and A. Lang, *J. Sci. Instruments*, 25, 405 (1948).

[3] Reference 1, pp. 204–206.

[4] W. O. Statton in *Handbook of X-Rays in Research and Analysis*, E. Kaelble (editor), McGraw-Hill, New York, 1967, Chapter 21.

[5] L. E. Alexander, S. Ohlberg, and G. R. Taylor, *J. Appl. Phys.*, **26**, 1068 (1955).

[6] L. Segal, J. J. Creely, and C. M. Conrad, *Rev. Sci. Instruments*, **21**, 431 (1950).

[7] L. Segal and C. M. Conrad, *Amer. Dyestuff Reporter*, August 26, 1957.

[8] J. J. Renton and W. L. Baun, U.S. Air Force Materials Laboratory, Wright-Patterson Air Force Base, Ohio, Tech. Doc. Report No. ASD–TDR–63–469, June 1963.

[9] P. H. Hermans, D. Heikens, and A. Weidinger, *J. Polymer Sci.*, **35**, 145 (1959).

[10] Reference 1, pp. 252 and 297.

[11] D. T. Keating and B. E. Warren, *Rev. Sci. Instruments*, **23**, 519 (1952).

[12] S. Ergun in *Chemistry and Physics of Carbon*, Vol. 3, P. L. Walker, Jr. (editor), Marcel Dekker, New York, 1968, pp. 224–233.

[13] Reference 1, pp. 211 ff. and 236–239.

[14] Commission on Crystallographic Apparatus of the International Union of Crystallography, *Acta Cryst.*, **9**, 520 (1956).

[15] H. Morimoto and R. Uyeda, *Acta Cryst.*, **16**, 1107 (1963).

[16] G. M. B. Dobson, *Proc. Roy. Soc. (London)*, **A104**, 248 (1923).

[17] H. S. Peiser, H. P. Rooksby, and A. J. C. Wilson (editors), *X-Ray Diffraction by Polycrystalline Materials*, The Institute of Physics, London, 1955, pp. 76–77, 620–634.

[18] L. V. Azaroff and M. J. Buerger, *The Powder Method in X-Ray Crystallography*, McGraw-Hill, New York, 1958, pp. 184–187.

[19] A Taylor, *X-Ray Metallography*, 2nd ed., Wiley, New York, 1961, pp. 275–278.

[20] K. C. Ellis and J. O. Warwicker, *J. Polymer Sci.*, Part A, **1**, 1185 (1963).

[21] M. J. Buerger, *The Photography of the Reciprocal Lattice*, Monograph No. 1, American Society for X-Ray and Electron Diffraction, 1944.

[22] M. J. Buerger, *The Precession Method in X-Ray Crystallography*, Wiley, New York, 1964.

[23] J. A. Howsmon and N. M. Walter in *Physical Methods in Chemical Analysis*, 2nd ed., Vol. I, W. G. Berl (editor), Academic Press, New York, 1960, pp. 154 and 161.

[24] M. V. King, *Acta Cryst.*, **21**, 629 (1966).

[25] D. R. Holmes, C. W. Bunn, and D. J. Smith, *J. Polymer Sci.*, **17**, 159 (1955).

[26] Y. Kinoshita, *Makromol. Chem.*, **33**, 21 (1959).

[27] E. S. Clark and L. T. Muus, *Z. Krist.*, **117**, 119 (1962).

[28] Reference 1, pp. 184–185 and 198.

[29] A. J. C. Wilson, *Rev. Sci. Instruments*, **20**, 831 (1949).

[30] Reference 1, pp. 326–328.

[31] M. J. Buerger, *Amer. Mineral.*, **21**, 11 (1936).

[32] M. J. Buerger, *J. Appl. Phys.*, **16**, 501 (1945).

[33] J. Mann and L. Roldan-Gonzalez, *J. Polymer Sci.*, **60**, 1 (1962).

[34] I. Fankuchen and H. Mark, *J. Appl. Phys.*, **15**, 368 (1944).

[35] N. Kasaї and M. Kakudo, *J. Polymer Sci.*, Part A, **2**, 1955 (1964).

[36] F. G. Chesley, *Rev. Sci. Instruments*, **18**, 422 (1947).

[37] H. Lipson, J. B. Nelson, and D. P. Riley, *J. Sci. Instruments*, **22**, 184 (1945).

[38] Reference 17, pp. 139–141.

[39] *International Tables for X-Ray Crystallography*, Vol. III, Kynoch Press, Birmingham, 1962, pp. 79–81.

[40] Reference 1, pp. 80–81.

[41] A. Guinier, *Compt. Rend.*, **204**, 1115 (1937); *Ann. Phys.* (Paris), **12**, 161 (1939).

[42] P. H. Hermans and A. Weidinger, *Makromol. Chem.*, **18–19**, 75 (1956).

[43] P. H. Hermans and A. Weidinger, *Textile Research J.*, **30**, 770 (1960).

[44] G. Farrow, *Polymer*, **2**, 409 (1961).

[45] P. M. de Wolff, *Acta Cryst.*, **1**, 207 (1948).

[46] T. Johansson, *Naturwiss.*, **20**, 758 (1932); Z. *Physik*, **82**, 507 (1933).

[47] O. E. A. Bolduan and R. S. Bear, *J. Appl. Phys.*, **20**, 983 (1949); R. S. Bear and O. E. A. Bolduan, *J. Appl. Phys.*, **22**, 191 (1951).

[48] (a) H. Kiessig, *Kolloid-Z.*, **98**, 213 (1942); (b) ibid., **152**, 62 (1957).

[49] C. R. Worthington, *J. Sci. Instruments*, **33**, 66 (1956).

[50] I. Fankuchen and M. H. Jellinek, *Phys. Rev.*, **67**, 201 (1945).

[51] A. Guinier and G. Fournet, *Small-Angle Scattering of X-Rays*, Wiley, New York, 1955, p. 109.

[52] P. Kaesberg, H. N. Ritland, and W. W. Beeman, *Phys. Rev.*, **74**, 71 (1948).

[53] O. Kratky, *Z. Elektrochem.*, **58**, 49 (1954).

[54] O. Kratky, *Kolloid-Z.*, **144**, 110 (1955).

[55] O. Kratky and A. Sekora, *Monatsh. Chem.*, **85**, 660 (1954).

[56] O. Kratky, *Z. Elektrochem.*, **62**, 66 (1958).

[57] O. Kratky and Z. Skala, *Z. Elektrochem.*, **62**, 73 (1958).

[58] O. Kratky in *Progress in Biophysics*, Vol. 13, Pergamon, New York, 1963, pp. 139–141.

[59] V. Luzzati and R. Baro, *J. Phys. Radium, Phys. Appl.*, **22**, 186A (1961).

[60] A. Franks, *Proc. Phys. Soc. (London)*, **B68**, 1054 (1955).

[61] A. Franks, *British J. Appl. Phys.*, **9**, 349 (1958).

[62] G. F. Elliott and C. R. Worthington, *J. Ultrastruct. Res.*, **9**, 166, 170 (1963).

[63] A. Elliott, *J. Sci. Instruments*, **42**, 312 (1965); *Hilger J.*, **11**, 38 (1968).

[64] H. E. Huxley and W. Brown, *J. Mol. Biol.*, **30**, 383 (1967).

[65] S. S. Pollack and L. E. Alexander, *J. Chem. Eng. Data*, **5**, 88 (1960).

[66] S. S. Pollack, W. H. Robinson, R. Chiang, and P. J. Flory, *J. Appl. Phys.*, **33**, 237 (1962).

[67] For example, see [1], pp. 268–269 and 305–311.

[68] H. Geiger and W. Müller, *Physik. Z.*, **29**, 839 (1928).

[69] Reference 58, pp. 105–173.

[70] V. Luzzati, J. Witz, and R. Baro, *J. Phys., Phys. Appl.*, **24**, 141A (1963).

[71] P. A. Ross, *J. Opt. Soc. Amer.*, **16**, 433 (1928).

[72] P. Kirkpatrick, *Rev. Sci. Instruments*, **10**, 186 (1939); **15**, 223 (1944).

[73] W. Ruland, *Acta Cryst.*, **14**, 1180 (1961).

[74] H. A. McKinstry and M. A. Short, *J. Sci. Instruments*, **37**, 178 (1960).

[75] A. R. Lang, *Rev. Sci. Instruments*, **27**, 17 (1956).

[76] H. T. Johann, *Z. Physik*, **69**, 185 (1931).

[77] P. M. de Wolff, *Acta Cryst.*, **13**, 835 (1960).

[78] W. A. Wooster, G. N. Ramachandran, and A. R. Lang, *J. Sci. Instruments*, **26**, 156 (1949).

[79] B. R. Banerjee, *Rev. Sci. Instruments*, **26**, 564 (1955).

[80] R. E. Ogilvie, *Rev. Sci. Instruments* (to be published).

[81] D. M. Koffman and S. H. Moll, *Norelco Reporter*, **11**, 95 (1964).

[82] S. Krimm and R. S. Stein, *Rev. Sci. Instruments*, **22**, 920 (1951).

[83] J. H. Wakelin, H. S. Virgin, and E. Crystal, *J. Appl. Phys.*, **30**, 1654 (1959).

[84] V. Luzzati, *Acta Cryst.*, **13**, 939 (1960).

[85] W. Ruland, *Norelco Reporter*, **14**, 12 (1967).

[86] C. A. Taylor, *J. Polymer Sci.*, Part C, **20**, 19 (1967).

[87] C. A. Taylor and H. Lipson, *Optical Transforms*, Cornell University Press, Ithaca, N.Y., 1965 (also G. Bell and Sons, London, 1964).

[88] H. Lipson and C. A. Taylor, *Fourier Transforms and X-Ray Diffraction*, G. Bell and Sons, London, 1958.

[89] For example, C. R. Berry, *Amer. J. Physics*, **18**, 269 (1950); C. W. Hooper, W. E. Seeds, and A. R. Stokes, *Nature*, **175**, 679 (1955); [87], pp. 54–56.

[90] For example, R. Bonart, Z. *Krist.*, **109**, 296 (1957); R. Hosemann, *Polymer*, **3**, 349 (1962).

[91] P. Predecki and W. O. Statton in *Small-Angle X-Ray Scattering*, H. Brumberger (editor), Gordon and Breach, New York, 1967, p. 131.

3

Degree of Crystallinity in Polymers

3-1 THE CONCEPT OF CRYSTALLINITY

As has already been explained in Chapter 1, the presence of both sharp and diffuse diffraction effects in the x-ray patterns of polymers was formerly accepted as evidence for a *two-phase concept* of polymer structure, which is to say that there exist relatively perfect crystalline domains (crystallites) interspersed with amorphous regions. However, this simplified picture of polymer structure is not fully compatible with a number of more recent experimental findings that are attested by a wealth of observational data accumulated principally during the past 10 years. These experimental results require that our earlier notions of polymer structure be modified to allow for the occurrence of such phenomena as polymer single crystals, chain folding, lamellar crystalline growths, and intermediate crystalline objects such as axialites and hedrites, lattice dislocations, and one- and two-dimensional ordering in drawn fibers. The new picture of polymer structure that is emerging may perhaps be designated the *crystal-defect concept* as opposed to the earlier *two-phase concept.* Present indications are that these concepts, though profoundly different, actually complement each other because each embodies truthful aspects of the nature of ordering in polymers.

An important consequence of the crystal-defect concept is that a portion of the x-ray scattering from the crystalline domains is diffuse and contributes to the so-called amorphous background, thus bringing into question the simple method of estimating degree of crystallinity by separating the diffraction pattern into sharp (crystalline) and diffuse (amorphous) components. It should also be mentioned that unequal numerical results are usually obtained when the percent crystallinity of a given polymer specimen is measured by different physical methods, such as x-ray diffraction, density, infrared absorption, or nuclear magnetic resonance (NMR). Thus it has become the practice to differentiate the x-ray crystallinity from the infrared crystallinity, NMR crystallinity, etc.

From these considerations we must conclude that the *degree of crystallinity* of a polymer is a concept that cannot be unambiguously defined, much less be subject to experimental measurement in an absolute sense. Thus numerical values of crystalline and amorphous contents obtained by well-established and purportedly absolute x-ray techniques are now generally regarded as of doubtful absolute significance. Increasing preference is shown for expressing the degree of three-dimensional order by some relative numerical quantity, which may be termed a *crystallinity index*.

The physical and mechanical properties of polymers are profoundly dependent on the degree of crystallinity, regardless of how we may choose to measure it or of the structural interpretation we may place on it; for example, the tensile strength and stiffness of polymer fibers are directly related to the degree of alignment of the molecular chains parallel to the fiber axis and hence to the degree of crystallinity. Likewise the stress at which the polymer starts to draw (yield stress) increases with the degree of crystallinity.

3-2 BASIC PRINCIPLES OF THE X-RAY METHOD

The intensity of the x-rays scattered over all angles by a given assemblage of atoms is independent of their state of order or disorder [1]. It follows that if the crystalline and amorphous scattering in the diffraction pattern can be separated from each other, the crystalline fraction is equal to the ratio of the integrated crystalline scattering to the total scattering, both crystalline and amorphous. In the ensuing treatment we shall adhere rather closely to the presentation given by Ruland[2]. We designate the magnitude of the reciprocal-lattice vector ρ_{hkl} by the symbol s, which is related by a factor 2π to S (see Sections 1-4.2 and 1-4.4):

$$s = |s| = \frac{2 \sin \theta}{\lambda} = \frac{S}{2\pi}.$$ (3-1)

Let $I(s)$ be the intensity of coherent x-ray scatter from a specimen at the point s in reciprocal space and $I_c(s)$ the part of the intensity at the same point that is concentrated in the crystalline peaks (reciprocal-lattice points). The integrals of these intensities over all reciprocal space are

$$\int_0^\infty I(s) \, dv_s = 4\pi \int_0^\infty s^2 I(s) \, ds \qquad (3\text{-}2)$$

and

$$\int_0^\infty I_c(s) \, dv_s = 4\pi \int_0^\infty s^2 I_c(s) \, ds. \qquad (3\text{-}3)$$

In these equations $I(s)$ and $I_c(s)$ refer to the intensities of scattering per unit solid angle at the radial distance s in reciprocal space. Thus

$$I(s) = \frac{1}{4\pi} \int_0^{4\pi} I(s) \, d\omega. \qquad (3\text{-}4)$$

To a first approximation the fraction of crystalline material in the specimen is given by

$$x_c = \frac{\displaystyle\int_0^\infty s^2 I_c(s) \, ds}{\displaystyle\int_0^\infty s^2 I(s) \, ds}, \qquad (3\text{-}5)$$

in which x_c tends to be smaller than the true crystalline fraction because part of the x-ray intensity that is scattered by the crystalline regions is lost from the peaks and appears as diffuse scatter in the background as a result of atomic thermal vibrations and lattice imperfections. (See also Chapter 7.) Allowance can be made for this loss of intensity by including in the intensity expression a lattice-imperfection factor D, as Ruland has shown [2]. This factor expresses the loss of intensity suffered by the crystalline reflections as a result of all departures of the atoms from their ideal positions.

We first define the weighted mean-square atomic-scattering factor of the polymer concerned,

$$\overline{f^2} = \frac{\sum N_i f_i^2}{\sum N_i}, \qquad (3\text{-}6)$$

N_i being the number of atoms of type i in the stoichiometric formula.

Equations 3-2 and 3-3 may then be further developed as follows:

$$4\pi \int_0^\infty s^2 I(s) \, ds = 4\pi \int_0^\infty s^2 \overline{f^2} \, ds, \tag{3-7}$$

$$4\pi \int_0^\infty s^2 I_c(s) \, ds = 4\pi x_c \int_0^\infty s^2 \overline{f^2} D \, ds. \tag{3-8}$$

Dividing (3-8) by (3-7) and solving for x_c, we find

$$x_c = \frac{\displaystyle\int_0^\infty s^2 I_c \, ds \;\; \int_0^\infty s^2 \overline{f^2} \, ds}{\displaystyle\int_0^\infty s^2 I \, ds \;\; \int_0^\infty s^2 \overline{f^2} D \, ds}. \tag{3-9}$$

The applicability of (3-9) to experimental scattering curves presupposes the following:

1. The orientation of the structural elements within the specimen is random.
2. The ordering in the crystalline regions is fully three-dimensional.
3. The integrals are valid when applied to finite angular ranges.
4. The crystalline peaks can be separated from the background in a precise and reproducible manner.
5. The imperfection factor D is known.

With regard to the foregoing requirements we may make the following comments, respectively:

1. If the diffraction pattern reveals some degree of preferred orientation in the specimen, it must be randomized in the sense of (3-4) by some method. Desper and Stein[3a] have given a critical discussion of randomization techniques, and Ruland and Dewaelheyns [3b] have invented an electronic device for the spherical averaging of fiber patterns.
2. When the diffraction pattern indicates that the degree of ordering is less than three-dimensional, the mathematical formulas must be modified as discussed on pp. 142–143.
3. The validity of the integrals will be good if relatively large angular ranges s_0 to s_p are employed.
4. The most reproducible method of separating the crystalline peaks from the background is to draw a smooth curve from one minimum point to the next, following the general curvature of the independent scattering curve $(s^2 \overline{f^2})$ as modulated by long-period

undulations. Since appreciable overlapping of adjacent lines begins only when the integral breadths exceed 0.02 to 0.03 Å⁻¹, this method of drawing the demarcation line has the effect of restricting the term "crystalline" to regions with a mean dimension larger than 30 to 50 Å or to regions characterized by paracrystalline distortions that are smaller than about 0.10† [4, 5].

5. Requirement 5 is satisfied inasmuch as Ruland's method provides analytical values of D and x_c simultaneously.

For lattice imperfections of the first kind,‡ including thermal vibrations, D can be approximated well with a Gaussian curve, whereas for imperfections of the second kind‡ the theory shows that D conforms more closely to the expression [5, 6]

$$D = \frac{2 \exp (-as^2)}{1 + \exp (-as^2)}. \tag{3-10}$$

However, Ruland found that calculated values of the coefficient

$$K = \frac{\int_{s_0}^{s_p} s^2 \overline{f^2} \, ds}{\int_{s_0}^{s_p} s^2 \overline{f^2} D \, ds}, \tag{3-11}$$

when plotted as a function of s_p, the upper angular limit of integration, gave approximately straight lines with the same slope regardless of whether D was assumed to be given by (3-10) or by a Gaussian function (see Figure 3-4 for typical plots of K versus s_p) [4]. This showed that all kinds of displacements of the atoms from their ideal positions can be lumped together in one Gaussian lattice-imperfection factor,

$$D = \exp (-ks^2), \tag{3-12}$$

in which k is the sum of three terms due to thermal motion (k_T) and to lattice imperfections of the first (k_I) and second (k_{II}) kinds [4]:

$$k = k_T + k_I + k_{II}. \tag{3-13}$$

It must be emphasized that (3-7), (3-8), and (3-9) are valid only when the ordering in the crystalline regions is three-dimensional. In that

†This fraction expresses the integral breadth of the distribution of first-neighbor distances relative to the average first-neighbor distance. See also Chapter 7 for a discussion of paracrystalline lattice distortions, including the meaning of distortions of the first and second kinds.
‡See Section 7-2.2.

case (3-8) may also be expressed as

$$4\pi \int_0^\infty s^2\, I_c(s)\; ds = \frac{x_c}{NV} \sum_h \sum_k \sum_{l}^{+\infty} F_D^2, \tag{3-14}$$

where V is the volume of the unit cell, N is the number of atoms in the unit cell, and F_D is the structure factor formulated so as to make allowance for deviations of the atoms from their ideal sites. When the ordering is exclusively two-dimensional, only reflections of the type $(hk0)$ appear sharply in the $s^2\, I(s)$ curve, and (3-14) must be modified as follows:

$$4\pi \int_0^\infty s^2\, I_c(s)\; ds = \frac{x_c}{NV} \sum_h \sum_{k}^{+\infty} F_D^2$$

$$= \frac{x_c|\mathbf{a}\times\mathbf{b}|}{V} 2\pi \int_0^\infty s^2[f^2]\, D\; ds. \tag{3-15}$$

If the ordering is purely one-dimensional in the c-direction, only reflections $(00l)$ appear in the $s^2\, I(s)$ curve, and (3-14) becomes

$$4\pi \int_0^\infty s^2\, I_c(s)\; ds = \frac{x_c}{NV} \sum_{l}^{+\infty} F_D^2$$

$$= \frac{x_c\mathbf{c}}{V} 2\pi \int_0^\infty [f^2]\, D\; ds. \tag{3-16}$$

In (3-15) and (3-16)

$$[f^2] = \frac{1}{N}\left[\sum_j f_j^2 + \sum_n \left(\sum_k f_k\right)^2\right], \tag{3-17}$$

in which for (3-15) j refers to atoms that do not coincide in the projection on the ab plane along the direction of the c-axis, and k refers to atoms that do coincide in this projection, forming n points of coincidence [7]. For (3-16) j refers to atoms that do not coincide in the projection on the c-axis parallel to the ab plane, and k denotes atoms that coincide in this projection. Also $\mathbf{a}$, $\mathbf{b}$, $\mathbf{c}$ are the three vector edges of the unit cell.

A consideration of (3-15), (3-16), and (3-17) shows that they cannot be applied without enough prior knowledge of the structure of the two- or three-dimensionally ordered phase to specify the numbers of atoms of types j and k. It must also be remembered in applying (3-15) and (3-16) that x_c specifies the weight fraction of the polymer that possesses two- or one-dimensional order rather than the fraction that has full three-dimensional (crystalline) order, as in the ideal case represented by (3-8), (3-9), and (3-14). Equation 3-9 shows that, when the ordering is three-dimensional, no knowledge of the structure is required but only the chemical composition as expressed by (3-6).

Ruland[4] also treats theoretically a structural model that has a continuous gradation from ordered to disordered character but at the same time is capable of producing reflections of sufficient sharpness to be regarded as crystalline. The specific model is a one-dimensional paracrystal in which the first-neighbor distribution can be split into a narrow and a wide distribution, which constitutes an extreme case of lattice imperfection inasmuch as only the interactions of first neighbors are taken into account. Limitations of space do not allow a presentation of the mathematics here, but the conclusion that was reached is a very significant one, at least insofar as it may apply to two- and three-dimensional structures. This conclusion is that *analyses of crystallinity can lead to crystalline fractions smaller than 0.8 only when there exists segregation into ordered and disordered domains* (a two-phase system). This also means that a *completely "homogeneous" disorder can be present only if crystalline fractions larger than 0.8 are found by Ruland's method of analysis.*

To set the stage for a description of some practical procedures for measuring crystallinity in polymers, we shall first discuss in a semiquantitative way Ruland's exceptionally thorough and precedent-making investigations of crystallinity in two polymer systems. These methods, although not readily adaptable to routine use, are of fundamental importance because they serve to set up guidelines and standards of reference for all workers in this field.

3-3 EXPERIMENTAL PROCEDURES

3-3.1 Crystallinity from the Integrated Crystalline and Amorphous Scatter, with Allowance for Lattice Imperfections (Method of Ruland [2])

Ruland's procedure conforms closely to the basic equations (3-6), (3-7), (3-8), and (3-9) and yields the crystalline fraction x_c and the coefficient k of the lattice-imperfection exponent [see (3-12)]. He applied the method to two kinds of polymer systems — the first, polypropene, being one in which the interchain forces are relatively weak (exclusively van der Waals); and the second, the polyamides nylon 6 and nylon 7, being systems in which the dominant interchain forces are relatively strong (hydrogen bonds). It could be anticipated that lattice imperfections might be more important in the second kind of system than in the first. This point should be resolved on the basis of the relative sizes of k.

Crystallinity and Lattice Imperfections in Polypropene. Four polypropene samples with the following prehistories were studied:

No. 1. Isotactic polypropene heated to its melting point and quenched in water at room temperature.

No. 2. The same, annealed 1 hr at 105°C.

No. 3. The same, annealed $\frac{1}{2}$ hr at 160°C.

No. 4. Heptane extract of crude polypropene, highly atactic.

The intensities were measured with a xenon-filled proportional counter using $CuK\alpha$ radiation monochromatized with Ross balanced filters of nickelous oxide and cobaltic oxide. The symmetrical-reflection geometry was employed, and the thickness of the flat specimen was made large enough to keep the absorption independent of θ. [See (2-10) and (2-14).] The experimental intensity curve was corrected for polarization, and the angular scale was converted from 2θ to s.

In order to subtract the calculated intensity of incoherent scattering from the experimental intensity curve it is necessary to fit the experimental curve to the overall theoretical intensity curve, which consists of both coherent and incoherent components. This can be done most accurately at large angles where the incoherent component exceeds the coherent and where the fluctuations of the experimental curve about the total independent scattering curve have become small, as can be seen, for example, in Figure 1-25. The incoherent scattering[†] as calculated with (1-6) is actually appreciably larger than that experimentally observed because the wavelength and absorption coefficient of the incoherent scattering are both somewhat larger than those of the coherent radiation. This results in significant reductions in the incoherent intensity because of absorption within the sample, in the air path between sample and counter, and in the balanced filters. Ruland evaluated the magnitudes of these corrections and applied them to the theoretical incoherent scattering of polypropene as calculated with (1-6) for the chemical composition $(CH_2)_n$. Figure 3-1 shows the incoherent-scattering curves as calculated (a) omitting the Breit-Dirac factor and absorption corrections, (b) including the Breit-

[†]It is suggested that the reader examine again Section 1-3.2.

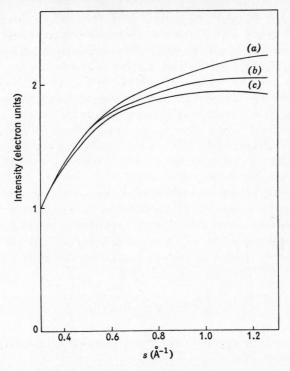

Figure 3-1 Compton scattering calculated for $(CH_2)_n$ in electron units per atom (*a*) omitting the Breit-Dirac factor and absorption corrections, (*b*) including the Breit-Dirac factor, and (*c*) including both the Breit-Dirac factor and absorption corrections. (Ruland[2].)

Dirac factor and (c) including both the Breit-Dirac factor and absorption corrections. Both factors are significant.

The coherently scattered intensity per atom at large values of s is given to a first approximation by $\overline{f^2}$, the mean-square atomic-scattering factor of the polymer, but a closer match with the experimental curve is given by modifying (T-19) of Appendix 1[8] with the aid of (3-6) to express the average scattered intensity per atom in the form

$$I(s) = \overline{f^2} + \frac{1}{\sum N_i} \sum_m^{m \neq n} \sum_n f_m f_n \frac{\sin 2\pi s r_{mn}}{2\pi s r_{mn}}, \qquad (3\text{-}18)$$

where the terms for $m = n$ combine to yield $\overline{f^2}$. The second term gives the oscillations of the actual coherent-scattering curve about the mean scattering curve, $\overline{f^2}$, resulting from the frequently occurring interatomic distances r_{mn}. Ruland substituted in (3-18) the known invariant intramolecular bonded distances and the mean intermolecular distance in linear polypropene to calculate the course of the coherent-scattering curve at large values of s. When this undulating curve was combined with the corrected incoherent-scattering curve, it gave the overall calculated intensity curve shown in Figure 3-2, which is seen to compare well in shape with the experimental scattering of the four polypropene specimens considering only their long-range oscillations. The experimental curves were normalized to the calculated curve by appropriate factors and the theoretical incoherent scattering was then subtracted, thus yielding experimental curves of coherent scattering in electron units for the four specimens.

Since the long-range undulations of the scattering curve are not influenced by the degree of crystallinity, an examination of the scattering curves of specimens differing in crystallinity makes it possible to choose several rather broad integration intervals extending from a fixed minimum angle s_0 to various maximum angles s_p in which the relationship of (3-7) is applicable; that is,

$$\int_{s_0}^{s_p} s^2 I(s) \, ds = \int_{s_0}^{s_p} s^2 \overline{f^2} \, ds. \tag{3-19}$$

Then (3-9) can be applied over these angular ranges to any given specimen in the form

$$x_c = \frac{\displaystyle\int_{s_0}^{s_p} s^2 I_c \, ds}{\displaystyle\int_{s_0}^{s_p} s^2 I \, ds} K(s_0, s_p, D, \overline{f^2}) = \text{constant}, \tag{3-20}$$

where K is defined by (3-11).

Figure 3-3 shows the curve of $s^2 I(s)$ versus s of one of the polypropene samples (No. 3). To determine x_c from the $s^2 I_c$ and $s^2 I$ curves of any specimen it is useful to prepare a nomogram of K versus s_p for a range of values of the coefficient k in the lattice-imperfection factor $D = \exp(-ks^2)$. Such a nomogram, calculated by using (3-11) for $s_0 = 0.1$ and for $\overline{f^2}$ corresponding to the chemical composition $(CH_2)_n$, is shown in Figure 3-4. It will be noticed that the curves of K versus s_p are nearly linear. For a given polypropene sample Ruland could read from this nomogram the optimal value of k to make x_c as nearly constant as possible irrespective of s_p. Table 3-1 shows the numerical results that were obtained for the four polypropene samples described above by using four angular ranges terminating at $s_p = 0.3$, 0.6, 0.9, and 1.25. For all four samples the most constant

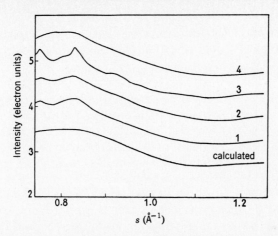

Figure 3-2 High-angle-scattering curve of polypropene calculated from the known most invariant interatomic distances and compared with the experimental scattering curves of samples 1, 2, 3, and 4. The latter curves have been arbitrarily displaced along the ordinate axis for clarity. (Ruland [2].)

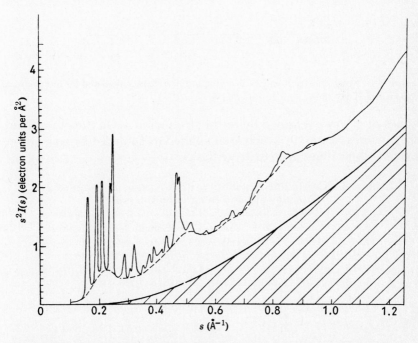

Figure 3-3 Curve of $s^2 I(s)$ versus s for polypropene sample No. 3. (Ruland [2].)

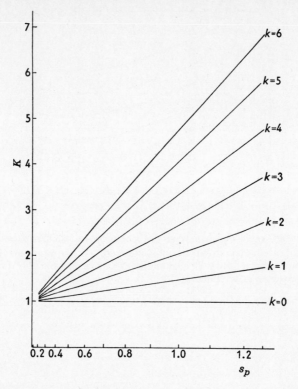

Figure 3-4 Nomogram of K values as a function of k and s_p calculated for the chemical composition $(CH_2)_n$ and $s_0 = 0.1$. (Ruland[2].)

values of x_c were obtained by setting k equal to 4. On the other hand, the x_c values for $k = 0$, which are included in Table 3-1 by way of contrast, exhibit pronounced drifts with s_p.

Table 3-1   Crystalline Fraction x_c in Polypropene Samples as a Function of k and Integration Interval[a]

Interval (s_0-s_p)	Sample 1 $k=0$	Sample 1 $k=4$	Sample 2 $k=0$	Sample 2 $k=4$	Sample 3 $k=0$	Sample 3 $k=4$	Sample 4 $k=0$	Sample 4 $k=4$
0.1–0.3	0.270	0.329	0.353	0.431	0.546	0.666	0.120	0.146
0.1–0.6	0.159	0.294	0.222	0.411	0.333	0.616	0.078	0.144
0.1–0.9	0.105	0.305	0.145	0.421	0.220	0.638	0.044	0.128
0.1–1.25	0.067	0.315	0.095	0.447	0.145	0.682	0.029	0.136
Mean x_c		0.31		0.43		0.65		0.14

[a]Data from [2].

It is perhaps surprising that one and the same value of k should characterize all four specimens despite their very different thermal prehistories. In interpreting this result it must be remembered that k includes the effects of both thermal vibrations and lattice imperfections of the first kind (sometimes referred to as frozen-in thermal motion). Since lattice imperfections would be expected to change with variations in thermal treatment, the constancy of k suggests that in polypropene such imperfections are very small in comparison with the thermal vibrations. This conclusion is also supported by crystal-structure studies performed by Natta and Corradini[9], which led to a mean isotropic thermal-vibration coefficient of $B = 7.5$ to 8.5 Å^{-2}, in good agreement with Ruland's result since $B = 2k$.† It should also be observed that Ruland's findings favor a two-phase system of crystalline and noncrystalline regions in polypropene, in which the crystalline regions are relatively well ordered and gradual transitions from the crystalline to the amorphous state cannot be present in significant amounts.

Crystallinity and Lattice Imperfections in Nylons 6 and 7 [4]. Bulk specimens of polycaprolactam (nylon 6) and poly-ω-heptanolactam (nylon 7) were subjected to varied thermal treatments, as in the case of polypropene, in an effort to produce specimens varying in crystallinity and degree of lattice imperfection. The experimental technique and method of deriving the final coherent-scattering curves were essentially the same as those described in the preceding section. Again the integral coherent intensities $s^2 I(s)$ and $s^2 I_c(s)$ were determined for four angular ranges, extending from $s_0 = 0.1$ to $s_p = 0.4$, 0.65, 0.95, and 1.25. The calculated values of k that are required to yield constant values of x_c for each sample are listed in Tables 3-2 and 3-3. It is seen that for nylon, unlike for polypropene, the numerical values of k depend on the thermal treatment. Thus seven specimens of nylon 6 required values of k that ranged from 3.0 to 5.6, and three specimens of nylon 7 called for k values that ranged from 3.6 to 7.7.

Bearing in mind that $k = k_T + k_I + k_{II}$ [see (3-13)], we now propose with Ruland[4] that the minimum value of k observed for the nylon specimens, viz., $k = 3.0$, is equal to or somewhat greater than k_T and that the amount by which a given k exceeds 3.0 is equal to or somewhat less than $k_I + k_{II}$, the sum of the lattice imperfections of the first and second kinds. This interpretation is reasonable because the crystalline regions in all the polyamide specimens would be expected

---

†By reference to p. 42 it is seen that the temperature factor is $\exp\left[-2B(\sin^2\theta)/\lambda^2\right]$, whereas the lattice-imperfection factor is $\exp\left[-4k(\sin^2\theta)/\lambda^2\right]$.

Table 3-2   Crystalline Fraction x_c in Samples of Nylon 6 as a Function of k and Integration Interval[a]

Interval (s_0-s_p)	Sample 3 $k=0$	$k=3.0$	Sample 4 $k=0$	$k=4.2$	Sample 6 $k=0$	$k=3.0$	Sample 7 $k=0$	$k=5.6$
0.10–0.40	0.260	0.338	0.169	0.242	0.216	0.281	0.214	0.340
0.10–0.65	0.175	0.306	0.102	0.216	0.139	0.242	0.119	0.308
0.10–0.95	0.129	0.327	0.072	0.238	0.105	0.265	0.076	0.322
0.10–1.25	0.091	0.330	0.050	0.245	0.075	0.273	0.054	0.341
Mean x_c		0.33		0.24		0.27		0.33

Interval (s_0-s_p)	Sample 8 $k=0$	$k=3.9$	Sample 9 $k=0$	$k=4.4$	Sample 10 $k=0$	$k=3.7$
0.10–0.40	0.232	0.324	0.253	0.367	0.241	0.331
0.10–0.65	0.143	0.289	0.143	0.312	0.154	0.302
0.10–0.95	0.101	0.314	0.101	0.346	0.106	0.317
0.10–1.25	0.070	0.318	0.072	0.367	0.076	0.329
Mean x_c		0.31		0.35		0.32

[a]Data from [4].

Table 3-3   Crystalline Fraction x_c in Samples of Nylon 7 as a Function of k and Integration Interval[a]

Interval (s_0-s_p)	Sample 2 $k=0$	$k=7.1$	Sample 3 $k=0$	$k=7.7$	Sample 4 $k=0$	$k=3.6$
0.10–0.40	0.331	0.594	0.233	0.440	0.169	0.231
0.10–0.65	0.172	0.541	0.108	0.366	0.111	0.214
0.10–0.95	0.108	0.577	0.071	0.409	0.073	0.211
0.10–1.25	0.075	0.598	0.050	0.433	0.055	0.234
Mean x_c		0.58		0.41		0.22

[a]Data from [4].

to exhibit similar thermal vibrations at any given temperature. Thus lattice imperfections should account for a contribution to k amounting to from 0 to 2.6 in the case of nylon 6 and from 0.6 to 4.1 in the case of nylon 7.

These numerical results lead to the conclusion that in these polyamide specimens the deviations from ideal lattice periodicity that result from lattice imperfections are of the same general magnitude as those that result from temperature vibrations. However, how these imperfections are apportioned between types I and II cannot be determined from the present method of analysis. In principle this distinction can be made on the basis of line-profile measurements as a function of s since lattice distortions of the first kind affect the intensities but not the breadths of the reflections, whereas those of the second kind cause increasing broadening with s. However, the success of the line-breadth analysis is contingent on the presence of at least three well-resolved orders of reflection from a given set of crystallographic planes, and polymer patterns seldom afford this degree of resolution. Furthermore, the analysis must adequately take into account the broadening that results from the finite size of the crystalline regions (crystallite-size broadening) as well as the broadening that is due to lattice imperfections of the second kind. This subject is discussed in some detail in Chapter 7.

3-3.2 Crystallinity from the Intensity of Amorphous Scatter, Lattice Imperfections Not Considered

This method is especially suited to the measurement of crystallinity in elastomers. It requires that (a) 100% amorphous polymer be available and (b) the principal amorphous halo be not seriously interfered with by reflections from the crystalline phase. The amorphous fraction is assumed to be proportional to the intensity of the amorphous halo, and the crystalline fraction is determined by difference,

$$x_c = 1 - x_a. \tag{3-21}$$

It is necessary that the absorption of x-rays and the scattering mass be constant for the various samples being compared or that corrections be applied for variations of these parameters. It is the usual practice to neglect the correction for the incoherent scattering, which does not introduce a serious error since the incoherent intensity is relatively small at the site of the principal amorphous halo. Furthermore, neglect of the incoherent scattering greatly simplifies the calculations since it is not necessary to reduce the scattered intensity to electron units.

A prototype of the photographic method was first devised by Field [10] and applied to natural rubber. Important improvements were introduced by Goppel and Arlman [11–13], including in particular a compound diffraction camera consisting of a miniature calibrating camera within the principal camera, as shown in Figure 3-5 [11]. The primary beam traverses the collimator (1) and, after penetrating the polymer specimen (2), impinges on a reference sample (3) in the cone-

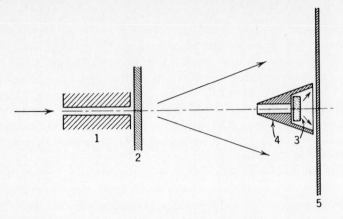

Figure 3-5 Diagram of Goppel's camera for simultaneously recording the diffraction patterns of rubber and the inner lines of the pattern of a standard sample. (Goppel [11].)

shaped miniature camera (4). The diffraction patterns of the polymer and standard are thus recorded simultaneously on the flat film (5) so that both patterns are affected similarly by any variations in the intensity of the primary beam. Furthermore, changes in the intensity of the standard reflection from one experiment to the next provide a means for correcting the polymer intensities for differences in the thicknesses of the specimens being compared.

Because the precision of the Goppel-Arlman method suffers from the usual limitations of photographic measurements, an equivalent but more precise diffractometric technique has been devised and applied to natural and synthetic rubbers [14]. In presenting the mathematical basis of these equivalent techniques we shall employ the following quantities:

I_0 = intensity of the incident x-ray beam.

I_a, I_a' = intensities of the amorphous halos of the completely and partially amorphous polymers, respectively.

I_R, I_R' = intensities of a reflection from the reference specimen corresponding to I_a and I_a', respectively.

μ, μ_R = linear absorption coefficients of the polymer and reference specimen, respectively.

t, t' = thicknesses of the 100% amorphous and partially amorphous specimen, respectively.

t_R = thickness of the reference specimen.

x_c, x_a = crystalline and amorphous fractions, respectively, of the partially amorphous polymer.

To a fair approximation the amorphous fraction may be expressed as

$$x_a = \frac{\int_0^\infty s^2 I_a'(s)\, ds}{\int_0^\infty s^2 I_a(s)\, ds}. \tag{3-22}$$

Furthermore, if only the amplitude (but not the shape) of the amorphous scattering curve changes with changing amorphous content, I_a' and I_a at any particular value of s (such as the peak of the amorphous halo) may be taken as proportional to the integrals in (3-22), which becomes simply

$$x_a = \frac{I_a'(s_i)}{I_a(s_i)}. \tag{3-23}$$

Such a measure of the intensity of the amorphous halo of the 100% amorphous specimen may be written

$$I_a = KI_0 t e^{-\mu t}, \tag{3-24}$$

in which we have deleted the (s_i) for clarity. Likewise, the intensity of a reflection from the reference specimen is

$$I_R = K' I_0 e^{-\mu t} e^{-\mu_R t_R}, \tag{3-25}$$

wherein we take account of the fact that the beam suffers absorption in both the polymer and standard specimens before being diffracted.

At this point we make the assumption that the linear absorption coefficients of the crystalline and amorphous polymers are identical and equal to μ, which is only approximately true since $\mu_a/\mu_c = \rho_c/\rho_a$ and the crystalline density is usually somewhat larger than the amorphous density. However, if we assume this approximation to be acceptable, we can write for the intensity of the amorphous halo of the partially amorphous polymer specimen

$$I_a' = x_a K I_0 t' e^{-\mu t'}, \tag{3-26}$$

whereas the intensity of the reference reflection with the partially amorphous polymer specimen in place is

$$I_R' = K' I_0 e^{-\mu t} e^{-\mu_R t_R}. \tag{3-27}$$

Division of (3-24) by (3-25) and of (3-26) by (3-27) gives

$$\frac{I_a}{I_R} = \frac{K}{K'} \frac{t}{e^{-\mu_R t_R}} \tag{3-28}$$

and

$$\frac{I_a'}{I_R'} = \frac{K}{K'} \frac{t'}{e^{-\mu_R t_R}} x_a,$$ (3-29)

whereupon combination of (3-28) and (3-29) followed by rearrangement leads to

$$x_a = \frac{(1/t')(I_a'/I_R')}{(1/t)(I_a/I_R)}.$$ (3-30)

It is important to make some estimate of the precision with which x_a of (3-30) can be determined as a function of the random errors in the measured values of its six component factors. The *relative* standard deviation σ_a is related to the σ's of the two t and four I factors as follows:

$$\sigma_a^2 = \sum_{i=1}^{2} \sigma_{t_i}^2 + \sum_{j=1}^{4} \sigma_{I_j}^2,$$

and if we postulate, perhaps somewhat optimistically, that optimal experimental techniques suffice to reduce σ_t and σ_I to 0.02 and 0.01, respectively, we find

$$\sigma_a^2 = 2 \times (0.02)^2 + 4 \times (0.01)^2 = 0.0012$$

and

$$\sigma_a = 0.035, \text{ or } 3.5\%.$$

Thus the ultimate limitation on the precision with which x_a can be determined is imposed by the t's, which are inherently difficult to measure with high precision or accuracy.

Figure 3-6 is a schematic diagram, and Figure 3-7 is a photographic perspective, of a counter apparatus used at Mellon Institute for measuring I_a, I_a', I_R, and I_R' of (3-30)[14]. The instrument consists of a standard Norelco wide-angle goniometer and a number of accessories. The direct x-ray beam from a copper-target tube is defined by the 0.25° divergence slit T and its axial divergence is limited additionally by the aperture A in the brass plate M, which shields the counter tubes G_x and G_s from unwanted radiation originating in the neighborhood of the x-ray-tube housing. The beam penetrates the polymer E, and x-rays scattered by the amorphous component are received by G_x. The specimen holder shown in Figure 3-7 permits extension of the sample in a vertical direction only, so that the amorphous intensity can be measured only on the meridian. Figure 3-8 shows the apparatus with an azimuthal specimen holder, which was required for determining the integrated intensity of the amorphous halo when

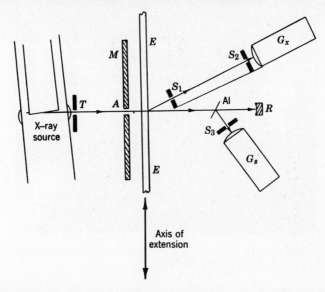

Figure 3-6 Schematic diagram of diffractometric apparatus for measuring crystallinity in elastomers. (Alexander, Ohlberg, and Taylor[14].)

preferred orientation was found to exist in certain extended amorphous specimens. With reference to Figures 3-6 and 3-7, the slits A and S_1 minimize the air-scattered x-rays that can be received by G_x. The direct beam, weakened by absorption in the specimen, penetrates

Figure 3-7 Perspective of apparatus shown in Figure 3-6. (Alexander, Ohlberg, and Taylor[14].)

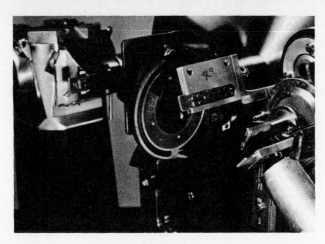

Figure 3-8   Close-up view of diffractometric apparatus showing azimuthal specimen holder. (Alexander, Ohlberg, and Taylor[14].)

a reference sample of 0.002-in. aluminum foil, Al, and is then intercepted by the beam stop R. A diffracted ray from the foil is received by the counter G_s, which provides a measure of the direct-beam intensity, I_R or I'_R, after transmission through the polymer. Slits S_2 and S_3 are both equipped with nickel β-filters.

The pulses from counter G_s are registered with the Norelco scaler and those from G_x, with an auxiliary scaler. An interconnecting timing circuit is incorporated to enable both units to record counts simultaneously for the same time interval. Normally the control switch is set to shut off the counters when G_s has counted a predetermined number of pulses (for example, 25,600), which effectively normalizes each recorded count N_a or N'_a to a constant direct-beam flux.

Figure 3-9 is a diagram of the simple device that was used to hold elastomer specimens at fixed extensions. Experience in this laboratory has shown that the thickness ratio t'/t of (3-30) is given with good accuracy by $\alpha^{-1/2}$, the square root of the extension ratio, provided that the specimen is supported in such a way as to permit unconstrained, and therefore proportionate, diminution in its thickness and width (*affine* extension).

If the lengths of an elastomer before and after extension are l_0 and l, the extension ratio is

$$\alpha = \frac{l}{l_0}. \tag{3-31}$$

The reader should distinguish between α and the fractional elonga-

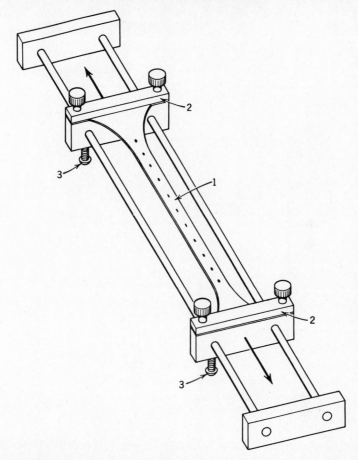

Figure 3-9 Device for holding elastomer specimens at fixed extensions during diffractometric measurements. 1—Stretched elastomer with fiducial marks; 2—clamps to grip specimen; 3—set screws to fix the degree of extension.

tion, or degree of elongation, ϵ, defined by

$$\epsilon = \frac{l-l_0}{l_0} = \alpha - 1. \tag{3-32}$$

The extension ratio can be determined by measuring the separation before and after extension of small India-ink fiducial marks placed a few millimeters apart along the longitudinal axis of the elastomer, as shown in Figure 3-10. When the azimuthal specimen holder is employed, the sample is first extended to the desired length in the

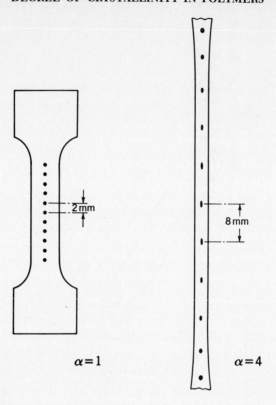

Figure 3-10 Scheme for determining the extension ratio on the basis of the separation of fiducial marks.

instrument shown in Figure 3-9, after which the central portion of the strip is clamped in the azimuthal holder and the unused ends then released. It is also possible to determine t and t' with the aid of a vernier caliper or, when the thickness is known to be very uniform, it can sometimes be calculated from a knowledge of the density and weight per unit area of surface:

$$t = \frac{w}{\rho A}. \tag{3-33}$$

Crystallinity in Natural Rubber[14]. The crystalline fraction x_c as a function of extension ratio was determined for a natural-rubber specimen at room temperature. The rubber studied was prepared from a sample of smoked sheet by purifying it and cross-linking it with 0.6% decamethylene *dis*-methylazodicarboxylate using essentially the method of Flory, Rabjohn, and Shaffer[15]. Measurements were made at a series of extension ratios ranging from 1 to 9, the values of α being determined by the method shown in Figure 3-10. Previous investigations of the x-ray scattering versus elongation of natural rubber failed to disclose any deviations from uniformity of the intensity of the amorphous halo with azimuth or changes in the radial position or profile of the halo[16]. Accordingly it was assumed in the present investigation that a peak measurement of the amorphous intensity at a single azimuth β and a single 2θ could be regarded as representative of the integrated amorphous scattering.

Nickel-filtered CuKα x-rays were employed and Geiger-Müller counters G_x and G_s were used to measure the peak counting rates N_a and N_R of the amorphous halo and standard reflection, respectively. The intensities I_a and I_R were derived by correcting N_a and N_R for the dead-time losses of the Geiger counter and subtracting the estimated components of intensity due to air scatter and general x-radiation. The first correction was made as explained in Section 2-4.2 with the formula

$$N = \frac{N_o}{1 - N_o \tau}, \tag{3-34}$$

τ being assigned the effective value 2.5×10^{-4} sec for the Norelco type 62019 Geiger-Müller tube used in conjunction with a full-wave rectified x-ray source. The air-scattered component N_A to be subtracted was taken to be

$$N_A = (N_A)_0 \frac{N_R}{(N_R)_0}, \tag{3-35}$$

$(N_A)_0$ and $(N_R)_0$ being, respectively, the counting rates corrected for dead time of counters G_x and G_s with no elastomer specimen, and N_R being the G_s counting rate with the elastomer in place.

Because of the low sensitivity of the counters to the general x-radiation, which consists principally of short wavelengths (see again Figure 2-40), the size of the correction to be applied was found by measuring the diffracted intensity from the specimen after transmission through 0.35 mm of aluminum, which removes more than 99% of the CuKα x-rays but only about 27% of the general radiation. It must be emphasized that results of higher accuracy could have been obtained by employing more monochromatic registration techniques, as discussed in Section 2-4.3. Thus the use of proportional or scintillation counters together with balanced filters and pulse-height discrimination would have eliminated the need for dead-time and general-radiation corrections.

The detailed experimental data that were used in the calculation of x_c versus α for natural rubber are presented in Table 3-4. With the thickness ratio t/t' set equal to $\alpha^{1/2}$, (3-30) assumes the form

$$x_a = \alpha^{1/2}\left[\frac{(I_a'/I_R')}{(I_a/I_R)}\right].$$ (3-36)

The measurements for $\alpha = 1, 2, 3, 3.5$, and 4 were made at one sitting with $I_a/I_R = 0.847$, whereas those for $\alpha = 5, 6, 7, 8$, and 9 were made at another time with $I_a/I_R = 0.803$. The final crystallinity values versus α are plotted as curve A of Figure 3-11, which may be compared with curve B of Goppel and Arlman[13] and curve C of Nyburg[17] for natural-rubber specimens of somewhat different types. The three curves agree in showing little or no crystallinity at extension ratios smaller than 4, but they suggest possible minor differences in the progress of crystallization at the higher extensions.

Crystallinity in Polybutadiene[14]. The specimens to be analyzed had been polymerized at 5°C without the aid of modern stereospecific catalysis and showed crystalline reflections only at temperatures lower than 0°C. Very similar polybutadiene polymers had been examined previously with x-rays by Beu et al.[18]. The diffraction

Table 3-4 Crystallinity x_c in Natural Rubber as a Function of Extension[a]

α	Observed Counting Rate N_a	N_R	Counting Rate Corrected for Dead Time N_a	N_R	White Radiation Plus Air Scatter	I_a'	I_R'	I_a'/I_R'	$\alpha^{1/2}\left(\dfrac{I_a'}{I_R'}\right)$	x_a	x_c
1	87.6	95.5	89.6	97.9	6.7	82.9	97.9	0.847	0.847	1.000	0.000
2	70.4	106.5	71.6	109.4	6.3	65.3	109.4	0.597	0.844	0.996	0.004
3	62.5	113.0	63.6	116.3	6.1	57.5	116.3	0.494	0.856	1.011	−0.011
3.5	58.1	114.0	58.9	117.3	6.0	52.9	117.3	0.451	0.844	0.996	0.004
4	54.8	117.5	55.5	121.0	6.0	49.5	121.0	0.409	0.818	0.966	0.034
5[b]	49.0	132.0	49.5	136.5	5.7	43.8	136.5	0.321	0.718	0.894	0.106
6[b]	44.6	134.5	45.1	139.2	5.5	39.6	139.2	0.284	0.696	0.867	0.133
7[b]	37.9	130.0	38.2	134.4	5.2	33.0	134.4	0.2455	0.650	0.809	0.191
8[b]	32.10	130.5	32.3	135.0	4.9	27.4	135.0	0.2030	0.574	0.715	0.285
9[b]	29.79	139.0	30.0	144.0	4.9	25.1	144.0	0.1743	0.523	0.651	0.349

[a]All counting rates are in counts per second.
[b]These results were obtained at a later sitting and are referred to a new value of the "constant" $\alpha^{1/2}(I_a/I_R) = 0.803$, rather than 0.847, for $\alpha = 1$.

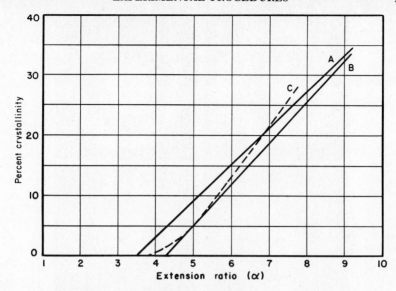

Figure 3-11 Crystallinity in natural-rubber specimens at room temperature as measured by (A) Alexander, Ohlberg, and Taylor[14], (B) Goppel and Arlman[13], (C) Nyburg[17].

pattern of the stretched material showed an axial repeat distance of about 5 Å and other features that identify the principal crystalline phase as the pseudohexagonal phase ascribed by Natta and co-workers to *trans* 1,4-polybutadiene [19, 20]. Figure 3-12 shows the x-ray diffraction patterns produced by the 5°C polybutadiene specimens in the present study at elongations of 0 and 535% and at temperatures of 25 and −15°C. The patterns of the extended samples show unmistakable evidence of preferred molecular orientation in the form of an intensification of the amorphous halo on the equator.

The counter apparatus used in the measurement of crystallinity in natural rubber was also applied to the study of polybutadiene. For the measurements at lowered temperatures the entire sample-supporting mechanism was enclosed in a windowless Styrofoam insulating box of such low density that the absorption of the direct and diffracted x-ray beams was very small. Air from the laboratory air line was passed successively through a drying column of silica gel, a flow meter, and a copper cooling coil immersed in a slush of dry ice and 1:1 chloroform-carbon tetrachloride in a Dewar flask, after which it was introduced into the cooling box through openings in both the top and bottom. By regulating the flow rate any temperature down to −40°C could be maintained for several hours with a mean variation

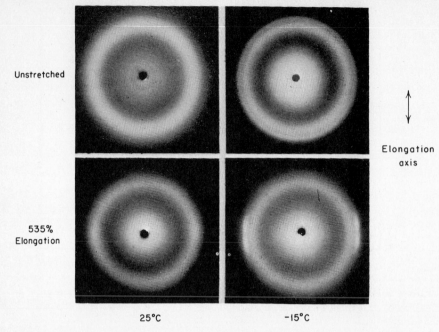

Figure 3-12 X-Ray patterns of a 5°C polybutadiene specimen at −15 and 25°C, and elongations of 0 and 535%. Nickel-filtered CuKα radiation. (Alexander, Ohlberg, and Taylor [14].)

not exceeding ±1°. The temperature of the specimen was measured with either a thermometer or a thermocouple. Nickel-filtered CuKα x-radiation was employed.

Preliminary measurements at room temperature were made of the radial (2θ) profiles of the amorphous halo at azimuths† ranging from 0 to 90° for a highly stretched specimen and compared with the radial profile of an unstretched specimen. The experimental intensities were corrected for general radiation and air scatter in the manner previously described, and then normalized to constant direct-beam flux and divided by the respective thicknesses. The ratio of these corrected and normalized intensities,

$$\frac{I_{\alpha=6.1}}{I_{\alpha=1}} = \frac{[(1/t)(I_a/I_R)]_{\alpha=6.1}}{[(1/t)(I_a/I_R)]_{\alpha=1}}, \tag{3-37}$$

when summed over all azimuths, was found to be equal to unity within

†In conformity with the convention adhered to in this monograph the azimuths of the meridian and equator are designated 0° and 90°, respectively. The reverse convention was employed in the original publication [14].

experimental error, thus verifying the constancy of the integrated intensity of the amorphous halo as a function of extension.

It was also found that in the extended state not only the peak intensity but also the integral breadth of the radial profiles (ω_i) and the radial angle of the point of maximum intensity ($2\theta_{max}$) showed a dependence on azimuth. The integral breadth of a radial profile is defined by

$$\omega_i = \frac{1}{I_{max}} \int I(2\theta)\, d(2\theta), \tag{3-38}$$

the intensities being the measured values above background. As explained in Section 6-1.5, for randomly oriented polymers the mean distance between adjacent molecular chains may be taken to be [21]

$$r = 1.22 \left(\frac{\lambda}{2 \sin \theta} \right) = 1.22\, d_{Bragg}, \tag{3-39}$$

which was considered to be applicable to both the extended and unextended samples in the present investigation.

The salient numerical data are presented in Table 3-5. As might be expected, for $\beta = 0°$ and $\alpha = 6.1$ the values of $2\theta_{max}$, r, and ω_i agree with their values for the unstretched elastomer. The highly oriented molecules are more densely packed than the unoriented ones, as evidenced by an average intermolecular distance of 5.49 Å for $\beta = 0°$ and 5.41 Å for $\beta = 90°$. The decrease in integral breadth of the radial

Table 3-5 Properties of Radial Sections through the Amorphous Halo of Polybutadiene for Extension Ratios of 1.0 and 6.1[a]

Parameter	$\alpha = 1.0$	$\alpha = 6.1$	
		$\beta = 0°$	$\beta = 90°$
$2\theta_{max}$(degrees)	19.75	19.75	20.10
Equivalent r between molecular chains (Å)	5.49	5.49	5.41
Integral breadth ω_i of radial profile (degrees)	11.1	11.1	9.6
Relative peak intensity I_{max}	1.00	0.98	1.62

[a]Data from [14]. Cu$K\alpha$ radiation ($\lambda = 1.542$ Å).

section through the halo at $\beta = 90°$ when extension occurs is indicative of a sharper distribution in the magnitudes of the intermolecular vectors.

For the determination of crystalline fraction versus temperature the following experimental procedure was employed. Prior to cooling the specimen was stretched to the desired extension ratio and mounted in the azimuthal holder, and its azimuth was then adjusted to some fixed value for the first series of measurements. The intensity ratio $(I'_a/I'_R)_{T_0}$ was measured at $T_0 = 25°C$, after which the specimen at the same extension and azimuth was enclosed in the insulating box and its temperature reduced to T_1 (about 10°C). An equilibration period of 30 min was allowed, and then the ratio $(I'_a/I'_R)_{T_1}$ was measured until it reached constancy, which occurred within a few minutes except at low extensions. The temperature of the specimen was next reduced to T_2 (for example, $-6°C$), a second equilibration period was allowed, and the ratio $(I'_a/I'_R)_{T_2}$ was measured, and so on. The experimental intensity ratios at an extension ratio of 5.55 and for three azimuths (0, 30, and 60°) are plotted in Figure 3-13 for temperatures between $+25$ and $-28°C$. It is seen that the three curves are in essential agreement above the crystallization temperature, but that below this temperature the curve for $\beta = 60°$ deviates appreciably from the other two. This can be attributed to interference by the strong near-equatorial reflections with $d = 3.97$ Å, which becomes increasingly severe

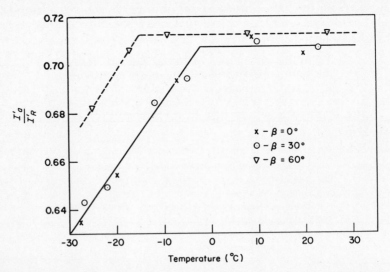

Figure 3-13 Amorphous-halo intensity of polybutadiene versus temperature at azimuths of 0, 30, and 60°. Extension ratio, 5.55. (Alexander, Ohlberg, and Taylor [14].)

as β approaches 90° (see again Figure 3-12) and precludes the possibility of reliable measurements of the amorphous halo in this region. The curves for $\beta = 0$ and 30° indicate that crystallization in this specimen at an extension ratio of 5.55 begins somewhat below 0°C and increases rather linearly with further reduction of the temperature.

The good agreement of the foregoing results at azimuths of 0 and 30° suggests that the intensity of the amorphous halo diminishes proportionately at all azimuths when crystallization takes place, although this conclusion cannot be regarded as rigorous in view of the difficulties that affect measurements at the lower azimuths. However, on the assumption that this conclusion was valid, calculations of the crystalline fraction in polybutadiene at reduced temperatures were made on the basis of simple meridional measurements of the ratios $(I_a'/I_R')_T$. Letting T_0 refer to a temperature at which the extended specimen is presumably completely amorphous (25°C in this study) and setting the thickness ratio corresponding to two temperatures equal to unity, we can apply (3-30) to the measurement of crystallinity resulting from reduction of temperature at constant elongation in the following form:

$$(x_a)_T = \frac{(I_a'/I_R')_T}{(I_a'/I_R')_{T_0}}. \tag{3-40}$$

Figure 3-14 gives plots of the percent crystallinity versus temperature for extension ratios of 4.38 and 6.13. The values of x_c are derived from (3-40) by the relation $x_c = 1 - x_a$. It is seen that the crystalline fraction increases with reduction of temperature and with increase of the extension ratio. In addition to the statistical errors of counting, the measured intensity values are affected by additional random and undetermined errors introduced by experimental difficulties associated with the low-temperature technique, with the result that the I' values can hardly have a precision of less than 3%, thus yielding a net standard deviation in x_a of the order of 6%. Hence the proposed linear form of the curves of Figure 3-14 is somewhat tentative, especially at the lower crystallinities.

3-3.3 Crystallinity from the Intensities of Both the Crystalline and Amorphous Scatter, Regression Curve, Lattice Distortions Not Considered

In the absence of the 100% amorphous polymer, this method is applicable provided that (a) samples of the polymer with a considerable range of crystallinity are available and (b) it is possible to

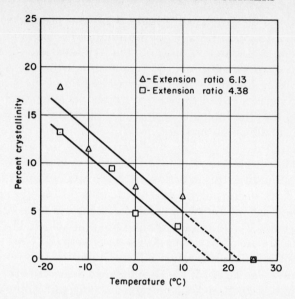

Figure 3-14   Percent crystallinity in 5°C polybutadiene as a function of temperature for extension ratios of 4.38 and 6.13. (Alexander, Ohlberg, and Taylor[14].)

draw an acceptable demarcation line between the crystalline and amorphous scattering over some angular interval that encompasses the principal amorphous halo. This procedure was first developed by Hermans and Weidinger for the determination of crystallinity in cellulose[22] and was subsequently applied by them to several synthetic polymers, including polyethylene[23], isotactic polypropene[24], and isotactic polystyrene (in collaboration with Challa) [25]. The method makes use of certain simplifying assumptions and procedures that render it well adapted to routine use without conflicting with the basic physical principles given early in this chapter. Two polypropene samples gave essentially the same numerical values of crystallinity by the method of Hermans and Weidinger[24] as were obtained by the much more rigorous and elaborate procedure of Ruland[2].

The Hermans-Weidinger method comprises the following steps:

1. The diffraction patterns (over a limited angular range) of a number of specimens of the polymer with various degrees of crystallinity are normalized to equal μt (x-ray optical density) and equal intensity of the direct beam.

2. A reasonable demarcation line is drawn between the crystalline and amorphous scattering.

3. Suitable peak or integral intensities proportional to the crystalline and amorphous fractions are selected and measured (I_c and I_a, respectively).

4. The regression line that gives the statistically optimum linear dependence of I_c on I_a is calculated.

5. Linearity of the I_c versus I_a curve to within experimental error provides confirmation that the method of measuring these quantities is in harmony with the theory (see below), whereupon the investigator may proceed to determine the crystalline fractions of any other specimens from their I_c or I_a values by reference to the established regression line.

An evaluation of the foregoing procedure in the light of the theoretical equations presented at the beginning of this chapter and in comparison with the rigorous procedure of Ruland[2] shows that certain intensity factors have been neglected—including the factor s^2 in (3-2), (3-3), and (3-5); the polarization factor; and the correction for incoherent scattering. These simplifications are permissible because the method of Hermans and Weidinger[22–24] depends only on the existence of *proportionality* between the experimentally measured crystalline intensity I_c and the crystalline fraction, and between the amorphous intensity I_a and the amorphous fraction. There is no requirement that complete integral intensities in the sense of (3-5) or (3-9) be measured.

With Hermans and Weidinger, then, we postulate that the crystalline fraction is proportional to I_c by some unknown proportionality constant p, or

$$x_c = pI_c, \qquad (3\text{-}41)$$

and that the amorphous fraction is proportional to I_a by an unknown factor q,

$$x_a = qI_a = 1 - x_c. \qquad (3\text{-}42)$$

Then for two samples that differ considerably in crystalline content

$$x_{c1} = pI_{c1} \quad \text{and} \quad x_{c2} = pI_{c2}$$

and

$$1 - x_{c1} = qI_{a1} \quad \text{and} \quad 1 - x_{c2} = qI_{a2},$$

from which it is possible to eliminate the constants p and q and solve for x_{c1} and x_{c2} in terms of the four intensities. Division of (3-41) by (3-42) leads to

$$\frac{I_c}{I_a} = \frac{q}{p} \frac{x_c}{(1 - x_c)}, \qquad (3\text{-}43)$$

and solution for x_c gives

$$x_c = \left(1 + \frac{qI_a}{pI_c}\right)^{-1}. \qquad (3\text{-}44)$$

If the measured values of I_c are plotted against the values of I_a, the points will in principle lie on a straight line that may be regarded as the regression line of the correlation between the quantities I_c and I_a. The points of intersection of the regression line with the coordinate axes give the values of I_c and I_a that correspond to a completely crystalline and a completely amorphous sample, respectively, values that we may denote by I_{100c} and I_{100a}. From (3-41) and (3-42) it can be seen that the numerical values of the proportionality constants are given by

$$p = \frac{1}{I_{100c}} \quad \text{and} \quad q = \frac{1}{I_{100a}}. \qquad (3\text{-}45)$$

It is helpful but not essential to the plotting of the regression line of I_c on I_a that the 100% amorphous polymer be available.

We now proceed to illustrate the procedure with the determination by Challa, Hermans, and Weidinger [25] of the crystalline fraction in isotactic polystyrene. The radiation used was CuKα from a highly stabilized x-ray source, and the diffracted intensities were measured with a proportional counter, thus eliminating the need for nonlinearity corrections. The specimens for x-ray analysis were molded under pressure into disks about 2 mm thick, every effort being made to minimize preferential orientation of the crystallites. The intensity measurements were made with the specimens in the symmetrical-transmission position (Figure 2-8c), and a crystal monochromator in the diffracted beam (Section 2-4.3) was employed to remove the general-radiation components. The polystyrene specimens were obtained as white powders by polymerization with a Ziegler-Natta-type catalyst. Entirely amorphous reference samples, Nos. 1–3, were prepared by heating the original polymer above its melting point and quenching. The remaining samples, Nos. 4–14, were given a variety of treatments in order to induce different degrees of crystallinity.

The experimental intensities as directly measured were first corrected for the air-scattering counting rate $j(\theta)$ and for the constant cosmic and radioactive background counting rate k. If we combine these correction terms with a factor P/P_R to normalize to constant direct-beam intensity, the counting rate to be subtracted at the Bragg angle θ for a specimen of x-ray optical density $D = \mu t$ is [23]

$$\frac{P}{P_R}(j + k) \exp\left(-D \sec\theta\right). \qquad (3\text{-}46)$$

For the experimental conditions of Challa, Hermans, and Weidinger [25] k had the fixed value of 0.185. The size of the correction given by (3-46) diminishes rapidly with increasing 2θ, becoming negligible above 20°. Variations in the intensity of the primary beam were determined immediately before and after each experiment by measuring the intensity diffracted by a standard Perspex plate, P_R being its accepted reference value and P its value corresponding to the experimental measurements in question.

After the raw intensities were corrected for air scatter, they were corrected for differences in effective scattering mass and for variations in primary-beam intensity. The effective scattering mass of a given polymer specimen is given by its absorption exponent, or optical density, $D = \mu t$. This quantity was determined experimentally for each sample by measuring the intensity of the monochromatic x-ray beam with and without the specimen in place, yielding respectively I and I_0, from which

$$\mu t = -\log_e \frac{I}{I_0}. \tag{3-47}$$

In this connection the reader may refer again to Section 1-3.1. According to (2-7) for the symmetrical-transmission geometry, the intensity scattered at the Bragg angle θ by a specimen of optical density D_1 is

$$I_1 = (I_0)_1 A t_1 \sec \theta [\exp(-D_1 \sec \theta)]. \tag{3-48}$$

But $(I_0)_1$ for a polymer of given chemical composition is proportional to its density of packing and hence to its linear absorption coefficient, which may be expressed as

$$(I_0)_1 = k\left(\frac{\mu}{\rho}\right)\rho' = k\mu',$$

in which ρ and ρ' are respectively the density of the solid polymer and the density including interstices, and μ' is the linear absorption coefficient including interstices,

$$\mu' = \frac{\mu\rho'}{\rho}.$$

Equation 3-48 then becomes

$$
\begin{aligned}
I_1 &= kA\mu_1' t_1 \sec \theta [\exp(-D_1 \sec \theta)] \\
&= kAD_1 \sec \theta [\exp(-D_1 \sec \theta)],
\end{aligned} \tag{3-49}
$$

and for a specimen of optical density $D_2 = \mu_2' t_2$

$$I_2 = kAD_2 \sec\theta \, [\exp{(-D_2 \sec\theta)}]. \tag{3-50}$$

Division of (3-49) by (3-50) gives

$$\frac{I_1}{I_2} = \frac{D_1 \exp{(-D_1 \sec\theta)}}{D_2 \exp{(-D_2 \sec\theta)}}, \tag{3-51}$$

which is the appropriate factor for normalizing intensities diffracted by specimen 2 to those of specimen 1 as a reference standard.

Now $D_1 \exp{(-D_1 \sec\theta)}$ in (3-51) may arbitrarily be assigned any constant value, such as unity, and if we drop the subscript 2 and include the correction for fluctuations in primary-beam intensity, P/P_R, we have the following expression that gives the normalized intensity I_N as a function of the raw intensity corrected for air scatter, I [23]:

$$I_N = \frac{P_R}{P} \frac{I}{D \exp{(-D \sec\theta)}}. \tag{3-52}$$

Figures 3-15 and 3-16 show the x-ray scattering curves over the angular range 12 to 30° of a partially crystalline and a fully amorphous polystyrene specimen, respectively. In addition to the principal amorphous halo centered at 18.4° this angular range encompasses two major crystalline reflections with the indices (220) and (211), as well as four weaker peaks. To resolve the crystalline and amorphous scattering Challa, Hermans, and Weidinger [25] proceed as follows. First, the angular range of measurement is chosen, whenever possible, in such a manner that at the limits A and D, or A' and D', the intensities of both the crystalline reflections and the principal amorphous halo may be regarded (to be true, somewhat arbitrarily) as zero. A straight line is then drawn from A to D, separating the intensity components *above*, which *are proportional* to the crystalline and amorphous scattering masses, from the intensity components *below*, which *are not proportional* (incoherent, thermal diffuse, some amorphous scatter).

The next step in the analysis of the polystyrene pattern is to select an intermediate point C (at 20°) at which the amorphous intensity CE is large and the crystalline intensity is assumed to drop to zero. The separation between the peaks (211) and (410/31$\bar{1}$) is 3°, which is regarded as large enough to justify this assumption. The line of demarcation BCD between the amorphous and crystalline scattering is now drawn in so as to be congruent with the profile of the 100% amorphous specimen in this region (Figure 3-16). The total area above the curve BCD in Figure 3-15 is now taken to be proportional

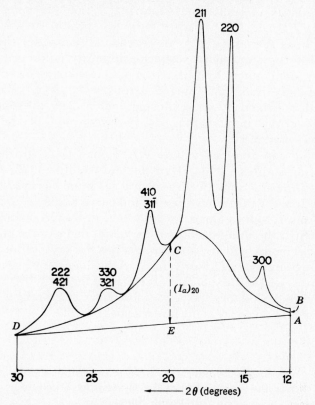

Figure 3-15 X-Ray scattering curve of a partially crystalline isotactic polystyrene specimen. (Challa, Hermans, and Weidinger [25].)

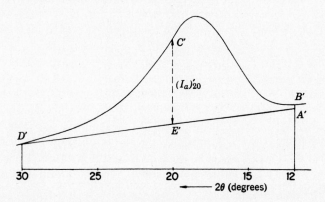

Figure 3-16 X-Ray scattering curve of a fully amorphous isotactic polystyrene specimen. (Challa, Hermans, and Weidinger [25].)

171

to the intensity of crystalline scattering, I_c, and the area between the lines AED and BCD is taken to be proportional to the intensity of amorphous scattering, I_a.† An alternative parameter that is proportional to the amorphous scattering intensity is the ordinate CE, which we shall denote by $(I_a)_{20}$. It is more convenient to determine experimentally than the integral intensity I_a since $(I_a)_{20}$ is not appreciably affected by air scattering, and it does not require construction of the demarcation line between the crystalline and amorphous scattering.

Using the experimental values of I_c, I_a, and $(I_a)_{20}$ for the several polystyrene samples, Challa, Hermans, and Weidinger[25] calculated the crystalline fractions in three ways: (a) from a regression curve based on I_c and I_a, (b) from a regression curve based on I_c and $(I_a)_{20}$, and (c) from I_a values alone, as described in the preceding section. Table 3-6 lists the numerical intensity values that were obtained from the 14 specimens and used in calculating the crystalline fractions. The average intensity $\langle I_a \rangle$ or $\langle (I_a)_{20} \rangle$ of the amorphous specimens 1, 2, and 3 was used as the standard of reference for the 100% amorphous state. Figure 3-17 shows the regression curve of I_c on I_a. A statistical analysis of this curve gives the linear equation

$$I_c = 1053 - 2.158 I_a, \tag{3-53}$$

Table 3-6 Numerical Values of Intensities Used
for the Calculation of the Crystalline Fractions
in Polystyrene Specimens[a]

Specimen Number	$I_c + I_a$	I_a	I_c	$(I_a)_{20}$	Air Scattering
$\langle 1, 2, 3 \rangle$		485		62.66	22.6
4	565	445	120	57.4	20.0
5	575	402	173	52.0	27.0
6	620	380	240	49.2	20.0
7	620	378	242	49.0	19.8
8	620	360	260	46.5	17.5
9	645	355	290	46.0	20.2
10	680	342	338	44.5	25.0
11	653	350	303	45.2	17.2
12	680	332	348	43.2	27.2
13	660	325	335	42.0	20.0
14	670	325	345	42.0	15.2

[a]Data from [25].

†It should be mentioned that the curve BCD is uncorrected for air scatter, for which reason its ordinate at $2\theta = 12°$ falls between the point A and the total intensity.

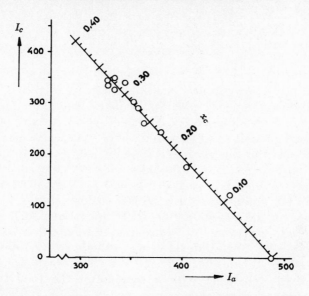

Figure 3-17 Regression of I_c on I_a for isotactic polystyrene specimens. Scale of crystalline fraction x_c is indicated along the regression line. (Challa, Hermans, and Weidinger [25].)

from which $I_c = 1053$ for a 100% crystalline specimen and $I_a = 488$ for a 100% amorphous specimen. By reference to (3-45) it is seen that these intensities correspond to the proportionality constants $p = 1/1053$ and $q = 1/488$. Substitution of (3-41) into (3-53) then gives

$$x_c = 1 - 0.00205 I_a. \tag{3-54}$$

Equation 3-54 was used to compute the scale of x_c that is seen on the regression line of Figure 3-17. With the aid of this scale the degree of crystallinity of any new polystyrene specimen can be read off directly as a function of either I_c or I_a. Equation 3-44 can also be used to calculate x_c from measured values of the intensity ratio I_a/I_c, which has the numerical coefficient $q/p = 2.158$. Thus

$$x_c = \left(1 + 2.158\frac{I_a}{I_c}\right)^{-1}. \tag{3-55}$$

If numerical values of $(I_a)_{20}$ rather than I_a are to be used as measures of the amorphous scattering, a new regression curve is found from which $(I_a)_{20} = 62.7$ for a 100% amorphous specimen. Then $q' = 1/62.7$, and (3-54) and (3-55) are replaced by

$$x_c = 1 - 0.0158(I_a)_{20} \tag{3-56}$$

and

$$x_c = \left[1 + 16.6\frac{(I_a)_{20}}{I_c}\right]^{-1}. \tag{3-57}$$

Table 3-7 compares the crystalline fractions in the 11 partially crystalline polystyrene samples as calculated from the intensity data of Table 3-6 in the three ways described above. In processing the experimental intensities the first step was to subtract the air scattering integrated over the angular range 12 to 30° [see (3-46)] from the area above the base line AED over the same angular range, measured with a planimeter, which is a measure of $I_c + I_a$. Normalization of this area with (3-52) then gave the correctly scaled value of $I_c + I_a$. The area below the demarcation line BCD and above AED was then measured with a planimeter, whereupon normalization with the aid of (3-52) gave I_a. The value of I_c was calculated by subtracting I_a from the total scattering above AED, $I_c + I_a$. This is simpler and more accurate than measuring with a planimeter the total crystalline scattering directly. The values of x_c that were determined by the three procedures are seen to be in very good agreement, lending encouragement to the use of methods involving only $(I_a)_{20}$ or only $(I_a)_{20}$ and I_c, which are more satisfactory for routine use than methods involving both I_a and I_c.

Inspection of the integral breadths [see (3-38)] of the prominent reflections (211) and (220) and of their relative integral intensities

Table 3-7   Crystalline Fractions in Polystyrene
Specimens Calculated in Three Ways[a]

Specimen Number	From I_a	From I_a and I_c	From $(I_a)_{20}$ and I_c
4	0.08	0.088	0.11
5	0.165	0.175	0.165
6	0.215	0.22	0.226
7	0.22	0.225	0.226
8	0.25	0.26	0.250
9	0.265	0.27	0.275
10	0.29	0.30	0.315
11	0.312	0.29	0.312
12	0.312	0.32	0.328
13	0.325	0.335	0.323
14	0.325	0.335	0.33

[a]Data from [25].

(Table 3-8) shows some significant variations among the specimens, particularly for Nos. 4 and 5. The differences in integral breadth may be attributed to differences in crystallite size or perfection or both, which in general would not be expected to introduce serious errors into the determination of the crystalline fraction as long as integrated crystalline intensities are measured. However, variations in the ratio $R = I(211)/I(220)$ are more serious because they are most likely to occur as the result of preferred orientation of the crystallites. Specimen 4 exhibits the greatest deviation, and it is undoubtedly significant that this specimen was subjected to unusually heavy compression and was found to possess a degree of planar orientation. If we suppose that the precision in the measurement of $I(211)$ and $I(220)$ is on the order of 4%, the standard deviation in the ratio R of these two intensities is $2^{1/2} \times 4\% = 5.7\%$, which is equivalent to ± 0.08 in the ratio itself. Excluding the ratio of No. 4 as being anomalously large, the mean of the remaining 10 ratios is 1.44, and it is seen that in no case do the individual deviations amount to 3σ, although for five specimens (Nos. 5, 8, 9, 12, and 14) the deviation exceeds 2σ. Thus only in the case of specimen 4 is the deviation of R from the mean, $\langle R \rangle$, unquestionably established from the statistical standpoint, which constitutes a potential source of error in the calculations of

Table 3-8   Observed Integral Breadths of the (211) and (220) Reflections of Polystyrene and Ratios of their Integral Intensities[a]

Specimen Number	Integral Breadths (degrees) (211)	(220)	$R = \dfrac{I(211)}{I(220)}$	$\lvert R - \langle R \rangle \rvert$
4	1.56	0.445	2.15	—
5	1.45	1.09	1.63	0.19
6	1.32	0.54	1.49	0.05
7	1.21	0.48	1.49	0.05
8	1.29	0.59	1.27	0.17
9	1.24	0.50	1.62	0.18
10	1.30	0.51	1.37	0.07
11	1.21	0.57	1.33	0.11
12	1.33	0.75	1.63	0.19
13	1.40	0.65	1.30	0.14
14	1.14	0.54	1.26	0.18
			$\langle R \rangle = 1.44$[b]	

[a] Data from [25].
[b] Excluding specimen 4.

x_c that involve I_c. That the error in this particular case is actually negligible, however, is shown by the good agreement between the values of x_c that were determined for specimen 4, first from I_a alone and second from I_a and I_c. (See Table 3-7.)

3-3.4 Crystallinity from Differential Intensity Measurements, Crystallinity Index

Wakelin, Virgin, and Crystal[26] devised a method, based exclusively on the measurement of intensity differences, that circumvents some of the experimental and interpretative difficulties inherent in the foregoing methods. Their procedure, which they applied to cotton cellulose, expresses numerically the degree of ordering in a given specimen relative to the minimum and maximum values that are observed in a sampling of specimens of the same polymer species. Such a crystallinity index constitutes a particularly appropriate device for characterizing the degree of ordering in polymer systems such as cellulose in which the completely ordered and completely disordered "reference standards" are difficult or impossible to procure.

This method conforms to the underlying principles presented in Section 3-2 insofar as the differential intensity measurements must encompass an angular range that is of sufficient size to include most of the crystalline peaks. However, it deserts the principles on which so-called absolute methods are based in that it does not require the drawing of a demarcation line separating the crystalline and amorphous scattering, a rather arbitrary operation at best. Instead it is necessary to measure only the overall scattering intensity at small angular increments over the entire range of 2θ in question and to compare these ordinates point by point with the corresponding ordinates of the scattering curves generated by the reference specimens of minimum and maximum crystallinity. It is important that the angular increment, generally not larger than $0.5°$ 2θ, be small enough to sample the relatively sharp crystalline peaks as well as the broad amorphous fluctuations.

Figure 3-18 illustrates these points with the intensity curves over the range $2\theta = 10$ to $45°$ of the "crystalline" and "amorphous" reference specimens chosen by Wakelin, Virgin, and Crystal[26]. The points indicated on the crystalline curve designate the angles at which the experimental intensities I_c and I_a were measured. For any "unknown" specimen of intermediate crystallinity experimental intensities I_u must be measured at the same sites.

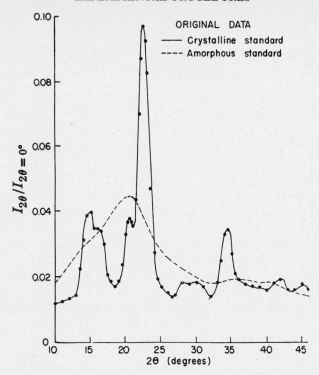

Figure 3-18 X-Ray intensity curves of the crystalline and amorphous standards used for the analysis of cotton cellulose. (Wakelin, Virgin, and Crystal [26].)

The numerical values of I_c, I_a, and I_u can be used to calculate an *integral* or a *correlation* crystallinity index. For either index the differential intensities $I_u - I_a$ and $I_c - I_a$ are first determined at all incremental points between the angular limits $2\theta_0$ and $2\theta_m$. The *integral* index C_i is calculated with the expression

$$C_i = \frac{\sum\limits_{2\theta_0}^{2\theta_m} (I_u - I_a)}{\sum\limits_{2\theta_0}^{2\theta_m} (I_c - I_a)}, \qquad (3\text{-}58)$$

which is seen to be a measure of the area between the curves of I_u and I_a divided by the area between the curves of I_c and I_a. The *correlation* index is the slope of the straight line obtained on plotting $I_u - I_a$ versus $I_c - I_a$ for all the data points, as illustrated by Figure 3-19 for a typical cotton sample. The data points fall in both the first and third quadrants since $I_u - I_a$ and $I_c - I_a$ are both positive when the

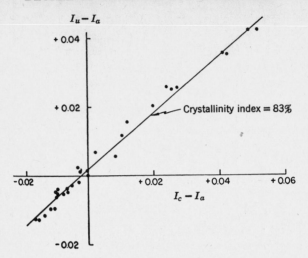

Figure 3-19 Plot of $I_u - I_a$ versus $I_c - I_a$ for a typical cotton sample (Rowden). (Wakelin, Virgin, and Crystal [26].)

crystalline curve is higher than the amorphous curve, whereas both quantities become negative at points where the amorphous curve is higher than the crystalline. The equation of the best straight line through the plotted points may be written

$$(I_u - I_a) = C_c(I_c - I_a) + B, \tag{3-59}$$

C_c being the correlation crystallinity index and B the ordinate intercept. Since B is theoretically zero, one criterion of the acceptability of experimental data is that B must be small. A large negative or positive value of B is indicative of systematic errors, which must be diagnosed and eliminated. Experience has shown the correlation index to be more reliable than the integral index, besides which the former is better suited to statistical evaluation. Hence in the more detailed discussion that follows we shall deal only with the *correlation* crystallinity index.

Practical Application of the Correlation Index. First we may observe that the experimental requirements for the determination of a correlation index are not stringent, a decided advantage for its routine use. In order to conform to the theoretical principles presented earlier the important requirements are the following:

1. The orientation of the crystalline and amorphous regions must be random. This is of course true of all the methods for determining crystallinity that have been described in this chapter.

2. The intensities I_a, I_c, and I_u must be corrected for variations in the primary-beam intensity and normalized to equal x-ray optical density, μt. (See again Section 3-3.3.)

3. For specimens of different μt it is necessary to apply absorption corrections appropriate to the geometry, as explained in Sections 2-2.1 and 2-2.5.

4. For specimens of similar μt it is not necessary to correct for air scattering since the air-scatter components of I_a, I_c, and I_u will be equal and hence will disappear in the computation of the differences $I_c - I_a$ and $I_u - I_a$. However, when μt differs from sample to sample, air scattering should be eliminated experimentally or corrected for by methods that have been already described. [See "Crystallinity in Natural Rubber" in Section 3-3.2, Section 3-3.3, and (3-46).]

5. If counter registration of intensities is employed, care must be taken to correct for losses due to nonlinear response at high counting rates. Likewise, in photographic recording the microdensitometer curves must be converted to intensity curves by reference to calibrated density scales [27].

We now take note of some customary experimental precautions that may be disregarded or de-emphasized in the determination of the correlation index. As a general rule it is not essential to employ highly monochromatic radiation because the white radiation components, like air scattering, are largely eliminated in subtracting I_a from I_c and I_u. However, monochromatization by means of a crystal monochromator or balanced filters is of advantage since the general-radiation contribution to I_a versus 2θ differs somewhat from the general-radiation contribution to I_c or I_u. Multiplication of $I(2\theta)$ by angular correction factors such as the reciprocal of the polarization factor or s^2, as is demanded by the general theory, is not necessary because these factors affect I_c, I_a, and I_u identically, and the ratios $(I_u - I_a)/(I_c - I_a)$ are not changed. Also there is no theoretical compulsion to convert the angular scale from 2θ to s since these variables are almost linearly related up to $2\theta = 60°$, which encompasses the significant portion of most polymer patterns.

Even more important, the incoherent scattering may be neglected since it is the same for both crystalline and amorphous regions and so is eliminated in the process of computing the differential intensities. An important consequence is that the intensities need not be reduced to absolute (electron) units. In actual practice useful values of the crystallinity index can be derived when only experimental requirements 1, 2, and 5 are satisfied.

The determination of the correlation index of a given set of data can be optimized by calculating the least-squares regression of $I_u - I_a$ on $I_c - I_a$. If we let $I_u - I_a = y$ and $I_c - I_a = x$, and denote the number of pairs of observations by n, and if we assume that the variability of y is large with respect to that of x, the slope of the least-squares regression of y on x is

$$C_c = \frac{\sum xy - (1/n) \sum x \sum y}{\sum x^2 - (1/n) \left(\sum x \right)^2}. \qquad (3\text{-}60)$$

The degree of scatter of the data about the least-squares-regression line can be expressed as $1 - r^2$, where r is a correlation coefficient defined by

$$r = \frac{\sum xy - (1/n) \sum x \sum y}{\left\{ \left[\sum x^2 - (1/n) \left(\sum x \right)^2 \right] \left[\sum y^2 - (1/n) \left(\sum y \right)^2 \right] \right\}^{1/2}}. \qquad (3\text{-}61)$$

Crystallinity Index of Cellulose [26]. For the crystalline standard Wakelin, Virgin, and Crystal chose a native cotton cellulose hydrolyzed in 1N HCl for 24 hr at 60°C, washed, and dried. The amorphous standard was a kiered† alcohol-extracted native cotton. (The x-ray scattering curves of these specimens are shown in Figure 3-18.) The samples that were selected for study included nine cottons representing a wide range of physical properties and six reference samples synthesized by mixing the amorphous and crystalline standards with the following percentages of the crystalline standard: 20, 33, 50, 67, 80, and 89. Diffractometric specimens of similar scattering mass and little or no preferred orientation were prepared by cutting the fibers with scissors to the shortest possible length and then pressing 125 mg into a compact disk 1.27 cm in diameter under a pressure of 75,000 kg/cm² in a Carver laboratory press. This procedure yielded sample disks with x-ray absorption exponents μt in the range 0.55 to 0.80 for CuKα radiation.

The curves of intensity versus 2θ were measured with the aid of the special diffractometer built by Krimm and Stein [28] for polymer studies. (See "Crystal Monochromators" in Section 2-4.3.) This instrument employed the transmission technique with perpendicular incidence of the beam from a crystal monochromator (Section 2-2.1), and the diffracted rays were received with a Geiger-Müller counter in the investigation presently described. A reflection from a small piece of 0.005-cm nickel foil located in the direct beam was measured

†Boiled in a chemical solution to bleach and remove gums.

by a monitor counter in order to compensate for variations in the output of the x-ray generator.

One objective of the investigation of Wakelin, Virgin, and Crystal was to determine the magnitude of the change in the correlation index that results when the *original intensities* are substituted for *fully corrected intensities* in the calculations. By "original intensities" we here refer to the measured counting rates modified only as follows:

1. Corrected for cosmic and radioactive background count.
2. Corrected for losses due to dead time of the Geiger-Müller counter.
3. Corrected for fluctuations in the intensity of the x-ray source.
4. Normalized to equal scattering mass (by dividing the counting rate at each angle by the integrated scattering over the angular range 5 to 50°).

By "fully corrected intensities" we mean the counting rates corrected by the above factors and in addition the following:

5. Polarization.
6. Diffractometer geometry.
7. Absorption by the sample.
8. Moisture content of the sample.
9. Incoherent radiation.

The reader is referred to the original paper for the details of the corrections [26].

As examples of the scattering curves of the cottons measured, Figures 3-20 and 3-21 show respectively plots of the *original intensities* of a highly crystalline and poorly crystalline cotton, each being compared with the intensity curve of the amorphous standard. The application of the various corrections and the subsequent calculation of C_c and r for each cotton specimen were performed with a computer.

Table 3-9 compares the values of the crystallinity index and correlation coefficient for nine cottons as calculated from the original intensities and from the fully corrected intensities. Table 3-10 compares the known weight fraction of crystalline standard in each synthetic mixture with the respective crystallinity index determined from the original and fully corrected intensity curves. The results support the conclusion that there is little justification for exhaustively correcting the raw intensity data prior to calculating the crystallinity index. The essential corrections are those needed to compensate for fluctuations in the intensity of the x-ray source and to normalize the observed intensities to equal scattering mass.

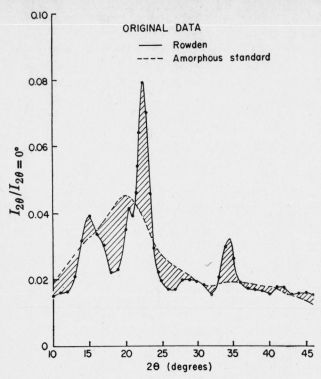

Figure 3-20   X-Ray intensity curve of Rowden cotton (solid line) compared with that of the amorphous standard (broken line). (Wakelin, Virgin, and Crystal [26].)

Table 3-9   Crystallinity Indices C_c and Correlation Co-efficients r for Nine Cottons[a]

Cotton Type	Original Intensities		Corrected Intensities	
	C_c	r	C_c	r
Acala	0.676	0.951	0.694	0.966
Deltapine	0.642	0.939	0.672	0.951
Montserrat	0.624	0.936	0.680	0.944
Rowden	0.728	0.949	0.724	0.961
Triple Hybrid	0.674	0.933	0.658	0.961
Watson Mebane	0.664	0.959	0.670	0.963
SXP	0.563	0.832	0.543	0.848
CO–5087	0.307	0.791	0.297	0.726
CO–5119	0.300	0.679	0.283	0.692

[a]Data from [26].

182

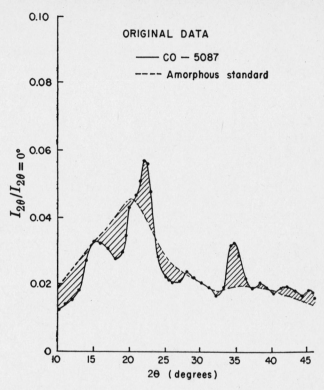

Figure 3-21 X-Ray intensity curve of cotton modified with anhydrous ethylamine (solid line) compared with that of the amorphous standard (broken line). (Wakelin, Virgin, and Crystal[26].)

Table 3-10 Comparison of Analyzed Crystalline-Index Values with Known Compositions of Reference Mixtures[a]

Percent Crystalline Standard in Mixture	Analyzed Crystalline Index C_c	
	Original Data	Corrected Data
20	18	24
33	25	29
50	48	55
67	65	71
80	78	83
89	79	85

[a]Data from [26].

183

Crystallinity Index of Poly(ethylene terephthalate) [29]. Statton has made use of photographic rather than counter data to determine the correlation crystallinity indices of a series of poly(ethylene terephthalate) specimens. The exposures were made with nickel-filtered CuKα radiation in an evacuated flat-film camera using a sample-to-film distance of 5 cm. (See Section 2-3.5.) Bundles of fibers mounted perpendicular to the direct beam were exposed for about 2 hr with a collimator containing two 0.025-in. pinholes separated by 3 in. To minimize graininess of the diffraction patterns fine-grained Eastman type AA film was used, and a fixed development time of 5 min and temperature of 68°F were observed in order to further ensure reproducible results. Instead of experimentally randomizing the orientations of the fibers, as was done in the analyses of cellulose fibers described above, Statton rotated the fiber specimens during the exposure around an axis coincident with the primary beam, which converted all arcs and spots in the pattern to circles. The films were measured radially with a Leeds and Northrup Knorr-Albers microphotometer.

Any particular scattering curve was normalized to constant exposure conditions by multiplying its ordinates by the ratio of the integral intensity of the standard specimen to that of the unknown, both integrated over the angular range $2\theta = 12°30'$ to $38°48'$. The intensities of the reference and unknown specimens were measured at angular increments of 21 minutes of arc over the same range. In Figure 3-22 the intensity curves of the crystalline and amorphous standards and of a typical poly(ethylene terephthalate) unknown over a portion of this angular range are shown. Also indicated in this diagram are some of the differences of ordinates, $I_u - I_a$ and $I_c - I_a$, which were employed in calculating the crystallinity index. Table 3-11 lists 43 pairs of $I_c - I_a$ and $I_u - I_a$ values taken from the curves of the standard specimen and a representative unknown over the range $2\theta = 12°30'$ to $38°48'$, and Figure 3-23 is a plot of these data together with the derived regression line. It is seen that the crystallinity index C_c as defined earlier is equal to 0.695, the slope of this line. In order to facilitate the calculation of the normalization factors and the data points of the regression plot Statton made use of a small (Bendix G–15) computer, although a desk calculator would have sufficed for the purpose.

Under some circumstances it may be desirable to substitute for the "slope" index C_c some proportional measure of the crystallinity on a different scale. Thus a "percentage relative crystallinity" can be deduced for any given sample by arbitrarily assigning the numerical

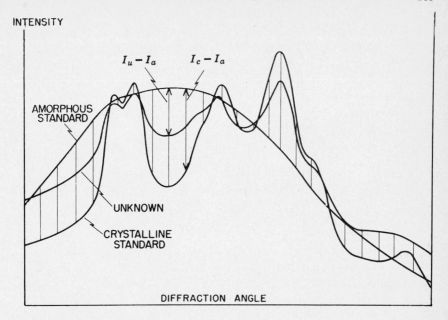

INTENSITY

$I_u - I_a$ $I_c - I_a$

AMORPHOUS
STANDARD

UNKNOWN

CRYSTALLINE
STANDARD

DIFFRACTION ANGLE

Figure 3-22 X-Ray intensity curves of an "unknown" and the standard specimens of poly(ethylene terephthalate) showing some of the intensity differences used in calculating the crystallinity index. (Statton [29].)

values 0 and 100 to the slopes of the regression lines corresponding respectively to the amorphous and crystalline standards. Statton's amorphous standard with a limiting index of 0 was an undrawn, melt-cast film of poly(ethylene terephthalate), and the same specimen annealed for 65 hr in air at 245°C in a small muffle furnace became the crystalline standard with an arbitrary index of 100.

A series of synthetic reference samples of poly(ethylene terephthalate) were prepared by sandwiching together varying numbers of the crystalline and amorphous standard films. Table 3-12 illustrates the reproducibility that was obtainable with Statton's procedure by the results from seven x-ray exposures of a given sample (windup No. 1) and from one exposure each of repeat windups Nos. 2 and 3. Table 3-13 compares the analyzed crystallinity-index values with the (approximately) known compositions of three synthetic reference specimens. A variety of thermal treatments was applied to amorphous, unoriented film and to drawn, slightly crystallized fiber in order to induce varying degrees of crystallinity. In Table 3-14 the normalized C_c values on a scale of 0 to 100 are presented as a function of the mode of treatment for 11 samples.

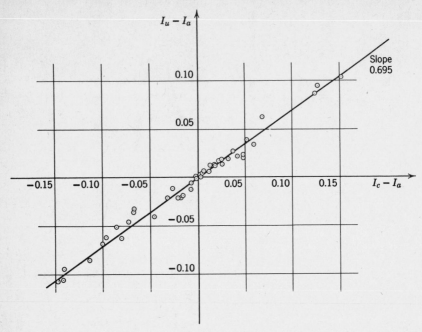

Figure 3-23 Plot of $I_u - I_a$ versus $I_c - I_a$ for a representative poly(ethylene terephthalate) specimen. The plotted data are taken from Table 3-11. (Statton [29].)

Table 3-11 Tabulation of $I_u - I_a$ and $I_c - I_a$. Data for a Representative Sample of Poly(ethylene terephthalate)[a]

$I_c - I_a$	$I_u - I_a$	$I_c - I_a$	$I_u - I_a$
−0.0681	−0.0359	0.1510	0.1030
−0.0854	−0.0518	0.1260	0.0942
−0.0956	−0.0620	0.0691	0.0607
−0.0725	−0.0459	0.0597	0.0333
−0.0278	−0.0124	0.0256	0.0135
−0.0062	−0.0058	0.0122	0.0073
−0.0071	−0.0128	0.0028	0.0000
−0.0335	−0.0205	−0.0018	−0.0005
−0.1000	−0.0594	−0.0020	0.0009
−0.1410	−0.0955	0.0042	0.0042
−0.1480	−0.1080	0.0119	0.0074
−0.1420	−0.1050	0.0187	0.0135
−0.1140	−0.0865	0.0246	0.0177
−0.0813	−0.0639	0.0233	0.0169

Table 3-11   (cont'd)

−0.0461	−0.0402	0.0220	0.0171
−0.0214	−0.0214	0.0188	0.0144
−0.0161	−0.0190	0.0137	0.0116
−0.0189	−0.0204	0.0384	0.0251
0.0055	0.0056	0.0480	0.0218
0.0515	0.0380	0.0485	0.0203
0.1240	0.0861	0.0430	0.0204
		0.0316	0.0184

[a]Courtesy of W. O. Statton.

Table 3-12   Reproducibility of Repetitive Measurements of Crystallinity Index[a] of Poly(ethylene terephthalate)[b]

Sample Description	Percentage Relative Crystallinity	Probable Error
Fiber		
Sample windup No. 1:		
Exposure 1	41.3	1.6
Exposure 2	41.4	1.6
Exposure 3	41.6	2.0
Exposure 4	40.7	1.4
Exposure 5	40.8	1.4
Exposure 6	42.2	1.5
Exposure 7	41.4	1.3
New sample windup No. 2	40.7	1.7
New sample windup No. 3	40.0	1.7
Film		
Unoriented, annealed; sample not rotated	100.0	Standard
Annealed; sample rotated	100.6	0.6

[a]Normalized to percentage relative crystallinity as explained on p. 184.
[b]Data from [29].

Table 3-13   Accuracy of Crystallinity-Index[a] Measurements for Poly(ethylene terephthalate)[b]

Approximate Composition	Sample No.	Percentage Relative Crystallinity	Probable Error
Group A:			
⅔ Crystalline standard	1	63.4	1.1
⅓ Amorphous standard	1	63.7	1.0
	2	59.7	0.5
	2	58.4	0.8
	3	62.5	0.8
Group B:			
⅓ Crystalline standard	1	30.3	0.6
⅔ Amorphous standard	2	25.0	0.9
	2	24.8	0.8
	3	32.2	0.9
	3	32.3	0.7
	4	29.9	0.8
	5	34.6	1.9
Group C:			
¼ Crystalline standard			
¾ Amorphous standard		23.3	0.7

[a]Nomalized to percentage relative crystallinity as explained on p. 184.
[b]Data from [29].

Table 3-14   Crystallinity Index[a] of Poly(ethylene terephthalate) as a Function of Pretreatment[b]

Sample	Pretreatment	Percentage Relative Crystallinity
Amorphous, unoriented film	Control	0
	Heated at 183°C for 10 sec in air	15
	Heated at 183°C for 65 hr in air	64
	Heated at 245°C for 10 sec in air	43
	Heated at 245°C for 65 hr in air	100
Drawn, slightly crystallized fiber	Control	20
	Boiled in H_2O for 30 min	25
	Heated taut in air at 150°C for 30 min	26
	Heated taut in air at 200°C for 30 min	39
	Heated taut in air at 220°C for 30 min	57
	Heated slack in oil at 220°C for 15 sec	46
	Heated taut in oil at 220°C for 15 sec	49
	Heated slack in oil at 240°C for 15 sec	51

[a]Normalized to percentage relative crystallinity as explained on p. 184.
[b]Data from [29].

3-4 CRYSTALLINITY FROM OTHER PHYSICAL MEASUREMENTS

Before concluding this chapter we shall discuss rather briefly three other physical methods that are reputed, at least under favorable circumstances, to yield information concerning the degree of crystallinity in polymers. Since these techniques, unlike the x-ray method, are not directly sensitive to the degree of crystallinity, we shall indicate the specific polymer property that each measures and whether or not a correlation has been found to exist between this property and degree of crystallinity. Comparisons will be made of numerical results obtained by these methods with results from x-rays, not because the latter have absolute accuracy but because a theoretical relationship between degree of crystallinity and the nature and magnitude of the x-ray scattering effects is well established. Thus in discussing the measurement of crystallinity it is unavoidable that we treat the x-ray method as the basis of reference.

3-4.1 Density

Crystallization in a typical polymer system is accompanied by an increase in density, or decrease in specific volume, of about 10%. From a knowledge of the crystalline and amorphous densities ρ_c and ρ_a, and of the density of the partially crystalline "unknown" specimen, ρ, the crystalline fraction of the unknown may be assumed to be given by the relationship [30]

$$x_c = \frac{\rho_c}{\rho}\left(\frac{\rho - \rho_a}{\rho_c - \rho_a}\right). \tag{3-62}$$

This formula can be derived as follows. For a partially crystalline specimen we first designate the following quantities:

$V_c, V_a =$ volumes of crystalline and amorphous components, respectively.

$W_c, W_a =$ weights of crystalline and amorphous components, respectively.

$$V = V_c + V_a = \text{total volume of specimen.} \tag{3-63}$$
$$W = W_c + W_a = \text{total weight of specimen.} \tag{3-64}$$

The weight fraction of crystalline material is

$$x_c = \frac{W_c}{W} = \frac{\rho_c V_c}{\rho V}; \tag{3-65}$$

but from (3-63)

$$\frac{V_c}{V} = 1 - \frac{V_a}{V}, \tag{3-66}$$

and since $\rho V = \rho_c V_c + \rho_a V_a$, we also have

$$\frac{V_a}{V} = \frac{1}{\rho_a}\left(\rho - \rho_c \frac{V_c}{V}\right). \tag{3-67}$$

Combining (3-66) and (3-67) and solving for V_c/V, we find

$$\frac{V_c}{V} = \frac{\rho - \rho_a}{\rho_c - \rho_a}, \tag{3-68}$$

which when substituted for V_c/V in (3-65) gives (3-62). Many investigators have not included the factor ρ_c/ρ in (3-62). This has the effect of yielding the volume fraction, rather than the weight fraction, of crystalline polymer.

The relationship between the weight and volume fractions of crystallinity, $x_{c,w}$ and $x_{c,v}$, respectively, is given in terms of ρ and ρ_c by

$$x_{c,w} = \frac{\rho_c}{\rho} x_{c,v} \tag{3-69}$$

and in terms of ρ_a and ρ_c by

$$\frac{1 - x_{c,w}}{x_{c,w}} = \frac{\rho_a}{\rho_c} \frac{1 - x_{c,v}}{x_{c,v}}. \tag{3-70}$$

The volume fraction of crystallinity is important principally in birefringence calculations, a topic that is discussed in Section 4-5.

In order to apply (3-62) it is necessary to calculate ρ_c from a knowledge of the structure of the crystalline phase and to determine experimentally ρ_a, the density of the polymer in the completely amorphous condition, and ρ, the density of the unknown. The last two quantities can be calculated from precise pyknometric measurements of the corresponding specific volumes v_a and v, since $v_a = 1/\rho_a$ and $v = 1/\rho$. Furthermore, the validity of (3-62) evidently rests on the assumptions that (a) the density of the amorphous polymer phase is the same regardless of the size of the amorphous regions or of their degree of preferred orientation and (b) no pores exist in the amorphous standard or unknown specimens. In actual practice neither assumption is acceptable under all conditions. In particular it is now known that the packing density of highly oriented molecular chains is greater than that of randomly oriented chains in the amorphous state (for example, see p. 163 and [31]); also it seems unlikely that the density of the amorphous regions in a quenched polymer would be the same as the density of amorphous regions in a slowly cooled polymer.

Farrow and Ward[31] determined the crystallinities of unoriented poly(ethylene terephthalate) films and fibers at several temperatures and of oriented fibers in which the degree of orientation was controlled by varying the draw ratio. Tables 3-15, 3-16, and 3-17 compare their numerical results by the density, infrared, and x-ray methods. It is seen that whereas a reasonable correlation exists between x-ray and density crystallinities for the unoriented specimens, there is no such correlation in the case of the oriented fibers. The tabulated density crystallinities were based on $\rho_a = 1.335$ g/cm^3 for an amorphous poly(ethylene terephthalate) specimen quenched rapidly from the melt and $\rho_c = 1.455$ g/cm^3 for the crystalline phase as calculated from the crystal structure. The experimental density of each specimen was determined by observing the point to which it sank in a graded density column[32]. Comparison of Farrow and Ward's x-ray crystallinities (Table 3-17) for heat-crystallized poly(ethylene terephthalate) yarns shows a very reasonable degree of agreement with the values obtained by Statton (Table 3-14) for fibers subjected to rather similar crystallizing conditions.

The last column in Tables 3-16 and 3-17 gives the density of the amorphous material in oriented fibers as calculated from the x-ray crystallinities and the calculated density of crystalline poly(ethylene terephthalate). The average value is seen to be about 1.355 g/cm^3, which is appreciably higher than that of the unoriented amorphous material.

Table 3-15 Crystallinities of Poly(ethylene terephthalate) Samples from X-Ray, Infrared, and Density Measurements — Unoriented Film and Yarn[a]

Sample	Temperature (°C)	Density (g/cm³)	Percent Crystallinity		
			Density	Infrared	X-Ray
0.005-in. film	117	1.359	20	61	29
	146	1.378	36	55	35
	186	1.385	41	59	37
	213	1.395	50	73	46
	227	1.391	48	73	48
0.006-in. yarn	117	1.356	18	41	31
	146	1.382	38	35	33
	186	1.384	41	37	38
	213	1.398	52	46	49
	227	1.358[b]	19[b]	54[b]	41[b]

[a]Data from [31].
[b]These results are inconsistent.

Table 3-16   Crystallinities of Poly(ethylene terephthalate) Samples from X-Ray, Infrared, and Density Measurements—Oriented Fibers[a]

| Sample | Draw Ratio | Density (g/cm³) | Percent Crystallinity | | | Density of Oriented Amorphous Material (g/cm³) |
			Density	Infrared	X-Ray	
Yarns drawn	3.0	1.356	18	50	Nil	1.356
without hot pin	3.5	1.361	22	59	2	1.359
or plate	4.0	1.369	28	72	8	1.359
	4.5	1.360	21	76	7	1.352
Yarns drawn	3.0	1.385	42	58	27	1.354
with hot pin	3.5	1.385	42	63	27	1.354
and plate	4.0	1.385	42	77	25	1.355
	4.5	1.382	39	75	24	1.354
	5.0	1.382	39	84	28	1.348

[a]Data from [31].

3-4.2   Infrared Absorption

In principle infrared absorption is *not* a good measure of long-range order, or crystallinity; rather, certain infrared bands of polymers can be specifically related to configurational states of the molecular chains, in particular to their tacticity. In the infrared spectrum of

Table 3-17   Crystallinities of Poly(ethylene terephthalate) Samples from X-Ray, Infrared, and Density Measurements—Heat Treated Oriented Fibers[a,b]

| Sample | Draw Ratio | Density (g/cm³) | Percent Crystallinity | | | Density of Oriented Amorphous Material (g/cm³) |
			Density	Infrared	X-Ray	
Yarns drawn	3.0	1.402	56	81	39	1.355
without hot pin	3.5	1.402	56	77	41	1.353
or plate	4.0	1.404	58	80	38	1.359
	4.5	1.409	62	85	38	1.364
Yarns drawn	3.0	1.408	61	75	40	1.359
with hot pin	3.5	1.412	64	80	39	1.365
and plate	4.0	1.412	64	86	41	1.365
	4.5	1.412	64	83	41	1.365
	5.0	1.406	59	86	41	1.356

[a]Heat treatment: fibers were heat crystallized for 30 min at 212°C under tension.
[b]Data from [31].

polypropene several bands have been identified that can be attributed to the isotactic chain segments that exist in the helical configuration [33, 34]. In effect this is believed to include virtually *all* isotactic segments, with the consequence that the infrared technique gives a measure of isotacticity in polypropene. Brader[33] selected the 8.57-micron band as the criterion of helical configuration and the 10.27-micron band as the measure of the density-thickness product, and the absorbance ratio $A_{8.57\mu}/A_{10.27\mu}$ was actually employed to deduce the amount of helical structure in a given specimen. It is to be expected that both small and large isotactic chain segments will contribute to such an infrared measurement of isotacticity, whereas only relatively large isotactic elements will contribute much to long-range ordering and crystallinity. Hence it can be anticipated that in general numerical values of a tactic fraction from infrared measurements will tend to be larger than values of the crystalline fraction determined with x-rays. Speaking more loosely, the infrared crystallinity will tend to be larger than the x-ray crystallinity.

More commonly investigators have had some success in relating the intensities of specific infrared bands to the *amorphous* fraction in polymers; for example, the crystalline regions in vinyl polymers would be expected to contain chain elements with a large proportion of left-left or right-right (isotactic) sequences of placement of the side groups, whereas the chain segments in amorphous regions would be expected to consist of a large proportion of random (atactic) sequences. Thus Miller and Willis[35] and Farrow and Ward[31] studied the use of the 898-cm^{-1} infrared band of poly(ethylene terephthalate) as a measure of the amorphous fraction. (See Figure 3-24.) Actually this band has been definitely assigned to a *gauche* configuration of the $-OCH_2CH_2O-$ group. When crystallization occurs as a result of heating or drawing, the intensity of this band diminishes as a result of a change of configuration of this group from *gauche* to *trans*. Reference to Tables 3-15, 3-16, and 3-17 will enable the reader to compare the crystallinities of a number of poly(ethylene terephthalate) samples as deduced from the intensity of the 898-cm^{-1} infrared band with the numerical values from x-rays and density. In accordance with the expectations cited above it is seen that the crystalline fractions derived by infrared absorption are larger than those from x-rays. On the other hand, in studies of polyethylene Hendus and Schnell[36] found that x-rays gave slightly higher crystalline fractions than infrared. It can undoubtedly be concluded that there is no consistent proportionality between the results obtained by the two methods.

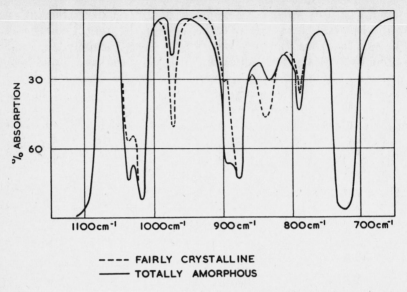

---- FAIRLY CRYSTALLINE
———— TOTALLY AMORPHOUS

Figure 3-24 Crystalline and amorphous infrared absorption bands in poly(ethylene terephthalate). Broken lines—fairly crystalline material; solid lines—totally amorphous material. (Miller and Willis [35].)

3-4.3 Nuclear Magnetic Resonance

The possibility of measuring degree of crystallinity by the nuclear-magnetic-resonance (NMR) method is based on the fact that when the frequency of molecular motion equals or exceeds approximately 10^4 sec^{-1} the resonance curve is narrowed. Thus it is necessary for the polymer in question that there be a temperature at which the average correlation time for molecular motion in the crystalline regions is $\tau_{cc} \gg 10^{-4}$ sec, whereas the correlation time for the amorphous regions is $\tau_{ca} \ll 10^{-4}$ sec [37, 38]. Under these conditions the NMR absorption consists of a superposition of a narrow resonance and a broad resonance. The narrow and broad components are resolved graphically, and the ratio of the areas of the narrow and broad components is taken as equal to the ratio of the amounts of rapidly moving polymer and relatively stationary polymer. On the assumption that the rapidly and slowly moving portions represent amorphous and crystalline regions, respectively, a numerical value can be assigned to the crystalline fraction.

Slichter and McCall [37] studied molecular motion in two poly-ethylenes, one highly branched (DYNK) and one linear (Marlex 50). It was found that for Marlex 50 a considerable range of temperature existed within which $\tau_{cc} \gg 10^{-4}$ sec and $\tau_{ca} \ll 10^{-4}$ sec, thus permitting a satisfactory value of the fraction of relatively stationary material to be determined. The value, 90%, agrees well with x_c for Marlex 50 as determined by x-rays. On the other hand, no temperature was found for DYNK at which the necessary conditions were satisfied for such an analysis.

Farrow and Ward[39] measured the NMR crystallinity of oriented and unoriented samples of poly(ethylene terephthalate) over a wide temperature range. Only above the glass-transition temperature, 110°C, was the necessary composite resonance curve obtained to permit the measurement of the mobile and relatively immobile fractions. Figure 3-25 shows a typical composite derivative curve of poly(ethylene terephthalate) with the sketched-in line separating the narrow and broad components. The actual intensities of the

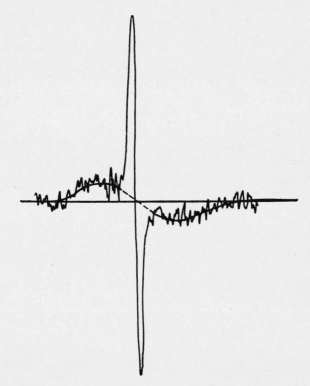

Figure 3-25 Typical NMR signal of poly(ethylene terephthalate) above the glass-transition temperature, showing the composite nature of the derivative curve. (Farrow and Ward[39]; reprinted by permission of The Institute of Physics and The Physical Society.)

narrow and broad resonances were found by calculating the integral of this derivative curve. Table 3-18 compares the crystalline fractions of the several specimens as measured by NMR and x-rays. It is seen that the NMR fractions are larger, which can possibly be attributed to portions of the polymer other than the truly crystalline regions being relatively stationary, because of, for example, numerous cross-links or chain entanglements.

It is fair to conclude that NMR can be of value in estimating crystallinity only when it is employed in conjunction with other methods of characterizing the polymer in question. As Slichter observes [38], "It is necessary to establish that there is a correspondence between degree of crystallinity, which depends on morphology, and the NMR measurement, which depends on the vigor of motion." In general, then, a correlation does not exist between crystallinities as determined by NMR and x-rays.

Table 3-18   Crystallinities of Poly(ethylene terephthalate) Samples from NMR and X-Ray Measurements[a]

Sample	NMR Crystallinity at 180°C (percent)	X-Ray Crystallinity (percent)	
		25°C	180°C
Normal unoriented chip	72	52	46
Unoriented chip (cooled slowly)	72	62	44
Oriented film	80	53	—
Poly(tetradeuteroethylene terephthalate)	74	60	—

[a]Data from [39].

GENERAL REFERENCES

R. Bonart, R. Hosemann, F. Motzkus, and H. Ruck, "X-Ray Determination of Crystallinity in High-Polymeric Substances," Norelco Reporter, 7, 81 (1960).

G. Farrow and D. Preston, "Measurement of Crystallinity in Drawn Poly(ethylene terephthalate) Fibres by X-Ray Diffraction," British J. Appl. Phys., 11, 353–358 (1960).

W. O. Statton, "Characterization of Polymers," in Handbook of X-Rays in Research and Analysis, E. Kaelble (editor), McGraw-Hill, New York, 1967, Chapter 21, pp. 5–9.

H. P. Klug and L. E. Alexander, X-Ray Diffraction Procedures, Wiley, New York, 1954, pp. 621–630.

W. Ruland, "X-Ray Determination of Crystallinity and Diffuse Disorder Scattering," Acta Cryst., 14, 1180 (1961).

J. H. Wakelin, H. S. Virgin, and E. Crystal, "Development and Comparison of Two

X-Ray Methods for Determining the Crystallinity of Cotton Cellulose," *J. Appl. Phys.*, **30**, 1654–1662 (1959).

SPECIFIC REFERENCES

[1] A. W. Coven, *Phys. Rev.*, **41**, 422 (1932); G. E. M. Jauncey and F. Pennell, *Phys. Rev.*, **43**, 585 (1932); **44**, 128 (1933).

[2] W. Ruland, *Acta Cryst.*, **14**, 1180 (1961).

[3] (a) C. R. Desper and R. S. Stein, *J. Polymer Sci.*, Part B, **5**, 893 (1967). (b) W. Ruland and A. Dewaelheyns, *J. Sci. Instruments*, **44**, 236 (1966).

[4] W. Ruland, *Polymer*, **5**, 89 (1964).

[5] W. Ruland, *Faserforsch. Textiltechnik*, **15**, 533 (1964).

[6] R. Hosemann and S. N. Bagchi, *Direct Analysis of Diffraction by Matter*, North-Holland, Amsterdam, 1962, p. 665.

[7] After A. J. C. Wilson, *Nature*, **150**, 151 (1942).

[8] H. P. Klug and L. E. Alexander, *X-Ray Diffraction Procedures*, Wiley, New York, 1954, p. 590.

[9] G. Natta and P. Corradini, *Nuovo Cimento*, **15**, Suppl. 1, 40 (1960).

[10] J. E. Field, *J. Appl. Phys.*, **12**, 23 (1941).

[11] J. M. Goppel, *Appl. Sci. Res.*, **A1**, 3 (1947).

[12] J. J. Arlman, *Appl. Sci. Res.*, **A1**, 347 (1949).

[13] J. M. Goppel and J. J. Arlman, *Appl. Sci. Res.*, **A1**, 462 (1949).

[14] L. E. Alexander, S. Ohlberg, and G. R. Taylor, *J. Appl. Phys.*, **26**, 1068 (1955).

[15] P. J. Flory, N. Rabjohn, and M. C. Shaffer, *J. Polymer Sci.*, **4**, 225 (1949).

[16] For example, S. Krimm and A. V. Tobolsky, *Textile Research J.*, **21**, 805 (1951).

[17] S. C. Nyburg, *British J. Appl. Phys.*, **5**, 321 (1954).

[18] K. E. Beu, W. B. Reynolds, C. F. Fryling, and H. L. McMurry, *J. Polymer Sci.*, **3**, 465 (1948).

[19] G. Natta, L. Porri, and P. Corradini, *Atti Accad. Naz. Lincei*, **20**, 728 (1956).

[20] G. Natta, *Nuovo Cimento*, **15**, Suppl. 1, 31 (1960).

[21] Reference 8, p. 632.

[22] P. H. Hermans and A. Weidinger, *J. Appl. Phys.*, **19**, 491 (1948); *J. Polymer Sci.*, **4**, 135 (1949); *J. Polymer Sci.*, **5**, 565 (1950).

[23] P. H. Hermans and A. Weidinger, *Makromol. Chem.*, **44–46**, 24 (1961).

[24] P. H. Hermans and A. Weidinger, *Makromol. Chem.*, **50**, 98 (1961).

[25] G. Challa, P. H. Hermans, and A. Weidinger, *Makromol. Chem.*, **56**, 169 (1962).

[26] J. H. Wakelin, H. S. Virgin, and E. Crystal, *J. Appl. Phys.*, **30**, 1654 (1959).

[27] Reference 8, pp. 364–376.

[28] S. Krimm and R. S. Stein, *Rev. Sci. Instruments*, **22**, 920 (1951).

[29] W. O. Statton, *J. Appl. Polymer Sci.*, **7**, 803 (1963).

[30] H. G. Kilian, *Kolloid-Z.*, **176**, 49 (1961).

[31] G. Farrow and I. M. Ward, *Polymer*, **1**, 330 (1960).

[32] H. J. Kolb and E. F. Izard, *J. Appl. Phys.*, **20**, 564 (1949).

[33] J. J. Brader, *J. Appl. Polymer Sci.*, **3**, 370 (1960).

[34] M. P. McDonald and I. M. Ward, *Polymer*, **2**, 341 (1961).

[35] R. G. J. Miller and H. A. Willis, *J. Polymer Sci.*, **19**, 485 (1956).

[36] H. Hendus and G. Schnell, *Kunststoffe*, **51**, 69 (1961).

[37] W. P. Slichter and D. W. McCall, *J. Polymer Sci.*, **25**, 230 (1957).

[38] W. P. Slichter, *Advances in Polymer Sci.*, **1**, 35 (1958).

[39] G. Farrow and I. M. Ward, *British J. Appl. Phys.*, **11**, 543 (1960).

4

Preferred Orientation in Polymers

When a linear polymer is subjected to mechanical deformation by means of cold work—such as drawing, stretching, or rolling—the molecular chains usually tend to align themselves parallel to the direction of deformation, although in some polymer species they may tend to assume some angle with respect to the axis of deformation. In either case the polymer is spoken of as *preferentially oriented* or, more commonly, simply as *oriented*. If we consider only the crystalline regions, orientation might be defined more specifically as the relationship of some crystallographic direction or directions to an external reference frame. In the present chapter we are concerned very largely with the use of x-ray diffraction techniques to reveal and characterize orientation modes in the crystalline regions of linear polymers. Although x-ray patterns under special conditions can reveal the presence of orientation of the amorphous regions [1], this application at present has little quantitative value and may even lead to wrong conclusions.

In contrast to x-ray diffraction, optical birefringence measurements give a measure of the overall preferred orientation that prevails in a polymer specimen, including both crystalline and amorphous regions. In a later section something will be said of the possibility of combining x-ray and optical birefringence data to give some idea of the extent of amorphous orientation.

Preferred orientation of linear polymers has great practical consequences, which arise from the fact that the primary bonds within the molecular chains are much stronger than the interchain forces.

Preferential orientation of the chains parallel to the direction of external stress can therefore result in optimal tensile, or break, strength. Geil[2] observes that on a weight basis many crystalline polymer fibers are stronger than steel! Numerous commercial applications of linear polymers utilize this anisotropy of molecular forces. It may also be pointed out that in various fibrous proteins preferred orientation of the molecular structure plays an important role during growth and in the subsequent functioning of the molecule in its environment. Thus the characterization of orientation constitutes one of the major functions of x-ray analysis as applied to polymers.

4-1 DIFFRACTION PATTERNS OF ORIENTED POLYMERS

It is suggested that the reader refer to Sections 1-4.5 and 1-4.6 for the discussion on the origin and nature of the x-ray patterns that are produced by *randomly* and *preferentially oriented* microcrystalline specimens. As illustrated by Figure 1-40, it was shown that the diffraction patterns of axially oriented polymers bear a resemblance to the patterns generated by rotating single crystals inasmuch as the constituent reflections are spots or arcs that lie on layer lines. It was also explained that specimens with a very high degree of c-axis (fiber-axis) orientation cannot produce reflections of type $(00l)$ with the incident beam perpendicular to the fiber axis because these reciprocal-lattice nodes lie on the axis of rotation and cannot intersect the sphere of reflection. Likewise (hkl) reflections with h and k very small relative to l are likely to be absent. When the c-axis orientation is less perfect, however, reflections of these classes may appear as faint, diffuse arcs. Figure 4-1 illustrates how the intensities of the $(00l)$ reflections can be enhanced by appropriate degrees of tilting of the fiber axis, thereby bringing these planes into optical diffracting position. Table 4-1 gives the relative intensities of the (002), (004), and (006) reflections versus tilt angles of 90, 82, 74, 67, and 59° with respect to the incident beam. The reader may note that the fiber patterns of Figure 4-1 are rich in reflections but show only a moderate degree of orientation. Hence the reflections are arcs of considerable length. Figure 4-2 shows a series of photographs of a much more highly oriented specimen of polyethylene prepared over a wide range of inclination angles between 90 and 15°. The reader's attention is directed in particular to the variation in intensity of the (002) reflection.

The foregoing considerations show that a correct understanding of the kind of preferred orientation that prevails in a selected polymer specimen cannot be obtained from a simple fiber pattern prepared

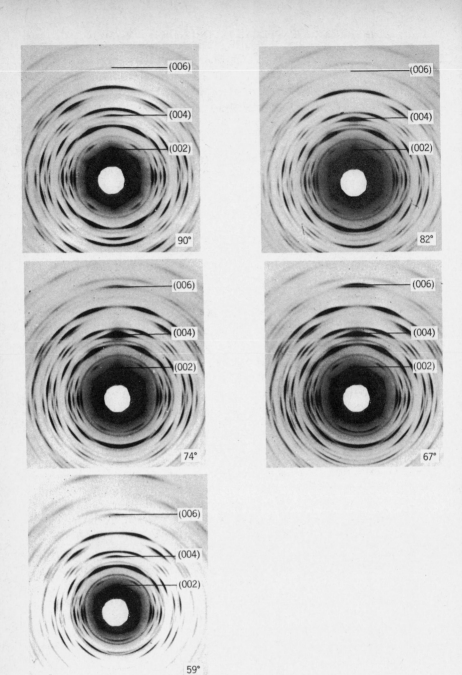

Figure 4-1 Fiber photographs prepared with the fiber axis c inclined at angles of 90, 82, 74, 67, and 59° to the incident beam. Mercuric-chloride complex of poly-(ethylene oxide); $CuK\alpha$ radiation, flat camera. (Courtesy of S. S. Pollack.)

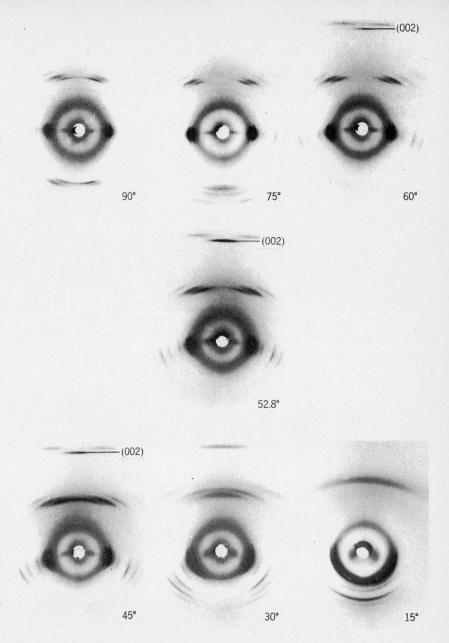

Figure 4-2 Fiber photographs prepared with the fiber axis c inclined at angles of 90, 75, 60, 52.8, 45, 30, and 15° to the incident beam. Highly oriented polyethylene; CuKα radiation, cylindrical camera. The angle of 52.8° is calculated to maximize the intensity of (002). (Courtesy of E. S. Clark.)

201

Table 4-1   Dependence of the Intensities of the Axial Reflections (002), (004), and (006) on Inclination with Respect to the Incident Beam[a]

	Intensities of Axial (Meridional) Reflections		
Inclination (degrees)	(002)	(004)	(006)
90	vs	m	vw
82	vs	s	w
74	s	vs	mw
67	m	vs	m
59	w	m	mw

Abbreviations: vs = very strong, s = strong, m = medium, mw = medium weak, w = weak, vw = very weak.

[a]Material: mercuric-chloride complex of poly(ethylene oxide); CuKα radiation, flat camera. (See Figure 4-1.)

with the fiber axis perpendicular to the direct beam. Instead the intensities of reflections with appropriate indices must be observed as a function of angle of tilt of the fiber axis. Furthermore, when the mode of preferred orientation is more complex than simple axial, the intensities of certain reflections are found to depend not only on the inclination but also on the azimuth of the specimen. Thus it has come to be realized in recent years that orientation states in linear polymers can be investigated most effectively by the methods of pole-figure analysis that have been applied successfully to the elucidation of textures in metals for many years.

4-2   REPRESENTATION OF PREFERRED ORIENTATION

We shall here confine ourselves to the crystalline phase and, as heretofore, assume that the molecular chains are parallel to the crystallographic axis c. Given the unit-cell dimensions and the mode of packing of the molecular chains within the cell, we are able to describe any given orientation mode in terms of the orientations in space of the normals, or poles, to selected crystallographic planes. If the polymer specimen were a single crystal, it would be possible to mark on the surface of a sphere concentric with it the locations of the points of intersection of all plane normals of interest, as illustrated in Figure 4-3. Such a construction constitutes the spherical projection of a crystal, and it provides a precise and unequivocal representation of its orientation. Because of the practical difficulties involved in

75

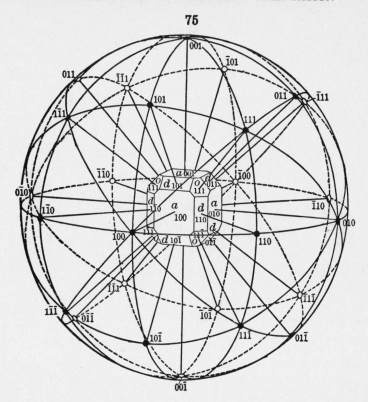

Figure 4-3 The spherical projection of a crystal. (Dana-Ford, *A Textbook of Mineralogy*, 4th ed., Wiley, New York, 1932.)

working with spherical maps it is customary to plot pole figures by means of planar, stereographic projections. As depicted in Figure 4-4, the stereographic projection can be derived from the northern hemisphere of a spherical polar map by drawing straight lines from the various (hkl) poles to the south pole of the sphere. The stereographic projection, then, is composed of the points of intersection of these lines with the equatorial plane of the sphere. Likewise, a stereographic projection of the southern hemisphere can be obtained by drawing lines from the crystallographic poles contained thereon to the north pole of the sphere.

The angular coordinates of a pole on a spherical projection can be conveniently expressed in terms of latitude (α) and longitude (β) as illustrated in Figures 4-5 and 4-29. Occasionally it is useful to specify the colatitude, $\phi = 90° - |\alpha|$. Such a net of latitude and longitude circles can be projected on a plane to form a stereographic net in two

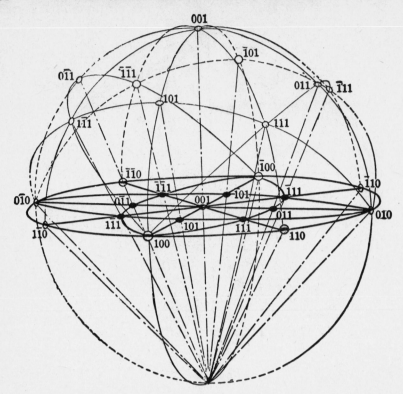

Figure 4-4 Derivation of the stereographic projection from the spherical projection. (Dana-Ford, *A Textbook of Mineralogy*, 4th ed., Wiley, New York, 1932.)

ways. When the projection is made from the north or south pole on the equatorial section of the sphere in the manner described in the preceding paragraph, a polar stereographic net such as shown in Figure 4-6 is obtained. When the projection is made from one end of a diameter in the equatorial section perpendicularly on a plane containing the north-south axis, a Wulff, or meridional, stereographic net results, as shown in Figure 4-7. The polar net is the more useful of the two for depicting polymer textures, especially when counter diffractometric techniques are used for the intensity measurements.

The sum total of all orientations—that is, the orientation distribution function—of a crystalline polymer is referred to as its *texture*. Of course the degree of preferred orientation that is attainable in a polymer specimen is far short of that manifested by a single crystal, with the result that the poles of a particular set of planes (*hkl*) occupy a distribution of orientations centering about the preferred direction.

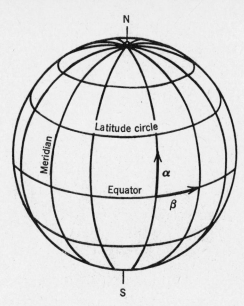

Figure 4-5   Globe ruled with latitude and longitude circles. (C. S. Barrett, *Structure of Metals*, 2nd ed., McGraw-Hill, New York, 1952.)

Often the distribution is very broad and diffuse, or the preferred direction may be ill-defined. For this reason it is the usual practice to portray in any one pole figure only the orientations of one set of crystallographic planes. Thus we may draw a (100), a (001), or a (110) pole figure, but we would not ordinarily map on one pole figure both the (100) and (001) planes. Figure 4-8 shows the relationship between a point S on a powder-diffraction circle produced by diffracting planes with pole P and the equivalent point P' on a stereographic projection. The incident and diffracted beams and the pole of the diffracting planes lie on a common plane that is turned through an angle β from the vertical. All the diffracting planes are inclined at an angle θ to the direct beam, and the poles of these planes make an angle $90° - \theta$ with the beam, from which it follows that the poles of all planes that diffract into a given Debye-Scherrer ring are also located on a circle in the stereographic projection (pole figure). However, as the radius of the Debye-Scherrer ring on the film increases, the radius of the ring on the pole figure decreases. The reader will also perceive from Figure 4-8 that when the photographic film and stereographic projecting plane are both perpendicular to the x-ray beam, the radial (2θ) and azimuthal (β) coordinates of a point on the diffraction pattern are

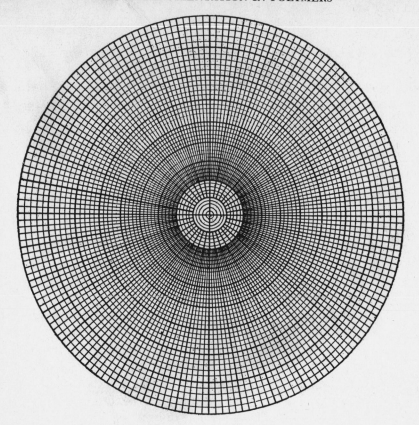

Figure 4-6   Polar stereographic net. (C. S. Barrett, *Structure of Metals*, 2nd ed., McGraw-Hill, New York, 1952.)

directly related to the radial ($\frac{1}{2}[90° - \theta]$) and azimuthal ($\beta$) coordinates of the equivalent point on the pole figure, thus providing the geometrical basis for plotting pole figures.

4-2.1 Classification of Orientation Modes

Sisson's[3] qualitative classification of the orientation modes observed in crystalline cellulose forms the basis for a classification scheme that may be applied to polymers in general. Heffelfinger and Burton[4] have modified and extended the treatment of Sisson, arriving at the six modes that are presented by means of idealized pole figures in Figure 4-9. More complex as well as mixed orientation modes are frequently observed, but they can usually be characterized as combinations of the basic modes of Figure 4-9.

Figure 4-7 Meridional stereographic net. (C. S. Barrett, *Structure of Metals*, 2nd ed., McGraw-Hill, New York, 1952.)

Any given mode of preferred orientation can be described by specifying the orientation of a crystallographic operator element (axis or plane) with respect to a set of orthogonal axes of reference in the polymer specimen, as shown in Figure 4-9a and b. The pole figures of Figure 4-9c presuppose monoclinic symmetry with the first setting of axes†, which conforms with the convention generally followed in this monograph that c is the unique axis. In this setting of axes the mono-clinic angle (here considered to be greater than 90°) is γ and is formed by the axes a and b. It should also be noted that the reciprocal-lattice

†The reader should refer to *International Tables for X-Ray Crystallography*, Vol. I, Kynoch Press, Birmingham, 1952, p. 55: "Other Data, No. 1" and the descriptions of monoclinic space groups, pp. 76–101.

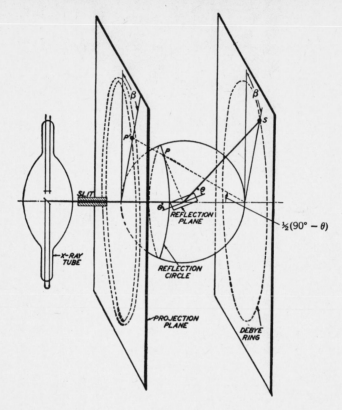

Figure 4-8 Relation between a point S on a powder-diffraction circle, pole P of the diffracting planes, and point P' on the stereographic projection. (C. S. Barrett, *Structure of Metals*, 2nd ed., McGraw-Hill, New York, 1952.)

axes a^* and b^* coincide with the poles of the planes (100) and (010), respectively. The reference axes in the sample, M, T, and N, designate respectively the machine (draw), transverse, and normal directions in the customary manner. The plane of projection of the pole figures is MT. A polar stereographic coordinate net is assumed with zero longitude (β) at M, zero latitude (α) corresponding to the periphery of the figures, and 90° latitude located at the center and corresponding to the direction N. For axial orientation (fiber texture) M is taken to be the fiber axis, or axis of draw or extrusion. In Figure 4-9c the orientation modes are specified by describing the orientation of (2) a crystallographic axis with respect to a reference plane (planar), (3) a plane with respect to a plane (uniplanar), (4) an axis with respect to an axis (axial), (5) a plane with respect to an axis (plan-axial), or (6) a combination of

(3) and (4) (uniplanar-axial). In uniplanar orientation the radius of the (100) ring is γ^*, which is less than 90° since $\gamma^* = 90° - \gamma$. On the other hand, in axial orientation with c parallel to M the longitude of the (100) and (010) poles is 90° since these poles correspond to the reciprocal-lattice axes a^* and b^*, which make angles of 90° with c (or c^*, since c and c^* are coincident in monoclinic geometry with the first setting of axes).

Figure 4-10[5] shows (200), (020), and (110) pole figures of polyethylene film at low elongation (100%). The specimen was cross-linked with a small dose of beta radiation, melted, stretched in the molten state, and then crystallized. The texture is seen to be approximately axial with the (200) poles (a-axis) preferentially oriented parallel to M and the (020) and (110) poles distributed with approximate cylindrical symmetry about M. Because of the orthogonality of the polyethylene unit cell, orthorhombic with $a = 7.40$, $b = 4.93$, $c = 2.534$ Å, the locus of the (020) poles is at right angles to (200), whereas the (110) poles lie at an intermediate angle. A second specimen prepared in the same way and then crystallized at 200% elongation gave the pole figures shown in Figure 4-11, which reveal that the approximate axial texture at low elongation has been replaced by a more complex texture in which the (200) poles are cylindrically distributed about M at an angle of about 50°, and the (020) poles tend to concentrate in the plane of the film near the transverse direction T, whereas the (110) vectors tend to concentrate in the plane of the film at an intermediate angle between M and T. Heffelfinger and Burton [4] examined the texture in poly(ethylene terephthalate) films subjected to a high one-way stretch and found it to be uniplanar-axial with c well oriented parallel to M and the (100) planes essentially parallel to the film plane MT.

4-3 PREPARATION OF POLE FIGURES

The diffraction intensities required for the preparation of a pole figure can be most conveniently and accurately obtained diffractometrically. The counter is fixed at the angle 2θ that corresponds to the chosen planes [for example, (200)], and the specimen is rotated about two perpendicular axes so as to permit the measurement of the intensities diffracted by these selected planes oriented at various angles within the specimen. The well-established procedures for preparing quantitative pole figures of metals [6] are also applicable to polymers, although with certain limitations that are imposed by the much smaller x-ray absorption coefficients and lower crystallographic

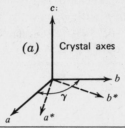

	ORIENTATION MODE		GEOMETRY		
	Heffelfinger – Burton	Sisson	Preferred Orientation	Crystal Operator Element	Reference Element
1	Random	Random	None	None	None
2	Planar	Uniplanar	A crystal axis *(c)* lying in a reference plane *(MT)*	Axis *c*	Plane *MT*
3	Uniplanar		A crystal plane (010) parallel to a reference plane *(MT)*	Plane (010)	Plane *MT*
4	Axial	Uniaxial	A crystal axis *(c)* parallel to a reference axis *(M)*	Axis *c*	Axis *M*
5	Plan – axial		A crystal plane (010) parallel to a reference axis *(M)*	Plane (010)	Axis *M*
6	Uniplanar-axial	Selective uniaxial	A given crystal axis *(c)* parallel to a reference axis *(M)* and a given plane (010) parallel to a reference plane *(MT)*	Axis *c* Plane (010)	Axis *M* Plane *MT*

Figure 4-9 Classification of orientation modes. (After Heffelfinger and Burton[4].)

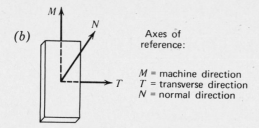

(b)

Axes of reference:

M = machine direction
T = transverse direction
N = normal direction

REPRESENTATIVE

POLE FIGURES

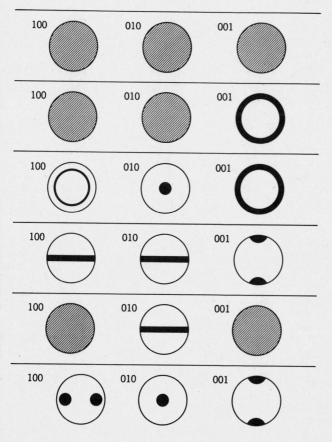

Figure 4-9 Classification of orientation modes. (After Heffelfinger and Burton[4].)

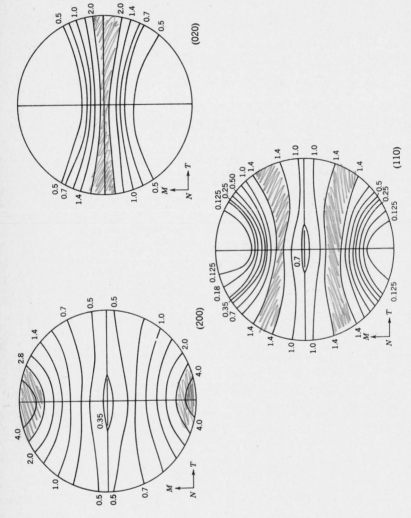

Figure 4-10 Pole figures of polyethylene film at 100% elongation. Symmetry approximately axial. M = machine direction, T = transverse direction, N = normal direction. (Desper and Stein [5].)

212

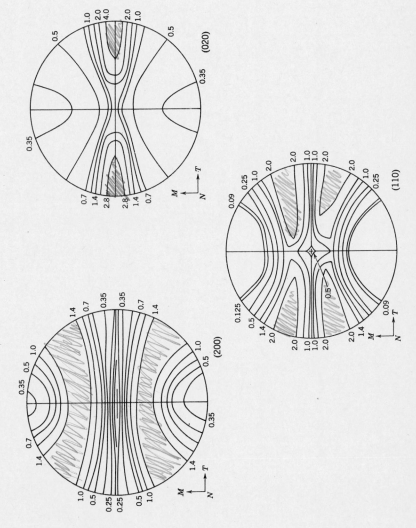

Figure 4-11 Pole figures of polyethylene film at 200% elongation. M = machine direction, T = transverse direction, N = normal direction. (Desper and Stein [5].)

213

symmetries of polymers. Thus the transmission technique is more important for polymers, whereas the reflection technique has important advantages for metals. For sheet specimens of polymers this has the consequence that the larger portion of the latitude range, from 0 to 60 or 70° in α, is most easily mapped by the transmission technique — and the relatively small zone remaining around $\alpha = 90°$, by the reflection technique. (See again Figures 4-5, 4-6, and 4-29.) The usual method of representation is by means of a polar stereographic net with the machine (M) and transverse (T) directions in the equator and the normal direction (N) at $\alpha = 90°$. (See Figure 4-9b.)

We wish to emphasize at this point that the intensity measurement that is required at any particular polar coordinates α, β is the integrated intensity of the reflection (hkl) above background, I_i. As indicated in Figure 4-12, this quantity can be measured by fixing α and β, scanning the reflection over the 2θ interval between two background positions $2\theta_{b_1}$ and $2\theta_{b_2}$, and then subtracting from the integrated counts the background counts (hatched zone). In actual practice it is often assumed that I_i is proportional to the overall peak intensity I_p minus the background intensity at $2\theta_p$ (determined as the average of I_{b_1} and I_{b_2}). This method presupposes that the 2θ distribution of intensity in the reflection is invariant, that is, independent of α and β and other systematic factors. A more reliable, although approximate, measure of I_i can be had by multiplying the peak height above background by the width at half-maximum intensity, $\omega_{1/2}$.

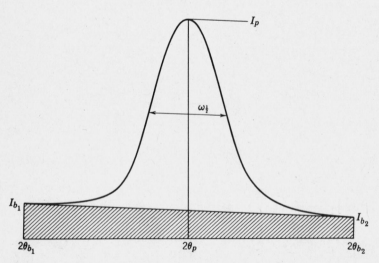

Figure 4-12   Profile of a reflection as a function of 2θ; α and β constant.

If highly monochromatic x-rays are used rather than β-filtered radiation, the background correction is considerably reduced and may even be justifiably neglected if it is very small in relation to I_p. Granted that this is the case, if it can also be shown that absorption corrections vary only slightly over the pertinent ranges of α and β, the directly measured peak intensities I_p may be regarded as roughly proportional to the pole densities and plotted directly on the pole figure. The nature and quantitative evaluation of the corrections for x-ray absorption by the specimen are discussed below in connection with a treatment of the transmission and reflection geometries. In concluding these introductory remarks it is to be emphasized that the investigator must not fail to take cognizance of the errors incurred if he employs any intensity measurements other than the integrated intensity above background, I_i.

4-3.1 Transmission Technique, Sheet Specimen

The geometrical arrangement of the transmission technique — which was first employed by Decker, Asp, and Harker[7] in studies of metals — is shown in relation to Cartesian axes in Figure 4-13. The direction Z coincides with the goniometer axis, and XY is the plane that contains the incident (s_0) and diffracted (s) x-ray beams and center of the polymer specimen. When $\alpha = 0°$, the plane of the specimen

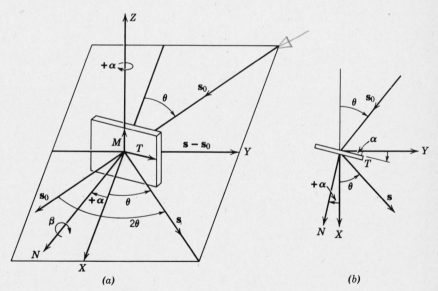

(a) (b)

Figure 4-13 Geometry of the transmission technique for pole-figure measurements.

coincides with the plane YZ, and we have the symmetrical-transmission arrangement, which was discussed in Section 2-2.1 and illustrated in Figure 2-8c. It is the usual practice to orient the machine (draw) direction M in the Z-direction. With $\alpha = 0°$ the poles of the diffracting planes, which are coincident with Y, lie in the plane of the specimen, and rotation of the specimen about its normal N (β-rotation), which coincides with X, has the effect of scanning the periphery of the pole figure (the $0°$ latitude circle). Circles at higher latitudes can be scanned by turning the specimen about the Z-axis to the desired value of α and then rotating the specimen about its normal N (β-rotation) as before. It can be seen that when $\alpha \neq 0°$ the transmission geometry is asymmetrical.

An inspection of Figure 4-13b shows that the transmission method fails when α approaches $90° - \theta$ because at this angle the diffracted beam is parallel to the specimen surface. Thus for a representative Bragg angle of $15°$ the *hypothetical limiting* value of α is $75°$ and the *largest practicable* angle is about $65°$. When $\alpha \neq 0°$ the departure from symmetrical-transmission geometry causes the overall length of the x-ray-beam path through the specimen to increase, thus calling for a thinner specimen than that specified by (2-9) to yield maximum diffracted intensity:

$$t_m = \frac{1}{\mu \sec \theta}. \qquad (2\text{-}9)$$

For polymers that consist exclusively of light elements (carbon, oxygen, nitrogen, hydrogen) the linear absorption coefficient μ for $CuK\alpha$ x-rays lies in the range of 5 to 8 cm^{-1}, from which (2-9) gives an optimal thickness of from 1.0 to 2.0 mm for moderate to small values of θ. However, in the present application the optimal thickness of such a "light" polymer is prescribed not so much by (2-9) as by the need to keep geometrical broadening of the diffracted ray to a practicable minimum. (In this connection the reader may wish to refer to Section 2-2.1.)

Not only the specimen thickness but also the equatorial divergence (in the XY plane) of the direct beam must be sufficiently limited so as not to greatly impair the resolution of the diffraction pattern. The widths of the reflections determine the resolution that can be attained; thus two reflections with similar intensity profiles cannot be resolved unless their angular separation exceeds the individual width at half-maximum intensity. An analysis of the factors that combine to generate the observed reflection profile was given some years ago for conventional powder diffractometry[8, 9]. Although a similar comprehensive investigation of the experimental conditions employed in pole-figure measurements would be valuable, a much simpler semi-quantitative approach involving some approximations will suffice to determine the choice of experimental conditions.

In the following treatment the symbols γ and ω denote angular distributions and widths, respectively, on the 2θ scale in degrees. Thus ω_x and ω_s are the angles subtended at the specimen by the widths of the x-ray source and receiving aperture, respectively. The divergence ϵ of the primary beam is the angle subtended by the primary-beam aperture at the x-ray source.

The observed reflection profile is compounded of several component profiles, the more important of which are the following:

γ_0 = pure profile of the reflection as determined by the crystallite-size distribution and lattice imperfections,

γ_λ = spectral dispersion,

γ_x = x-ray source,

γ_t = thickness of (sheet) specimen,

γ_ϵ = divergence of the incident beam,

γ_s = dimensions of the receiving aperture.

The observed reflection profile is given by the convolution:†

$$\gamma = \overbrace{\gamma_0 \cdot \gamma_\lambda \cdot \gamma_x \cdot \gamma_t \cdot \gamma_\epsilon \cdot \gamma_s}. \tag{4-1}$$

If the individual profiles, γ_i, of (4-1) are Gaussian in shape with overall widths ω_i and half-maximum widths $(\omega_i)_{1/2}$, their convolution γ has the overall width

$$\omega = \sum_i \omega_i \tag{4-2}$$

and half-maximum width

$$\omega_{1/2} = \left[\sum_i (\omega_i)_{1/2}^2 \right]^{1/2}. \tag{4-3}$$

Of the elements in (4-3) the breadth $(\omega_\lambda)_{1/2}$ at relatively small Bragg angles is negligible in relation to the other ω_i's. Thus we need to estimate $(\omega_{0xt\epsilon})_{1/2}$ for representative experimental conditions and determine the corresponding receiving-aperture widths that are compatible with satisfactory resolution. Although several of these elementary profiles differ greatly from Gaussian, in particular γ_t and γ_ϵ, it can be shown that (4-3) still defines a generally valid *lower limit*

†The convolution, or fold, of two functions is defined in Appendix 1. The commutative law applies to multiple convolutions such as (4-1), which are calculated sequentially:

$$\gamma_{0\lambda} = \overbrace{\gamma_0 \cdot \gamma_\lambda},$$
$$\gamma_{0\lambda x} = \overbrace{\gamma_{0\lambda} \cdot \gamma_x},$$
$$\gamma_{0\lambda xt} = \overbrace{\gamma_{0\lambda x} \cdot \gamma_t}, \text{ etc.}$$

of their joint broadening effect. Also, to a good approximation the half-maximum widths of the three geometrical elements, x, t, and ϵ, may be taken to be respectively equal to one-half their overall widths:

$$(\omega_x)_{1/2} = \tfrac{1}{2}\omega_x, \qquad (\omega_t)_{1/2} = \tfrac{1}{2}\omega_t, \qquad (\omega_\epsilon)_{1/2} = \tfrac{1}{2}\omega_\epsilon. \qquad (4\text{-}4)$$

An analysis similar to that which has been applied to the measurement of intensities from single crystals [10] shows that ω_t and ω_ϵ are given to a good first approximation by

$$\omega_t = 57.3 \, \frac{t}{d} \frac{\sin 2\theta}{\cos(\theta - \alpha)} \quad \text{degrees}, \qquad (4\text{-}5)$$

$$\omega_\epsilon = \epsilon \left[1 + \frac{\cos(\theta + \alpha)}{\cos(\theta - \alpha)} \right] \quad \text{degrees}. \qquad (4\text{-}6)$$

In these equations ϵ is the divergence permitted by the direct-beam aperture, t is the specimen thickness, and d is the goniometer radius (d = source-to-specimen distance = specimen-to-receiving-aperture distance).

The considerations given above permit a reasonable estimate to be made of the width of the reflection profile

$$\gamma_{0\lambda x t \epsilon} = \overbrace{\gamma_0 \cdot \gamma_\lambda \cdot \gamma_x \cdot \gamma_t \cdot \gamma_\epsilon}. \qquad (4\text{-}7)$$

To determine the appropriate receiving-aperture width we may refer to an earlier study of the broadening of reflection profiles by a rectangular receiving aperture [8]. This analysis showed that the broadening of a reflection of half-maximum width $\omega_{1/2}$ amounts to 6 and 12% for Gaussian [$\exp(-k^2\zeta^2)$] and Cauchy [$1/(1 + k^2\zeta^2)$] profiles, respectively, if the receiving-aperture width ω_s is set equal to $\tfrac{1}{2}\omega_{1/2}$. Since the profile $\gamma_{0\lambda x t \epsilon}$ tends to assume an intermediate form, we may safely conclude that, if $\omega_s \leq \tfrac{1}{2}\omega_{1/2}$, the broadening will be less than 10% and good resolution of the diffraction pattern is assured.

Figures 4-14 and 4-15 show plots of ω_s versus α for narrow and wide x-ray sources, respectively. In all cases the width of the profile at half-maximum intensity prior to the action of the receiving aperture is taken to be twice the plotted value of ω_s for the same experimental conditions. Each figure consists of parts a and b corresponding to $(\omega_0)_{1/2} = 0$ and $0.50°$. Each of the four families of curves includes one curve for $\theta = 15°$, $t = 1$ mm, $\epsilon = 0.25°$ and three other curves corresponding to separate increases in each of these parameters to $\theta = 30°$, $t = 2$ mm, $\epsilon = 0.50°$. A fixed goniometer radius of 145 mm has been assumed. A narrow, or "line," source is required for highest resolution and is obtained by viewing the focal spot laterally at a take-off

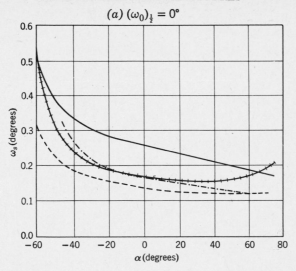

(a) $(\omega_0)_{\frac{1}{2}} = 0°$

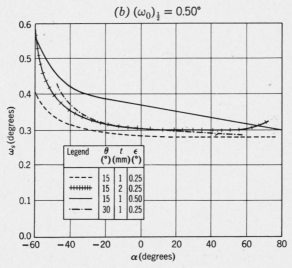

(b) $(\omega_0)_{\frac{1}{2}} = 0.50°$

Legend	θ (°)	t (mm)	ϵ (°)
- - - -	15	1	0.25
++++++	15	2	0.25
———	15	1	0.50
·—·—	30	1	0.25

Figure 4-14 Maximum receiving-aperture width (ω_s) as a function of Bragg angle θ, specimen thickness t, and beam divergence ϵ. Transmission geometry, goniometer radius 145 mm, narrow x-ray source: $(\omega_x)_{1/2} = 0.05°$, $\omega_0 =$ width of pure diffraction profile.

angle of 2 or 3°, as is done in conventional powder diffractometry. The length of the x-ray focal spot when viewed laterally corresponds to the axial, or Z, dimension of Figure 4-13. Since this length is 10 mm or more, it is (a) necessary to restrict the axial divergence of the direct

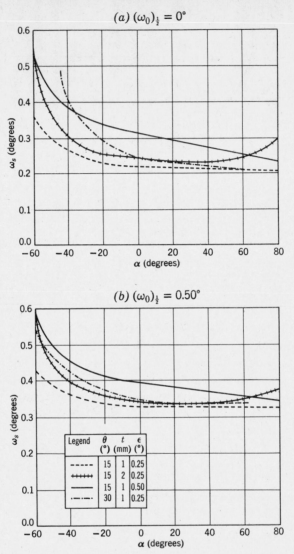

Figure 4-15 Maximum receiving-aperture width (ω_s) as a function of Bragg angle θ, specimen thickness t, and beam divergence ϵ. Transmission geometry, goniometer radius 145 mm, wide x-ray source: $(\omega_x)_{1/2} = 0.35°$, $\omega_0 = $ width of pure diffraction profile.

and diffracted beams by means of Soller-slit collimators [11] and (b) desirable for good intensity that the axial dimension of the specimen equal or exceed the length of the source.

An evaluation of Figures 4-14 and 4-15 in the light of the related text above leads to the following conclusions:

1. From the standpoints of resolution and effective accessible angular range the $+\alpha$ region is superior to the $-\alpha$ region, which is important since in nearly all textures the $+\alpha$ and $-\alpha$ hemispheres are symmetrically equivalent, obviating the need to make measurements in both regions.†

2. The method of calculation yields a good approximation of the experimental profile width or, under some conditions, its lower limit. Thus the resolution to be anticipated might be somewhat poorer, but not better, than that corresponding to the calculated ω_s.

3. The significant variation of ω_s (and therefore ω) with α demonstrates that integrated rather than peak intensity measurements are mandatory in accurate work.

4. Aperture widths that are smaller than those indicated will provide virtually no improvement in resolution and at the same time reduce the intensity. Conversely, under circumstances where optimal resolution is not required, greater intensity can be had by employing apertures that are wider than the values indicated.

5. The use of a line source is effective in promoting resolution only when the other components, in particular $(\omega_0)_{1/2}$, are also small. This is readily seen by comparing Figures 4-14a and b, and to a less striking degree by comparing the four curves of Figure 4-14a.

No well-defined minimum dimension of the receiving aperture in the axial (Z) direction exists. When a line source is used, it should be somewhat larger than the length of the source, hence ordinarily 10 to 15 mm. When the focal spot is viewed longitudinally so as to give a "spot" focus in projection, a reasonable compromise between resolution and intensity can be realized by relating the height of the receiving aperture in angular units, ω_s', to the axial divergence of the primary beam, ω_ϵ', and the Z-dimension of the focal spot in angular units, ω_x', by the Gaussian combination principle as stated by (4-3). Thus we may write

$$\omega_s' = [(\omega_\epsilon')_{1/2}^2 + (\omega_x')_{1/2}^2]^{1/2}, \tag{4-8}$$

which gives for representative values of $(\omega_\epsilon')_{1/2} = 0.25$ and $(\omega_x')_{1/2} = 0.40°$ the result: $\omega_s' = 0.47°$. This method is meant to give only the approximate dimension required; for higher resolution ω_s' may be reduced, and for higher intensity it may be increased.

The intensity diffracted when $\alpha \neq 0°$ will differ from that at $\alpha = 0°$ because of increased x-ray path length through the specimen. Decker, Asp, and Harker [7] have derived a correction formula that can be

†The sense of positive α-rotation adopted in this book corresponds to the simplest mathematical interpretation for actual working conditions. However, the reader should note that it is opposite to the convention of Decker, Asp, and Harker [7].

applied to intensities measured at nonzero values of α. If we make allowance for the change of convention with respect to the sense of positive and negative α, this formula is

$$\frac{I_{0°}}{I_\alpha} = \frac{\mu t \exp(-\mu t/\cos\theta)}{\cos\theta} \times \frac{[\cos(\theta-\alpha)/\cos(\theta+\alpha)] - 1}{\exp[-\mu t/\cos(\theta-\alpha)] - \exp[-\mu t/\cos(\theta+\alpha)]}$$
$$\text{(transmission technique).} \qquad (4\text{-}9)$$

In order to use (4-9) it is necessary to know the μt of the specimen being investigated. An approximate value can be calculated from the measured thickness t and known chemical composition of the specimen; however, for accurate calculations it is better to determine μt experimentally by measuring the diminution in intensity of a monochromatic x-ray beam on passing perpendicularly through the specimen. A reflection from a single crystal or strongly diffracting crystalline powder may serve as the monochromatic beam. Equation 4-9 has been used to calculate the values in Table 4-2 of the correction factor to be applied corresponding to representative values of μt and θ, and over the range of α that is accessible at the Bragg angle concerned.

More commonly it is the practice to circumvent the absorption corrections by normalizing all intensities from the oriented polymer specimen to the intensities diffracted at the same angles by a randomly oriented polymer specimen of the same kind. These normalized intensities can be plotted directly on the pole figure as relative pole densities. Numbers greater than unity indicate greater than average

Table 4-2 Correction Factors To Be Applied to Intensities Measured by the Transmission Technique; Dependence on μt, α, and θ

	$\theta = 15.0°$			$\theta = 30.0°$			$\theta = 45.0°$		
	μt			μt			μt		
α (°)	0.25	0.50	1.00	0.25	0.50	1.00	0.25	0.50	1.00
−60	0.395	0.563	1.052	—	—	—	—	—	—
−45	0.593	0.674	0.864	0.405	0.527	0.798	—	—	—
−30	0.766	0.802	0.878	0.628	0.679	0.782	0.465	0.566	0.754
−15	0.902	0.921	0.938	0.828	0.842	0.869	0.737	0.761	0.807
0	1.000	1.000	1.000	1.000	1.000	1.000	1.000	1.000	1.000
+15	1.030	1.051	1.072	1.131	1.151	1.187	1.275	1.317	1.397
+30	1.043	1.092	1.196	1.255	1.358	1.564	1.734	2.112	2.811
+45	1.027	1.169	1.494	1.511	1.964	2.976	—	—	—
+60	1.078	1.536	2.871	—	—	—	—	—	—

pole densities, whereas numbers smaller than unity denote smaller than average densities. Jones[12] has pointed out that this approach is subject to errors arising from possible differences in the degree of crystallinity between the oriented and random samples. A scaling factor can be determined by measuring the integrated intensity of crystalline scattering from the oriented and random specimens, $\int I_{or}$ and $\int I_{ra}$, from which the factor to be applied to the intensities of the random specimen is $\int I_{or}/\int I_{ra}$.

The method of normalization to a random sample still poses the problem that it is not easy to procure truly random specimens with the same thickness and absorption coefficient as the oriented sample. Wilchinsky[13] in a study of preferred orientation in cold-rolled polypropene made use of a reference specimen consisting of tightly packed, fine polypropene powder. Lindenmeyer and Lustig[14] in an investigation of extruded polyethylene found it difficult to eliminate preferred orientation entirely but were able to prepare samples with minimum orientation in the form of lightly pressed briquettes of finely precipitated polymer. Higher degrees of compression introduced distinct orientation. Desper and Stein[5] prepared a randomly oriented sample by casting powdered polyethylene in epoxy resin.

4-3.2   Reflection Technique, Sheet Specimen

The reflection technique with counter recording was first described by Schulz[15]. In the reflection geometry (Figure 4-16) the α-axis

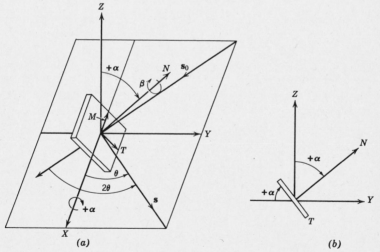

Figure 4-16   Geometry of the reflection technique for pole-figure measurements.

coincides with the Cartesian axis X. When $\alpha = 0°$ the plane of the specimen lies in the equatorial plane XY and its normal N coincides with Z. Ordinarily the machine direction is set parallel to X. From the figure it is evident that the most favorable geometry for the reflection method prevails in the general neighborhood of $\alpha = 90°$ (N coincident with Y).

If the outer region of the pole figure has been constructed from intensities measured by the transmission technique over the latitude range $\alpha = 0$ to $60°$, the remaining central zone can be filled in by using reflection intensities. The most convenient experimental procedure is to set the longitude β to successive fixed values, scanning at each β by means of the α-motion along the given longitude line between $\alpha = 30$ and $90°$. Intensities measured in the region of overlap $\alpha = 30$ to $60°$ are then available for normalizing the transmission and reflection intensities to a common basis.

In the reflection arrangement the path of the x-rays through the specimen is a minimum for $\alpha = 90°$ and increases as α diminishes. When $\alpha = 90°$ the criterion for minimum thickness to yield maximum diffracted intensity is given by (2-10),

$$t \geq \frac{3.2}{\mu} \frac{\rho}{\rho'} \sin \theta. \tag{2-10}$$

For "light" nonporous polymers with μ in the range of 5 to 8 cm^{-1} this equation prescribes a minimum thickness of the order of 1.5 to 2.0 mm for a representative Bragg angle of $18°$, although for $\alpha \ll 90°$ the thickness can be reduced. As in the transmission technique, it is preferable in the interest of preserving good resolution to restrict t to about 1 mm, even at the expense of some loss of intensity for α near $90°$.

When α is less than $90°$ defocusing occurs because of the appreciable height (Z-dimension) of the incident x-ray beam, which results in a broadening of the diffracted beam at the receiver. Hence both the specimen thickness and the axial divergence of the incident beam must be limited to preserve good resolution. As in the transmission case, the observed reflection profile can be expressed as a convolution,

$$\gamma = \widehat{\gamma_0 \gamma_\lambda \gamma_x \gamma_t \gamma_\alpha \gamma_\epsilon \gamma_s}. \tag{4-10}$$

In this equation all factors have the meanings given in (4-1) except for the additional factor γ_α, which represents the added broadening that is caused by tilting the specimens to angles other than $\alpha = 90°$. As in the transmission case, $(\omega_\lambda)_{1/2}$ may be neglected, and in addition

$(\omega_\epsilon)_{1/2}$ is vanishingly small because of the basic parafocusing geometry. In a manner parallel to that followed in the calculation of $(\omega_{0xt\epsilon})_{1/2}$ and ω_s for the transmission method, we now proceed to calculate $(\omega_{0xta})_{1/2}$ and ω_s for representative experimental conditions pertinent to the reflection technique.

The profiles γ_0 and γ_x that were used in the transmission analysis are immediately applicable again, with the one exception that a narrow (line) source cannot be employed in the reflection method since it leads to prohibitively great broadening of the diffracted beam for values of α that deviate much from $90°$. Accordingly we shall assign to the source only the width $(\omega_x)_{1/2} = 0.35°$. By reference to (2-11) it is seen that to a first approximation ω_t can be expressed as

$$\omega_t = 114.6 \frac{t}{d} \cos\theta \quad \text{degrees.} \qquad (4\text{-}11)$$

To a good approximation ω_α is given by

$$\omega_\alpha = \omega_\epsilon' \cos\theta \cot\alpha, \qquad (4\text{-}12)$$

in which ω_ϵ' is defined as in (4-8). Proceeding in the same way as for transmission geometry [see (4-4)], we postulate that

$$(\omega_x)_{1/2} = \tfrac{1}{2}\omega_x, \qquad (\omega_t)_{1/2} = \tfrac{1}{2}\omega_t, \qquad (\omega_\alpha)_{1/2} = \tfrac{1}{2}\omega_\alpha, \qquad (4\text{-}13)$$

and we again specify that ω_s shall not exceed $\tfrac{1}{2}\omega_{1/2}$ in order to limit the broadening to 10%.

Figures 4-17a and b give two families of curves portraying ω_s versus α for $(\omega_0)_{1/2} = 0$ and $0.50°$. Each family of curves consists of one curve for $\theta = 15°$, $t = 1\,mm$, $\omega_\epsilon' = 0.25°$, and three curves corresponding to separate increases in each of these parameters to $\theta = 30°$, $t = 2\,mm$, $\omega_\epsilon' = 0.50°$. Two parameters have constant values: $(\omega_x)_{1/2} = 0.35°$, $d = 145\,mm$. Comparison of Figures 4-15 and 4-17 shows that when a broad source (spot focus) is employed, the optimal receiving-aperture width is of the same order of size for both the transmission and reflection techniques. Thus when the pure diffraction profile is very narrow $[(\omega_0)_{1/2} \to 0°]$ the indicated width is between 0.2 and 0.3°, whereas for $(\omega_0)_{1/2} = 0.50°$ it falls between 0.3 and 0.4°. It will also be noted that the receiving-aperture width exhibits a stronger dependence on sample thickness for the reflection technique than for the transmission technique. As in the transmission geometry, (4-8) can be used to estimate the minimum axial (Z) dimension of the receiving aperture for the reflection method, and likewise the height (but not the width) of the aperture may be increased to enhance the intensity without an undue sacrifice of resolution.

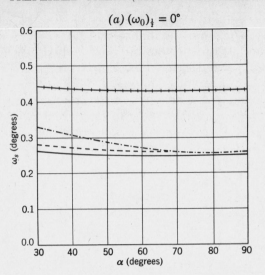

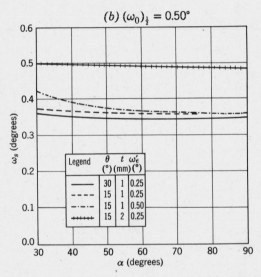

Figure 4-17 Maximum receiving-aperture width (ω_s) as a function of Bragg angle θ, specimen thickness t, and axial beam divergence ω'_ϵ. Reflection geometry, goniometer radius 145 mm, wide x-ray source: $(\omega_x)_{1/2} = 0.35°$, $\omega_0 =$ width of pure diffraction profile.

Schulz[15] has shown that the following correction formula should be applied to intensities measured at values of α other than 90°:

$$\frac{I_{90°}}{I_\alpha} = \frac{1 - \exp\left(-2\mu t/\sin\theta\right)}{1 - \exp\left(-2\mu t/\sin\theta \sin\alpha\right)} \quad \text{(reflection technique).} \quad (4\text{-}14)$$

Numerical values calculated for typical experimental conditions with this equation are given in Table 4-3. It will be noted that for small values of θ and larger values of μt the correction may be neglected except when exceptionally accurate results are required. As in the transmission geometry, the general practice is to avoid making such corrections by normalizing the measured intensities to the corresponding intensities scattered by a "random" specimen. (See p. 222.)

4-3.3 Other Techniques

Heffelfinger and Burton[4] in their studies of poly(ethylene terephthalate) made use of a General Electric single-crystal orienter (a three-circle goniometer), the **X**- and **Φ**-circles being fitted with synchronous motors to permit the intensities to be registered with a potentiometer strip-chart recorder. The orientation of the M, T, and N axes corresponded to the transmission geometry (Figure 4-13), so that the **Φ**- and **X**-circles of the instrument furnished the α- and β-motions, respectively. In order to eliminate the need for absorption corrections [see again (4-9) and Table 4-2] Heffelfinger and Burton prepared laminated specimens in the form of small $\frac{1}{16} \times \frac{1}{16} \times \frac{1}{4}$-in. rectangular prisms. To test the efficacy of this approach they measured the intensities that were diffracted by a specimen of unoriented, well-crystallized poly(ethylene terephthalate) over a range of 90° in **Φ** and **X**. As shown in Figure 4-18, the intensity of the (100) reflection was very constant as a function of **Φ** but showed a small decrease as **X** was varied from 0 to 90°. Laminated specimens of the required thickness were built up as follows. A number of $\frac{1}{4} \times 1$-in. strips of

Table 4-3 Correction Factors To Be Applied to Intensities Measured by Reflection Technique; Dependence on μt, α, and θ

α (°)	$\theta = 15.0°$ μt 0.25	0.50	1.00	$\theta = 30.0°$ μt 0.25	0.50	1.00	$\theta = 45.0°$ μt 0.25	0.50	1.00
90	1.000	1.000	1.000	1.000	1.000	1.000	1.000	1.000	1.000
80	0.995	0.999	1.000	0.991	0.995	0.999	0.990	0.994	0.997
70	0.981	0.995	1.000	0.965	0.982	0.996	0.958	0.973	0.990
60	0.958	0.990	1.000	0.923	0.960	0.992	0.909	0.941	0.978
50	0.930	0.985	1.000	0.867	0.933	0.987	0.841	0.898	0.965
40	0.895	0.981	1.000	0.801	0.905	0.984	0.760	0.851	0.953
30	0.873	0.979	1.000	0.731	0.881	0.982	0.670	0.804	0.944

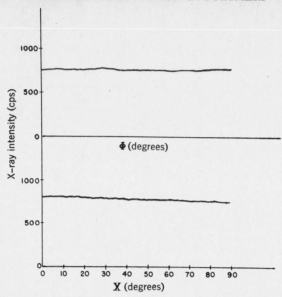

Figure 4-18 Intensity versus single-crystal-orienter coordinates Φ and X diffracted by a small rectangular prism of unoriented poly(ethylene terephthalate). (100) planes; Φ and X equivalent to pole-figure coordinates α and β, respectively. (Heffelfinger and Burton[4].)

polymer film were cut and stacked to a depth of about $\frac{1}{16}$ in. in a vise-type clamp with a $\frac{1}{4}$-in. depression. The clamp was then closed and the laminations trimmed to the thickness of the clamp with a razor blade, after which Duco cement was applied to the trimmed edges, binding the laminations together. After drying, the $\frac{1}{16} \times \frac{1}{16} \times \frac{1}{4}$-in. rectangular prismatic specimen could be removed and mounted for x-ray analysis. The device for compressing and holding laminated specimens described in Chapter 2(see Figure 2-5) is also useful for pole-figure measurements.

For studying axial textures in cylindrical specimens Cullity and Freda[16] introduced a transmission technique that encompasses the entire range of α and β. Although it is not readily applicable to most polymer specimens, the method has been adapted in a special way by Wilchinsky[17] to the measurement of axial orientation in polypropene films. He prepared a rod-shaped specimen by rolling the polymer film into a tight cylindrical rod about 1 mm in diameter, with the fiber axis parallel to the cylinder axis. Obviously this method is capable of revealing only an average axial orientation component of more complex textures. For further details the reader may consult the references cited.

Whatever the geometrical arrangement that is employed for scanning the diffracted intensity as a function of α and β, the usual procedure for registering the intensities is to channel the counter output into a pulse-integrating circuit, which in turn actuates a strip-chart recorder. (See also Section 2-4.1.) By relating the chart speed to the angular scanning speed it is easy to determine the pole coordinates α, β corresponding to any point on the recorder chart. Alternatively the investigator may choose to record the counts manually point by point over the desired range of α and β, which is somewhat more accurate but much more tedious. In either case, however, he will need to scan the 2θ profile of the reflection and to determine the magnitude of the background correction.

4-3.4 Special Instrumentation

Two manufacturers offer pole-figure (texture) goniometers with counter recording.† Figure 4-19 shows such a commercial goniometer with a transmission specimen holder (1) in place. For the transmission technique the large vertical circle (2) that bears the specimen holder (1) rotates on ball bearings about a horizontal axis within an outer, stationary, toothed ring (3), providing the β-rotation about the specimen normal N. (See again Figure 4-13.) To make measurements at higher latitudes the entire vertical-circle assembly is turned about its vertical (Z) axis to the desired α-setting and the β-scan is performed as before. A synchronous motor drive and appropriate gear trains permit either the α- or β-circles to be turned individually.

For the reflection technique the vertical circle provides the α-motion; and the small horizontal azimuth circle (4), the β-motion. Figure 4-20 is a view of the goniometer with the specimen (1') in the reflecting position at $\alpha = 90°$. Also visible in the figure are the direct- and diffracted-beam collimating systems, (5) and (6), respectively, and the drive motor (7). By turning the α- and β-circles simultaneously it is possible to scan the pole figure along a spiral path, as indicated in Figure 4-21 for the reflection technique. This method of scanning was first used by Holden[18] in the study of metal textures. According to the precision of counting or degree of resolution desired, the Siemens goniometer can be adjusted so as to provide different scanning speeds or spirals of different pitches, respectively; for example, with a spiral pitch of 5° (5° increase in

†General Electric Company, Analytical Measurement Business Section, West Lynn, Massachusetts 01905. Siemens Aktiengesellschaft, D-75 Karlsruhe 21, Postfach 21 1080, Germany. (Available in the United States through Eastern Scientific Sales Company, 43 East Main Street, Marlton, New Jersey 08053.)

Figure 4-19 Siemens pole-figure goniometer with transmission-specimen holder. 1—transmission-specimen holder; 2—vertical (β) circle; 3—toothed outer rim that houses vertical circle; 4—azimuthal circle; 5—direct-beam collimator; 6—diffracted-beam collimator; 7—drive motor. (Courtesy of Siemens Aktiengesellschaft.)

α per 360° rotation of β) the speed of scan can be selected so as to map the range of α from 0 to 75° in either 90 or 180 min; or when a more detailed pole figure is needed, a pitch of 2.5° can be employed and the 0 to 75° scan completed in 180 min. Figure 4-22 shows the recorder trace of a cold-rolled polyethylene film scanned over the α-range 15 to 90° with a pitch of 5°. It is seen that fifteen 360° β-cycles are traversed within this α-range. This instrument provides rectangular apertures for both the primary and diffracted beams that are continuously adjustable from 0 to 10 mm both horizontally and vertically. Chirer[19] has invented a "texturograph" for use with the Siemens texture goniometer that registers the intensities directly on a polar chart, thus eliminating the rather tedious operation of plotting the intensities from the strip chart on the spiral stereographic blank.

Geisler[20] devised a turntable-type recorder that can be directly coupled to a pole-figure goniometer, thereby automatically producing the pole figure. The apparatus synchronizes the α- and β-coordinates, respectively, of the specimen orientation with the α-coordinate of a

Figure 4-20 Siemens pole-figure goniometer with reflection-specimen holder. 1′—reflection-specimen holder in $\alpha = 90°$ position; 2—vertical circle that provides the α-motion; 5—direct-beam collimator; 6—diffracted-beam collimator; 7—drive motor. (Courtesy of Siemens Aktiengesellschaft.)

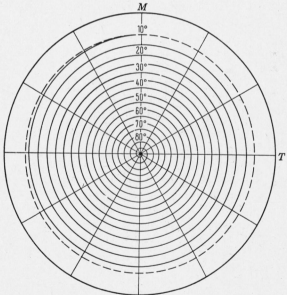

Figure 4-21 Spiral pole-figure pattern corresponding to a simultaneous scan of α and β. Five-degree pitch; M = machine (draw) direction; T = transverse direction.

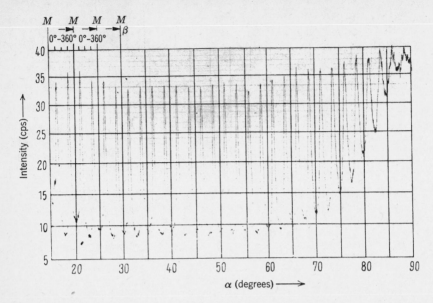

Figure 4-22   Recorder trace of cold-rolled polyethylene film. Reflection technique, spiral scan with 5° pitch, 15 β-cycles in α-range 15 to 90°. Machine direction (M) corresponds to $\beta = 0$°. [H. Neff, *Siemens Z.*, **31**, 23 (1957).]

printer arm and the angle of rotation, β, of the turntable, which bears a pole-figure chart. In an improved version marketed by the General Electric Company (Figure 4-23) the counter output of the pole-figure goniometer is fed to an X-Y recorder, which prints out the pole figure. The X-motion is provided by the printer head; and the Y-motion, by the chart paper. The rotation angle β of the specimen is resolved into $\sin^2 \beta$ and $\cos^2 \beta$ signals, which respectively actuate the X- and Y-motions so as to produce one circular traverse of the printing head for each 360° rotation of the specimen. The amplitudes of the sine and cosine signals are related to the latitude setting of the specimen in such a way that the diameter of a given β-circle on the chart corresponds to the colatitude ϕ of the specimen.

In the General Electric unit the numerical printer is of the multipoint type with inks of 10 different colors, making it possible to register 10 levels of diffracted intensity. An aluminum cam cut to the proper shape for the x-ray absorption of the particular specimen in question modulates the voltage applied to the printer so as to effect the absorption corrections required for the transmission or reflection technique. [See (4-9) and (4-14).] Thus the colored numbers that

Figure 4-23 Automatic XY pole-figure recorder (A) coupled to a diffractometer with pole-figure goniometer (B). Also visible are counter (C), x-ray tube (D), cam for absorption compensation (E), mode of operation control (F), and counting and scaling units (G). (Courtesy of General Electric Company, Analytical Measurement Business Section, and I. Cohen, Westinghouse Bettis Atomic Power Laboratory.)

are printed on the pole-figure chart are, except for background corrections, the pole densities on an arbitrary scale.

Four-circle single-crystal diffractometers[21] are well suited to the measurement of intensity data for pole figures, the only instrumental modifications required being the construction of beam collimators of sufficiently large aperture and suitable transmission and reflection specimen holders. The X- and ω-circles provide the β- and α-motions, respectively, for the transmission technique, whereas the X- and Φ-circles provide the α- and β-motions, respectively, for the reflection technique. For transmission measurements it is highly desirable that the X-circle be of the 360° (full-circle) rather than 90° type. Even more valuable for pole-figure measurements are single-crystal diffractometers under automatic control by means of computers, punched cards, or paper tape. Such an apparatus can be programmed to collect pole-figure intensities according to any prearranged scheme, and the output data are punched directly on cards or tape for subsequent computer processing[12, 22]. Figure 4-24 shows a four-circle single-crystal diffractometer with a crystal mono-chromator in the direct beam.†

4-3.5 Illustrative Preparation of a Pole Figure

We shall describe the steps in the preparation of a pole figure beginning with the measurement of the intensities. For this purpose we shall make use of intensity data from the (040) planes of poly-propene obtained by Wilchinsky[17].‡

The specimen was prepared by parallel stacking of layers of oriented polymer film to form a pad approximately 1 mm thick. Nickel-filtered CuKα x-rays generated at 45 kV (peak) and 20 mA

†Picker Instruments, 1020 London Road, Cleveland, Ohio 44110.
‡Experimental data supplied by Z. W. Wilchinsky in a private communication. The specimen is polypropene film D of reference [17].

Figure 4-24 Four-circle single-crystal diffractometer. 1 — Goniometer head for mounting specimen; 2 — telemicroscope; 3 — x-ray tube; 4 — crystal-monochromator housing; 5 — 360° X-circle; 6 — Φ-circle; 7 — direct-beam collimator; 8 — diffracted-beam collimator; 9 — scintillation counter. (Courtesy of Picker Corporation.)

were employed, and the intensities were measured with a Norelco Geiger-counter diffractometer equipped with a manual pole-figure device. A line source obtained by viewing the target at an angle of 3° was used for both the transmission and reflection measurements, the height of the beam being reduced for the reflection geometry. Table 4-4 gives the orientation of the reference axes in the specimen and the corresponding values of the coordinates α, ϕ, and β.

The intensities were registered on a strip-chart recorder at 10° intervals over the angular ranges $0° \leqslant \alpha \leqslant 90°$ and $0° \leqslant \beta \leqslant 360°$. The intensity of the background as a function of α was determined by means of 2θ scans at successive fixed values of α, and the intensities above background were then read directly from the chart record, giving the quantities I_{exp} in columns 2 and 5 of Table 4-5. This table illustrates the workup of the raw intensity data at $\alpha = -40°$ including the calculation of a factor to scale the transmission intensities to the level of the reflection intensities. Intensity data were taken by reflection at $|\alpha| = 40, 50, 60, 70, 80$, and 90° and by transmission at $|\alpha| = 0, 10, 20, 30$, and 40°, the intensities being measured by both techniques at $|\alpha| = 40°$ to permit determination of the scaling factor.

The absorption of the specimen for $CuK\alpha$ x-rays was measured experimentally as explained on p. 222 (see also Section 1-3.1), yielding $\mu t = 0.242$. Absorption corrections were calculated as a function of θ and α, using (4-9) for transmission and (4-14) for reflection. For the (040) reflection of polypropene $\theta = 8.44°$, which results in the following correction factors at $\alpha = -40°$:

$$\left(\frac{I_0}{I_\alpha}\right)_T = 0.7124; \qquad \left(\frac{I_{90°}}{I_\alpha}\right)_R = 0.9688. \qquad (4\text{-}15)$$

Table 4-4 Axes of Reference in Polypropene Specimen and Corresponding Values of the Co-ordinates α, ϕ, and β[a]

Reference Axis	Direction in Specimen	α (degrees)	ϕ (degrees)	β (degrees)
X	T	0	90	90, 270
Y	M	0	90	0, 180
Z	N	90	0	—

[a]Analysis of Wilchinsky[17]. Experimental data supplied in a private communication.
Symbols: α — latitude, ϕ — colatitude, β — longitude; T — transverse, M — machine, N — normal.

Table 4-5   Processing of Experimental Data for Polypropene[a,b]

(1)	Transmission			Reflection		(7)	(8)
	(2)	(3)	(4)	(5)	(6)		
		$I'_T =$	$I_T =$		$I_R =$	$I(\phi, \beta) =$	
β (deg.)	I_{exp}	$2.38\,I_{exp}$	$1.387\,I'_T$	I_{exp}	$3.22\,I_{exp}$	$\frac{1}{2}(I_T + I_R)$	$0.207\,I(\phi, \beta)$
0	2.50	5.53	7.67	2.85	8.77	8.22	1.7
10	1.65	3.92	5.44	2.25	7.24	6.34	1.3
20	0.75	1.78	2.47	1.10	3.54	3.01	0.6
30	0.50	1.19	1.65	0.60	1.93	1.79	0.4
40	0.25	0.59	0.82	0.35	1.12	0.97	0.2
50	0.25	0.59	0.82	0.20	0.64	0.73	0.2
60	0	0	0	0.05	0.16	0.08	0
70	0.38	0.90	1.25	0.03	0.10	0.68	0.1
80	0.25	0.59	0.82	0	0	0.41	0.1
90	0.25	0.59	0.82	0.06	0.19	0.51	0.1
100	0.25	0.59	0.82	0.16	0.51	0.67	0.1
110	0.25	0.59	0.82	0.09	0.29	0.56	0.1
120	0.50	1.19	1.65	0	0	0.83	0.2
130	0.25	0.59	0.82	0.06	0.19	0.51	0.1
140	0.25	0.59	0.82	0.10	0.32	0.57	0.1
150	0.25	0.59	0.82	0.16	0.51	0.67	0.1
160	0.50	1.19	1.65	0.20	0.64	1.15	0.2
170	1.25	2.97	4.12	1.00	3.21	3.67	0.8
180	2.18	5.18	7.18	2.30	7.40	7.29	1.5
190	1.90	4.52	6.27	2.95	9.49	7.88	1.6
200	0.85	2.02	2.80	1.60	5.15	3.98	0.8
210	0.15	0.36	0.50	0.70	2.25	1.38	0.3
220	0	0	0	0.30	0.96	0.48	0.1
230	0	0	0	0.20	0.64	0.32	0.1
240	0.40	0.95	1.32	0.15	0.48	0.90	0.2
250	0	0	0	0.13	0.42	0.21	0
260	0.30	0.71	0.98	0.10	0.32	0.65	0.1
270	0.25	0.59	0.82	0.06	0.19	0.51	0.1
280	0	0	0	0.05	0.16	0.08	0
290	0	0	0	0.06	0.19	0.10	0
300	0.25	0.59	0.82	0.25	0.80	0.81	0.2
310	0.25	0.59	0.82	0.06	0.19	0.51	0.1
320	0.10	0.24	0.33	0	0	0.17	0
330	0.35	0.83	1.15	0	0	0.58	0.1
340	0.65	1.54	2.14	0.40	1.28	1.71	0.4
350	1.30	3.09	4.28	1.05	3.38	3.83	0.8

[a]Data from [17] and private communication by Z. W. Wilchinsky. Polypropene specimen D.

[b]Experimental conditions: (040) planes, $\theta = 8.44°$, $\alpha = -40°$.

The conversion of the experimental intensities (columns 2 and 5 of Table 4-5) to corrected intensities (columns 3 and 6) was accomplished by using the factors

$$\frac{S_T}{I_d}\left(\frac{I_0}{I_\alpha}\right)_T \quad \text{and} \quad \frac{S_R}{I_d}\left(\frac{I_{90°}}{I_\alpha}\right)_R, \tag{4-16}$$

S being an adjustable instrumental factor (2, 4, 8, or 16) that determines the counting rate corresponding to full-scale deflection of the recorder pen, and I_d being a measure of the direct-beam intensity that is obtained by counting a diffracted ray from a standard sample. For the transmission measurements $S_T = 2$ (corresponding to approximately 100 counts per second full-scale deflection) and $I_d = 59.95$ in arbitrary units, whereas for reflection $S_R = 2$ and $I_d = 60.2$, giving the following numerical factors for the two geometries:

$$I_T' = 2.38 I_{\text{exp}} \quad \text{(transmission)}$$

and

$$I_R = 3.22 I_{\text{exp}} \quad \text{(reflection)}. \tag{4-17}$$

For computational convenience these factors have been multiplied by 100.

Application of (4-17) to the data in columns 2 and 5 of Table 4-5 gave the numerical results shown in columns 3 and 6. A factor for scaling the transmission intensities was derived from the quantities in columns 3 and 6 as follows:

$$\frac{\sum_\beta I_R}{\sum_\beta I_T'} = \frac{62.66}{45.19} = 1.387. \tag{4-18}$$

Multiplication of the intensities I_T' in column 3 by this factor gave the scaled values I_T in column 4.

A further step preliminary to the construction of the (040) pole figure was the determination of the mean (040) intensity averaged over the entire orientation sphere. In terms of the corrected intensities $I(\phi, \beta)$ this may be expressed as follows:

$$\langle I \rangle = \frac{\displaystyle\int_0^{\pi/2} \int_0^{2\pi} I(\phi, \beta) \sin \phi \, d\beta \, d\phi}{\displaystyle\int_0^{\pi/2} \int_0^{2\pi} \sin \phi \, d\beta \, d\phi}. \tag{4-19}$$

The reciprocal of $\langle I \rangle$ serves as a normalizing factor to convert all intensities $I(\phi, \beta)$ to relative intensities, which are identical with relative pole densities at the same ϕ, β coordinates. Table 4-6 outlines the numerical calculation of this factor from the (040) intensities. The integrals of (4-19) have been approximated by summations at 2° intervals, values of $I(\phi)$ at angles intermediate between 0, 10, 20°, etc., being read from a plot of $I(\phi)$ versus ϕ. For economy of space the numerical data have been reproduced in full only for $0° \geq \phi \geq 10°$. When the normalization factor,

$$\frac{1}{\langle I \rangle} = 0.207, \tag{4-20}$$

is applied to the average corrected intensities $I(\phi, \beta)$ of column 7 in Table 4-5, the relative pole densities in column 8 are obtained.

In the same manner the I_{exp} intensities that were measured by transmission for $|\alpha| = 0$, 10, 20, and 30° and by reflection for $|\alpha| =$

Table 4-6   Calculation of Factor To Relate Individual Pole Densities to Average Densities (Polypropene, (040) planes)[a]

α (degrees)	ϕ (degrees)	$I(\phi) = \sum\limits_{\beta} I(\phi, \beta)$	$\sin \phi$	$I(\phi) \sin \phi$
90	0	4428.0	0.0000	000.0
88	2	4000.0	0.0348	139.2
86	4	3600.0	0.0697	250.9
84	6	3175.0	0.1045	331.8
82	8	2750.0	0.1391	382.5
80	10	2331.0	0.1736	404.7
70	20	390.0	0.3420	133.4
60	30	137.6	0.5000	68.8
50	40	77.0	0.6428	49.5
40	50	62.7	0.7660	48.0
30	60	66.5	0.8660	57.6
20	70	72.5	0.9397	68.1
10	80	68.0	0.9848	67.0
0	90	53.5	1.0000	53.5
			$\sum = 29.0917$	5060.5

$$\langle I \rangle = \frac{\sum\limits_{\phi=0°}^{90°} \sum\limits_{\beta=0°}^{360°} I(\phi, \beta) \sin \phi \, \Delta\beta \, \Delta\phi}{\sum\limits_{\phi=0°}^{90°} \sum\limits_{\beta=0°}^{360°} \sin \phi \, \Delta\beta \, \Delta\phi} = 4.83$$

$$\frac{1}{\langle I \rangle} = 0.207$$

[a]Experimental data supplied by Wilchinsky in private communication. Polypropene specimen D of reference [17].

50, 60, 70, 80, and 90° were converted to $I(\phi, \beta)$ values on the same scale as the data of column 7 in Table 4-5. These intensities plotted as a function of β over the range $0° \leq \beta \leq 360°$ are shown by the 10 curves of Figure 4-25. These $I(\phi, \beta)$ values were then scaled by the

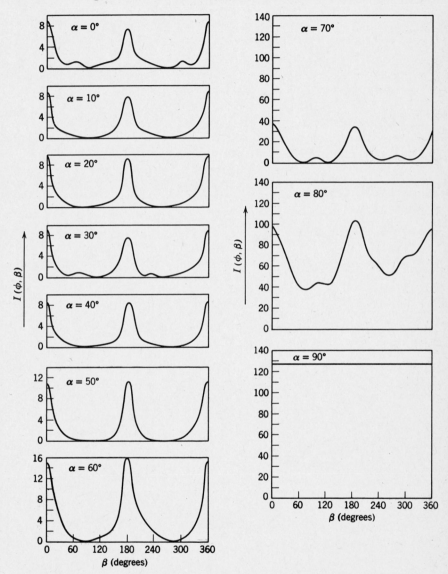

Figure 4-25   Plots of $I(\phi, \beta)$ versus β for $|\alpha| = 0, 10, 20, 30, \ldots, 90°$ for (040) reflection of oriented polypropene film. (After Wilchinsky [17].)

factor 0.207 to supply the relative pole densities required to construct the (040) pole figure, Figure 4-26. Direction N is normal to the plane of the diagram, which coincides with MT, M being vertical and T horizontal. It will be seen that the (040) pole density is well below average (unity) over most of the pole figure, especially in the transverse regions. A ridge of density somewhat less than 2 units in height extends in the machine direction, and a very high concentration of poles exists in the immediate neighborhood of the normal to the polymer film. This means that the great majority of the polymer crystallites are oriented with their (040) planes nearly parallel to the plane of the polymer film. To a fair approximation the pole figure possesses horizontal and vertical lines of symmetry. The apparent deviations from this symmetry may probably be attributed to experimental error.

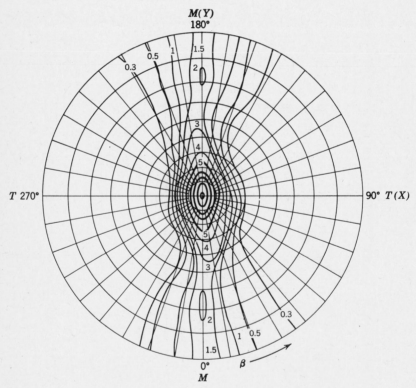

Figure 4-26 The (040) pole figure of oriented polypropene film derived from the intensity data plotted in Figure 4-25. Thick contours near center indicate relative densities of 5, 10, 15, 20 and 25. (Experimental data supplied by Wilchinsky.)

4-4 ANALYTICAL DESCRIPTION OF PREFERRED ORIENTATION

4-4.1 Axial Orientation

The concept of an orientation function to specify quantitatively the degree of axial orientation in crystalline fibers was originated by P. H. Hermans and co-workers[23]. Such a function takes the form

$$f_\phi = \tfrac{1}{2}(3 \langle \cos^2 \phi \rangle - 1) = 1 - \tfrac{3}{2}\langle \sin^2 \phi \rangle, \tag{4-21}$$

where $\langle \cos^2 \phi \rangle$ and $\langle \sin^2 \phi \rangle$ represent respectively the mean-square cosine and sine (averaged over all the crystallites) of the angle between a given crystal axis and the fiber axis, which serves as the reference direction. The nature of this function implies that other crystallographic axes as well as the one specified are arranged with cylindrical symmetry about the reference direction.

Stein[24] extended Hermans' treatment for axial orientation by setting up three equations like (4-21) to specify the degree of orientation of each of the crystal axes with respect to the reference direction Z:

$$f_{a,z} = \tfrac{1}{2}(3\langle \cos^2 \phi_{a,z} \rangle - 1),$$
$$f_{b,z} = \tfrac{1}{2}(3\langle \cos^2 \phi_{b,z} \rangle - 1), \tag{4-22}$$
$$f_{c,z} = \tfrac{1}{2}(3\langle \cos^2 \phi_{c,z} \rangle - 1).$$

The corresponding geometry is shown in Figure 4-27. For the discussion of axial orientation we may ignore the azimuthal angles σ_a, σ_b, and σ_c, which would display uniform distributions between 0 and 2π. As shown in Table 4-7, $\langle \cos^2 \phi \rangle$ for a given crystallographic axis assumes a value of 1 for perfect alignment with Z, $\tfrac{1}{3}$ for random orientations, and 0 for precise perpendicularity to Z. At the same time f assumes the respective values 1, 0, and $-\tfrac{1}{2}$.

If the crystallographic axes are orthogonal (refer to Table 1-1),

$$\cos^2 \phi_{a,z} + \cos^2 \phi_{b,z} + \cos^2 \phi_{c,z} = 1, \tag{4-23}$$

from which it follows that

$$f_{a,z} + f_{b,z} + f_{c,z} = 0. \tag{4-24}$$

This result shows that only two of the orientation functions are needed to specify the orientation, which may be represented as a point on a ternary diagram, as shown in Figure 4-28. The limitations on the ranges of the orientation functions, which are presented in Table 4-7, confine all points $(f_{c,z}, f_{a,z})$ within the right-triangular region outlined in the figure. The origin corresponds to the randomly

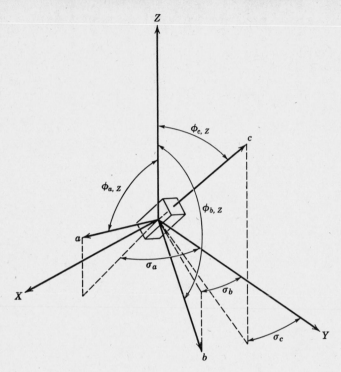

Figure 4-27 Stein's coordinate system for specifying orientation modes in orthorhombic polymers. (After Stein[24].)

oriented state, and each apex corresponds to complete alignment of
one crystal axis parallel to Z and each side of the triangle to one
crystal axis perpendicular to Z. It may also be noted that a straight
line proceeding from the origin to one of the apices corresponds to
the process of the given axis becoming preferentially oriented

Table 4-7 Values of $\langle \cos^2 \phi \rangle$ and f for
Three States of Orientation

	Orientation with Respect to Reference Direction Z		
Parameter	Parallel	Random	Perpendicular
$\langle \cos^2 \phi \rangle$	1	$\frac{1}{3}$	0
f	1	0	$-\frac{1}{2}$

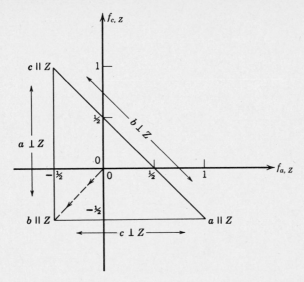

Figure 4-28   Representation of axial orientation by a point within a right triangle. (After Stein [24].)

parallel to Z while the other crystallographic axes become randomly oriented about this axis (axial orientation). Thus in Figure 4-28 the broken line denotes progressive orientation of b toward parallelism with Z. More generally, variations in preferred orientation that occur during mechanical or thermal treatment of a polymer may be specified in terms of migration of a point within the triangle [24].

Experimental Evaluation of $\langle cos^2 \phi \rangle$. Numerical values of the mean-square cosines of (4-22) can be calculated from the distribution of plane normals in an appropriate pole figure or directly from the fully corrected intensity distribution $I(\phi, \beta)$ from which the pole figure was derived. With reference to Figure 4-29, it will be seen that in the case of axial orientation about Z the total number of (hkl) plane normals oriented at a given latitude α or colatitude ϕ is proportional to the circumference of the circle of radius r, which is given by $\sin \phi$. Therefore in order to obtain $\langle \cos^2 \phi_{hkl,z} \rangle$ averaged over the entire surface of the orientation sphere it is necessary to weight $I_{hkl}(\phi, \beta)$ by $\sin \phi$. Thus [25, 26]

$$\langle \cos^2 \phi_{hkl,z} \rangle = \frac{\displaystyle\int_0^{\pi/2} I(\phi) \sin \phi \cos^2 \phi \, d\phi}{\displaystyle\int_0^{\pi/2} I(\phi) \sin \phi \, d\phi}, \tag{4-25}$$

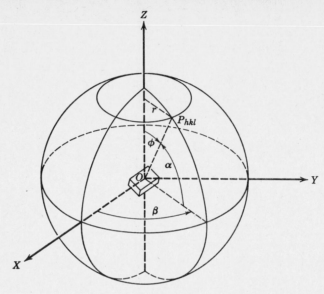

Figure 4-29

where

$$I(\phi) = \int\limits_{0}^{2\pi} I(\phi, \beta)\, d\beta. \tag{4-26}$$

In most oriented specimens the symmetry of the texture is such as to require integration of the diffracted intensity over only one octant; that is, $0 \le \phi \le \pi/2$ and $0 \le \beta \le \pi/2$.

The diffracted intensities from axially oriented specimens are independent of β, thus simplifying the intensity measurements and requiring only the integration (4-25) to be performed. For more complex modes of orientation the pole-figure intensities $I(\phi, \beta)$ that correspond to the particular planes (hkl) concerned must first be integrated over β according to (4-26) and then over ϕ by (4-25). The reader's attention is also directed again to the technique of rolling a polymer film into a tight cylindrical rod in order to measure a desired axial orientation component of more complex textures. (See Section 4-3.3.) We emphasize that the above procedure for the determination of particular $\langle \cos^2 \phi \rangle$ functions treats all systems as if they are axially oriented, regardless of the actual complexity of the texture. The process of integrating over the azimuthal coordinate β washes out the higher orientation information, leaving only the average orientation of the crystal axis relative to the specified reference direction [27].

If the crystallographic symmetry is such that there are no other directions in the polymer crystals that are equivalent to the particular direction under consideration [a plane normal (hkl) or one of the crystallographic axes a, b, c], such a direction is termed *unique*. Only when the direction is unique will the pertinent $\langle \cos^2 \phi \rangle$ and f functions assume the ideal values given in Table 4-7 for parallel and perpendicular orientation. However, since in polymers we are nearly always interested in unique crystallographic directions, the following development will avoid unnecessary complexity by treating only such directions. Table 4-8 lists the unique directions that are present in the various crystal systems. For a discussion of non-unique directions in crystals and their multiplicities the reader is referred to the literature [28].

Stein's model for axial orientation [24] as expressed in Figure 4-27 and (4-22), (4-23), and (4-24) was developed primarily for application to polyethylene, which is orthorhombic. Wilchinsky [26, 29] has modified the treatment so as to make it applicable to nonorthogonal crystal systems and shown how a desired $\langle \cos^2 \phi \rangle$ function can be determined even in the absence of reflecting planes normal to the chosen crystallographic direction. With reference to Figure 4-30: Z is the reference direction in the specimen; a, b, and c are the crystallographic axes; $N(hkl)$ is the normal to reflecting planes with intercepts m, n, p; U, V, c are Cartesian coordinate axes, of which c coincides with the crystallographic direction whose orientation is of interest. Except for this last restriction the crystal may be oriented in any arbitrary manner with respect to the axes U and V. Now let Z and N be unit vectors in the Z and N directions. Let e, f, and g be the direction cosines of N with respect to the axes U, V, and c, re-

Table 4-8   Unique Directions in the Six Crystal Systems[a]

Crystal System	Unique Directions
Triclinic	All directions
Monoclinic:	
$b \perp ab$ plane	a, b, c axes and all directions in ac plane
$c \perp ab$ plane	a, b, c axes and all directions in ab plane
Orthorhombic	a, b, c axes
Tetragonal	c axis
Hexagonal	c axis
Cubic	None

[a]Data from [26].

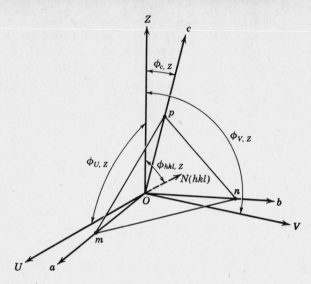

Figure 4-30 Wilchinsky's generalized model for specifying orientation modes in crystalline polymers. (After Wilchinsky [26, 29].)

spectively, and let $\mathbf{i}$, $\mathbf{j}$, and $\mathbf{k}$ be unit vectors along these axes. Then the vectors $\mathbf{Z}$ and $\mathbf{N}$ may be expressed as follows:

$$\mathbf{Z} = (\cos \phi_{U,z})\mathbf{i} + (\cos \phi_{V,z})\mathbf{j} + (\cos \phi_{c,z})\mathbf{k}, \qquad (4\text{-}27)$$

$$\mathbf{N} = e\mathbf{i} + f\mathbf{j} + g\mathbf{k}. \qquad (4\text{-}28)$$

Their scalar product is

$$\mathbf{N}\cdot\mathbf{Z} = \cos \phi_{hkl,z} = e \cos \phi_{U,z} + f \cos \phi_{V,z} + g \cos \phi_{c,z}, \qquad (4\text{-}29)$$

from which for a distribution of orientations the mean-square cosine of $\phi_{hkl,z}$ is given by the important equation

$$\begin{aligned}
\langle \cos^2 \phi_{hkl,z} \rangle = {}& e^2\langle \cos^2 \phi_{U,z} \rangle + f^2\langle \cos^2 \phi_{V,z} \rangle + g^2\langle \cos^2 \phi_{c,z} \rangle \\
& + 2ef\langle \cos \phi_{U,z} \cos \phi_{V,z} \rangle \\
& + 2fg\langle \cos \phi_{V,z} \cos \phi_{c,z} \rangle \\
& + 2eg\langle \cos \phi_{c,z} \cos \phi_{U,z} \rangle. \qquad (4\text{-}30)
\end{aligned}$$

The quantity that is of primary interest in (4-30) is $\langle \cos^2 \phi_{c,z} \rangle$, which expresses the degree of alignment of the crystallographic direction c (by convention parallel to the molecular chains) with the reference direction Z. The quantity $\langle \cos^2 \phi_{hkl,z} \rangle$ is determined from measurements of the intensity diffracted by the (hkl) planes following methods described earlier in this chapter, and e, f, and g for the same planes can be calculated from a knowledge of the crystal structure. The three squared cosine functions and three cosine products are the six unknowns that must be determined. In general they can be evaluated from measurements of $\langle \cos^2 \phi_{hkl,Z} \rangle$ for six sets of planes. Substitution in (4-30) of these six $\langle \cos^2 \phi_{hkl,Z} \rangle$ functions together with their respective sets of e, f, and g values provides six simultaneous equations that can be solved for the six unknowns. Actually these are reduced to five by the orthogonality relationship:

$$\langle \cos^2 \phi_{U,Z} \rangle + \langle \cos^2 \phi_{V,Z} \rangle + \langle \cos^2 \phi_{c,Z} \rangle = 1. \qquad (4\text{-}31)$$

Additional reductions in the number of unknowns result from the presence of crystallographic mirror planes and n-fold axes of symmetry. These simplifications are presented in Table 4-9. Their introduction into (4-30) reduces the number of independent (hkl) planes that are required to determine $\langle \cos^2 \phi_{c,Z} \rangle$ to the figures given in Table 4-10 for different classes of reflections and for the five crystal systems important in polymer structure. We shall illustrate the application of (4-30) with two analyses reported by Wilchinsky[17, 26].

Polypropene Film[17, 26]. Isotactic polypropene is monoclinic with $a = 6.65$, $b = 20.96$, and $c = 6.50$ Å; $\beta = 99.3°$ (b is the monoclinic axis)[30]. The molecular chains are parallel to c, with the result that $\langle \cos^2 \phi_{c,Z} \rangle$ is a direct measure of the molecular orientation relative to some reference direction Z in the specimen. Since there

Table 4-9 Simplifications Applicable to Equation 4-30 Resulting from Crystallographic Symmetry[a]

Symmetry Condition	Resultant Simplification
Monoclinic system: $\quad b \perp ac$ plane $\quad c \perp ab$ plane	$\langle \cos \phi_{U,Z} \cos \phi_{V,Z} \rangle = \langle \cos \phi_{V,Z} \cos \phi_{c,Z} \rangle = 0$ $\langle \cos \phi_{V,Z} \cos \phi_{c,Z} \rangle = \langle \cos \phi_{c,Z} \cos \phi_{U,Z} \rangle = 0$
Orthorhombic system	All cross-product averages are zero
Tetragonal and $\quad$ hexagonal systems	All cross-product averages are zero and $\qquad \langle \cos^2 \phi_{U,Z} \rangle = \langle \cos^2 \phi_{V,Z} \rangle$
$(hk0)$ planes	$g = 0$
$(00l)$ planes, $c \perp a$ $\quad$ and $c \perp b$	$e = f = 0, \quad g = 1$
Randomness about $\quad c$-axis	All cross-product averages are zero and $\qquad \langle \cos^2 \phi_{U,Z} \rangle = \langle \cos^2 \phi_{V,Z} \rangle$

[a]Data from [26].
Note: $\phi_{c,Z}$ is the angle between the c-axis and the reference direction Z.

Table 4-10   Number of Independent (hkl)
Planes Needed to Determine $\langle \cos^2 \phi_{c,z} \rangle$[a]

Crystal System	hkl	$hk0$	$h0l$	$00l$	Randomness about c-Axis
Triclinic	5	3	5	—	1
Monoclinic:					
$b \perp ac$	3	2	3	—	1
$c \perp ab$	3	3	2	1	1
Orthorhombic	2	2	2	1	1
Hexagonal	1	1	1	1	1
Tetragonal	1	1	1	1	1

[a]Data from [26].

are no strong reflections of the type $(00l)$, it was necessary to determine $\langle \cos^2 \phi_{c,z} \rangle$ indirectly by means of intensity measurements from strongly diffracting planes of another type. Three suitable $(hk0)$ reflections — (040), (130), and (110) — were available, of which only two were required (refer to Table 4-10, monoclinic symmetry with two-fold axis b perpendicular to the plane of ac). The planes (040) and (110) were selected for the evaluation of $\langle \cos^2 \phi_{c,z} \rangle$. By reference to Table 4-9 it will be seen that the following simplifications result for this case:

For $(hk0)$ planes: $g = 0$, and so $e^2 + f^2 = 1$.

For $b \perp ac$ plane: $\langle \cos \phi_{U,z} \cos \phi_{V,z} \rangle = 0$.

Substitution of these conditions into (4-30) gives

$$\langle \cos^2 \phi_{hk0,z} \rangle = e^2 \langle \cos^2 \phi_{U,z} \rangle + f^2 \langle \cos^2 \phi_{V,z} \rangle$$

$$= e^2 \langle \cos^2 \phi_{U,z} \rangle + (1 - e^2) \langle \cos^2 \phi_{V,z} \rangle. \qquad (4\text{-}32)$$

We may then write for two reflections

$$\langle \cos^2 \phi_{1,z} \rangle = e_1^2 \langle \cos^2 \phi_{U,z} \rangle + (1 - e_1^2) \langle \cos^2 \phi_{V,z} \rangle, \qquad (4\text{-}33)$$

$$\langle \cos^2 \phi_{2,z} \rangle = e_2^2 \langle \cos^2 \phi_{U,z} \rangle + (1 - e_2^2) \langle \cos^2 \phi_{V,z} \rangle. \qquad (4\text{-}34)$$

These two equations and the orthogonality relationship (4-31) may now be solved simultaneously for the desired root, $\langle \cos^2 \phi_{c,z} \rangle$, with the result

$$\langle \cos^2 \phi_{c,z} \rangle = 1 + \frac{(2e_2^2 - 1) \langle \cos^2 \phi_{1,z} \rangle - (2e_1^2 - 1) \langle \cos^2 \phi_{2,z} \rangle}{e_1^2 - e_2^2}. \qquad (4\text{-}35)$$

Let (110) and (040) be respectively reflections 1 and 2 of (4-35). With reference to Figure 4-30, for monoclinic symmetry with $b \perp ac$

the b crystallographic axis may be taken to be coincident with V. But the (040) plane normal coincides with b, and therefore also with V, so that its direction cosines are

$$e_2 = 0, \quad f_2 = 1, \quad g_2 = 0.$$

From the geometry of the unit cell it is found that the (110) plane normal makes an angle of 72.5° with b (and V) so that for these planes

$$e_1 = \sin 72.5° = 0.9537,$$

$$f_1 = \cos 72.5° = 0.3007,$$

$$g_1 = 0.$$

Substitution of these numerical values of e_1 and e_2 into (4-35) gives

$$\langle \cos^2 \phi_{c,z} \rangle = 1 - 1.099 \langle \cos^2 \phi_{110,z} \rangle - 0.901 \langle \cos^2 \phi_{040,z} \rangle. \quad (4\text{-}36)$$

Table 4-11 summarizes the calculation of $\langle \cos^2 \phi_{040,z} \rangle$ from the (040) intensities with (4-25). Equation 4-26 shows that the value of $I(\phi)$ for any given ϕ is given by the area under the curve of $I(\phi, \beta)$ versus β over the range 0 to 360°. The 10 pertinent plots for $\phi = 0, 10, 20, \cdots, 90°$ are given in Figure 4-25. In Table 4-11 the areas have been approximated by the summations

$$I(\phi) = \sum_{\beta=0°}^{360°} I(\phi, \beta),$$

taken at 5° intervals; for example, at $\phi = 50°$ (latitude $|\alpha| = 40°$) the sum of the $I(\phi, \beta)$ data (column 7 of Table 4-5) is 62.7. The next step is to determine the sums $\Sigma I(\phi) \sin \phi$ and $\Sigma I(\phi) \sin \phi \cos^2 \phi$. For this purpose the calculated values listed in columns 4 and 5 of Table 4-11 are plotted against ϕ as shown in Figure 4-31 and the numerical values of the two functions are read off at 2° intervals, thus permitting reasonable approximations to the integrals of (4-25) to be determined. The result is

$$\langle \cos^2 \phi_{040,z} \rangle = \frac{\sum\limits_{0°}^{90°} I(\phi) \sin \phi \cos^2 \phi}{\sum\limits_{0°}^{90°} I(\phi) \sin \phi} = \frac{3570}{5020} = 0.711.$$

A parallel treatment of the (110) intensities gives

$$\langle \cos^2 \phi_{110,z} \rangle = 0.245.$$

Substitution of these numerical values into (4-36) leads to the result

$$\langle \cos^2 \phi_{c,z} \rangle = 0.090.$$

Table 4-11   Calculation of $\langle \cos^2 \phi_{040,Z} \rangle$ for Polypropene[a]

$\lvert \alpha \rvert$ (degrees)	ϕ (degrees)	$I(\phi)$	$I(\phi) \sin \phi$	$I(\phi) \sin \phi \cos^2 \phi$
90	0		0	0
85	5	3658.0	318.8	316.4
80	10	2331.0	404.7	392.5
70	20	390.0	133.4	117.8
60	30	137.6	68.8	51.6
50	40	77.0	49.5	29.1
40	50	62.7	48.0	19.8
30	60	66.5	57.6	14.4
20	70	72.5	68.1	8.0
10	80	68.0	67.0	2.0
0	90	53.5	53.5	0

$$\sum_{0°}^{90°} I(\phi) \sin \phi = 5020,$$

$$\sum_{0°}^{90°} I(\phi) \sin \phi \cos^2 \phi = 3570,$$

$$\langle \cos^2 \phi_{040,Z} \rangle = \frac{3570}{5020} = 0.711.$$

[a]Experimental data supplied by Wilchinsky in private communication. Polypropene specimen D of reference [17].

Fiber of Poly-4-methylpentene-1 [26].   In this crystalline polymer the molecular chains are parallel to the c-axis of a tetragonal unit cell with the dimensions $a = 18.50$, $c = 13.76$ Å [31]. We wish to determine the orientation of the c crystallographic axis with respect to the fiber (Z) axis of the specimen. By reference to Table 4-9 we see that for the tetragonal system (4-30) simplifies to

$$\langle \cos^2 \phi_{hkl,z} \rangle = (e^2 + f^2) \langle \cos^2 \phi_{U,z} \rangle + g^2 \langle \cos^2 \phi_{c,z} \rangle$$
$$= (1 - g^2) \langle \cos^2 \phi_{U,z} \rangle + g^2 \langle \cos^2 \phi_{c,z} \rangle. \tag{4-37}$$

A second r̶ ̶ ̶ ̶ship is that of orthogonality, which may be written

$$1 = 2 \langle \cos^2 \phi_{U,z} \rangle + \langle \cos^2 \phi_{c,z} \rangle \tag{4-38}$$

$\langle \cos^2 \phi_{V,z} \rangle$ for the tetragonal system. These two ved for $\langle \cos^2 \phi_{c,z} \rangle$, with the result

$$\phi_{c,z} \rangle = \frac{1 - g^2 - 2 \langle \cos^2 \phi_{hkl,z} \rangle}{1 - 3g^2}. \tag{4-39}$$

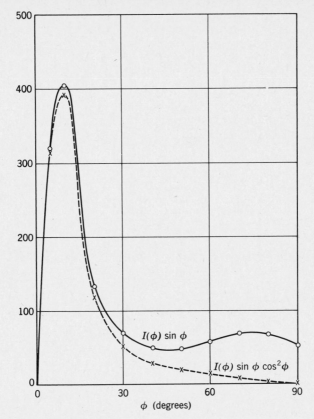

Figure 4-31 Plots of $I(\phi) \sin \phi$ and $I(\phi) \sin \phi \cos^2 \phi$ used in the calculation of $\langle \cos^2 \phi_{040,z} \rangle$ for polypropene (see Table 4-11). (Experimental data supplied by Wilchinsky in a private communication.)

Thus it is necessary to measure the diffracted intensities from only one set of planes, which is also stated in Table 4-10. Wilchinsky chose the strong reflection from the (200) planes, for which $g = 0$. (See again Table 4-9.) Thus (4-39) reduces further to

$$\langle \cos^2 \phi_{c,z} \rangle = 1 - 2\langle \cos^2 \phi_{200,z} \rangle, \tag{4-40}$$

which, then, expresses the orientation of the c-axis with respect to the fiber axis as a function of the orientation of the (200) poles. The specimen used for the I_{200} intensity measurements consisted of a sheet of parallel fibers. Because of the axial symmetry of the fibers it was not necessary to vary the azimuth β. From the $I(\phi)$ data plots of $I(\phi) \sin \phi$ and $I(\phi) \sin \phi \cos^2 \phi$ were prepared (Figure 4-32) and

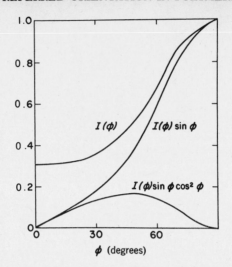

ϕ (degrees)

Figure 4-32   Plots of functions used in the calculation of $\langle \cos^2 \phi_{200,z} \rangle$. Fiber of poly-4-methylpentene-1. (Wilchinsky [26].)

the areas used to compute $\langle \cos^2 \phi_{200,z} \rangle$ with (4-25). The numerical result was 0.232, from which by (4-40)

$$\langle \cos^2 \phi_{c,z} \rangle = 0.536.$$

This may be compared with values of 1 for perfect alignment of c with Z and $\frac{1}{3}$ for completely random orientations of the c-axes. (See again Table 4-7.)

4-4.2   Higher Orientation Modes

No general agreement exists regarding the concepts of orientation modes that are more complex than simple axial nor of the most satisfactory terminology for describing them. With particular reference to optical birefringence Desper and Stein [5] designate as *biaxial* all orientation distributions that possess three mutually perpendicular mirror planes of symmetry (corresponding to *MN*, *MT*, and *NT* of Figure 4-9*b*), a broad category that includes all of Heffelfinger and Burton's modes 2, 3, 4, 5, and 6. An example of lower symmetry is furnished by polyethylene film unidirectionally recrystallized under the influence of a pronounced temperature gradient in one direction [5]. The (200), (020), and (110) pole figures from the work of Desper and Stein, shown in Figure 4-33, display only the symmetry plane *MN*. When the symmetry is lower than

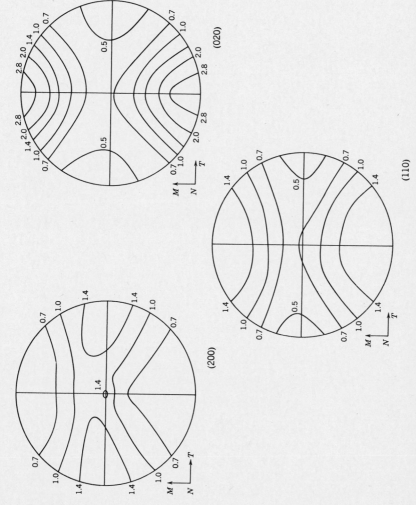

Figure 4-33 Pole figures of unidirectionally recrystallized polyethylene film showing only a single plane of symmetry (MN). M = machine direction, T = transverse direction, N = normal direction. (Desper and Stein[5].)

253

biaxial, as defined above, the equations used in birefringence calculations are no longer valid. (See Section 4-5.)

The term "biaxial" is also used in two more restrictive connotations. In industrial polymer technology it is sometimes used to describe the strains to which a specimen has been subjected. Thus a biaxially stretched film would be one that is stretched in both the machine and transverse directions. "Biaxial" is also occasionally used to indicate a texture in which there are two maxima in the polymer chain distribution, one in the machine and the other in the transverse direction [32].

From a consideration of the idealized pole figures of Figure 4-9 it will be realized that any kind of complex texture can be described by a number of properly chosen pole figures, commonly three; for instance, (100), (010), and (001) pole figures are usually sufficient, although certain textures may be more effectively portrayed by pole figures of planes with more than one index, such as (110), (101), or (111). It must be remembered also that the preparation of a pole figure corresponding to any particular indices (hkl) is contingent on the existence of an x-ray reflection of sufficient intensity from these planes; for example, the analysis of c-axis orientation in polypropene film described in the preceding section could not be carried out directly from a $(00l)$ pole figure because of the absence of $(00l)$ reflections of adequate intensity. In contrast to axial orientation, more complex textures are much more difficult to specify analytically by means of $\langle \cos^2 \phi \rangle$ or f orientation functions; nevertheless both Stein and Wilchinsky have developed methods for dealing with certain modes.

Stein has treated biaxial orientation in orthorhombic polymers [33, 34]. In addition to the uniaxial angles $\phi_{a,z}$, $\phi_{b,z}$, and $\phi_{c,z}$ of Figure 4-27 it is now necessary to consider the azimuthal angles σ_a, σ_b, and σ_c that specify the angles between the Y-axis and the projections of a, b, and c in the XY plane. Three new orientation functions can be defined as follows:

$$f_{\sigma_a} = 2\langle \cos^2 \sigma_a \rangle - 1,$$
$$f_{\sigma_b} = 2\langle \cos^2 \sigma_b \rangle - 1, \qquad (4\text{-}41)$$
$$f_{\sigma_c} = 2\langle \cos^2 \sigma_c \rangle - 1.$$

With reference to Figure 4-27 and Table 4-12, the orientation functions (4-41) assume values between 1 and −1 as a function of the orientation of the crystallographic axes. These orientation functions are interrelated to each other and to $f_{a,z}$, $f_{b,z}$, and $f_{c,z}$ [see (4-22)] as

Table 4-12   Values of σ, $\cos^2 \sigma$, and f_σ as a Function of Axial Orientation

(X, Y, Z represent N, T, M directions, respectively)

Orientation of Axis a, b, or c	σ (degrees)	$\cos^2 \sigma$	f_σ
In plane of polymer film YZ	0	1	1
Random with respect to plane of film YZ	45	$\frac{1}{2}$	0
In XZ plane	90	0	−1

a result of the following restrictions:

$$\sin^2 \phi_{a,z} \cos^2 \sigma_a + \sin^2 \phi_{b,z} \cos^2 \sigma_b + \sin^2 \phi_{c,z} \cos^2 \sigma_c = 1,$$
$$\sin^2 \phi_{a,z} \sin^2 \sigma_a + \sin^2 \phi_{b,z} \sin^2 \sigma_b + \sin^2 \phi_{c,z} \sin^2 \sigma_c = 1. \tag{4-42}$$

If there is no relationship between the degrees of biaxial and axial orientation, which means that there is no correlation between $f_{a,z}$ and f_{σ_a}, between $f_{b,z}$ and f_{σ_b}, and between $f_{c,z}$ and f_{σ_c}, then (4-42) lead to the result

$$f_{\sigma_a}(1-f_{a,z}) + f_{\sigma_b}(1-f_{b,z}) + f_{\sigma_c}(1-f_{c,z}) = 0. \tag{4-43}$$

Furthermore, because of the orthogonality of the orthorhombic crystallographic axes

$$f_{a,z} + f_{b,z} + f_{c,z} = 0, \tag{4-44}$$

as in the case of axial orientation [see (4-24)].

The numerical evaluation of the orientation functions $f_{\sigma_a}, f_{\sigma_b}$, and f_{σ_c} from x-ray intensities is a complex and tedious task, with the result that to date no illustrative analyses have appeared in the literature. For this reason and because of the special limitations that affect the foregoing theory, which are expressed by (4-43) and (4-44), we shall now discuss the analysis of higher orientation modes following the more empirical but generalized approach of Wilchinsky [13, 17].

Starting with a given $I(\phi, \beta)_Z$ intensity (or pole) distribution, we can determine orientation functions in directions other than the original reference direction Z, in particular the orthogonal reference directions X and Y corresponding to T and M, respectively. Thus we can specify the mean orientation of the normals to some particular family of planes (hkl) with respect to X, Y, and Z by three squared direction cosines [13, 22]:

$$\langle \cos^2 \phi_{hkl,X} \rangle, \qquad \langle \cos^2 \phi_{hkl,Y} \rangle, \qquad \langle \cos^2 \phi_{hkl,Z} \rangle.$$

Because of the orthogonality of the coordinate axes, only two of the three cosines are independent, the third being fixed by the relationship

$$\langle \cos^2 \phi_{hkl,X} \rangle + \langle \cos^2 \phi_{hkl,Y} \rangle + \langle \cos^2 \phi_{hkl,Z} \rangle = 1. \qquad (4\text{-}45)$$

Any two of these squared cosines may be used to plot the state of orientation on an equilateral ternary diagram, as shown in Figure 4-34 [5]. The location of the point (*hkl*) depends both on the numerical values of *h*, *k*, and *l* — and on the state of orientation. The points 1, 2, 3, and 4 represent four limiting cases of orientation of the (*hkl*) normals as follows [5]:

Point 1. At apex X. Perfect orientation parallel to X.

$$\langle \cos^2 \phi_{hkl,X} \rangle = 1,$$
$$\langle \cos^2 \phi_{hkl,Y} \rangle = \langle \cos^2 \phi_{hkl,Z} \rangle = 0. \qquad (4\text{-}46)$$

Point 2. On side YZ. Orientation in Y, Z plane and perpendicular to X.

$$\langle \cos^2 \phi_{hkl,X} \rangle = 0,$$
$$\langle \cos^2 \phi_{hkl,Y} \rangle + \langle \cos^2 \phi_{hkl,Z} \rangle = 1. \qquad (4\text{-}47)$$

Point 3. Center of triangle. Random orientation.

$$\langle \cos^2 \phi_{hkl,X} \rangle = \langle \cos^2 \phi_{hkl,Y} \rangle = \langle \cos^2 \phi_{hkl,Z} \rangle = \tfrac{1}{3}. \qquad (4\text{-}48)$$

Point 4. On line bisecting angle at vertex Z. Axial orientation about Z-direction.

$$\langle \cos^2 \phi_{hkl,X} \rangle = \langle \cos^2 \phi_{hkl,Y} \rangle = \tfrac{1}{2}(1 - \langle \cos^2 \phi_{hkl,Z} \rangle). \qquad (4\text{-}49)$$

Polypropene Film [17]. We shall now illustrate the evaluation of *c*-axis orientation with respect to reference directions other than Z by returning to Wilchinsky's analysis of polypropene film. In Section 4-4.1 the numerical result $\langle \cos^2 \phi_{c,Z} \rangle = 0.090$ was obtained by analyses of the intensity distributions of the (040) and (110) planes. It is possible to cross-plot the pole-figure data of Figure 4-26, which were obtained with respect to the Z-axis of reference ($\phi_Z = 0°$), so as to yield the distribution of poles with respect to the X- or Y-axis. We now consider X as the new reference axis of the pole orientation sphere. In Figure 4-35 the horizontal solid lines and circular broken lines represent latitude lines with respect to Z and X, respectively. By means of coordinate transformations the values of ϕ_Z and ϕ_X can

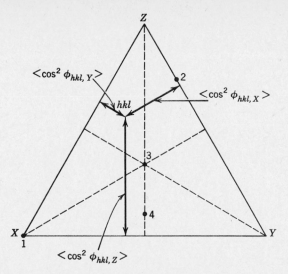

Figure 4-34 Representation of Wilchinsky's generalized orientation functions by a point within an equilateral triangle. (Desper and Stein[5].)

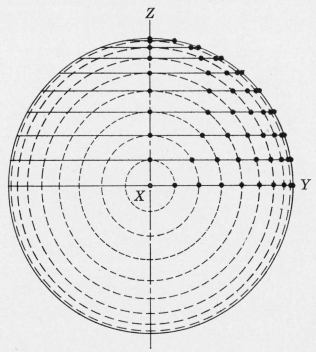

Figure 4-35 Method of cross-plotting pole densities to transform coordinates ϕ_Z, β_Z to ϕ_X, β_X. (Wilchinsky[17].)

be calculated at each point of intersection of the two sets of latitude lines. Values of $I(\phi_Z, \beta_Z)$ are then read off the original curves of Figure 4-25, and new curves of $I(\phi_X, \beta_X)$ versus β_X are constructed. In the same way curves of $I(\phi_Y, \beta_Y)$ can be prepared.

The I_{040} and I_{110} data thus re-expressed in coordinates relative to the X- and Y-axes are now substituted in equations analogous to (4-35) to yield $\langle \cos^2 \phi_{c,X} \rangle$ and $\langle \cos^2 \phi_{c,Y} \rangle$, the numerical values being 0.09 and 0.83, respectively. Figure 4-36 portrays the orientation of the c-axes of Wilchinsky's polypropene specimen D by means of an equilateral-triangle plot after the fashion of Figure 4-34. In plotting the point C, which specifies the mean orientation of the c-axes, the coordinate $\langle \cos^2 \phi_{c,Y} \rangle$ has been adjusted from 0.83 to 0.82 in order to conform to the orthogonality condition,

$$\langle \cos^2 \phi_{c,X} \rangle + \langle \cos^2 \phi_{c,Y} \rangle + \langle \cos^2 \phi_{c,Z} \rangle = 1. \qquad (4\text{-}50)$$

It will be seen that the experimental sum is

$$0.09 + 0.83 + 0.09 = 1.01,$$

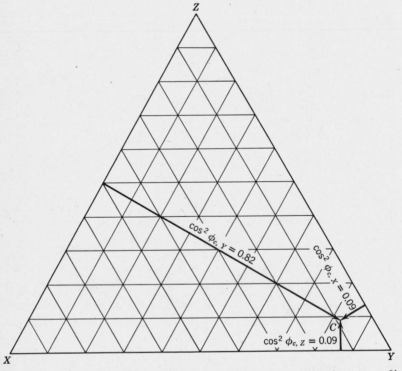

Figure 4-36 Triangular representation of c-axis orientation in polypropene film. (Experimental data supplied by Wilchinsky in private communication.)

which agrees well with the theory. Figure 4-36 shows clearly that the c-axes are preferentially oriented close to Y, the machine direction, and that they lie nearly perpendicular to X and Z, the transverse and normal directions, respectively.

4-4.3   Other Methods of X-Ray Analysis

Method of Sack [35].   Lack of space will permit us to comment only briefly on important analytical treatments of preferred orientation published by Sack [35], by Roe and Krigbaum [36–38], and more recently by Williams [39]. Sack's analysis of axial orientation, although equivalent to Wilchinsky's in essentials, embraces certain additional features, which may be summarized as follows:

1. Sack derives simplifications in the formulas that apply under particular symmetry conditions.

2. His treatment makes allowance for overlapping or imperfectly resolved reflections.

3. With reference to relationships among the $\langle \cos^2 \phi \rangle$ or f functions of the general form of (4-30) Sack shows how greater accuracy in the determination of a sought parameter can be realized by over-determining the solution with the aid of standard statistical methods, such as least squares, when more than the minimum number of experimental data are available. (See again Table 4-10.)

4. He relates the accuracy of the derived $\langle \cos^2 \phi \rangle$ or f functions to the statistical accuracy of the experimental measurements.

Sack applied these methods to a re-evaluation of the results published by Hermans and co-workers [23b] on preferred orientation in regenerated cellulose. He showed that the new results differed very little from those of Hermans numerically but that, at the same time, they possessed higher inherent accuracy as a consequence of the overdetermination effect.

Method of Roe and Krigbaum [36–38].   The method of Roe and Krigbaum represents an extension of the treatments of Wilchinsky and Sack but is a more general and elegant development. In effect it transforms pole-figure distributions $I_{hkl}(\phi, \beta)$† to an inverse pole figure, which is to say that the crystallite-orientation distribution is derived from a number of plane-normal orientation distributions. The mathematical procedure involves expansion of the $I_{hkl}(\phi, \beta)$ distributions in series of Legendre polynomials and expansion of the crystallite-orientation function in a series of spherical harmonics. The

†The reader may refer again to Section 4-3.5 and Tables 4-5 and 4-6.

coefficients of the Legendre polynomials are then determined from the experimental I_{hkl} (ϕ, β) data, and the coefficients of the series of spherical harmonics are found by solving simultaneous equations involving the Legendre coefficients for different planes, thus leading to the numerical solution of the crystallite-orientation distribution.

Roe and Krigbaum [36] first developed the method for the case of axial orientation, after which Roe generalized the treatment for application to any anisotropic polycrystalline material [38]. Perhaps the most useful way of portraying the Roe-Krigbaum crystallite-orientation distribution functions is in the form of inverse pole figures on stereographic charts. In contrast to a direct pole figure, which depicts the orientation distribution of given plane normals with respect to the reference axes, the inverse pole figure [40] depicts the orientation distribution of a given reference direction (for example, the fiber axis) with respect to the crystallographic axes. For polymer crystals of orthorhombic or higher symmetry it is necessary to portray only one quarter of the stereographic projection, as shown in Figure 4-37, since the remaining quarter sections are equivalent. In the orthogonal systems the a, b, and c crystallographic axes may conveniently be oriented to coincide with the three corners of the diagram, as indicated in Figure 4-37b by the poles (200), (020), and (002).

Although $I(\phi, \beta)$ distributions for as few as four sets of planes will normally suffice to reveal the qualitative characteristics of an axial orientation distribution, the inclusion of data from a larger number of planes has the effect of increasing the accuracy of detail and resolution of the orientation diagram. Roe and Krigbaum expedited the collection of the very extensive amount of experimental intensity data by using an automatic diffractometer, and they employed a high-speed digital computer for the complex and lengthy calculations. With such excellent facilities for collecting and processing the data, they found it possible to completely characterize the crystallite orientation in an axially oriented polymer specimen to a high state of accuracy within a period of three days.

The Roe-Krigbaum method of fiber-texture analysis embodies direct analogies with the well-known methods of single-crystal structure analysis. Thus the representation of the probability density of crystallite orientation by a spherical harmonic series corresponds to the representation of the electron-density distribution by a Fourier series. Also in both methods the presence of special symmetry elements results in systematic absences of certain classes of coefficients in the corresponding series.

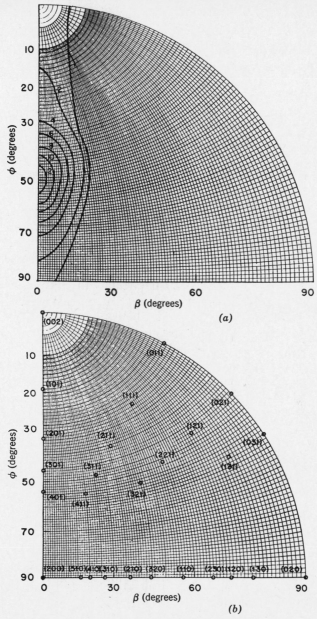

Figure 4-37 Meridional stereographic representation of axial texture in cross-linked polyethylene at an extension ratio of 4.58. Colatitude ϕ and longitude β. (*a*) Distribution function of the fiber axis; (*b*) corresponding locations of plane normals. (Krigbaum and Roe [37].)

261

Krigbaum and Roe[37] illustrated their method by applying it to axial textures in strained samples of cross-linked polyethylene at extension ratios of 2.20, 3.32, 4.58, and 6.67. It was observed that with increasing extension the plane normals of the type $(h0l)$ oriented most strongly parallel to the fiber axis, the h/l ratio of best alignment decreased continuously as the extension increased. At an extension ratio of 4.58 the most highly oriented plane normals were observed to be (301) and (401), as shown in Figure 4-37. This figure and those for extension ratios of 2.20, 3.32, and 6.67 also showed that the (020) normals tended to orient perpendicularly to the direction of extension and that the orientation sharpened with increasing extension.

Williams[39] has recently published a generalized series method for representing the distribution function for an inverse pole figure. From statistical theory the method should be superior to that based on Legendre polynomials, besides being much easier to apply to actual problems and yielding a correct evaluation of the error in the results.

Milberg[41] has developed an analytical method, based on a difference cylindrical-distribution function, that makes use of the total scattering, both crystalline and amorphous; it is particularly useful in the study of specimens with only a moderate degree of axial orientation. Since the primary value of the method is its capacity to reveal structural information, it is appropriate to present this topic in Chapter 6. (See Section 6-1.5.)

Method of Azimuthal Breadth.   For such purposes as comparison of relative degrees of axial orientation in a series of related polymer specimens the methods described in the foregoing sections are frequently entirely too laborious and their use is hardly justifiable. Instead a given specimen can be rather easily characterized by some kind of orientation index related to the intensity distribution in an azimuthal scan of an appropriate diffraction arc. (See Fig. 4-38.) That there is a correspondence between the intensity in a diffraction arc at a particular angle to the meridian and the number of crystallites oriented at the same critical angle to the fiber axis was first pointed out by Weissenberg[42], and early applications of this relationship were made by Sisson and Clark[43] to the study of cellulose fibers. The theoretical relationships were investigated in detail by J. J. Hermans et al.[23b], and extensive studies of orientation in cellulose were carried out by P. H. Hermans [25, 44].

When the fiber axis is perpendicular to the incident beam, it can be shown by spherical trigonometry that [23b]

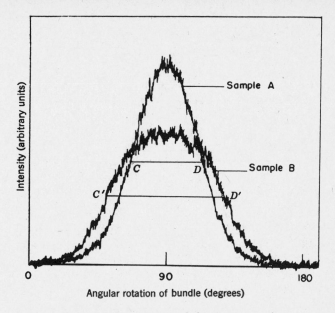

Figure 4-38 Azimuthal intensity tracings of the (002) arc of cotton samples with relatively high (sample A) and low (sample B) degrees of crystallite orientation. [J. J. Creely et al., *Textile Research J.*, **26**, 789 (1956).]

$$\cos \beta = \frac{\cos \phi}{\cos \theta} , \qquad (4\text{-}51)$$

a relationship first derived by Polanyi[45]. From this relationship it can be seen that for moderate to small Bragg angles the angle of inclination of the plane normal to the fiber axis, ϕ, is practically equal to the azimuthal angle β as directly measured from the meridian on the film. If a meridional (00l) reflection of adequate intensity can be obtained,† the degree of alignment of the molecular chains, taken to be coincident with the c-axis, is given directly by the azimuthal extension of the (00l) reflection. Since, however, meridional reflections are frequently weak, near-meridional reflections, if resolvable, may also be used to almost equally good effect[46], and numerous investigators have used the azimuthal intensity profile of an equatorial reflection as an empirical index of degree of alignment. Granted that $\cos \theta$ in (4-51) may be set equal to unity, the distribution of intensity versus β in a meridional reflection measures directly the *density* of

†The reader's attention is again directed to the conditions governing the generation of (00l) reflections as given in Section 4-1.

(00*l*) plane normals (number of poles per unit solid angle) as a function of ϕ. On the other hand, as has been pointed out in Section 4-4.1 (see "Experimental Evaluation of $\langle\cos^2\phi\rangle$"), the total *number* of plane normals oriented at a given inclination ϕ is proportional to $I(\phi)$ multiplied by $\sin\phi$ [or $I(\beta)$ multiplied by $\sin\beta$]. For comparative work such as we are now concerned with, the correction $\sin\beta$ is usually neglected, so that pole densities rather than total numbers of poles are compared. The most common degree-of-orientation index is the azimuthal breadth at half-maximum intensity, or 50% intensity breadth, although the 40% breadth and integral breadth are also employed. The reciprocals of these breadths are preferred by some investigators since the magnitude of the reciprocal increases with increasing parallelism of alignment[47, 48].

The unit cell of native cellulose (cellulose I) is monoclinic, with $a = 8.20$, $b = 10.34$, and $c = 7.84$ Å; and $\beta = 82°$[49]. The molecular chains are parallel to b (the unique axis), and it is known that the crystallites tend to orient spirally about the fiber axis[50]. Sisson and Clark[43] used a photographic method to determine the degree of orientation in various native cellulose fibers from the azimuthal distribution of intensity in the strongest equatorial reflection, (002). The Sisson-Clark technique did not resolve the doubling of the reflections that are produced by the spiral structure in the cotton specimens examined, although compression-wood cellulose did reveal a minimum in the orientation curve parallel to the fiber axis. (See Figure 4-39*a* and *b*.) The degree of alignment was expressed in several ways, a particularly useful method being the percentages of the total crystallites showing inclinations within successive 5° angular ranges. Multiplication of each percentage by the mean cosine of the angular range involved followed by addition of the products then yielded a weighted cosine index for each orientation distribution, which varied from 100 for perfect parallelism with the fiber axis to 50 for totally random orientation. Typical cottons examined by Sisson and Clark gave such weighted cosine indices in the range of 85 to 89, signifying a high degree of alignment.

Segal, Greely, and Conrad[51, 52] converted the photographic method of Sisson and Clark to a diffractometric technique for evaluating the degree of orientation in cotton fibers from the azimuthal widths of equatorial reflections. A key component of their apparatus is a device for mounting a bundle of fibers with a high degree of parallelism and under constant tension. This device can be rotated at a controlled speed so as to permit the preparation of a strip-chart record of the azimuthal intensity distribution in a given reflection.

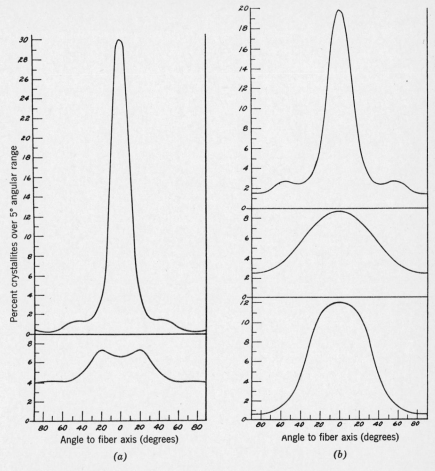

Figure 4-39 Typical orientation curves of cellulose fibers, (002) data: (*a*) upper—ramie, lower—compression wood; (*b*) upper—cotton mercerized with tension, center—cotton mercerized without tension, lower—unmercerized cotton. (Sisson and Clark[43].)

Scanning the (002) equatorial reflection with this apparatus, these investigators evaluated the relative degrees of orientation in native cottons and various cotton-amine complexes, taking the width at 50% maximum intensity as the index of orientation. Using this same technique of intensity measurement, DeLuca and Orr [53, 54] made use of the (002) intensity profile to determine the degree of crystallite orientation and the spiral angle in various native, decrystallized, and mercerized cottons. They resolved the intensity profile into two identical peaks of Gaussian shape of half-maximum breadth $2\alpha'$, which are separated by twice the spiral angle ϕ', as shown in Figure 4-40. Table 4-13 summarizes the results obtained by DeLuca and Orr for seven native cotton specimens. Included are the azimuthal angles corresponding to 50 and 40% of maximum intensity, and the orientation and spiral angles, α' and ϕ', respectively. In a later publication [55] these investigators discussed a graphical method for determining the amorphous background correction prior to the analysis of the azimuthal profile and measurement of crystallinity.

Dumbleton and Bowles [46] have studied perfection of orientation in fibers and films of poly(ethylene terephthalate) by using the azimuthal profile of the near-meridional reflection $(\bar{1}05)$. The unit cell of poly(ethylene terephthalate) is triclinic, with $a = 4.56$, $b = 5.94$, $c = 10.75\,\text{Å}$, $\alpha = 98.5°$, $\beta = 118°$, $\gamma = 112°$, and with the molecular chains parallel to c [56]. The diffracted intensities were measured with an automatic diffractometer using nickel-filtered CuKα radiation. With 2θ fixed at $43°$ azimuthal scans were made over the X range 0 to $180°$, equivalent to $-90° \leqslant \beta \leqslant 90°$. The $(\bar{1}05)$ profile is composed of two overlapping components situated at $10°$ on either side of the meridian. The individual components were resolved on the assumption that they

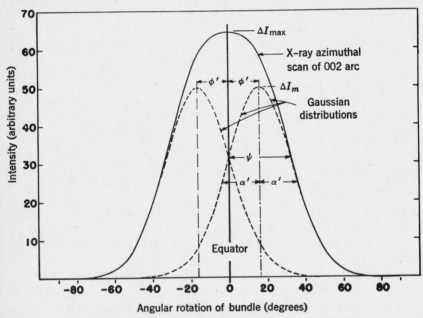

Figure 4-40   Resolution of the (002) azimuthal profile of cotton into two Gaussian distributions. (DeLuca and Orr [53].)

Table 4-13   Crystallite-Orientation and Spiral Angles of
Seven Native Cottons[a,b]

| Type of Cotton | (002) Arc | | Orientation Angle α' | Spiral Angle ϕ' |
	50% Intensity Angle	40% Intensity Angle		
Belgian Congo	35.8	39.5	21.9	17.7
Bobshaw No. 2	33.2	36.6	20.1	16.4
Bobshaw No. 1	33.0	36.3	19.7	16.3
Rowden	29.4	32.9	19.9	14.2
Hopi Acala	27.3	30.5	18.2	13.2
Interspecies	24.8	28.1	18.6	11.4
Strain 330	24.7	27.7	17.0	11.9

[a] Data from [53].
[b] All units are degrees.

were identical and symmetrical. The integral breadth was employed as the orientation index, and it was determined as one-half the total area under the joint ($\bar{1}05$) profile divided by the height of one resolved component. Calculations showed that broadening due to the geometry of the collimator could be neglected for integral breadths larger than $10°$.

An inspection of Figure 4-41 shows that the two components of the ($\bar{1}05$) profile become increasingly well resolved as the draw ratio α increases. Table 4-14 presents the integral breadths of the resolved ($\bar{1}05$) peak versus α as determined in two ways — first from the resolved peak and second from the intensity of the unresolved profile at $\mathbf{X} = 80°$. From the decrease of the integral breadth with draw ratio it is clear that the drawing process increases the degree of alignment of the molecules. Although both methods give similar results, the values from the resolved profile are more useful because they are more sensitive to changes in orientation at high elongations.

The method of azimuthal breadth analysis has been applied to numerous other polymers, of which we shall cite only two examples. Holmes and Palmer[57] evaluated the degree of orientation in polyethylene films prepared by the tubular extrusion process. They determined the alignment of the a- and b-axes from the azimuthal intensity profiles of the (200), (110), and (020) reflections. Dismore and Statton[58], using the yarn holder and camera described in Chapter 2 (see Sections 2-2.4 and 2-3.5), characterized the degree of orientation of axially oriented and annealed nylon 66 filaments by the widths at 50% maximum intensity of the (100) and (010) reflections.

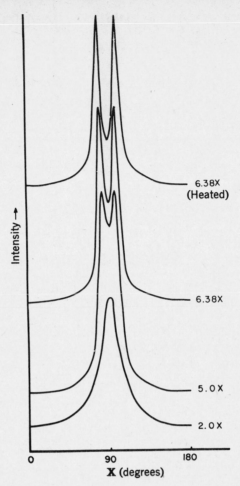

Figure 4-41   Effect of draw ratio and heat treatment on the azimuthal profile of the $(\bar{1}05)$ reflection of poly(ethylene terephthalate). (Dumbleton and Bowles [46].)

In nylon 66 [poly(hexamethylene adipamide)] the molecular chains are parallel to c in the triclinic cell, which has the dimensions $a = 4.9$, $b = 5.4$, and $c = 17.2$ Å; $\alpha = 48.5°, \beta = 77°, \gamma = 63.5°$ [59].

4-5   OPTICAL BIREFRINGENCE

The birefringence of a polymer specimen may be defined as the difference between the refractive indices in any two selected directions. Birefringence gives a useful measure of the overall state of preferred orientation, including both the crystalline and amorphous

Table 4-14 Crystallite-Orientation Angles of Poly(ethylene terephthalate) Fibers[a]

	Integral Breadth of $(\overline{1}05)$ Reflection (degrees)	
Draw Ratio	From Resolved Peak	From Intensity at $\mathbf{X} = 80°$
2.0	37	23
3.0	32	20
4.0	27	15
5.0	18	16
5.6	17	15
6.38	15	14
6.38[b]	14	14

[a]Data from [46].
[b]Fibers heated for 2 hr at 100°C.

regions. Since unoriented amorphous polymers are isotropic, the presence of birefringence in amorphous specimens indicates the existence of stresses. In general the optical character of an anisotropic specimen can be described by three refractive indices, n_x, n_y, n_z, along three orthogonal axes. Whereas all three indices are equal in an unoriented amorphous polymer, in a specimen that is axially oriented by drawing, the refractive indices parallel and perpendicular to the direction of draw are unequal, and the birefringence may be expressed as follows:

$$\Delta = n_{\parallel} - n_{\perp}.$$

The magnitude of Δ increases with the degree of orientation and with the optical anisotropy of the monomeric units of which the polymer is composed.

4-5.1 Axial Orientation

If we assume the applicability of the two-phase model of polymer morphology, we may express the measured birefringence of an axially oriented specimen as the sum of three terms arising from *crystalline*, *amorphous*, and *form* birefringence (subscripts c, a, and f, respectively) [27, 60]:

$$\Delta = x_{c,v}\Delta_c + x_a\Delta_a + \Delta_f$$

$$= x_{c,v}\Delta_c + (1 - x_{c,v})\Delta_a + \Delta_f. \tag{4-52}$$

Here $x_{c,v}$ is the volume fraction of crystalline polymer, which may be obtained from the weight fraction by means of (3-69). The crystalline and amorphous birefringences Δ_c and Δ_a may also be expressed in terms of the intrinsic birefringence of a single polymer crystal, Δ_c^0, and the birefringence of the perfectly oriented amorphous phase, Δ_a^0, respectively, which converts (4-52) to the form

$$\Delta = x_{c,v} f_c \Delta_c^0 + (1 - x_{c,v}) f_a \Delta_a^0 + \Delta_f. \tag{4-53}$$

In (4-53) f_c and f_a are orientation functions of the kind described in Section 4-4.1. Form birefringence Δ_f results from distortion of the electric field of the light wave on passing through a refractive-index boundary between two phases; for example, crystalline and amorphous or solid polymer and microvoid. The magnitude of Δ_f is usually small enough in relation to the other two terms to be neglected.

If f_c and $x_{c,v}$ are determined by appropriate x-ray measurements and if the refractive indices of the crystalline polymer and therefore also Δ_c^0 are known, it is possible to calculate the contribution of the crystalline phase to the birefringence, $x_{c,v} \Delta_c = x_{c,v} f_c \Delta_c^0$. Then, if the form birefringence can be calculated (or neglected), the amorphous contribution to the total birefringence Δ can be determined by difference with (4-53). For a low-density polyethylene sample Δ_f was found to amount to 5 to 10% of Δ [60].

4-5.2 Biaxial Orientation [5]

We may apply (4-52) to biaxially oriented films (see again Section 4-4.2) if we specify two independent birefringences:

$$\Delta_X = n_Z - n_Y \quad \text{(in-plane birefringence)}, \tag{4-54}$$

$$\Delta_Y = n_Z - n_X \quad \text{(out-of-plane birefringence)}. \tag{4-55}$$

By comparison with (4-52), Δ_X and Δ_1 may also be written as follows:

$$\Delta_X = x_{c,v} \Delta_{c,X} + (1 - x_{c,v}) \Delta_{a,X} + \Delta_{f,X}, \tag{4-56}$$

$$\Delta_Y = x_{c,v} \Delta_{c,Y} + (1 - x_{c,v}) \Delta_{a,Y} + \Delta_{f,Y}, \tag{4-57}$$

in which $\Delta_{c,X}$ and $\Delta_{c,Y}$ are two crystalline birefringences defined by

$$\Delta_{c,X} = n_{c,Z} - n_{c,Y}, \tag{4-58}$$

$$\Delta_{c,Y} = n_{c,Z} - n_{c,X}. \tag{4-59}$$

In these equations $n_{c,X}$, $n_{c,Y}$, and $n_{c,Z}$ should be understood as the indices of refraction that would be observed if the sample were 100% crystalline, form birefringence being neglected.

Desper and Stein [5] show that the crystalline birefringences may be expressed in terms of the $\langle \cos^2 \phi \rangle$ distributions and n_a, n_b, and n_c, the refractive indices of the crystals for light polarized along three mutually perpendicular (principal) directions in the crystals:

$$\Delta_{c,X} = (n_a - n_c)(\langle \cos^2 \phi_{a,Z} \rangle - \langle \cos^2 \phi_{a,Y} \rangle) + (n_b - n_c)(\langle \cos^2 \phi_{b,Z} \rangle - \langle \cos^2 \phi_{b,Y} \rangle)$$

(4-60)

and

$$\Delta_{c,Y} = (n_a - n_c)(\langle \cos^2 \phi_{a,Z} \rangle - \langle \cos^2 \phi_{a,X} \rangle) + (n_b - n_c)(\langle \cos^2 \phi_{b,Z} \rangle - \langle \cos^2 \phi_{b,X} \rangle).$$

(4-61)

In orthorhombic crystals the directions of n_a, n_b, and n_c coincide with the crystallographic axes; in monoclinic crystals one principal direction coincides with the unique axis (b), and the other two lie somewhere in the ac plane; in the triclinic system there are no restrictions on the orientation of n_a, n_b, and n_c with respect to the crystallographic axes.

4-5.3 Experimental Measurement of Birefringence

Figure 4-42 shows the experimental arrangement used by Stein [61] for the measurement of birefringence in polymers by the compensation method. The basic components of the optical system include a mercury light source, a filter for isolating the green line at 5461 Å, a Polaroid polarizer with its axis set at 45° to the vertical, a Babinet compensator (quartz wedge with its slow axis vertical), and a Nicol-prism analyzer with its polarizing axis set at 45° to the vertical and at 90° to the axis of the polarizer. With the light beam polarized at 45° to two of the three principal axes of a birefringent film and incident perpendicularly on the film, the light transmission is

$$T = \sin^2 \left(\frac{\delta}{2} \right),$$

(4-62)

where δ is the retardation, given by

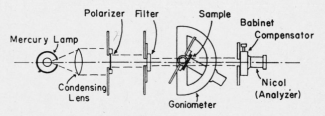

Figure 4-42 Experimental arrangement for measuring birefringence. (Stein [61].)

$$\delta = \frac{2\pi\Delta d}{\lambda}.$$ (4-63)

In (4-63) Δ is the birefringence, d is the thickness of the film, and λ is the wavelength of the light.

In the compensation method the birefringence of the specimen is balanced by moving the calibrated quartz wedge to the position giving equal and opposite retardation. According to (4-62) the light transmission is a maximum or minimum depending on whether δ is an odd or even multiple of π, respectively, which produces a pattern of light and dark fringes. Removal of the balanced specimen causes a shift in the fringe pattern, and the movement of the wedge required to restore the pattern to its original position measures the retardation δ. The two birefringence values of biaxially oriented films are obtained by measurements with the specimen tilted at several different angles to the perpendicular[61]. Limitations of space preclude a more complete treatment of experimental techniques. For detailed expositions the reader is referred to the literature [5, 61–63].

4-5.4   Illustrative Applications

Amorphous Birefringence of Axially Oriented Polyethylene [62]. Stein and Norris calculated the crystalline birefringence Δ_c of (4-52) from the average polarizabilities per unit volume parallel and perpendicular to the c-axis, $\alpha_\parallel$ and $\alpha_\perp$. Values of $\alpha_\parallel$ and $\alpha_\perp$ applicable to polyethylene can be determined from a knowledge of the principal refractive indices of crystals of high-molecular-weight n-paraffins. Substituting the refractive indices for $C_{36}H_{74}$ obtained by Bunn and Daubeny[64],

$$n_a = 1.514, \qquad n_b = 1.519, \qquad n_c = 1.575,$$

in the Lorenz-Lorentz equation [65],

$$\alpha = \frac{3}{4\pi}\frac{n^2-1}{n^2+2},$$ (4-64)

Stein and Norris obtained the following numerical values of the specific polarizabilities:

$$\alpha_\parallel = \alpha_c = 0.0790,$$

$$\alpha_\perp = \tfrac{1}{2}(\alpha_a + \alpha_b) = 0.0723.$$

Then the crystalline birefringence is [65]

$$\Delta_c = \frac{2}{9}\pi\frac{(n^2+2)^2}{n}(\alpha_\parallel - \alpha_\perp)_c f_c = 0.0572 f_c,$$ (4-65)

the average refractive index of polyethylene, n, having been assigned the value 1.50.

Numerical values of the orientation function f_c were determined as a function of elongation from photographic x-ray measurements. The volume fraction of crystallinity, $x_{c,v}$, was determined to be 0.56 on the basis of density measurements. (See again Section 3-4.1.) With these numerical data and with neglect of the form birefringence it was then possible to calculate the amorphous birefringence from measurements of the overall birefringence Δ by using (4-52) in the form

$$\Delta_a = \frac{\Delta - x_{c,v}\Delta_c}{1 - x_{c,v}}. \qquad (4\text{-}66)$$

Figure 4-43 shows the variation of the total birefringence and $x_{c,v}\Delta_c$, the calculated contribution of the crystalline phase to the birefringence, with percent elongation. To a first approximation the two curves vary proportionately with elongation, the crystalline birefringence accounting for about two-thirds of the total. The data for the

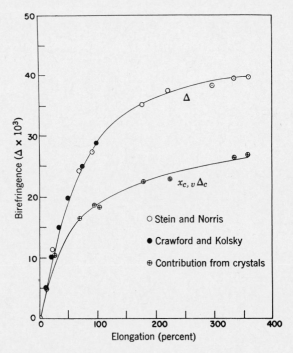

Figure 4-43 Variation with elongation of the measured birefringence, Δ, and the calculated contribution of the crystalline phase to the birefringence, $x_{c,v}\Delta_c$. (Stein and Norris [62].)

total birefringence obtained by Stein and Norris[62] are compared with the results of Crawford and Kolsky[66]. Figure 4-44 shows Stein and Norris' curve of Δ_a versus elongation as determined with (4-66) and compares it with a curve calculated by using the theory of birefringence of rubberlike networks proposed by Kuhn and Grün[67] and Treloar[68]. It is seen that whereas Stein and Norris' curve asymptotically approaches a value of 0.03 at high elongations, the slope of the Kuhn-Treloar curve increases with elongation.

Crystalline and Amorphous Birefringence of Biaxially Oriented Polyethylene Films[5, 69]. To illustrate the application of birefringence analysis to higher orientation modes we return to the polyethylene films described in Section 4-2.1, which were cross-linked with beta radiation, melted, stretched, and crystallized while elongated 100 and 200%. The pole figures of these specimens, which are

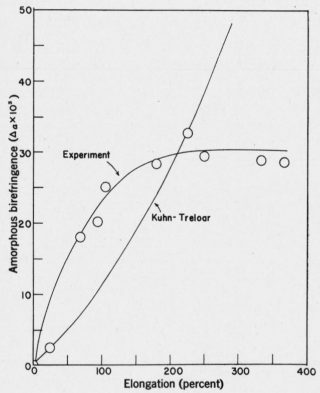

Figure 4-44 Variation with elongation of the calculated amorphous birefringence and comparison with the curve predicted by the Kuhn-Treloar theory. (Stein and Norris[62].)

shown in Figures 4-10 and 4-11, were prepared from x-ray intensities measured with an automated diffractometer[22]. The triangular orientation diagram of these specimens, Figure 4-45, clearly shows the approximately axial orientation in the low-elongation sample and the considerable degree of biaxial orientation in the high-elongation sample.

Orientation functions of the a, b, c crystallographic axes with respect to the orthogonal reference directions X, Y, Z (N, T, and M, respectively) were derived from the pole-figure data and are listed in Table 4-15. The $f_{a,q}$ and $f_{b,q}$ functions were first calculated directly from the (200) and (020) pole figures, respectively; the $f_{b,q}$ and $f_{c,q}$ functions were then deduced from the (110) and (200) data with the aid of equations equivalent to those derived by Wilchinsky[17, 26], that is, (4-32) to (4-35). Because of the lower experimental accuracy of the (020) relative to the (110) and (200) intensity measurements, Desper and Stein used only the latter two sets of data in the preparation of the orientation diagram of Figure 4-45 and in the calculation of the crystalline and amorphous birefringences. Substitution into (4-60) and (4-61) of the six $\langle\cos^2\phi\rangle$ values corresponding to the f's of Table 4-15 gave the crystalline birefringences $\Delta_{c,X}$ and $\Delta_{c,Y}$ in Table 4-16. The volume fractions of crystalline polymer were determined by x-rays and found to be 0.41 and 0.38 for the samples elongated 100 and 200%, respectively, and measurements of the in-plane birefringence gave $\Delta_X = -2 \times 10^{-4}$ and 186×10^{-4}, respectively, for these samples. The out-of-plane birefringence Δ_Y of the specimen at 200%

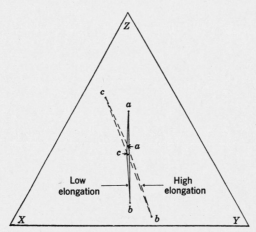

Figure 4-45 Orientation-function diagram of polyethylene film at low and high elongation. (Desper and Stein [5].)

Table 4-15   Crystalline-Orientation Functions of Polyethylene Film[a]

Function	100% Extension			200% Extension		
	$q = X$	Y	Z	$q = X$	Y	Z
$f_{110,q}$	0.0667	0.0791	−0.1459	0.0476	0.2462	−0.2938
$f_{a,q}$	−0.1468	−0.1469	0.2937	−0.0220	−0.0344	0.0564
$f_{b,q}$	0.1394	0.0933	−0.2327	−0.0123	0.2329	−0.2207
$f_{b,q}^{(b)}$	0.160	0.177	−0.341	0.0785	0.371	−0.450
$f_{c,q}^{(b)}$	−0.014	−0.033	0.047	−0.057	−0.337	0.394

[a]Data from [69].
[b]Values derived from the (110) and (200) data.

elongation was measured as 235×10^{-4}; the thickness of the specimen at 100% elongation was too great to permit measurement of Δ_Y. Finally with the aid of (4-56) and (4-57) the numerical amorphous birefringences $\Delta_{a,X}$ and $\Delta_{a,Y}$ of Table 4-16 were calculated, the terms $\Delta_{f,X}$ and $\Delta_{f,Y}$ being neglected.

The relatively large amorphous orientation that is found in these biaxially oriented specimens is thought to be a consequence of the cross-linking, which interferes with disorientation of the amorphous regions. The negative value of $\Delta_{a,X}$ that characterizes the specimen at 100% elongation indicates amorphous orientation *away from* the direction of extension, in agreement with previous findings of Judge and Stein[70]. Contrariwise, the high positive $\Delta_{a,Y}$ value found for the high-elongation specimen reveals a large degree of amorphous orientation parallel to the plane of the polymer film. It may also be

Table 4-16   Birefringence Values
of Polyethylene Film $(\times 10^4)^{[a]}$

Elongation	100%	200%
Crystalline Fraction $x_{c,v}$	0.41	0.38
Δ_X	−2	186
$\Delta_{c,X}$	48	300
$\Delta_{a,X}$	−37	115
Δ_Y	[b]	235
$\Delta_{c,Y}$	5	185
$\Delta_{a,Y}$	[b]	266

[a]Data from [69].
[b]Measurement not possible because of sample thickness.

pointed out that the smaller, but also positive, value of $\Delta_{a,x}$ at the higher elongation shows the most preferred direction of amorphous orientation to be the stretch direction and the least preferred the normal to the film.

In concluding this chapter special mention must be made of the very fruitful investigations of submicroscopic deformation morphology in polypropene by Samuels[71, 72], who demonstrated the potency of x-ray diffraction methods in combination with other physical techniques, including light scattering, optical birefringence, interference microscopy, infrared dichroism, and sonic modulus measurements.

GENERAL REFERENCES

1965 Book of ASTM Standards, American Society for Testing and Materials, Philadelphia, 1965, Part 31, pp. 151–167, Standard Method E81–63.

C. R. Desper and R. S. Stein, "The Measurement of Pole Figures, Orientation Functions, Birefringence and Infrared Dichroism for Polyethylene Films Prepared by Unidirectional Crystallization and Crystallization in the Oriented State," J. Appl. Phys., 37, 3990 (1966).

J. W. Jones, "The Preparation of Pole Figures for Polymers by Computer Techniques," in Advances in X-Ray Analysis, Vol. 6, Plenum Press, New York, 1963, p. 223.

R.-J. Roe and W. R. Krigbaum, "Description of Crystallite Orientation in Polycrystalline Materials Having Fiber Texture," J. Chem. Phys., 40, 2608 (1964).

R.-J. Roe, "Description of Crystallite Orientation in Polycrystalline Materials, III. General Solution to Pole Figure Inversion," J. Appl. Phys., 36, 2024 (1965).

R. A. Sack, "Indirect Evaluation of Orientation in Polycrystalline Materials," J. Polymer Sci., 54, 543 (1961).

· R. J. Samuels, "Evaluation of Orientation in Polycrystalline Polymeric Materials," Norelco Reporter, 10, 101 (1963).

R. S. Stein, "Optical Methods of Characterizing High Polymers," Chapter 4 in Newer Methods of Polymer Characterization, B. Ke (editor), Interscience, New York, 1964.

Z. W. Wilchinsky, "Measurement of Orientation in Polypropene Film," J. Appl. Phys., 31, 1969 (1960).

Z. W. Wilchinsky, "Recent Developments in the Measurement of Orientation in Polymers by X-Ray Diffraction," in Advances in X-Ray Analysis, Vol. 6, Plenum Press, New York, 1963, p. 231.

SPECIFIC REFERENCES

[1] L. E. Alexander and E. R. Michalik, Acta Cryst., 12, 105 (1959).

[2] P. H. Geil, Chem. Eng. News, August 16, 1965, p. 83.

[3] W. A. Sisson, J. Phys. Chem., 40, 343 (1936).

[4] C. J. Heffelfinger and R. L. Burton, J. Polymer Sci., 47, 289 (1960).

[5] C. R. Desper and R. S. Stein, J. Appl. Phys., 37, 3990 (1966).

[6] 1965 Book of ASTM Standards, American Society for Testing and Materials, Philadelphia, 1965, Part 31, pp. 151–167, Standard Method E81–63.

[7] B. F. Decker, E. T. Asp, and D. Harker, *J. Appl. Phys.*, **19**, 388 (1948).
[8] L. E. Alexander, *J. Appl. Phys.*, **21**, 126 (1950).
[9] Ibid., **25**, 155 (1954).
[10] L. E. Alexander and G. S. Smith, *Acta Cryst.*, **17**, 447 (1964).
[11] H. P. Klug and L. E. Alexander, *X-Ray Diffraction Procedures*, Wiley, New York, 1954, pp. 239–242.
[12] J. W. Jones, *Advances in X-Ray Analysis*, Vol. 6, Plenum Press, New York, 1963, p. 223.
[13] Z. W. Wilchinsky, *J. Appl. Polymer Sci.*, **7**, 923 (1963).
[14] P. H. Lindenmeyer and S. Lustig, *J. Appl. Polymer Sci.*, **9**, 227 (1965).
[15] L. G. Schulz, *J. Appl. Phys.*, **20**, 1030 (1949).
[16] B. D. Cullity and A. Freda, *J. Appl. Phys.*, **29**, 25 (1958).
[17] Z. W. Wilchinsky, *J. Appl. Phys.*, **31**, 1969 (1960).
[18] A. N. Holden, *Rev. Sci. Instruments*, **24**, 10 (1953).
[19] E. G. Chirer, *J. Sci. Instruments*, **44**, 225 (1967).
[20] A. H. Geisler, *Trans. Amer. Soc. Metals*, **45A**, 131 (1953).
[21] U. W. Arndt and B. T. M. Willis, *Single Crystal Diffractometry*, University Press, Cambridge, 1966.
[22] C. R. Desper, Project NR:056–378, Contract Nonr 3357(01), University of Massachusetts ONR Tech. Reports Nos. 78 and 80 (1965); *Advances in X-Ray Analysis*, Vol. 12, Plenum Press, New York, 1968.
[23] (a) P. H. Hermans and P. Platzek, *Kolloid Z.*, **88**, 68 (1939); (b) J. J. Hermans, P. H. Hermans, D. Vermaas, and A. Weidinger, *Rec. trav. chim. Pays-Bas*, **65**, 427 (1946).
[24] R. S. Stein, *J. Polymer Sci.*, **31**, 327 (1958).
[25] P. H. Hermans, *Contribution to the Physics of Cellulose Fibers*, Elsevier, Amsterdam, 1946, Appendix III (J. J. Hermans).
[26] Z. W. Wilchinsky, *Advances in X-Ray Analysis*, Vol. 6, Plenum Press, New York, 1963, p. 231.
[27] R. J. Samuels, *Norelco Reporter*, **10**, 101 (1963).
[28] Reference 11, pp. 150–152.
[29] Z. W. Wilchinsky, *J. Appl. Phys.*, **30**, 792 (1959).
[30] G. Natta, P. Corradini, and M. Cesari, *Atti Accad. Naz. Lincei*, **21**, 365 (1956); G. Natta and P. Corradini, *Nuovo Cimento*, **15**, Suppl. 1, 40 (1960).
[31] M. Litt, *J. Polymer Sci.*, Part B, **2**, 1057 (1964).
[32] For example, see Y. Nishijima, Y. Onogi, and T. Asai, *J. Polymer Sci.*, Part C, **15**, 237 (1966).
[33] R. S. Stein, *J. Polymer Sci.*, **31**, 335 (1958).
[34] R. S. Stein, *J. Polymer Sci.*, **50**, 339 (1961).
[35] R. A. Sack, *J. Polymer Sci.*, **54**, 543 (1961).
[36] R.-J. Roe and W. R. Krigbaum, *J. Chem. Phys.*, **40**, 2608 (1964).
[37] W. R. Krigbaum and R.-J. Roe, *J. Chem. Phys.*, **41**, 737 (1964).
[38] R.-J. Roe, *J. Appl. Phys.*, **36**, 2024 (1965).
[39] R. O. Williams, *J. Appl. Phys.*, **38**, 4029 (1967).
[40] L. K. Jetter, C. J. McMargue and R. O. Williams, *J. Appl. Phys.*, **27**, 368 (1956).
[41] M. E. Milberg, *J. Appl. Phys.*, **33**, 1766 (1962).
[42] K. Weissenberg, *Z. Physik*, **8**, 20 (1961).
[43] W. A. Sisson and G. L. Clark, *Chem. Eng. News, Anal. Ed.*, **5**, 296 (1933).
[44] P. H. Hermans, *Physical Chemistry of Cellulose Fibers*, Elsevier, New York, (1949).

[45] M. Polanyi, Z. *Physik*, **7**, 149 (1921).

[46] J. H. Dumbleton and B. B. Bowles, *J. Polymer Sci.*, Part A, **4**, 951 (1966).

[47] W. Kast and A. Prietzschk, *Kolloid-Z.*, **114**, 23 (1949).

[48] K. C. Ellis and J. O. Warwicker, *J. Polymer Sci.*, Part A, **1**, 1185 (1963).

[49] D. G. Fisher and J. Mann, *J. Polymer Sci.*, **42**, 189 (1960).

[50] W. L. Balls, *Proc. Roy. Soc. (London)*, **B95**, 72 (1923).

[51] L. Segal, J. J. Creely, and C. M. Conrad, *Rev. Sci. Instruments*, **21**, 431 (1950); J. J. Creely and C. M. Conrad, *Textile Research J.*, **35**, 863 (1965).

[52] L. Segal and C. M. Conrad, *Amer. Dyestuff Reporter*, August 26, 1957.

[53] L. B. DeLuca and R. S. Orr, *J. Polymer Sci.*, **54**, 457 (1961).

[54] L. B. DeLuca and R. S. Orr, *J. Polymer Sci.*, **54**, 471 (1961).

[55] L. B. DeLuca and R. S. Orr, *Textile Research J.*, **32**, 243 (1962).

[56] R. de P. Daubeny, C. W. Bunn, and C. J. Brown, *Proc. Roy. Soc. (London)*, **A226**, 531 (1954).

[57] D. R. Holmes and R. P. Palmer, *J. Polymer Sci.*, **31**, 345 (1958).

[58] P. F. Dismore and W. O. Statton, *J. Polymer Sci.*, Part C, **13**, 133 (1966).

[59] C. W. Bunn and E. V. Garner, *Proc. Roy. Soc. (London)*, **A189**, 39 (1947).

[60] R. S. Stein in *Newer Methods of Polymer Characterization*, B. Ke (editor), Interscience, New York, 1964, Chapter 4.

[61] R. S. Stein, *J. Polymer Sci.*, **24**, 383 (1957).

[62] R. S. Stein and F. H. Norris, *J. Polymer Sci.*, **21**, 381 (1956).

[63] Reference 60, pp. 175–176.

[64] C. W. Bunn and R. de P. Daubeny, *Trans. Faraday Soc.*, **50**, 1173 (1954).

[65] For example, see reference 60, p. 174.

[66] S. M. Crawford and H. Kolsky, *Proc. Phys. Soc. (London)*, **B64**, 119 (1951).

[67] W. Kuhn and F. Grün, *Kolloid-Z.*, **101**, 248 (1942); W. Kuhn, *J. Polymer Sci.*, **1**, 380 (1946).

[68] L. R. G. Treloar, *Trans. Faraday Soc.*, **42**, 83 (1946); L. R. G. Treloar, *The Physics of Rubber Elasticity*, Clarendon Press, Oxford, 1949, p. 147.

[69] C. R. Desper, *A Study of Crystallization and Orientation Mechanisms in Polyethylene*, Ph.D. thesis, University of Massachusetts, August 1966.

[70] J. T. Judge and R. S. Stein, *J. Appl. Phys.*, **32**, 2357 (1961).

[71] R. J. Samuels, *J. Polymer Sci.*, Part A, **3**, 1741 (1965).

[72] R. J. Samuels, *J. Polymer Sci.*, Part C, **20**, 253 (1967).

5

Macrostructure From
Small-Angle Scattering[†]

An attempt has been made in this chapter to assemble the principal theoretical formulations of small-angle scattering that have underlying significance or direct practical application in the investigation of polymers. This effort has encountered two formidable difficulties: (a) the need to recognize, and where possible to reconcile, dissimilarities in the mathematical expressions that have issued from the four major schools of theoretical investigation (Guinier, Kratky and Porod, Debye, Luzzati); and (b) the need to systematize to some degree the truly prolific nomenclature of the small-angle-scattering literature without at the same time conflicting unnecessarily with the basic terminology established thus far in the present monograph, especially in Chapter 1.

In order to preclude as far as possible confusion stemming from differences in nomenclature the presentation in this chapter, except when otherwise noted, adheres to the conventions of Kratky and Porod. Most expressions of practical import are presented first in the notation of Kratky and Porod and are numbered with the prefix K, followed immediately by the alternative form of Luzzati numbered with the prefix L. Included are some K and some L equations that are not common to both treatments.[‡] Topics and mathematical expressions

[†]Prior to the reading of this chapter it is suggested that the reader review Section 1-5. Instrumentation for small-angle scattering is described in Sections 2-3.5 and 2-4.

[‡]In addition to the employment of a different angular variable, Luzzati's expressions differ from Kratky's in that they are based on intensity defined as energy per square s

that are specifically due to Guinier, Debye, or others are plainly designated as such in the body of the text.

In an additional effort to systematize the nomenclature of this chapter a table has been appended (Table 5-7) that lists every non-trivial and nontransient symbol together with its concise definition and dimensional units. Unavoidably some trivial and transient symbols will be found to duplicate the basic symbols of Table 5-7, but in virtually every instance it will be self-evident that this is the case.

Small-angle x-ray scattering from an inhomogeneous system may consist of either diffuse or discrete effects. We shall defer the subject of discrete interferences to a later section of this chapter and devote our attention first to diffuse and continuous small-angle scattering. Such scattering is essentially particulate in origin but modulated more or less by interparticle interferences. Guinier [1–3] first showed that a single particle of colloidal dimensions can produce a character-istic diffuse small-angle-scattering pattern and that many identical particles at sufficiently large and irregular distances from each other will produce a total scattering that is simply the sum of the scattering of the individual particles. This is the concept of a *dilute system*, which produces pure *particulate scattering*.

The angular limit within which the scattering occurs is inversely proportional to the size of the inhomogeneities in the electron-density distribution within the specimen. Thus the diffuse scattering from a loose aggregation of oriented particles will have its largest extension in the direction of the smaller particle dimension and its smaller extension in the direction of the larger dimension, which is to say that the shape of the diffuse scattering depicts the particle shape in reverse. This is illustrated in Figure 5-1*b* by the scattering of dry-spun fibers of polyacrylonitrile, which is principally attributable to microvoids elongated parallel to the fiber axis (vertical direction in the photograph)[4]. This pattern may be compared with that of the unoriented polymer in Figure 5-1*a*.

It is of prime importance to realize that a diffuse small-angle-scattering pattern is never subject to an unequivocal structural interpretation on the basis of the x-ray data taken alone. However, specimens that yield identical scattering curves may be designated

---

unit $[s = (2 \sin \theta)/\lambda]$ rather than energy per square centimeter. Thus

$$I \text{ (Luzzati)} = (a^2 \lambda^2) \times I \text{ (Kratky)},$$

where a is the specimen-to-receiver distance in centimeters.

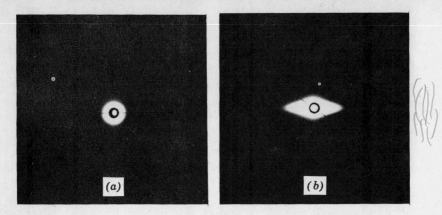

Figure 5-1   Diffuse small-angle scattering of polyacrylonitrile: (a) unoriented polymer; (b) experimental dry-spun fibers drawn 8× in water at 100°C. (Statton[4].)

scattering-equivalent systems[5]. In principle the following types of ambiguity must always be reckoned with in the interpretation of diffuse small-angle scattering:

1. As a consequence of the Babinet principle of reciprocity in optics it is not possible to distinguish the scattering by a system of particles in space from the scattering by a complementary system of micropores in a solid continuum. However, the choice between these alternatives can usually be made on the basis of independent evidence bearing on the nature of the system concerned.

2. It is not possible to differentiate without some degree of uncertainty the effects due to *particle shape* from those due to *polydispersity*. That is to say, scattering equivalence may result from a special particulate shape and a certain distribution of particle sizes.

3. It is frequently difficult to determine to what extent the scattering curve is determined by the isolated particles and to what extent by the interparticle interferences. Whereas *dilute systems* yield only particulate scattering, in *dense systems* interparticle interferences cannot be neglected. In addition to the dominant effect of *dilution*, *polydispersity*[6–8] and pronounced *anisotropy of particulate shape* [9] tend to favor the generation of pure particle scattering.

From the foregoing considerations we might easily conclude that the hope of achieving a valid interpretation of any given diffuse small-angle-scattering diagram is futile. Actually the prospects are not nearly this bleak because the investigator can usually resolve some of the ambiguities by bringing to bear on the problem additional information on the chemistry, physical properties, or history of the specimen in question.

5-1 PRELIMINARY CONSIDERATIONS

5-1.1 Angular Nomenclature

We define the angle of scattering—that is, the angle between the incident and diffracted beams—as 2θ, which is consistent with its meaning in wide-angle x-ray diffraction. For the sake of consistency with the literature sources it unfortunately becomes necessary to employ any one of three principal angular functions, †

$$s = |\boldsymbol{\rho}_{hkl}| = \frac{2\sin\theta}{\lambda}, \tag{5-1}$$

$$S = 2\pi s = \frac{4\pi\sin\theta}{\lambda}, \tag{5-2}$$

$$m = 2a\sin\theta. \tag{5-3}$$

Because of the smallness of the angles involved, these equations may be written without appreciable error as

$$s = \frac{2\theta}{\lambda}, \tag{5-4}$$

$$S = \frac{4\pi\theta}{\lambda}, \tag{5-5}$$

$$m = 2a\theta = a\lambda s. \tag{5-6}$$

In (5-6) m is the linear distance between the incident and scattered rays in the plane of registration. (See Figure 5-2.)

5-1.2 The Intensity of Small-Angle Scattering

It was explained in Chapter 1 [see (1-20)] that four factors modify the ideal intensity, or squared structure factor $|F(hkl)|^2$, diffracted by a crystalline substance. Of these four factors, all but the Lorentz factor may be neglected in small-angle-scattering studies. Because of the narrow range and small magnitude of the scattering angles involved, the absorption factor A is the same for the direct and diffracted rays, and the polarization factor P is effectively constant and equal to unity. Furthermore the multiplicity j may be set equal to

†The reader may refer again to (1-15) and p. 44.

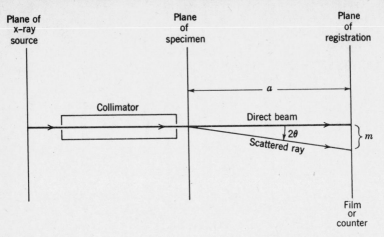

Figure 5-2   Schematic plan of small-angle-scattering geometry.

unity since the concepts of regularity in atomic arrangement and Bragg reflecting planes are not in general relevant in a description of the larger scale structure that gives rise to small-angle scattering. The nature of these inhomogeneities of colloidal dimensions may be expressed for spherically symmetrical scattering by the structure factor $F(S)$ rather than by $F(hkl)$, which describes the structural features of atomic dimensions. Accordingly for small-angle scattering we may write in place of (1-20)

$$I(S) = \overline{L \cdot F(S)^2},$$

in which L plays a role that is somewhat analogous to that of the Lorentz factor in classical x-ray crystallography.

The form of L as used in the above equation depends on the experimental conditions employed, the particular form of the inhomogeneities (particle shape) being studied, and the specific parameter sought; for example, in the evaluation of the integrated intensity of small-angle scattering it is proportional to S^{-2} for point collimation and S^{-1} for collimation by infinitely long, narrow slits. (See Section 5-2.2.) On the other hand, when either pinhole or slit collimation is employed, L is proportional to S^{-1} when the cross section of a rodlike particle is determined, whereas it is proportional to S^{-2} when the thickness of a platelike particle is sought.

In the mathematical treatment of small-angle scattering that is presented in this chapter we conform to the practice of most authorities in the field of not explicitly expressing the Lorentz factor L as such but, rather, including it implicitly in the various equations that are appropriate for specific experimental conditions and for the calculation of particular length or mass parameters.

5-1.3 Absolute-Intensity Measurements

Intensity measurements that are compared quantitatively to the intensity of the primary x-ray beam are referred to as *absolute intensities*. Whereas intensities measured on only a relative scale permit the determination of a number of geometric (length, area, and volume) parameters of a polymer system, absolute-intensity measurements open the way to important additional parameters related to mass—in particular the molecular weight and the electron-density distribution. The various intensity formulas given in this chapter are based exclusively on measurement of both the incident and scattered x-rays in the plane of registration. (See Figure 5-2.) The measurement of absolute intensities is a formidable experimental assignment because of the fact that the intensities of the incident and scattered x-rays differ by several orders of magnitude. In the three experimental techniques described below emphasis is placed on diffractometric measurements. Nevertheless, photographic recording offers an advantage in bridging the great intensity difference between the direct and diffracted beams since the direct beam needs to be exposed only momentarily, whereas the scattered beam requires an exposure of several hours. Weakening of the primary beam by a known factor with the aid of a rotating sectorial diaphragm permits the relationship between the exposures of the direct and scattered beams to be ascertained with the necessary accuracy [10].

Attenuation by Means of Filters. Luzzati [11, 12] measured the primary-beam intensity with the aid of very accurately calibrated absorbing filters. This method is acceptable only if the primary beam is strictly monochromatic because of the strong dependence of x-ray absorption on wavelength. Radiation from a copper-target tube, operated below 30 kV (peak) to eliminate the harmonic wavelength $\lambda/3$, was reflected from a quartz monochromator to yield the $CuK\alpha_1$ wavelength λ and the first harmonic $\lambda/2$. Calibrated nickel filters were used as attenuators. Nickel is especially suited to this purpose because the intensity ratio $I_\lambda/I_{\lambda/2}$ is not changed as a result of absorption at these copper wavelengths. For a fuller description of Luzzati's apparatus the reader is referred to Section 2-4.3.

Attenuation by Means of a Perforated Rotor [13–15]. By interposing between the x-ray tube and counter an x-ray-opaque rotating disk containing a few small transmitting holes, the intensity of the primary beam can be weakened by an accurately known factor. Kratky and co-workers constructed such a device with four holes. The counting rate measured with the rotor in operation was first corrected for resolving-time losses. (See Section 2-4.2.) The counting rate that corresponded to the full intensity of the primary beam was then obtained by multiplying this corrected rotor-attenuated counting rate by the factor UL/f_4, U being the circumference of the circular path described by the holes, L the cross-sectional area of the primary beam at the point of its intersection with the rotor, and f_4 the sum of the areas of the four holes.

Reference to a Standard Sample.   A third method of determining the absolute intensity of x-ray scattering from an "unknown" is to compare it quantitatively with the intensity scattered by a standard sample (for example, a metal sol)[16]. As an inert, stable primary standard Kratky and co-workers[10, 17] have employed a gold sol, the scattering power of which has remained invariant for many years and which can be precisely calculated with the expression (see Section 5-3.1)

$$(\rho_1 - \rho_0)^2 \, w_1 w_0.$$

In this formula ρ_1 and ρ_0 are respectively the electron densities of the solute and solvent in moles of electrons per cubic centimeter, and w_1 and w_0 are the respective volume fractions. As an actual working standard a special polyethylene specimen† in the form of a sheet 2 mm thick has been employed in place of the standard gold sol[18]. Extensive tests[19] of this material as a working standard have demonstrated its high resistance to irradiation and constancy of scattering power over long periods of time. In contrast to metal sols the polyethylene platelets are convenient to handle and always in instant readiness for use. In the calibration of the working standard with respect to the primary gold sol care must be taken to make the appropriate correction for their different absorptions of the x-ray beam.

5-1.4   Slit versus Pinhole Collimation

The general theory of small-angle scattering was originally developed for primary beams of pointlike cross section, which in practice can be closely realized only by collimating the beam with very small

†Lupolen 1811M, supplied by the Badische Anilin- und Soda-fabrik, Ludwigshafen am Rhein, Germany.

pinholes. The compelling need for greater intensity has, however, led to widespread use of slit collimators, which in turn has necessitated a modified mathematical treatment of the experimental data. Two ways of handling the slit-distorted, or "smeared," data have been devised: (a) formulas have been derived to correct the scattered intensities for the collimation error (to "unsmear" the intensity curve) and (b) a special theoretical framework has been developed to obtain directly from the slit-distorted intensity curve the principal geometric and mass parameters that were previously obtainable only from the point-collimator intensities. The first method suffers from two disadvantages: the unsmearing formulas always involve certain approximations and the calculations themselves are laborious, although the second objection has been largely negated in recent years by the increasing availability of high-speed computers. The second method of handling the slit-distorted intensities has one important deficiency in that the theory does not yield the complete scattering curves for particles of various specific shapes and for different size distributions. Thus when information about particle shape or polydispersity is sought it is still necessary to have recourse to intensities obtained with pinhole collimators or by correction of data obtained with slit collimators in order to make the necessary comparisons of the experimental and theoretical scattering curves. We shall therefore give a synoptical treatment of the topic of collimator-error corrections, indicating the important literature references that contain detailed descriptions of the methods involved.

5-1.5 The Correction of Collimation Errors

Infinitely Long and Narrow Slits. For slits of arbitrary length but negligible width the experimental intensity curve $\tilde{I}(S)$ is related to the equivalent intensity function $I(S)$ for point collimation by the equation [20]

$$\tilde{I}(S) = \int_0^\infty W(\phi)\, I[(S^2 + \phi^2)^{1/2}]\, d\phi, \tag{5-7}$$

where $W(\phi)$ is a weighting function the form of which depends on the collimating system and ϕ is an auxiliary variable of integration with the same dimensions as S. For infinitely long and narrow slits $W(\phi) = 1$, and (5-7) has the solution [21, 22]

$$I(S) = -\frac{2}{\pi} \int_0^\infty \frac{\tilde{I}'[(S^2 + t^2)^{1/2}]\, dt}{(S^2 + t^2)^{1/2}}. \tag{5-8}$$

In this equation $\tilde{I}'$ is the first derivative of the experimental intensity curve with respect to S.

The unsmearing process according to the method of Guinier and Fournet[21] [represented by (5-8)] consists of three steps: (a) differentiation of the experimental curve $\tilde{I}(S)$, (b) division through by the scattering angle S, and (c) smearing of the resultant curve. Kratky, Porod, and Skala[23] have described a graphical procedure for performing the differentiation and a mechanical analog device for the smearing operation of step c. In general, methods based on the Guinier-Fournet equation (5-8) tend to result in appreciable errors as a result of the differentiation of the experimental curve, which is ordinarily known only to an accuracy of several percent.

Narrow Slits of Arbitrary Height.   When the weighting function $W(\phi)$ of (5-7) can be assumed to have the form

$$W(\phi) = \frac{2p}{\pi^{1/2}} \exp\left(-p^2\phi^2\right), \tag{5-9}$$

Kratky, Porod, and Kahovec[17] showed that $I(S)$ is given by

$$I(S) = -\frac{\exp\left(p^2 S^2\right)}{p\pi^{1/2}} \int\limits_0^\infty \frac{N'[(S^2+t^2)^{1/2}]\,dt}{(S^2+t^2)^{1/2}}, \tag{5-10}$$

in which

$$N(S) = \tilde{I}(S) \exp\left(-p^2 S^2\right) \tag{5-11}$$

defines the intensity function of which N' is the first derivative.

Schmidt and Hight[24] showed how the accuracy of the determination can be increased by rearranging (5-10) so as to express $I(S)$ in terms of $\tilde{I}(S)$ itself rather than its derivative. Their equation has the form

$$I(S) = \frac{1}{p\pi^{1/2}}\left(\frac{\tilde{I}(S)}{S} + \int\limits_S^\infty \frac{du}{(u^2-S^2)^{3/2}}\left\{S\,\tilde{I}(S) - u\tilde{I}(u)\exp\left[-p^2(u^2-S^2)\right]\right\}\right). \tag{5-12}$$

An additional advantage of (5-12) compared with (5-10) is that it facilitates the numerical calculations. Schmidt and Hight[24] describe in some detail a scheme for the calculations, including the use of a digital computer. More recently an improvement of the Schmidt-Hight technique has been presented[25]. It has the advantage of being less sensitive to random errors in the input data besides being considerably more accurate when such errors in the input are very small.

Slits of Finite Breadth. As a general rule the breadth of the beam can be kept so small in relation to the length that a breadth correction can be neglected[23]. The principal exception is for intensities recorded at very small angles. Thus Kratky and Miholic[26] in their investigation of air-swollen, regenerated cellulose applied breadth corrections only at angles that corresponded to Bragg spacings larger than 1000 Å.

In summarizing the treatment of breadth corrections given by Kratky, Porod, and Skala[23] we may note that the actual experimental intensity curve $\tilde{I}(S)$ in the plane of measurement (normal to the axes of the collimator slits) is the convolution of the pure small-angle-scattering curve of the specimen, $I(S)$, and the profile of the primary beam in the plane of measurement, $Q(S)$,

$$\tilde{I}(S) = \int_{-\infty}^{+\infty} I(S-x)Q(x)\,dx. \tag{5-13}$$

Expansion of $I(S-x)$ in a Taylor's series about the abscissa S gives

$$\tilde{I}(S) = \int_{-\infty}^{+\infty} \left[I(S) - I'(S)x + I''(S)\frac{x^2}{2} - \cdots \right] Q(x)\,dx$$

$$= I(S) - I'(S)\bar{x} + I''(S)\frac{\overline{x^2}}{2} - \cdots \tag{5-14}$$

The origin is next shifted to $x = \bar{x}$, the center of gravity of the primary-beam profile, whereupon additional transformations lead to the solution

$$I(S) = \tilde{I}(S) - \left[\frac{\tilde{I}(S+\xi) + \tilde{I}(S-\xi)}{2} - \tilde{I}(S) \right], \tag{5-15}$$

in which $\xi = (\overline{x^2})^{1/2}$. The authors then outline the actual steps in unsmearing a point in the $\tilde{I}(S)$ curve at S as follows:

1. Determine $\bar{x}$, the center of gravity of the primary-beam profile.
2. Shift the origin on the abscissa axis from 0 to $\bar{x}$.
3. Determine graphically the root-mean-square abscissa $\xi = (\overline{x^2})^{1/2}$.
4. Determine the ordinates of the $\tilde{I}(S)$ curve at the distance ξ to the right and left of the point S, and connect the two points with a chord.
5. The separation of the midpoint of the chord from the measured point $\tilde{I}(S)$ is the indicated correction, which is then applied *in the reverse direction* to the measured point.

In regard to the application of corrections for the collimation error it should be emphasized that the corrections for length and breadth errors are independent of each other and can be performed in any desired sequence. The only underlying assumption is that the cross-sectional profile of the primary beam is uniform over its entire length—a condition that is sufficiently well fulfilled in practice. The reader is referred to the literature for additional papers devoted to the theory of the correction of slit-collimation errors [27, 28] and to computer techniques for performing the calculations [29–31].

It is apropos at this point to define more exactly what is meant by measurement of the intensity of the primary beam [26]. When collimation-error corrections are not applied, the primary-beam intensity is the number of photons per second in the whole beam. If corrections are applied for the length collimation error, its intensity is the number of photons per second corresponding to unit length of the beam cross section. Designating the primary-beam intensity thus defined (and measured in the plane of registration) as P_0 and the scattered-beam intensity as I, we then define the *absolute intensity* I_n as the ratio I/P_0. It should be emphasized that, when the corrections for the length collimation error are applied, the ratio I/P_0 is unchanged if both the scattered- and direct-beam counting rates are divided by the length of the receiving slit, which is then a valid procedure.

In what follows we shall denote by I *and* I_n *relative and absolute intensities corresponding to idealized point collimation and by* $\tilde{I}$ *and* $\tilde{I}_n$ *intensities corresponding to collimation by infinitely long and narrow slits.*

5-2   GENERALIZED SYSTEMS

5-2.1   Scattering Power†

If no assumptions whatever are made about the nature of the inhomogeneities that give rise to small-angle scattering from a specimen, the only parameter that can be determined is the *mean-square fluctuation* of the *electron density*, $\overline{(\rho - \bar{\rho})^2}$, which may be

---

†In his publications Luzzati employs *normalized absolute intensities*, i_n and $\tilde{i}_n$, rather than I_n and $\tilde{I}_n$. These quantities are related as follows:

$$i_n = \frac{I_n}{\nu\eta} = \frac{I_n}{\lambda^2 i_e \eta},$$

$$\tilde{i}_n = \frac{\tilde{I}_n}{\nu\eta} = \frac{\tilde{I}_n}{\lambda^2 i_e \eta}.$$

We shall use i_n and $\tilde{i}_n$ only in Section 5-4.7, where it is desirable for consistency with the original literature.

called the scattering power of the system [11, 32]. With Kratky and Miholic [26] we define the following quantities:

m = distance between the incident and scattered rays in the plane of registration in centimeters.

a = specimen-to-receiver distance in centimeters.

d = specimen thickness in centimeters.

λ = wavelength of x-rays in centimeters.

$i_e = 7.9 \times 10^{-26}$, the Thomson-scattering constant of a free electron.

N = Avogadro's number.

Then in terms of mole electrons per cubic centimeter

$$\overline{(\boldsymbol{\rho} - \overline{\boldsymbol{\rho}})^2} = \left(\frac{4\pi}{i_e N^2}\right)\frac{1}{da\lambda^3} \int\limits_0^\infty m^2\, I_n(m)\, dm. \qquad \text{(K5-16)}$$

If η is the thickness of the specimen expressed in electrons per square centimeter of surface and $\nu = \lambda^2 i_e$ (λ expressed in angstroms), the relative mean-square fluctuation of the electron density in electrons per cubic angstrom may be written [11, 12]

$$\frac{\overline{(\rho - \overline{\rho})^2}}{\overline{\rho}} = \frac{4\pi}{\nu\eta} \int\limits_0^\infty s^2\, I_n(s)\, ds = \frac{2\pi}{\nu\eta} \int\limits_0^\infty s\, \tilde{I}_n(s)\, ds. \qquad \text{(L5-16)}$$

In relating (K5-16) and (L5-16) it may be noted that $m = a\lambda s$ and that

$$\boldsymbol{\rho} = \frac{\rho}{N} \times 10^{24} \quad \text{and} \quad d = \frac{\eta}{\overline{\rho}N}.$$

5-2.2   The Invariant Q or $\tilde{Q}$

Integral scattering intensities of the form $\int\limits_0^\infty s^2\, I(s)\, ds$ or $\int\limits_0^\infty s\, \tilde{I}(s)\, ds$ have been termed the *invariant* Q, or $\tilde{Q}$, by Porod [33] and shown by him to have very useful properties. With intensities measured in absolute units, I_n or $\tilde{I}_n$, the invariant permits the determination for any system of the mean-square fluctuation of the electron density by means of (5-16). It will be shown in a later section that for a system of particles in a solvent, each phase being of uniform electron density, the invariant based on relative intensities permits the calculation of the average volume of a particle.

In terms of the angular variable s, then, we define the invariant for fine-pinhole and infinitely-long-slit collimation by the respective expressions

$$Q_s = \int\limits_0^\infty s^2 I(s) \, ds \qquad \text{(pinholes)}, \qquad (5\text{-}17)$$

$$\tilde{Q}_s = \int\limits_0^\infty s \, \tilde{I}(s) \, ds \qquad \text{(slits)}. \qquad (5\text{-}18)$$

The quantities Q_s and $\tilde{Q}_s$ as well as Q and $\tilde{Q}$ expressed in terms of any other angular variable are conveniently related by

$$\tilde{Q} = 2Q. \qquad (5\text{-}19)$$

By means of equations parallel in form to (5-17) and (5-18) the invariant may be defined in terms of S, θ, m, etc. Thus

$$Q_m = \int\limits_0^\infty m^2 I(m) \, dm \qquad (5\text{-}20)$$

and

$$\tilde{Q}_m = \int\limits_0^\infty m \, \tilde{I}(m) \, dm. \qquad (5\text{-}21)$$

In transforming the invariant in terms of one variable to the invariant expressed in terms of another due care must be exercised that the constants of proportionality relating s, S, m, θ, etc., are taken account of. [See again (5-4), (5-5), and (5-6).] Thus we find

$$
\begin{aligned}
\frac{Q_m}{Q_s} &= (a\lambda)^3, & \frac{\tilde{Q}_m}{\tilde{Q}_s} &= (a\lambda)^2, \\[2mm]
\frac{Q_m}{Q_\theta} &= (2a)^3, & \frac{\tilde{Q}_m}{\tilde{Q}_\theta} &= (2a)^2, \\[2mm]
\frac{Q_\theta}{Q_s} &= \left(\frac{\lambda}{2}\right)^3, & \frac{\tilde{Q}_\theta}{\tilde{Q}_s} &= \left(\frac{\lambda}{2}\right)^2, \\[2mm]
\frac{Q_S}{Q_s} &= (2\pi)^3, & \frac{\tilde{Q}_S}{\tilde{Q}_s} &= (2\pi)^2.
\end{aligned}
\qquad (5\text{-}22)
$$

The experimental evaluation of the invariant is difficult because the scattered intensities at both very small and very large angles must be determined with acceptable accuracy. In the absence of particulate clustering or a large degree of polydispersity it is normally possible to determine the very-low-angle portion of the curve by extrapolating a plot of $m^2 I(m)$ or $m \, \tilde{I}(m)$ versus m to zero angle. The high-angle portion (tail end) of the intensity curve likewise poses special problems of measurement. In the next section it is shown that the accuracy of this portion of the curve can be improved by making use of a theoretical finding of Porod[33] that the course of the tail end should conform to a constant limiting value of $m^4 I(m)$ or $m^3 \tilde{I}(m)$.

5-3 SYSTEMS OF TWO PHASES, EACH OF UNIFORM ELECTRON DENSITY

5-3.1 Scattering Power

It is assumed that the electron densities in the two regions are uniform and separated by a sharp boundary. To facilitate the treatment one phase, of electron density ρ_0 and volume fraction w_0, will be termed the dispersant (solvent); and the other, of electron density ρ_1 and volume fraction w_1, the disperse phase (solute). For a system of two uniform phases the expression for the scattering power is

$$\overline{(\rho - \overline{\rho})^2} = (\rho_1 - \rho_0)^2 \, w_1 w_0 = (\Delta \rho)^2 \, w_1 w_0, \qquad \text{(K5-23)}$$

with $w_1 + w_0 = 1$. The validity of this expression does not depend on the degree of dispersion (dilution) within the system but only on $\Delta \rho$. We see that w_1 and w_0 can be calculated from a knowledge of $\Delta \rho$ and the scattering power, $\overline{(\rho - \overline{\rho})^2}$. On the other hand, the scattering power of the system can be precisely calculated from a knowledge of ρ_1, ρ_0, w_1, and w_0—as, for example, in the important application cited at the end of Section 5-1.3 of the preparation of a standard gold sol [10, 17].

Returning to (L5-16) with electron densities expressed in electrons per cubic angstrom, we can show that for the two-phase system as presently defined [11, 12]

$$\frac{\overline{(\rho - \overline{\rho})^2}}{\overline{\rho}} = \frac{4\pi}{\nu\eta} \int\limits_0^\infty s^2 I_n(s) \, ds = c_e (1 - \rho_0 \psi) \left[\rho_1 - \frac{\rho_0}{1 - c_e(1 - \rho_0 \psi)} \right]. \qquad \text{(L5-24)}$$

In this equation two new quantities are introduced:

c_e = electron concentration expressed as the ratio of the number of electrons in the solute to the number in the solution.

ψ = electronic partial specific volume, the volume of solution occupied by one electron of the solute (cubic angstroms per electron).

Since ρ_0, c_e, and ψ are presumed known, ρ_1 can be determined from experimental measurements of Q_s (or $\tilde{Q}_s$) in absolute-intensity units with the aid of (L5-24).

5-3.2   Rule of Constancy of $s^4 I(s)$ or $s^3 \tilde{I}(s)$

As already mentioned in the preceding section, for the two-phase situation Porod[33] enunciated the important principle that in theory the tail end of the scattering curve should conform to the asymptotic course s^{-4} for collimation by very small pinholes and to s^{-3} for collimation by infinitely long narrow slits. This is equivalent to saying that $s^4 I(s)$ and $s^3 \tilde{I}(s)$ [or $m^4 I(m)$ and $m^3 \tilde{I}(m)$, etc.] should assume constant values. In general this relationship can be most readily verified experimentally for systems of granular particles. For elongated or platelike particles the theoretical asymptotic behavior recedes into the extreme end of the tail portion of the scattering curve where the intensity is extremely weak and commonly distorted by other background factors[34].

5-3.3   Specific Inner Surface O_s

When it can be verified experimentally that the intensity in the tail of the scattering curve follows the theoretical course, an important structural parameter can be obtained therefrom — namely, the specific inner surface O_s, which we shall define as the ratio of the area of the phase interface O to the volume occupied by the disperse phase V_1. For collimation by infinitely long and narrow slits we can then calculate O_s in terms of the asymptotic intensity and invariant from the equation [33–35]

$$O_s = \frac{O}{V_1} = \frac{8\pi w_0}{a\lambda} \frac{\lim \, [m^3 \tilde{I}(m)]}{\tilde{Q}_m}, \qquad \text{(K5-25)}$$

and for ideal pinhole collimation [35]

$$O_s = \frac{2\pi^2 w_0}{a\lambda} \frac{\lim \, [m^4 I(m)]}{Q_m}. \qquad \text{(K5-26)}$$

If λ is expressed in angstroms, O_s is obtained in square angstroms per cubic angstrom (or Å^{-1}). In terms of $\tilde{I}(s)$, $\tilde{Q}_s$, ρ_0, ρ_1, c_e, and ψ Luzzati [11, 12] has derived the following expression, which can be shown to be equivalent to (K5-25):

$$O_s = \frac{8\pi \lim \, [s^3 \tilde{I}(s)]}{\tilde{Q}_s} \left[1 - \left(\frac{\rho_0}{\rho_1 - \rho_0} \right) \frac{c_e(1 - \rho_0 \psi)}{1 - c_e(1 - \rho_0 \psi)} \right], \qquad \text{(L5-27)}$$

yielding O_s expressed in Å^{-1} if, as is usually the case, s is expressed in Å^{-1}. It should be noted that absolute intensities are not required for the determination of the specific surface.

Figure 5-3 shows plots of $s^3 \tilde{i}_n(s)$ versus s^3 for six saline solutions of the protein lysozyme at different concentrations ranging from 1.01 to 4.95%[36]. They exhibit a satisfactory approximation to straight horizontal lines for large s, which confirms the applicability of the model of uniform electron density in the disperse and dispersant phases, and justifies the use of (K5-25) or (L5-27) for the determination of the specific surface.

Failure of the intensity curve to conform to s^{-4} or s^{-3} at higher angles indicates a lack of homogeneity in the electron density of the solute, of the solvent, or of both. The asymptotic portion of the curve occurs at the higher scattering angles, which is just where electron-density variations of atomic dimensions would be expected to begin to affect the scattering curve. Luzzati, Witz, and Nicolaieff[36] showed that such electron-density fluctuations lead to an additional constant term in the $\tilde{I}$ scattering curve. Combining this effect with Porod's rule of the decrease of $\tilde{I}$ with the third power of the scattering angle, we can write for the overall form of the tail of the intensity curve

$$\tilde{I}(m) = K_1 m^{-3} + K_2$$

or

$$m^3 \cdot \tilde{I}(m) = K_1 + m^3 \cdot K_2. \tag{K5-28}$$

Thus a plot of $m^3 \cdot \tilde{I}(m)$ versus m^3 should give a straight line with positive slope K_2. This slope, then, determines the correction, $m^3 K_2$, which must be subtracted from the scattering curve, $m^3 \cdot \tilde{I}(m)$. We

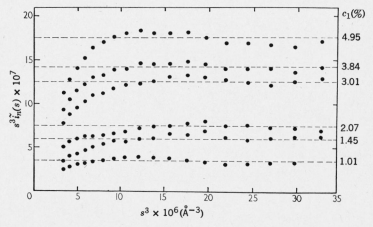

Figure 5-3 Saline solutions of lysozyme. Plots of $s^3 \tilde{i}_n(s)$ versus s^3 for six concentrations ranging from 1.01 to 4.95% as indicated. (Luzzati, Witz, and Nicolaieff[36].)

proceed in a similar way for intensities expressed in terms of other angular variables, $I(s)$, $I(\theta)$, etc. For further information on this correction the reader is referred to the literature [10, 36–38].

5-3.4 Correlation Function

Of particular importance in the study of dense two-phase systems is the concept of the *correlation function* (Debye et al. [39, 40]) or *characteristic* (Porod [33, 41]) defined by

$$\gamma(r) = \frac{\overline{(\Delta\rho)_1 \cdot (\Delta\rho)_2}}{(\Delta\rho)^2}, \qquad (5\text{-}29)$$

where $(\Delta\rho)_1$ and $(\Delta\rho)_2$ are the local deviations of the electron density from the average value $\bar{\rho}$ at points 1 and 2 separated by a distance r. The correlation function has the physical meaning of an average distribution of scattering matter in the vicinity of an origin which assumes all locations within the system with equal probability. It is therefore a model with continuously varying electron density that is *scattering equivalent* to the particular inhomogeneous system under consideration [34].

The correlation function is related to the scattering curve by a Fourier transformation, the same relationship that exists between the Patterson function and the distribution of (hkl) intensities of conventional crystal-structure analysis. (See Appendix 1.) For a *random* distribution of two phases, each of uniform electron density (ρ_1 and ρ_2) – such as holes in a solid or particles interspersed with voids – Debye, Anderson, and Brumberger [40] have shown that

$$\gamma(r) = \exp\left(\frac{-r}{\bar{l}_c}\right) \qquad (5\text{-}30)$$

and

$$\frac{O}{V} = \frac{4w_0 w_1}{\bar{l}_c}. \qquad (5\text{-}31)$$

The term O/V is the specific internal surface defined in terms of the overall volume of the system, V, and $\bar{l}_c$ is the *correlation distance* or *length*, a measure of the size of the inhomogeneities. From (5-30) it is easy to show that $\bar{l}_c$ is the integral breadth of the correlation function. Porod and associates [33, 34, 41] also refer to $\bar{l}_c$ as the *reduced inhomogeneity length*.

Somewhat easier to visualize are the *transversal*, or *inhomogeneity*, lengths $\bar{l}_0$ and $\bar{l}_1$, given by [34, 37]

$$\bar{l}_0 = \frac{4w_0}{O/V} \quad \text{and} \quad \bar{l}_1 = \frac{4w_1}{O/V}. \tag{5-32}$$

By comparison with (5-31) it can be seen that

$$\frac{1}{\bar{l}_c} = \frac{1}{\bar{l}_0} + \frac{1}{\bar{l}_1}. \tag{5-33}$$

To visualize $\bar{l}_0$ and $\bar{l}_1$ we imagine the system to be pierced in all directions at random by rays, as suggested by Figure 5-4. Then the mean length of all the line segments intercepted by the dispersant phase (white portions of arrows) is $\bar{l}_0$, and the mean length of all the segments intercepted by the disperse phase (black portions of arrows) is $\bar{l}_1$.

5-4 DILUTE PARTICULATE SYSTEMS – TWO PHASES, EACH OF UNIFORM ELECTRON DENSITY

5-4.1 General Principles

Among the various inhomogeneous systems that can scatter x-rays at small angles, dilute particulate systems are most amenable to definitive theoretical interpretation. In the light of this favorable out-

Figure 5-4   Physical concept of the *transversal*, or *inhomogeneity*, length. (Kratky [37]; reprinted from *Pure and Applied Chemistry* by permission of the International Union of Pure and Applied Chemistry and Butterworths Scientific Publications.)

look it seems unfortunate that the application of the well-developed theory of particulate scattering to polymer problems is restricted principally to polymers composed of large, discrete molecules (such as proteins), which can be dispersed in suitable solvents to an essentially monomolecular state. In solid polymers, such as synthetic and natural fibers, the primary particles are almost never well defined and at best are elongated, rod-shaped or platelike in form, and rather densely packed. In some instances such fibers, particularly the natural varieties, can be sufficiently swollen with a solvent or air to permit the theory of scattering by dilute systems to be applied with some degree of validity.

We wish to emphasize that experimental parameters free of any ambiguity can be derived from the small-angle-scattering pattern only if all the particles in the system are identical in size and shape (monodisperse) and separated sufficiently to eliminate interparticle interferences. Even if such idealized conditions exist, the shape must be known if the size is to be specified unequivocally, and vice versa. If the system is polydisperse to a significant degree, resolution of the shape and size parameters is not possible and only an average, or effective, parameter can be deduced.

Law of Guinier.   Guinier[1–3] first showed that for a dilute, monodisperse system in which the particles assume all orientations with equal probability the scattered intensity can be described by a power series as follows:

$$I(S) = I(0)\{1 - \tfrac{1}{3}S^2R^2 + \cdots\}. \tag{5-34}$$

The function $I(0)$ is the intensity scattered at zero angle, which is proportional to the total number of electrons in the system irradiated, and R is the electronic radius of gyration, which is the root mean square of the distances of all the electrons from the electronic center of gravity of the representative particle[42],

$$R = \left(\frac{\sum_k f_k r_k^2}{\sum_k f_k} \right)^{1/2}. \tag{5-35}$$

In (5-35) f_k is the scattering factor of the kth electron, which is a distance r_k from the electronic center of gravity. To a good approximation (5-34) may be written in the form

$$I(S) = I(0) \exp{(-\tfrac{1}{3}S^2R^2)} \tag{5-36}$$

or in terms of the angular variables m and s as

$$I(m) = I(0) \exp\left(-\frac{4}{3}\frac{\pi^2 m^2}{a^2 \lambda^2} R^2\right), \tag{K5-36}$$

$$I(s) = I(0) \exp\left(-\tfrac{4}{3}\pi^2 s^2 R^2\right), \tag{L5-36}$$

which is the familiar law of Guinier. In logarithmic form (5-36) becomes

$$\log_e I(S) = \log_e I(0) - \tfrac{1}{3}S^2 R^2, \tag{5-37}$$

and (K5-36) and (L5-36) assume corresponding forms.

When the law of Guinier is closely obeyed by an actual system (5-36) states that a plot of the intensity curve $I(S)$ versus S will be Gaussian, and (5-37) states that a plot of $\log_e I(S)$ versus S^2 will be linear over a large angular range with an ordinate intercept of $\log_e I(0)$ and a slope of $-\tfrac{1}{3}R^2$. These characteristics are illustrated in Figure 5-5. The law of Guinier is obeyed most closely over a large angular range by monodisperse systems of approximately spherical particles. However, we emphasize that when the logarithmic plot becomes nonlinear with increasing angle as a result of departures from spherical shape or monodispersity, the limiting slope as S approaches zero is still related to the mean radius of gyration of the particles by this fundamental law. In this connection Guinier and Fournet[43] observe that "the radius of gyration is the only precise parameter which can be determined by small-angle-scattering experiments without invoking supplementary hypotheses."

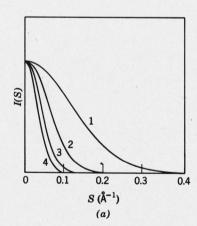 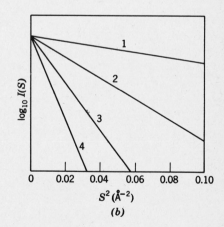

Figure 5-5 Small-angle-scattering curves conforming to the law of Guinier: (a) $I(S)$ versus S for particles with dimensions in the ratio $1 : 2 : 3 : 4$; (b) $\log_{10} I(S)$ versus S^2 corresponding to the four curves in (a).

Internal Solvation [26].   If the solute particles are swollen by absorption of solvent so that the volume fraction of unswollen solute w_1 becomes w_1', the internal solvation may be defined by a homogeneous swelling factor

$$q = \frac{w_1'}{w_1}.$$ (K5-38)

In view of the definition of Q_m expressed by (5-20), we may write (K5-16) in the form

$$\overline{(\boldsymbol{\rho} - \bar{\boldsymbol{\rho}})^2} = \frac{4\pi Q_m}{i_e N^2 da\lambda^3 P_0},$$ (K5-39)

which becomes for a dilute, two-component system of solute particles and solvent

$$Q_m = \left(\frac{1}{4\pi}\right) i_e N^2 da\lambda^3 P_0 \{ (\boldsymbol{\rho}_1 - \boldsymbol{\rho}_0)^2 \, w_1 w_0 \}.$$ (K5-40)

In (K5-39) and (K5-40) P_0 is the primary-beam intensity as defined in Section 5-1.5. For the solution containing the solute swollen by the factor q, w_1 of (K5-40) becomes $w_1 q$, and at the same time the electron-density difference $(\boldsymbol{\rho}_1 - \boldsymbol{\rho}_0)$ is decreased by the factor $1/q$. Hence the quantity in braces becomes

$$\frac{(\boldsymbol{\rho}_1 - \boldsymbol{\rho}_0)^2}{q^2} \cdot w_1 q \cdot w_0 = \frac{1}{q} \{ (\boldsymbol{\rho}_1 - \boldsymbol{\rho}_0)^2 \, w_1 w_0 \},$$

which shows that the experimental value of Q_m will be reduced from that given by (K5-40) by the factor $1/q$. Thus the ratio of Q_m calculated with (K5-40) (Q_{calc}) and its experimental value (Q_{exp}) is simply

$$q = \frac{Q_{\text{calc}}}{Q_{\text{exp}}},$$ (K5-41)

which provides the basis for an experimental determination of the internal solvation. For a solution of globular particles Luzzati [12, 36] has defined a related function α, the internal-solvation ratio. (See p. 304.)

5-4.2   Globular Particles

The law of Guinier is fulfilled most closely by globular particles; that is, particles none of whose dimensions is negligible with respect to the others. Table 5-1 gives the calculated radii of gyration of a

Table 5-1 Specific Forms of Radius of Gyration for
Bodies of Particular Shapes[a]

Body or Section	R^2
Ellipse, semiaxes a and b	$\frac{1}{4}(a^2+b^2)$
Sphere, radius r	$\frac{3}{5}r^2$
Hollow sphere with radii r_1 and r_2	$\frac{3}{5}(r_1^5-r_2^5)/(r_1^3-r_2^3)$
Ellipsoid, semiaxes a, b, c	$\frac{1}{5}(a^2+b^2+c^2)$
Prism, edges A, B, C	$\frac{1}{12}(A^2+B^2+C^2)$
Elliptic cylinder of height h and semiaxes a and b	$\frac{1}{4}(a^2+b^2+\frac{1}{3}h^2)$
Prisms and elliptic cylinders, in terms of height h and radius of gyration R_q of the basal area	$R_q^2+\frac{1}{12}h^2$
Hollow cylinder of height h and radii r_1 and r_2	$\frac{1}{2}(r_1^2+r_2^2)+\frac{1}{12}h^2$

[a]Data from [10].

number of bodies of simple and regular shape. It can be shown that a solid sphere has the smallest possible radius of gyration for a given volume. In terms of its radius r_{sp} its radius of gyration is

$$R_{sp} = r_{sp}\left(\frac{3}{5}\right)^{1/2}. \tag{5-42}$$

Thus an anisotropy, or shape, factor of a particulate system can be defined by the ratio of the experimental radius of gyration to that of a sphere of equal volume [10]:

$$f_a = \frac{R_{exp}}{R_{sp}}. \tag{K5-43}$$

The larger the value of f_a, the more anisotropic the shape of the particle; and the closer its approach to unity, the more nearly spherical is its shape. Clearly, f_a can be calculated only if the volume of the particle in question can be ascertained in some way; for example, by x-ray measurements as explained on p. 304 or from a knowledge of the density and an independent determination of the molecular weight (for a dispersion of discrete molecular particles such as a globular protein).

For scattering curves of dilute globular systems measured with infinitely long narrow slits, a plot of $\log_e \tilde{I}(m)$ versus m^2 is still linear except that the ordinate intercept and limiting slope assume new values,

$$\tilde{I}(0) \quad \text{and} \quad -\frac{4}{3}\frac{\pi^2}{a^2\lambda^2}\tilde{R}^2,$$

respectively. It can be shown that the theoretical expressions (K5-36) and (L5-36) must be replaced by

$$\tilde{I}(m) = \tilde{I}(0)\,\exp\left(-\frac{4}{3}\frac{\pi^2 m^2}{a^2\lambda^2}\tilde{R}^2\right) + Z(m), \qquad \text{(K5-44)}$$

$$\tilde{I}(s) = \tilde{I}(0)\,\exp\left(-\tfrac{4}{3}\pi^2 s^2 \tilde{R}^2\right) + \phi(s). \qquad \text{(L5-44)}$$

Thus $\tilde{I}(m)$ can be decomposed into two terms—a Gaussian term and a function $Z(m)$ that must be evaluated and subtracted from the experimental intensity curve $\tilde{I}(m)$ before the logarithmic plot for the extraction of $\tilde{R}$ is prepared. To determine $Z(m)$ the experimental curve $\tilde{I}(m)$ is plotted as $\log_e \tilde{I}(m)$ versus m^2; from its limiting slope as m approaches zero the equivalent *Gaussian intensity curve* is calculated and is then subtracted from the original $\tilde{I}(m)$ curve. The *residual curve* is the function $Z(m)$. The function $\phi(s)$ of (L5-44) can be derived in an analogous manner. The term $\tilde{R}$ is an *apparent* radius of gyration. The *true* radius of gyration can be derived from $\tilde{R}$, $\tilde{I}(0)$, and $Z(m)$ by means of the following relationship[10]:

$$R^2 = \frac{\tilde{R}^2 + \left[\dfrac{9\lambda^3 a^3}{16\pi^3}\left(\dfrac{3}{\pi}\right)^{1/2}\displaystyle\int_0^\infty m^{-4} Z(m)\,dm\right][\tilde{I}(0)\cdot\tilde{R}]^{-1}}{1 - \left[\dfrac{\lambda a}{2\pi}\left(\dfrac{3}{\pi}\right)^{1/2}\displaystyle\int_0^\infty m^{-2} Z(m)\,dm\right][\tilde{I}(0)\cdot\tilde{R}]^{-1}}\cdot \qquad \text{(K5-45)}$$

In terms of $\phi(s)$ the equivalent expression is[11, 12]

$$R^2 = \frac{\tilde{R}^2 + \left[\dfrac{9(3\pi)^{1/2}}{16\pi^4}\displaystyle\int_0^\infty s^{-4}\phi(s)\,ds\right][\tilde{I}(0)\cdot\tilde{R}]^{-1}}{1 - \left[\dfrac{(3\pi)^{1/2}}{2\pi^2}\displaystyle\int_0^\infty s^{-2}\phi(s)\,ds\right][\tilde{I}(0)\cdot\tilde{R}]^{-1}}\cdot \qquad \text{(L5-45)}$$

If the intensity extrapolated to zero angle is measured on an absolute scale, the mass of one particle in molecular-weight units can be determined with the aid of the relation[10, 17]

$$M = \frac{I_n(0)\,a^2}{i_e N d c_1 (z_1 - \bar{v}_1\rho_0)^2}\cdot \qquad \text{(K5-46)}$$

When a particle can be considered identical with one molecule, as in a dilute dispersion of a globular protein, M is the molecular weight. Equation K5-46 introduces the following new quantities:

c_1 = concentration of solute in grams per cubic centimeter of solution,

z_1 = number of mole electrons in 1 g of solute,

$$= \frac{\text{sum of atomic numbers}}{\text{sum of atomic weights}}$$

$$= \frac{\sum\limits_i n_i Z_i}{\sum\limits_i n_i A_i},$$

n_i = stoichiometric proportion of ith solute element of atomic number Z_i and atomic weight A_i,

$\bar{v}_1$ = partial specific volume of solute in cubic centimeters per gram.

We may note that the quantity $(z_1 - \bar{v}_1 \rho_0)$ represents the excess mole electrons, or contrast in electron density between the solute and solvent, which is responsible for the intensity of the small-angle scattering. An expression that is equivalent to (K5-46) but gives the number of electrons in one solute particle, m_1,

$$m_1 = \frac{I_n(0)}{\nu \eta c_e (1 - \rho_0 \psi)^2}, \tag{L5-46}$$

has been derived by Luzzati [11, 12]. For a monodisperse solute the molecular weight is then

$$M = \frac{m_1}{z_1}. \tag{5-47}$$

If the intensity at zero angle is obtained by extrapolation of intensities measured with slits rather than pinholes, $\tilde{I}_n(0)$ must be converted to $I_n(0)$ prior to its use in (K5-46) or (L5-46). The relationship is

$$I_n(0) = 2\left(\frac{\pi}{3}\right)^{1/2} \tilde{I}_n(0) \frac{\tilde{R}}{a\lambda} - \frac{1}{\pi} \int\limits_0^\infty m^{-2} Z(m)\, dm \tag{K5-48}$$

or

$$I_n(0) = 2\left(\frac{\pi}{3}\right)^{1/2} \tilde{I}_n(0) \tilde{R} - \frac{1}{\pi} \int\limits_0^\infty s^{-2} \phi(s)\, ds. \tag{L5-48}$$

The function $Z(m)$ or $\phi(s)$ is evaluated as described earlier except in terms of absolute rather than relative intensities.

From the intensity at zero angle and the invariant another important quantity can be derived—namely, the volume of one particle,

$$v_1 = \frac{I(0)}{2\pi \int_0^\infty s\,\tilde{I}(s)\,ds} = \frac{I(0)}{2\pi\tilde{Q}_s}, \tag{L5-49}$$

which may also be expressed in terms of Q_s:

$$v_1 = \frac{I(0)}{4\pi Q_s}. \tag{L5-50}$$

It is to be noted that absolute intensities are not required for the determination of the particle volume.

Two additional parameters can be determined from the intensity curve in absolute units: ρ_1, the electron density of the solute particles; and α, the internal-solvation ratio, which Luzzati defines as the ratio of the number of electrons swelling one solute particle to the number of electrons in the unswollen particle. The theoretical expressions are [11, 36]

$$\rho_1 - \rho_0 = \frac{2\pi \int_0^\infty s\,\tilde{I}_n(s)\,ds}{\nu\eta c_e(1-\rho_0\psi)} + \rho_0 c_e\,(1-\rho_0\psi), \tag{L5-51}$$

$$\alpha = \rho_0 \frac{1-\rho_1\psi}{\rho_1-\rho_0}. \tag{L5-52}$$

Equation L5-51 derives from the more general expression (L5-24) on the assumption that c_e is small. The assumption of monodispersity is not required for the determination of ρ_1 and α. It is worthwhile to note that Luzzati's α and Kratky's q [see (K5-38)] are related as follows:

$$q = \alpha + 1. \tag{5-53}$$

5-4.3 Rodlike Particles

It is assumed that the rods are rigid and have a uniform cross section and that the length l may be variable but, in any case, very large (specifically $ls \gg 1$ within the experimental range of s). It is also assumed that correlations in position and orientation among the particles are negligible. Three parameters can be determined—

namely, the radius of gyration of the cross section, R_c; the cross-sectional area A_c; and the mass per unit length of rod, which will be expressed in number of electrons μ or number of molecular-weight units μ per angstrom. Relative intensities suffice for the determination of R_c, but absolute intensities are required for the mass per unit length.

For ideal point collimation the scattered intensity in normalized absolute units may be written [11, 44]

$$\frac{I_n(s)}{\nu\eta} = \frac{1}{2s}(1 - \rho_0\psi)^2 c_e\mu(1 - 2\pi^2s^2R_c^2 + \cdots), \qquad (\text{L5-54})$$

and for "infinite" slit collimation [11]

$$\frac{\tilde{I}_n(s)}{\nu\eta} = \tfrac{1}{2}(1 - \rho_0\psi)^2 c_e\mu \exp(-\pi^2s^2R_c^2)K_0(\pi^2s^2R_c^2). \qquad (\text{L5-55})$$

In these expressions μ is the number of electrons per angstrom and $K_0(\pi^2s^2R_c^2)$ is a modified Bessel function of zeroth order [45, 46], the value of which decreases monotonically with increase in the argument [47]. Numerical values of this Bessel function have been published [48].

In order to determine R_c from *relative* intensities measured with fine pinholes (L5-54) may be simplified to

$$(s \cdot I)_s = (s \cdot I)_0 \exp(-2\pi^2s^2R_c^2), \qquad (5\text{-}56)$$

or, in logarithmic form,

$$\log_{10}(s \cdot I)_s = \log_{10}(s \cdot I)_0 - \frac{2\pi^2s^2R_c^2}{2.303}. \qquad (5\text{-}57)$$

From the magnitude of the slope of a plot of $\log_{10}(s \cdot I)_s$ versus s^2 the radius of gyration of the cross section is given by

$$R_c = \frac{1}{\pi}[\tfrac{1}{2} \times 2.303 \times (\text{slope})]^{1/2}, \qquad (5\text{-}58)$$

or, from a plot of $\log_{10}(m \cdot I)_m$ versus m^2, by

$$R_c = \frac{a\lambda}{2\pi}[2 \times 2.303 \times (\text{slope})]^{1/2}. \qquad (\text{K5-58})$$

The ordinate intercepts yield the products $(s \cdot I)_0$ and $(m \cdot I)_0$, respectively.

Luzzati solved (L5-55) for $(1 - \rho_0\psi)^2 c_e\mu$ and R_c simultaneously by plotting $\log_{10}(\tilde{I}_n(s)/\nu\eta)$ versus $\log_{10} s$ and then determining the magnitude of the two-dimensional translation required to make the experimental curve coincide with the curve calculated for the reference function,

$$\log_{10}\left[\tfrac{1}{2}\exp\left(-\pi^2 s^2\right)K_0(\pi^2 s^2)\right],$$

plotted against $\log_{10} s$. The translation required along the abscissae is proportional to R_c; and that along the ordinates, to $(1-\rho_0\psi)^2 c_e\mu$, from which the absolute values of R_c and μ can be calculated. In the process of attaining a coincidence of the two curves no rotation is allowed, only a pure translation.

The mass per unit length of rod can be found more simply from pin-hole-collimated data (or slit-collimated data corrected for the collimation error). Kratky and Porod[49] and Kratky[10, 44] have shown that the molecular-weight units per angstrom can be expressed in terms of the angular variable m as

$$\mu = \frac{2}{i_e N\lambda}\frac{(m\cdot I_n)_0 a}{dc_1(z_1-\bar{v}_1\rho_0)^2}. \tag{K5-59}$$

The reader may compare this expression with (K5-46) for the molecular weight of a globular particle. The product $(m\cdot I_n)_0$ is obtained from the ordinate intercept of a plot of $\log_{10}(m\cdot I)_m$ versus m^2, as mentioned earlier. The absolute intensity I_n is based on the direct-beam intensity measured in the plane of registration. (See again Section 5-1.3.) The coefficient $(2/i_e N\lambda)$ has the numerical value 27.3 for $CuK\alpha$ radiation.

Analogous to (L5-49) and (L5-50) relating v_1 to $I(0)$ for globular particles is the following equation relating to A_c to $(s\cdot I)_0$ for very elongated rodlike particles [10]:

$$A_c = \frac{(s\cdot I)_0}{2\pi Q_s} = \frac{(s\cdot I)_0}{\pi \tilde{Q}_s}. \tag{5-60}$$

In terms of the angular coordinate m (5-60) becomes [10]

$$A_c = \frac{(m\cdot I)_0}{Q_m}\cdot\frac{(a\lambda)^2}{2\pi} = \frac{(m\cdot I)_0}{\tilde{Q}_m}\cdot\frac{(a\lambda)^2}{\pi}. \tag{K5-60}$$

It will be seen that only relative intensities are needed for the determination of A_c.

5-4.4   Lamellar Particles [10]

When the thickness is very small in relation to the dimensions in the plane of the lamallae, it is possible to determine the mass per unit area and the thickness. In analogy to (K5-46) and (K5-59) for globular and rodlike particles, Kratky has shown that for lamellar particles the molecular-weight units per square angstrom are given by

$$\frac{M}{\text{Å}^2} = \frac{55.6\,(m^2 \cdot I_n)_0}{dc_1\,(z_1 - \bar{v}_1\rho_0)^2}.$$ (K5-61)

For lamellar particles a plot of $\log_{10}\,(m^2 \cdot I_n)$ versus m^2 is linear as m approaches zero, and the ordinate intercept yields the numerical value of $(m^2 \cdot I_n)_0$.

In analogy to (L5-49) for the volume of a globular particle and (K5-60) for the cross-sectional area of a rodlike particle the equation

$$t = \frac{(m^2 \cdot I)_0}{Q_m} \cdot \frac{a\lambda}{2} = \frac{(m^2 \cdot I)_0}{\tilde{Q}_m} \cdot a\lambda$$ (K5-62)

yields the thickness t of a lamellar particle in terms of the ratio $(m^2 \cdot I)_0/Q_m$ or $(m^2 \cdot I)_0/\tilde{Q}_m$. Equations K5-61 and K5-62 may of course be transformed to equivalent equations in terms of s or any other desired angular variable.

5-4.5   Linear Polymers in Solution [37]

Beginning with the random-flight model of an isolated linear polymer molecule [50], it is natural to proceed to the concept of polymer chains with persistence of direction, for which a reasonable theoretical groundwork has been laid relating the small-angle-scattering curve to the parameter called *persistence length* [51–55]. We visualize a chain consisting of n straight elements of length z, resulting in a fully extended length of $nz = l$. Now we postulate that each successive element makes an angle of Y with the direction of its predecessor but with random azimuth, as shown in Figure 5-6. A model of such a chain with $\cos Y = 0.8$ and $n = 100$ is shown in Figure 5-7. It can be shown that the entire change of direction of the final (nth) element with respect to the first is given by

$$(\cos Y)^n = (\cos Y)^{l/z}.$$

In the limiting case of a continuously curved (wormlike) chain $z \to 0$ and $\cos Y \to 1$, so that $(\cos Y)^{l/z}$ may be replaced by the negative exponential $\exp(-l/y)$, where y is defined as the *persistence length*.

The magnitude of y is a measure of *the tendency to maintain a given direction*. It can be shown on statistical grounds that the mean-square end-to-end distance of a chain of extended length l and persistence

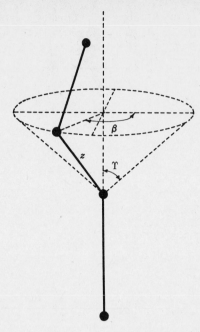

Figure 5-6 Portion of a chain consisting of three straight links. Length of link z, direction angle Υ, azimuth β. (Kratky[37]; reprinted from *Pure and Applied Chemistry* by permission of the International Union of Pure and Applied Chemistry and Butterworths Scientific Publications.)

length y is

$$\overline{r^2} = 2y\,(l - y + y e^{-l/y}).\tag{5-63}$$

In relation to our original link-chain model with straight segments z, the number of segments for a very long chain is related to the persistence length by

$$n = \frac{y}{z} = \frac{1}{2}\left(\frac{1 + \cos \Upsilon}{1 - \cos \Upsilon}\right).\tag{5-64}$$

From (5-63) it can be seen that for l very large

$$\overline{r^2} = 2\,yl.\tag{5-65}$$

The radius of gyration can then be obtained in terms of y and l:

$$R = \frac{(2yl)^{1/2}}{6^{1/2}} = (\tfrac{1}{3}\,yl)^{1/2}.\tag{5-66}$$

Theoretical considerations show that the scattering curves of

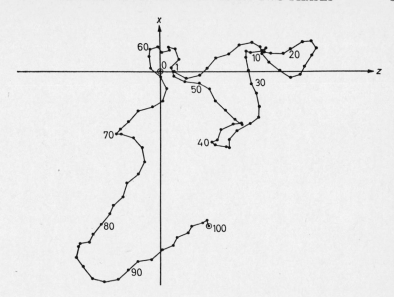

Figure 5-7 Chain of 100 straight links with $\cos Y = 0.8$. Projection in the xz plane. (Kratky[37]; reprinted from *Pure and Applied Chemistry* by permission of the International Union of Pure and Applied Chemistry and Butterworths Scientific Publications.)

statistically coiled linear polymer molecules consist of four zones, of which the three pertinent to the present discussion are shown in Figure 5-8. Proceeding from the smallest to larger angles, we have consecutively

Zone 1. I versus θ Gaussian, $\log I$ versus θ^2 linear.
Zone 2. I versus θ^{-2} linear, $I \cdot \theta^2$ versus θ constant.
Zone 3. I versus θ^{-1} linear, $I \cdot \theta$ versus θ constant.

The persistence length can be related in a direct manner to θ^*, the angle of transition between zones 2 and 3, by

$$y = \frac{3\lambda}{2\pi^2\theta^*} = \frac{\lambda}{6.57\theta^*}. \tag{5-67}$$

This transition angle θ^* can be most clearly defined by means of a plot of $I \cdot \theta^2$ versus θ, as illustrated in Figure 5-8b, in which it is the intersection point of the horizontal line of zone 2 and the straight line of zone 3 (which extrapolates to the origin). In terms of the Bragg spacing d^* corresponding to θ^* (5-67) becomes

$$y = \frac{d^*}{3.29}. \tag{5-68}$$

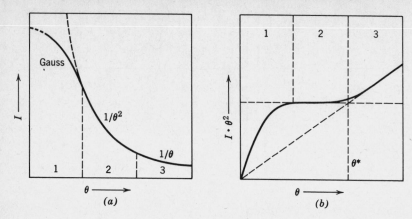

Figure 5-8   The three zones that comprise the low-angle portion of the scattering curve of a coiled linear polymer. (Kratky[37]; reprinted from *Pure and Applied Chemistry* by permission of the International Union of Pure and Applied Chemistry and Butterworths Scientific Publications.)

A more thorough statistical treatment of the scattering curve [54] gives in place of (5-68)

$$y = \frac{d^*}{2.73}, \tag{5-68'}$$

which is to be preferred in the quantitative interpretation of experimental intensity curves.

Figure 5-9 shows the $I \cdot m^2$ versus m scattering curves of solutions of high-molecular-weight nitrated cotton in acetone at concentrations of 6.65, 2.0, and 1.0%. The intensity zone 1 occurs at inaccessibly small angles, but zones 2 and 3 appear as theoretically expected. As the concentration diminishes, the transition point shifts toward lower angles, corresponding to an increase of the persistence length. Calculation of y according to (5-68'), followed by extrapolation to zero concentration, leads to $y = 110$ Å. It may be remarked that the numerical results depend somewhat sensitively on both the concentration and the nature of the solvent. "Good" solvents, like low concentrations, tend to result in larger values of y.

The theory of scattering by rodlike particles can also be applied to coiled linear polymers in solution, permitting the derivation of R_c, A_c, and μ with (5-58), (K5-60), and (K5-59), respectively. However, since the molecules consist only of single chains, R_c, A_c, and μ will be relatively small, and their numerical values will depend on zone 3 rather than the innermost portion of the scattering curve. Thus

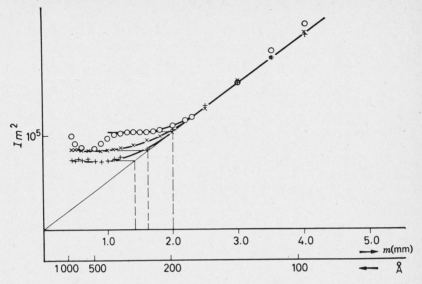

Figure 5-9 Plots of $I \cdot m^2$ versus m for solutions of high-molecular-weight nitrated cotton in acetone. Concentrations: o $= 6.65$, $\times = 2.0$, and $+ = 1.0\%$. (Kratky[37]; reprinted from *Pure and Applied Chemistry* by permission of the International Union of Pure and Applied Chemistry and Butterworths Scientific Publications.)

the Guinier plot of $\log(s \cdot I(s))$ versus s^2 based on this zone will have a very small slope, and a plot of $s \cdot I(s)$ versus s will be practically horizontal in zone 3 but rise steeply in zone 2, as illustrated in Figure 5-10. It also follows that the average value of $s \cdot I(s)$ in zone 3 may be taken directly as $(s \cdot I)_0$, and extrapolation to zero angle is not required.

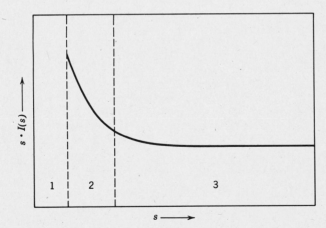

Figure 5-10 Plot of $s \cdot I(s)$ versus s for a linear polymer in solution.

When zone 1 is experimentally accessible, as in the case of solutions of low-molecular-weight cellulose nitrate in acetone, the slope of a Guinier plot of log $I(s)$ versus s^2 leads to the overall radius of gyration of the three-dimensionally coiled molecule [see (5-37)]. The limiting slopes of the four Guinier plots shown in Figure 5-11 yield R values of 83.5, 115.5, 89.0, and 133.0 Å. The results for these cellulose nitrate solutions are summarized in Table 5-2.

Calculations of the complete small-angle-scattering curves corresponding to wormlike (continuously curved) molecules of infinite

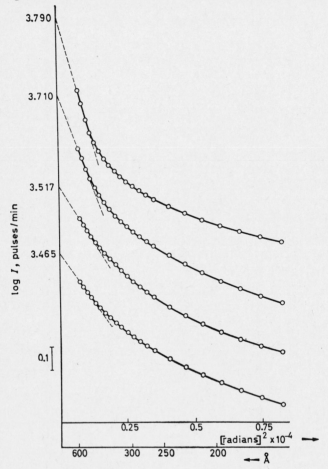

Figure 5-11   Guinier plots of the scattering curves of four low-molecular-weight cellulose nitrates in acetone solution. Concentration 0.5%. (Kratky [37]; reprinted from *Pure and Applied Chemistry* by permission of the International Union of Pure and Applied Chemistry and Butterworths Scientific Publications.)

Table 5-2   Persistence Lengths and Radii of
Gyration of Low-Molecular-Weight Cellulose
Nitrates in Acetone Solution[a]

Specimen	Percent Nitrogen	Persistence Length y (Å)	l/y	R (Å)
1	11.08	54.3	7.4	83.5
2	11.19	49.5	15.2	115.5
3	11.96	56.5	8.2	89.0
4	12.09	59.5	13.6	133.0

[a]Data from [37].

length have been carried out by Kratky and Porod[51, 52] and
Hermans and Ulmann[56], and for molecules of finite length by Heine
et al.[55, 57] and Luzzati and Benoit[58], among others. In view of the
fact that the theory of scattering by linear polymers in dilute solution
is currently in a state of rapid flux, it is not expedient to expand further
on this topic here. However, we may cite in closing recent important
contributions of Kirste and Wunderlich[59, 60] relative to the
determination of molecular conformation.

5-4.6   Polydispersity [61]

The impossibility, in general, of separating the influences of the
particle shape and size distribution on the small-angle-scattering
curve was noted at the beginning of this chapter. Thus the problem of
determining the statistical size distribution from the scattering curve
does not have a unique solution; rather the solution depends on the
particular hypotheses that are adopted.

If it can be assumed (a) that all the particles have the same shape,
(b) that the scattering curve obeys the exponential law of Guinier
(Gaussian shape), and (c) that the frequency distribution of the radii of
gyration assumes the form of a Maxwellian distribution, it becomes
possible to calculate the size distribution in a number of ways, among
which we may mention in particular the methods of Hosemann
[6, 8, 62] and Shull and Roess[27]. Roess and Shull[63] made very
accurate calculations of the scattering curves to be expected from
Maxwellian distributions of spheres and oblate ellipsoids as a function of three parameters – the particle shape v, the particle dimension
a, and the size distribution n. The results showed graphically that
variations in these parameters had relatively little effect on the form
of the scattering curves and that different combinations of n and v

values resulted in identical curves. When these findings are considered in relation to the limits of accuracy that affect the experimental measurement of the scattering curve, it must be concluded that the size-distribution parameter n can at best be determined with only poor accuracy.

The effect of the drastic limitations imposed by the present state of the theory has been to tend to discourage experimental studies of polydispersity by small-angle scattering, so that very few papers on this subject have appeared in recent years. Under these circumstances it is hardly worthwhile to expand further on the topic at this time.

5-4.7   Globular Particles: Example of Lysozyme [36]†

Specimens.   Lysozyme from white of hen's egg. Purity verified by sedimentation. Dispersed in 0.15 M aqueous saline solution in six concentrations ranging from 1.01 to 4.95%; concentrations determined by densitometry; pH $\simeq 5.7$.

Experimental Conditions.   Counter diffractometer as described in Section 2-4.3. CuKα radiation. Very long narrow slit collimation. Absolute intensities. Counting-rate curve registered with a strip-chart recorder. Very slow scanning speeds (as low as 1° in 8 hr). Solutions contained in cell with parallel windows of mica for transmission of the direct and scattered x-ray beams. Specimen thickness d measured with traveling microscope.

Information Known in Advance.   Solutions monodisperse and solute morphology globular [64]. From chemical analysis $M = 14,700$ [65] and 14,100 [66]; from light scattering $M = 14,800$ [67]. From chemical composition of the solvent (saline solution) $\rho_0 = 0.335$ electrons per cubic angstrom. Partial specific volume from accurate density measurements [68] 0.7138 cm^3/g; $\Sigma n_i A_i / \Sigma n_i Z_i = 1/z_i = 1.88$ [65].

Correction for Instrumental Blank.   Measure $\tilde{I}(s)$ curve of solution in cell. Divide by η, ν, and P_0 to give the *normalized absolute-intensity curve* in the nomenclature of Luzzati. (See footnote on p. 290.) Repeat procedure for the cell containing pure solvent to give the normalized absolute-intensity curve of the blank. The normalized absolute-intensity curve of the specimen is then

$$\tilde{i}_n(s) = \left[\frac{\tilde{I}(s)}{\nu \eta P_0} \right]_{\text{sample}} - \left[\frac{\tilde{I}(s)}{\nu \eta P_0} \right]_{\text{blank}}. \tag{5-69}$$

†To be consistent with the original literature we employ here Luzzati's *normalized* absolute intensities, i_n and $\tilde{i}_n$, which were introduced in Section 5-2.1.

In (5-69) η is calculated from d and $\bar{\rho}$ for the first term, and from d and ρ_0 for the second.

Determination of Parameters.

The procedure is as follows:

1. Plot $s^3 \bar{i}_n(s)$ versus s^3. (See Figure 5-3.) At large values of s the six solutions give satisfactory linear horizontal lines, as required by the theory. Read off the six limiting values of $s^3 \bar{i}_n(s)$. (See Table 5-3, row 3.)

2. Prepare Guinier plots of $\log_{10} \bar{i}_n(s)$ versus s^2 (Figure 5-12) and determine the $\bar{i}_n(0)$ values from the ordinate intercepts and the $\tilde{R}$ values from the limiting slopes. From (L5-44) converted to logarithms to the base 10

$$\tilde{R} = \frac{1}{2\pi} (3 \times 2.303 \times \text{slope})^{1/2}. \tag{L5-70}$$

The contributions of the term $\phi(s)$ in (L5-44) to $\bar{i}_n(s)$ and $\bar{i}_n(0)$ were less than 1% and considered to be negligible. The numerical values of $\bar{i}_n(0)$ are given in row 1 of Table 5-3.

3. Having determined the function $\phi(s)$ for each experimental curve as described in Section 5-4.2, calculate R from $\tilde{R}$ with (L5-45) and $i_n(0)$ from $\bar{i}_n(0)$ with (L5-48). Because of the negligible contribution of $\phi(s)$, these equations simplify respectively to

$$R = \tilde{R} \quad \text{and} \quad i_n(0) = 2 \left(\frac{\pi}{3} \right)^{1/2} \bar{i}_n(0) \, \tilde{R}.$$

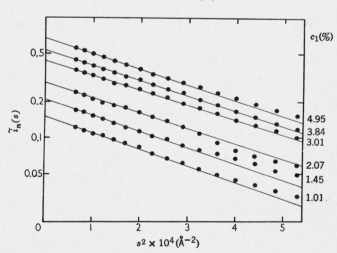

Figure 5-12 Saline solutions of lysozyme. Plots of $\log_{10} \bar{i}_n(s)$ versus s^2 for six concentrations ranging from 1.01 to 4.95% as indicated. (Luzzati, Witz, and Nicolaieff[36].)

The numerical results are presented in rows 2 and 5 of Table 5-3.

4. From the $\tilde{i}_n(s)$ curve calculate the invariant $\tilde{Q}_s$ with the aid of (5-18). Numerical results are listed in row 4 of Table 5-3.

5. Calculate ψ from $\bar{v}_1$ and z_1:

$$\psi = \frac{\bar{v}_1}{z_1 N} = \frac{0.7138 \times 10^{24} \times 1.88}{6.025 \times 10^{23}} = 2.227 \text{ Å}^3/\text{electron}.$$

We now continue the calculation of parameters employing the first solution ($c_1 = 1.01\%$) as an example.

6. Calculate c_e from c_1, the gravimetric concentration of the solute:

$$c_e = \frac{0.957\, c_1}{1 - 0.45\, c_1} = \frac{0.957 \times 0.0101}{1 - 0.45 \times 0.0101} = 0.00971.$$

7. Calculate m_1 and M with (L5-46) and (5-47):

$$m_1 = \frac{i_n(0)}{c_e(1 - \rho_0 \psi)^2} = \frac{4.63}{0.00971\,(1 - 0.335 \times 2.227)^2} = 7420 \text{ electrons}$$
$$\text{per particle};$$

$$M = \frac{m_1}{z_1} = 7420 \times 1.88 = 13{,}940. \text{ (Table 5-3, row 6.)}$$

8. Calculate $\rho_1 - \rho_0$ with (L5-51):

$$\rho_1 - \rho_0 = \frac{2.19 \times 10^{-6}}{0.00971\,(1 - 0.335 \times 2.227)} + 0.335 \times 0.00971\,(1 - 0.335 \times$$
$$2.227)$$

$$= 0.0900 \text{ electrons/Å}^3. \text{ (Table 5-3, row 7.)}$$

$$\rho_1 = 0.0900 + 0.335 = 0.425 \text{ electrons/Å}^3.$$

9. Calculate v_1 and O_s with (L5-49) and (L5-27):

$$v_1 = \frac{\tilde{I}(0)}{2\pi \tilde{Q}_s} = \frac{4.63}{219 \times 10^{-6}} = 21{,}100 \text{ Å}^3,$$

$$O_s = \frac{16 \times 3.1416^2 \times 3.50 \times 10^{-7}}{219 \times 10^{-6}} \times$$

$$\left[1 - \frac{0.335}{0.0900} \times \frac{0.00971\,(1 - 0.335 \times 2.227)}{1 - 0.00971\,(1 - 0.335 \times 2.227)} \right] = 0.250 \text{ Å}^{-1}.$$
$$\text{(Table 5-3, row 8.)}$$

10. Calculate α with (L5-52):

$$\alpha = \rho_0 \frac{1 - \rho_1 \psi}{\rho_1 - \rho_0} = 0.335 \left(\frac{1 - 0.425 \times 2.227}{0.0900} \right) = 0.20.$$

Table 5-3   Experimental and Calculated Values of Parameters for Lysozyme[a]

Parameter	Concentration, c_1 (percent)					
	1.01	1.45	2.07	3.01	3.84	4.95
$\tilde{i}_n(0)$	0.149	0.209	0.288	0.440	0.537	0.670
$R\,(\text{Å})$	15.2	15.4	15.2	14.9	15.1	15.3
$[\lim s^3\,\tilde{i}_n(s)] \times 10^7$	3.50	6.00	7.50	12.50	14.25	17.50
$2\pi\left[\displaystyle\int_0^\infty s\tilde{i}_n(s)\,ds\right] \times 10^6$	219	340	440	771	842	1042
$i_n(0)$	4.63	6.58	8.96	13.41	16.59	20.97
M	13,940	13,810	13,170	13,550	13,150	12,890
$\rho_1 - \rho_0$ (electrons/Å^3)	0.0900	0.0975	0.0900	0.0996	0.0933	0.0905
$O_s\,(\text{Å}^{-1})$	0.250	0.275	0.262	0.271	0.258	0.253

[a]Data from [36].

The final "best" parameters of the lysozyme particle are summarized in Table 5-4. With reference again to Table 5-3, it will be seen that O_s and $\rho_1 - \rho_0$ show no systematic dependence on the concentration, in accordance with the theory. Their least-squares mean values are respectively 0.260 Å^{-1} and 0.093 electrons per cubic angstrom. On the other hand, the final values of M and R have been obtained by least-squares extrapolations of the individual values to $c_1 = 0$. The molecular weight from small-angle scattering agrees well with reported values based on chemical methods, 14,700[65] and 14,100 [66]. Ritland, Kaesberg, and Beeman[69] obtained $R = 16.0$ Å for

Table 5-4   Parameters of the Lysozyme Particle from Small-Angle X-ray Scattering[a,b]

Parameter	Numerical Value	
$R = \tilde{R}\,(\text{Å})$	15.2	(± 0.2)
M	14,100	(± 200)
$\rho_1 - \rho_0$ (electrons/Å^3)	0.093	(± 0.003)
$O_s\,(\text{Å}^{-1})$	0.260	(± 0.007)
$v_1\,(\text{Å}^3)$	20,400	
α	0.17	

[a]Data from [36].
[b]Figures in parentheses are standard deviations from least-squares calculations.

lysozyme in aqueous solution—in good agreement with the result of the present study, 15.2 Å.

Although an unambiguous determination of the shape of the lysozyme particle cannot be made from the x-ray data, the relative dimensions (anisotropy) of scattering-equivalent particles can be deduced if some representative habit is assumed. On the reasonable assumption of ellipsoids of revolution Luzzati calculated the degree of prolateness or oblateness equivalent to (a) the experimental particulate volume and radius of gyration, and (b) the experimental particulate surface and radius of gyration. The results are summarized in Table 5-5, in which a is the semiaxis of rotation, b is the cross-sectional semiaxis, and $p = a/b$ is a measure of the anisotropy. In terms of p and b the volume of the ellipsoid of revolution is $\frac{4}{3} \pi p b^3$. For prolate and oblate ellipsoids $p > 1$ and $p < 1$, respectively, and for the limiting isotropic case of a sphere $p = 1$. As indicated, the experimental volume and surface functions could be satisfied either by a prolate ellipsoidal model with p of the order of 2 or 3, or by an oblate model with p equal to about 0.3 or 0.4. Substantially the same results were obtained earlier by Ritland, Kaesberg, and Beeman[69].

These results may be compared with the anisotropy factor of Kratky[10], $f_a = R_{exp}/R_{sp}$ [see (K5-43)]. From the volume of the lysozyme particle, 20,400 Å^3, the radius of the equivalent sphere is

$$r_{sp} = \left(\frac{3V}{4\pi} \right)^{1/3} = 16.9 \text{ Å},$$

Table 5-5 Models of Ellipsoids of Revolution Conforming to the Experimental Data for Lysozyme[a]

Volume Function $\frac{3v_1}{4\pi R^3}$	Surface Function RO_s	Parameter[b]		
		p	a (Å)	b (Å)
1.39		0.38	8.8	23.3
		2.2	28.7	13.0
	3.95	0.29	6.8	23.6
		3.3	31.3	9.5

[a]Data from [36].
[b]Semiaxis of rotation, a; cross-sectional semiaxis, b; anisotropy, p ($p = a/b$).

from which by (5-42)

$$R_{sp} = r_{sp} \left(\frac{3}{5}\right)^{1/2} = 13.1 \text{ Å}.$$

Then the anisotropy factor is

$$f_a = \frac{15.2}{13.1} = 1.16.$$

It can be shown that by their definitions p and f_a are related as follows:

$$f_a^2 = \frac{p^2 + 2}{3p^{2/3}}. \tag{5-71}$$

5-4.8   Rodlike Particles: Example of Air-Swollen Cellulose [26]

Specimens.   Regenerated cellulose, air-swollen by a previously described technique [70, 71] that separated the individual fibrils and randomized their alignment sufficiently to warrant use of the theory of dilute systems with certain limitations. Swelling factors of specimens ranged from $q = 1.00$ to 5.98.

Experimental Conditions.   Kratky U-bar apparatus as described in Section 2-3.5; CuKα radiation monochromatized with balanced filters. Geiger-Müller counter measurements; absolute intensities; P_0 measured with a perforated-rotor attenuator. (See Section 5-1.3.) Path of scattered x-rays evacuated. Minimum accessible angle 0.45 minutes of arc, equivalent to a Bragg spacing of 12,000 Å; maximum angle equivalent to 8 Å. Because of the great range of scattered intensities (five orders of magnitude), the angular range was covered in 10 sections, each with different widths of the entrance and receiving slits. Figures 5-13 and 5-14 show the 10 partial intensity curves for one of the specimens.

Processing of the Intensity Data.   The 10 partial intensity curves for each specimen as directly measured were first corrected for collimation errors by the methods described in Section 5-1.5. Only the innermost portions of those partial curves corresponding to Bragg spacings larger than 1000 Å were corrected for the breadth-collimation error [23]. The 10 partial angular ranges overlapped sufficiently to permit the 10 partial intensity curves to be blended into one final curve after corrections for collimation errors. This curve for the specimen with $q = 5.98$ is plotted in the form Im^2 versus m in Figures 5-15 and 5-16.

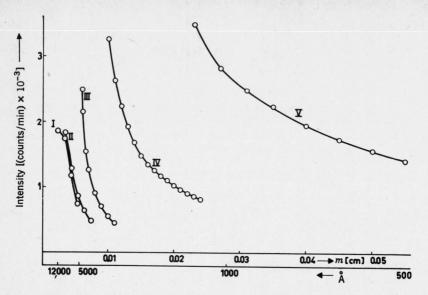

Figure 5-13 Five partial intensity curves for air-swollen-cellulose specimen for which $q = 5.98$. Uncorrected for collimation errors. Width of entrance slit varied from 0.006 to 0.06 mm. Width of counter slit varied from 0.006 to 0.1 mm. Range 12,000 to 500 Å; $a = 18.2$ cm. (Kratky and Miholic [26].)

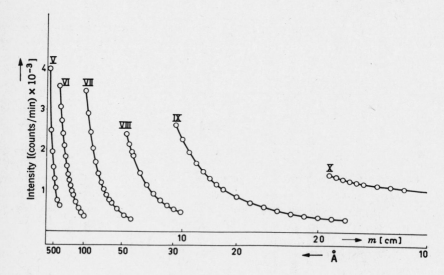

Figure 5-14 Six partial intensity curves for air-swollen-cellulose specimen for which $q = 5.98$. Uncorrected for collimation errors. Width of entrance slit varied from 0.06 to 0.3 mm. Width of counter slit varied from 0.1 to 0.8 mm. Range 500 to 10 Å; $a = 18.2$ cm. (Kratky and Miholic [26].)

320

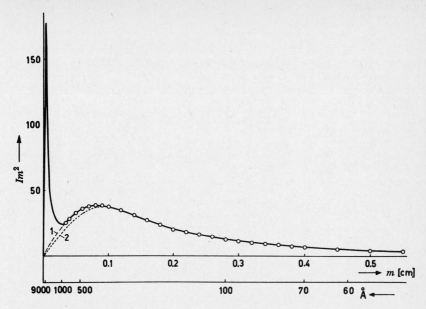

Figure 5-15 Curve of Im^2 versus m for air-swollen-cellulose specimen for which $q = 5.98$. Broken line 1 is a continuation of the right-hand portion of the curve. Broken line 2 separates the scattering of the disperse fraction from that of the cluster fraction. (Kratky and Miholic [26].)

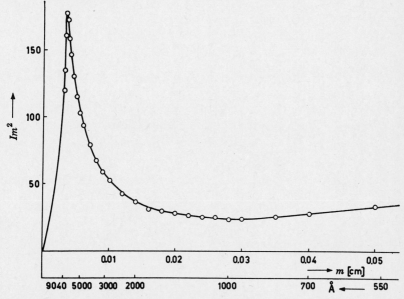

Figure 5-16 Inner portion of Figure 5-15 with abscissa scale enlarged tenfold. (Kratky and Miholic [26].)

Cluster fraction f[26]. In a system of air-swollen fibrils it is assumed that in general a fraction f remains aggregated (clustered) while a fraction $1-f$ is dispersed. From (K5-40) the product (w_1w_0) is given in terms of Q_m by

$$(w_1w_0)_{Q_m} = \frac{4\pi Q_m}{i_e N^2 \lambda^3 da\rho_1^2 P_0},$$ (K5-72)

where ρ_0 has been neglected in relation to ρ_1. However, the scattering due to the clustered fraction f may be assumed to occur at immeasurably small angles, so that Q_m in its entirety is not measurable. Nevertheless, for the macroscopic swelling factor q_m the quantity (w_1w_0) corresponding to the total invariant can be calculated from

$$(w_1w_0)_{q_m} = \frac{1}{q_m}\left(1 - \frac{1}{q_m}\right).$$ (K5-73)

It follows that $(w_1w_0)_{Q_m}$ determined experimentally will be smaller than $(w_1w_0)_{q_m}$ when clustering is present. The quotient

$$r = \frac{(w_1w_0)_{Q_m}}{(w_1w_0)_{q_m}} = \frac{Q \,(\text{disperse fraction})}{Q \,(\text{entire specimen})}$$ (K5-74)

represents the experimentally accessible fraction of the scattering power.

Since the clusters are not swollen, for a macroscopic swelling factor q_m the actual swelling of the dispersed fraction must be q', which is larger than q_m and is given by

$$q' = \frac{q_m}{r}.$$ (K5-75)

It is then easy to show that

$$q_m = f + q'(1-f).$$ (K5-76)

Solving (K5-76) for f and substituting q_m/r for q', we obtain

$$f = \frac{q_m(1-r)}{q_m - r}.$$ (K5-77)

The effective thickness d' of the specimen corresponding to the disperse fraction $(1-f)$ can then be expressed in terms of the overall thickness d as

$$d' = \frac{d(q_m - 1)}{q_m - r} = d\left(\frac{q_m - f}{q_m}\right).$$ (K5-78)

Determination of Parameters, Specimen with $q_m = 2.2$.† Because of space limitations we shall outline the numerical analysis of only this one specimen. Preliminary numerical information is as follows:

$d = 0.095$ cm.

$a = 24.0$ cm.

$P_0 = 2.01 \times 10^7$.

Solute has a density of 1.615 g/cm^3 and contains 0.53 mole electrons per gram. Therefore

$\rho_1 = 0.53 \times 1.615 = 0.86$ mole electrons/cm^3.

$Q_m = \int_0^\infty m^2 I(m) \, dm = 3.58 \times 10^4$, from area under curve for $q_m = 2.2$ (Figure 5-17).

1. Calculate $(w_1 w_0)_{Q_m}$ with (K5-72) (The constant $4\pi/i_e N^2 \lambda^3$ has the value 120.0 for Cu$K\alpha$ radiation.):

$$(w_1 w_0)_{Q_m} = \frac{120 \times 3.58 \times 10^4}{0.095 \times 24 \times (0.86)^2 \times 2.01 \times 10^7} = 0.127.$$

2. Calculate $(w_1 w_0)_{q_m}$ with (K5-73) and r with (K5-74):

$$(w_1 w_0)_{q_m} = \frac{1}{2.2}\left(1 - \frac{1}{2.2}\right) = 0.248,$$

$$r = \frac{0.127}{0.248} = 0.512.$$

3. From (K5-77) the cluster fraction is

$$f = \frac{2.2(1 - 0.512)}{2.2 - 0.512} = 0.636$$

and the disperse fraction is

$$1 - f = 0.364.$$

From (K5-75) the swelling factor of the disperse fraction is

$$q' = \frac{2.2}{0.512} = 4.30.$$

---

†Because of some computational errors in the results as originally published[26], the numerical parameters that follow are not in agreement in all cases. Only for d' and μ (M/A. of the original paper) are the differences appreciable. Thus in Table 1 of Kratky and Miholic's paper the M/A. values for series 1, 2, 3, and 4 should be 7740, 6770, 6930, and 7370, respectively, rather than 8480, 8310, 8480, and 8300.

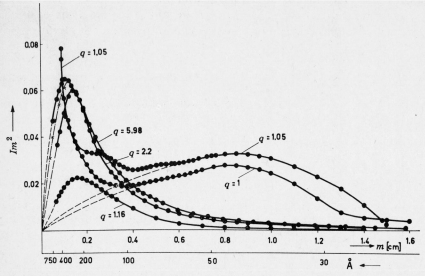

Figure 5-17   Curves of Im^2 versus m for five air-swollen-cellulose specimens. Swelling factors q: 1.00, 1.05, 1.16, 2.2, 5.98. (Kratky and Miholic[26].)

The effective thickness corresponding to the disperse fraction is, from (K5-78),

$$d' = \frac{0.095(2.2-1)}{2.2-0.512} = 0.0676 \text{ cm}.$$

The volume fraction of void space in the unclustered portion is

$$w_0' = 1 - \frac{1}{q'} = 1 - \frac{1}{4.30} = 0.767.$$

4. Prepare Guinier plots of $\log_{10}(I \cdot m)$ versus m^2 (Figure 5-18). From the slope we calculate with the aid of (K5-58) the radius of gyration of the cross section:

$$R_c = 47 \text{ Å}.$$

Note that the innermost portion of the curve bends downward below the straight-line extrapolation. The deviation of this portion of the curve from the law of Guinier may be attributed either to interparticle interferences or to finite particle length.

5. From (K5-60) we find for the cross-sectional area of a particle

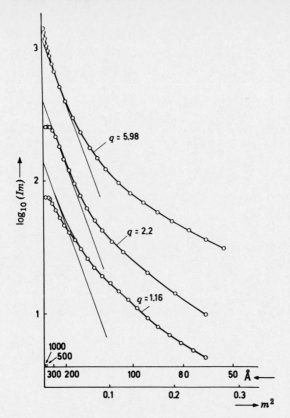

Figure 5-18   Air-swollen-cellulose specimens with swelling factors 1.16, 2.2, 5.98. Guinier plots of the cross-section factors [$\log_{10}(Im)$ versus m^2]. Scale of m normalized to a = 24 cm for all samples. (Kratky and Miholic [26].)

$$A_c = \frac{(m \cdot I)_0}{Q_m} \cdot \frac{(a\lambda)^2}{2\pi}$$

$$= \frac{1.48 \times 10^6}{3.58 \times 10^4} \cdot \frac{(24 \times 1.54)^2}{6.28} = 9000 \text{ Å}^2.$$

6. Determine the specific surface of the disperse phase. First it is necessary to verify that the $I(m)$ curve follows the asymptotic course m^{-4}, or $\tilde{I}(m)$ the course m^{-3}, at the larger angles. (See again Section 5-3.2.) The log log plots of these functions in Figure 5-19 verify that the theory is obeyed over a considerable angular range. Equation K5-26, with w_0 replaced by $w_0' = 0.767$, gives the result

$$O_s = 0.0465 \text{ Å}^{-1}.$$

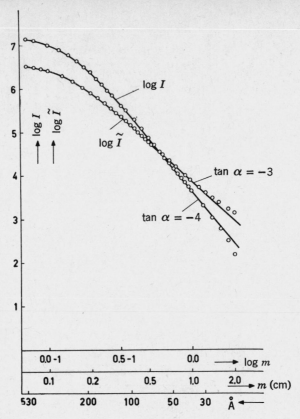

Figure 5-19 Plots of $\log_{10} I$ and $\log_{10} \tilde{I}$ versus $\log_{10} m$ of the air-swollen-cellulose specimen for which $q = 2.2$ ($a = 18.2$ cm). (Kratky and Miholic [26].)

This result is not very precise because of difficulty in selecting the proper value of the limiting constant $\lim [m^4 I(m)]$ from a plot of $[m^4 I(m)]/Q_m$ versus m.

7. Determine the mass per unit length with (K5-59). For CuKα radiation $(2/i_e N\lambda) = 27.3$. For the unclustered fraction the concentration of the solute is $c_1 = 1.615/4.29 = 0.378$ g/cm^3 and the effective sample thickness is 0.0676 cm. Then

$$\mu = \frac{27.3 \times 1.48 \times 10^6 \times 24.0}{0.0676 \times 0.378 \times (0.53)^2 \times 2.01 \times 10^7}$$

$$= 6770 \text{ molecular-weight units per angstrom.}$$

5-5 DENSE SYSTEMS

In general, dense polymeric systems (solid polymers) produce discrete interferences, diffuse scattering, or both at small angles. Discrete interferences are most likely to be observed in the patterns of oriented specimens such as fibers and films in addition to more or less diffuse scattering. On the other hand, solid polymers may yield diffuse scattering without any discrete "reflections." As was mentioned at the beginning of this chapter, in interpreting diffuse scattering at small angles from dense systems the interparticle interference effects cannot be neglected.

5-5.1 Diffuse Scattering and Voids

Prior to the mid 1950s there was extensive controversy as to the nature of the inhomogeneities in solid polymers that give rise to diffuse small-angle x-ray scatter. The first hypothesis to gain widespread support was that the scatter was caused by the presence of crystalline regions, or crystallites, of relatively high electron density, imbedded in an amorphous polymer matrix of lower electron density. This view led to conflicting interpretations of the small-angle scattering from cellulose, which was the first polymeric material to be extensively studied. As early as 1951 Porod[33, 41] presented theoretical arguments that microvoids must be involved in the generation of the diffuse small-angle scattering of x-rays by cellulose. Over a period of years Kratky and associates[72], Heyn[73], and others published experimental scattering data that were subsequently recognized to support the microvoid picture. Independent experimental evidence from the degradation, dyeing, and impregnation of fibers led Frey-Wyssling[74] to propose a micellar picture of fiber microstructure consisting of a continuous network of dense matter pierced by elongated voids. In 1956 Statton[75] reviewed the evidence accumulated to that date and concluded that the diffuse small-angle scattering of cellulose must be produced very largely by microvoids and that these voids are normally elongated parallel to the fiber axis, as is indicated by the greatest elongation of the scatter in the equatorial direction. (See Figure 5-20 for a typical example.) Important pieces of evidence that voids rather than crystalline regions produce the diffuse scatter are [75] (a) increasing steepness of the log I versus s^2 curve of rayon with swelling; (b) presence of a discrete maximum or inflection in the scatter from dried fibers, which is indicative of interparticulate interferences and requires the co-existent diffuse scatter to be generated by voids in accordance with

Figure 5-20 Small-angle diffuse scattering from ramie fibers. Pinhole collimation, fiber axis vertical. (Statton [75].)

Babinet's reciprocity law; and (c) the high intensity of the diffuse scatter from cellulose acetate, which is known to be amorphous.

It remained for Hermans, Heikens, and Weidinger[16, 76] to establish quantitatively that the observed scattering power of cellulose fibers is much higher than would be expected from mixed crystalline and amorphous regions and that it is, in fact, in agreement with the intensity calculated for a volume fraction of the order of 1% microvoids in a solid matrix. These investigators made use of (K5-16) modified for slit-collimated intensities,

$$\overline{(\Delta\rho)^2} = \frac{C}{dP_0} \int_0^\infty \theta\,\tilde{I}(\theta)\,d\theta, \qquad (5\text{-}79)$$

in which $\Delta\rho$ is expressed in electrons per cubic angstrom and C contains all the experimental and normalization constants. A vacuum camera was used in connection with crystal-monochromatized $CuK\alpha$ x-rays from a 4-kVA rotating-anode tube. The cellulose fibers to be studied were wound parallel to each other on the holder described in Section 2-2.4 (see also Figure 2-6), and the effective thickness of each specimen was determined by a quantitative measurement of the attenuation that was suffered by a monochromatic beam reflected from tungsten on transmission through the specimen. The mounted specimens were contained in a vacuum-tight cell, permitting them to be maintained air dry, soaked in water, or conditioned above salt-hydrate mixtures of known vapor pressure during the x-ray measurements. The intensity of the direct beam P_0 was ascertained by measuring the intensity of scattering from a secondary standard, a film of poly(methyl methacrylate), which was calibrated with respect to the known absolute intensity of scattering by a standard gold or silver sol calculated with the aid of (K5-23). The intensities were recorded on photographic films which were measured with a recording microphotometer capable of giving output readings directly proportional to the intensities. The low-angle limit of reliable measurement was less than 0.28°.

Table 5-6 presents the experimental findings for 16 rayon specimens in the air-dry and water-swollen states. In addition to the scattering power S_v the table gives the gravimetric density in the air-dry state and the fractional contribution f to the intensity integral $\int \theta \bar{I}(\theta) \, d\theta$ resulting from extrapolation of the $\theta \bar{I}(\theta)$ curve below $\theta = 0.25°$. The S_v values for the air-dry samples may be compared with a reference value of 4×10^{-4} calculated for a dry specimen without voids on the assumption that the scattering power is due only to the difference between the electron densities of the crystalline and amorphous fractions. For this purpose the crystalline fraction was taken to be 0.40, the dry-fiber density 1.52, and the crystalline density 1.59 g/cm³. With the exception of specimen 662 the scattering powers are seen to be from 2 to 5 times larger than that of the reference model without voids. Calculations of the scattering power of specimens containing 0.75 and 1.5% voids by volume give $S_v = 21 \times 10^{-4}$ and 38×10^{-4}, respectively, which may also be compared with the experimental data.

Not only is the hypothesis of microvoids in harmony with the experimental scattering-power data but it is supported by the plot of gravimetric density against scattering power for the air-dry samples shown in Figure 5-21. With increasing void content the scattering power would be expected to increase rapidly while the density decreases slowly, as is observed. If specimen 662 with the highest density of 1.499 g/cm³ and lowest scattering power of 4.1×10^{-4} is presumed to have zero void content, specimen 641 with the lowest density of 1.448 g/cm³ and highest scattering power must be assigned a microvoid content of 3.5% by volume on the basis of the density difference. On the other hand, when the difference in scattering power is used to calculate the volume fraction of microvoids on the basis of a three-phase system (crystalline cellulose, amorphous cellulose, and voids), a volume fraction of only 0.78% is obtained. The large disparity between these numerical results can be understood if only one-fifth of the total void space consists of sufficiently small voids to affect the accessible small-angle-scattering range.

**Table 5-6   Experimental Values of the Scattering
Power of Rayon Specimens in the Air-Dry and
Water-Swollen States[a]**

Specimen No.	Air-Dry Density (g/cm³)	Air-Dry		Water-Swollen	
		$S_r \times 10^4$	f	$S_v \times 10^4$	f
638	1·4860	7·6	0·10	30·5	0·05
639	1·4932	13·3	0·13	21·3	0·015
640	1·4887	9·2	0·24	18·3	0·06
641	1·448	19·9	0·13	25·4	0·015
642	—	12·7	0·09	25·5	0·03
643	1·491	8·7	0·07	22·1	0·035
644	1·4920	5·8	0·27	19·9	0·025
645	1·4853	11·7	0·19	26·7	0·075
646	1·471	14·6	0·16	17·7	0·04
647	1·4863	9·4	0·20	31·0	0·05
648	1·474	13·5	0·20	27·3	0·03
649	1·4848	11·8	0·09	28·7	0·025
661	1·4939	9·4	0·08	34·1	0·035
662	1·4990	4·1	0·17	19·1	0·04
663	1·4688	16·3	0·15	27·5	0·035
664	1·470	11·9	0·15	24·5	0·03

[a]Data from [16].

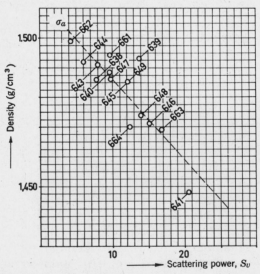

Figure 5-21   Regression curve of density versus scattering power for air-dry-rayon specimens. (Hermans, Heikens, and Weidinger [16].)

The majority of the experimental S_v values for the swollen specimens (see Table 5-6) are considerably larger than the value of 18×10^{-4} calculated for rayon without microvoids swollen to near saturation. A possible explanation for this effect is that at saturation existing voids are enlarged or new ones are formed and filled with water, resulting in an increase in the third phase consisting of micro-islands of water distributed among the noncrystalline regions. For further details of the experimental procedure and numerical calculations the reader is referred to the original paper[16]. Similar absolute measurements of some 60 rayons by Kratky and Sekora [37] showed nearly tenfold variations in scattering power and yielded microvoid contents ranging from near 0 to 0.7 vol%[37].

In recent reviews of the subject Statton[4, 77] shows that all the evidence supports the existence of microvoids in synthetic as well as natural fibers as the primary source of the observed diffuse small-angle scattering. In particular Statton makes the following points:

1. It is normal to find intense diffuse scatter in the small-angle patterns of synthetic as well as natural fibers, and very weak scattering is the exception rather than the rule.

2. The degree of crystallinity is unrelated to the diffuse scatter as is shown by large variations in the amounts of diffuse scatter with no change in the crystalline fraction and, conversely, by observations of large changes in the crystalline fraction with no appreciable change in the intensity of the diffuse scatter.

3. A fiber treatment that produces a decrease in the diffuse scatter is accompanied by an increase in density.

4. Compared with the void contents deduced for rayons (see above), experimental synthetic fibers display a range of diffuse-scatter intensity that extends from much lower to much higher equivalent void contents.

5. Most commercial fibers show a low level of diffuse scatter.

6. Microvoids are a logical consequence of the fact that long-chain polymers cannot efficiently fill all space. Thus the voids may be unfilled spaces around polymer chains, around fibrils, or between lamellae.

The amount of the diffuse small-angle scatter is strongly dependent on the history of formation of the fiber, as illustrated in Figure 5-22 with the example of nylon 66. Large void contents are promoted by those fiber-spinning procedures that result in rapid precipitation of the fibers from solution; for example, *wet*, *dry*, and *dispersion* spinning. Such a high level of voids can be subsequently reduced by the annealing effects of thermal after-treatments. Relatively small numbers of microvoids are produced by solidification of a viscous polymer from its melt (*melt* spinning), by slow evaporation of a polymer solution, or by spinning procedures that permit slower precipitation from solution.

In general as-spun fibers show a diamond-shaped diffuse scattering pattern, with the longer dimension parallel to the equator. According to the reciprocity between the dimension of an inhomogeneity and the angular extent of the small-angle scattering explained at the beginning of this chapter, this indicates some elongation of the microvoids in the meridional direction. Extension of polymer fibers by drawing processes has the effect of increasing the equatorial dimension of the scattering at the expense of the meridional dimension, thus signifying an elongation of the voids parallel to the direction of stretch (meridian). An illustration is provided by Figure 5-1b. Occasionally a drawn fiber also gives a meridional streak, indicative of elongated voids, or fissures, transverse to the fiber axis. For more comprehensive discussions of these topics the reader is referred to the excellent treatments by Statton [4, 77] and to a paper by Statton and Godard [78].

5-5.2   Discrete Interferences

If discrete interferences appear in the small-angle-scattering patterns of oriented polymers, they are usually observed in only one direction, most commonly on the meridian. Meridional reflections denote some degree of large-scale periodic character in the structure parallel to the fiber axis. In terms of the old Hess-Kiessig model [79] (see again Section 1-2.6) the long spacing derived from the first-order meridional reflection with the aid of Bragg's law can be most directly interpreted as the mean distance between the centers of successive crystalline regions along the fiber axis. It now seems probable that this simple interpretation is *not* invalidated by the newer concept of chain folding within the crystalline regions.

Reflection-Rich Meridional Patterns. Meridional interference patterns that exhibit a large number of sharp reflections at small angles were obtained from certain natural fibers, notably collagen and keratin, as early as 25 years ago by Kratky and Sekora [80] and by Bear [81]. Representative pinhole-collimated patterns of collagen and keratin from the work of Bear and co-workers are shown in Figure 5-23. In addition to some diffuse scatter the small-angle diagram of collagen consists of upward of 20 meridional orders diffracted by a

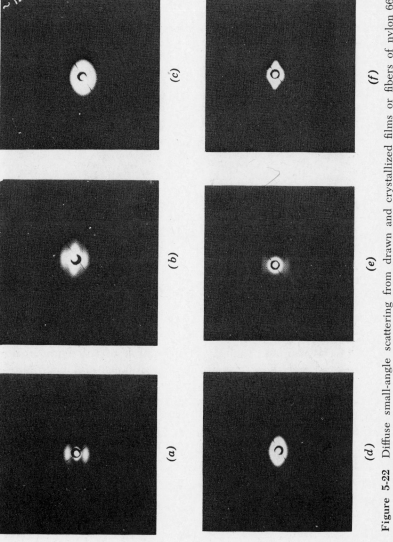

Figure 5-22 Diffuse small-angle scattering from drawn and crystallized films or fibers of nylon 66 prepared in a variety of ways; (*a*) melt pressed; (*b*) solvent cast; (*c*) wet spun; (*d*) interfacial spun; (*e*) melt spun, drawn in air (density = 1.132); (*f*) melt spun, drawn in acetone (density = 1.014). (Statton[77].)

333

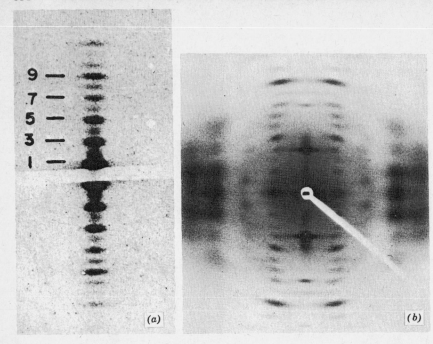

Figure 5-23 Small-angle x-ray diagrams containing numerous discrete meridional reflections: (*a*) collagen [Bear, Bolduan, and Salo, *J. Amer. Leather Chemists Assoc.*, **46**, 107 (1951)]; (*b*) keratin [Bear and Rugo, *Ann. N.Y. Acad. Sci.*, **53**, 627 (1951)].

fibril macroperiod that varies from 600 to 680 Å, depending on the condition of the sample. Such a large number of reflection orders reveals the existence of an essentially one-dimensional macrolattice with a rather regular period. Under these conditions the longitudinal spacings that correspond to the various orders of reflection can be calculated properly with the aid of Bragg's law. Bear and Bolduan[82, 83] derived expressions for determining the mean fibril diameter and the state of imperfection of the macrostructure transverse and parallel to the fiber axis from the corresponding profiles of the reflections of different orders. In spite of their striking appearance, such rich meridional small-angle patterns are very much the exception in the field of linear polymers and will not be discussed further here.

Isolated Diffuse Meridional Reflections. We shall now discuss a small-angle interference phenomenon that is characteristic of most synthetic fibers — namely a single somewhat diffuse reflection on the meridian, such as shown in Figure 1-20*a*. Hess and Kiessig[79, 84, 85] were the first to report their observation of such reflections in the course of x-ray studies of polyamide and polyester fibers. Subsequent investigators discovered similar meridional interferences in the small-angle patterns of numerous other synthetic fibers, and it was even found possible to evoke reflections of this kind from cellulose and silk by appropriate pretreatments [86, 87].

These characteristic sparse meridional interferences are best studied with pinhole collimation and have the following properties:

1. Most commonly only the first order is present, although occasionally one or two higher orders are detectable. When this is true, the several orders usually correspond to somewhat different fundamental Bragg spacings, an indication that a considerable degree of disorder is involved in the genesis of these interferences. From the theory of paracrystalline diffraction only the first order will be visible when the relative deviation from strict periodicity parallel to the fiber axis is about 25 to 35% [88].

2. The Bragg spacing that is calculated from the reflection or reflections is not related to the chemical or crystal-structure period in the meridional direction, which rules out a superlattice as the cause of the reflection.

3. The reflection actually has the form of a horizontal streak rather than an arc, showing that it is not the consequence of a range of orientation about the fiber axis, as discussed in Chapter 4. In Figure 5-24 the streaklike character of the interferences and a faint second order are both evident in the small-angle pattern of a linear polyethylene specimen drawn at 115°C to an extension ratio of 6:1. The length of the streak is inversely related to the transverse dimension of the ultimate scattering units, presumably fibrils, as shown schematically in Figure 5-25. This dependence was strikingly demonstrated by Hess and Kiessig [85] in studies of rolled Perlon L (polycaprolactam) fibers.

It is interesting to compare the long period derived from the meridional reflection with Bragg's law and a "crystallite length" obtained from the longitudinal breadth of a wide-angle meridional reflection with the aid of the Scherrer equation [78, 89],

$$L = \frac{K\lambda}{\beta_0 \cos \theta}. \tag{5-80}$$

In (5-80) L is the crystallite dimension, β_0 is the breadth of the reflection corrected for instrumental broadening and K is approximately unity. Figure 5-26 shows a plot of crystallite length versus long period for linear polyethylene from the work of Statton [90].

Figure 5-24 Small-angle diagram of linear polyethylene drawn to an extension ratio of approximately 6:1 at 115°C. Nickel-filtered Cu$K\alpha$ radiation; specimen-to-film distance, 400 mm; fiber axis, b_3. (Courtesy of R. Bonart.)

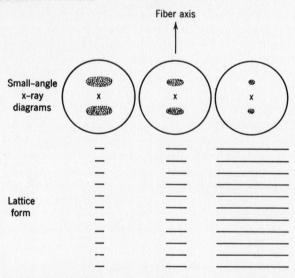

Figure 5-25 Relationship between the length of the meridional streak and the transverse dimension of the diffracting source. (After Hess and Kiessig[79] and Statton[77].)

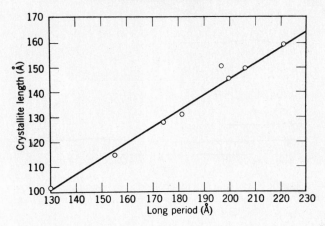

Figure 5-26 Comparison of crystallite length and long period for linear polyethylene fibers. (Statton[90].)

It will be noticed that the values of the long period are larger than the crystallite lengths. Thus subtraction of the crystallite length from the long period will yield a deduced amorphous length, provided that a model consisting of alternating crystalline and amorphous regions in the direction of the fiber axis is assumed. It may be empirically useful to compare different polymer specimens in terms of these parameters, but it is difficult to arrive at a satisfying physical interpretation. This is emphasized by comparing the results for fibers with parallel results for aggregates of polymer single crystals.

It is easy to prepare mats of parallel solution-grown single crystals (see Section 1-2.3) of sufficient thickness so that a small-angle-scattering pattern of several distinct meridional reflections is obtained, as illustrated in Figure 5-27a. For this purpose the direct beam collimated by pinholes is made to impinge on the mat parallel to the layering plane. It can be seen that this single-crystal-mat pattern resembles rather closely the pattern that is produced by a drawn fiber of the same polymer (linear polyethylene), as shown in Figure 5-27b. In the case of the mat the breadth of the wide-angle meridional reflections yields a value of the crystallite length (actually *crystallite thickness* under these conditions) that agrees well both with the long period from the small-angle reflection and the average thickness of the single crystals as determined by electron microscopy. On the other hand, the *crystallite width* that is calculated with (5-80) from the length (dimension $\Delta R_1'$ of Figure 7-6) of the meridional reflection is very much smaller than the width of the crystal platelets directly observed with the electron microscope, which leads to the

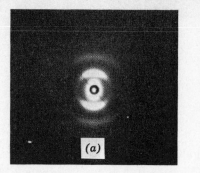 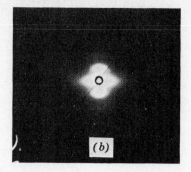

Figure 5-27   Typical small-angle x-ray patterns of linear polyethylene: (*a*) mat of parallel-stacked single crystals with *c*-axis of crystals vertical; (*b*) drawn fiber with fiber axis vertical. (Statton[77].)

inference that the crystals have a mosaic structure. These somewhat divergent interpretations relative to the meridional and equatorial dimensions of the crystals lead us to the strange conclusion that in the fiber direction the Scherrer equation is to be accepted as conveying the crystallite size in the conventional way, whereas in the transverse direction it must be interpreted in relation to crystal perfection[90]!

Four-Point Diagrams.   With suitable pretreatments a *four-point*, or *quadrant*, diagram can be obtained from most synthetic fibers. (See again Section 1-2.6 and Figure 1-20*b*.) The first recorded observations of such reflections were by Arnett, Meibohm, and Smith[91, 92] during studies of nylon 6·10 and polyethylene. Belbéoch and Guinier[93] obtained similar patterns from polyethylene fibers that had been cold drawn drastically enough to produce necking and then relaxed at various temperatures. Statton and Godard[78] and Bonart and Hosemann[88*a*] have demonstrated that certain rolled polymers or stretched polymer films generate meridional spots in one direction and a four-point diagram in a perpendicular direction. From cold-drawn and laterally pressed linear polyethylene Bonart and Hosemann obtained the usual meridional reflections with the incident beam parallel to the direction of compression—but a four-point diagram with the beam perpendicular to the compression direction (Figures 5-28*a* and *b*, respectively). Subsequent annealing of the specimen at 110°C for 12 hr intensified the four-point pattern and also increased the long period (Figure 5-28*c*). Pollack and Magill† found the opposite effect in studies of doubly-oriented nylon 66. Meridional reflections were

† Private communication, 1968.

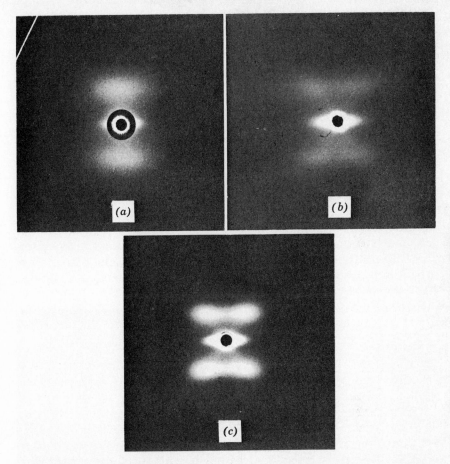

Figure 5-28 Small-angle x-ray diagrams of cold-drawn and laterally pressed linear polyethylene: (a) incident beam parallel to direction of compression; (b) incident beam perpendicular to direction of compression; (c) annealed 12 hr at 110°C, incident beam perpendicular to direction of compression. Fiber axis vertical. (Bonart and Hosemann[88a]; reprinted by permission of Dr. Dietrich Steinkopff, Darmstadt, Germany, *Kolloid-Z., Z. Polymere.*)

obtained with the incident beam perpendicular to the compression direction and a four-point diagram with the beam parallel.

With the aid of optical-diffraction analog apparatus Bonart and Hosemann[88a] and Predecki and Statton[94] have shown that a four-point diagram can be produced by a simple inclination of the individual linelike elements comprising a quasiperiodic scattering sequence in one dimension. Figure 5-29a shows such a two-dimensional model consisting of somewhat wavy lines arranged in a very roughly periodic manner in the direction of the fiber axis but individually oriented perpendicular to that axis *on the average*, and Figure 5-29*b* shows its light-optical diffraction pattern, consisting of two diffuse meridional reflections. The model of Figure 5-30*a* differs from that of Figure 5-29*a* only by an average inclination of the scattering elements to the right and left, and its diffraction pattern (Figure 5-30*b*) is seen to be a four-point diagram.

The foregoing results demonstrate that four-point as well as meridional small-angle patterns can be generated by essentially linear lattices that have little if any extension in the second or third dimension. Although this finding seems convincing, we must not fail to note that four-point diagrams can also be produced by rather different models, such as a continuous helix[93] or a somewhat irregularly staggered arrangement of scattering elements in adjoining fibrils as illustrated by Figure 1-21*b* or 1-22. We conclude, then, that a truly unequivocal explanation of the origin of four-point diagrams has not yet been arrived at.

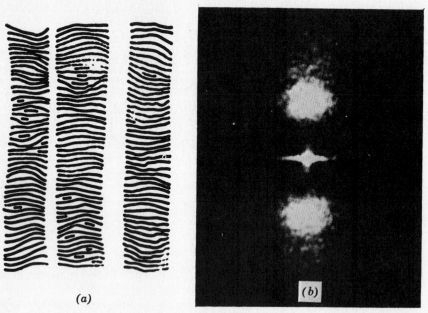

(a)   (b)

Figure 5-29 (a) Two-dimensional structure model with scattering elements consisting of quasiperiodic wavy lines with average orientation perpendicular to the fiber axis. (b) Light-optical diffraction pattern corresponding to (a). Fiber axis vertical. (Bonart and Hosemann[88a]; reprinted by permission of Dr. Dietrich Steinkopff, Darmstadt, Germany, *Kolloid-Z., Z. Polymere.*)

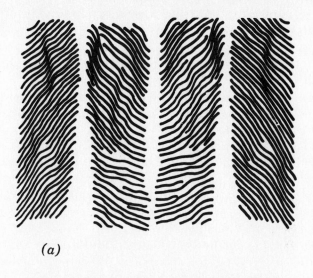

(a)

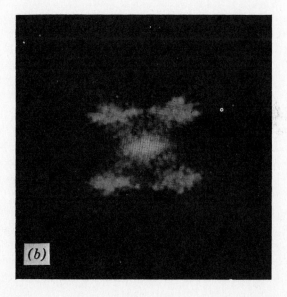

(b)

Figure 5-30 (a) Two-dimensional structure model similar to Figure 5-29a except that scattering elements have an average inclination with respect to the fiber axis. (b) Light-optical diffraction pattern corresponding to (a). Fiber axis vertical. (Bonart and Hosemann [88a]; reprinted by permission of Dr. Dietrich Steinkopff, Darmstadt, Germany, *Kolloid-Z., Z. Polymere.*)

Effect of Annealing on Long Spacings. Annealing of polymer fibers leads to the following observations:

1. The long spacing increases irreversibly.

2. The reflection sharpens somewhat and intensifies, often dramatically.

3. The length of the reflection (meridional streak) decreases.

4. Annealing in air results in a slow increase in period that is linear with the logarithm of the time.

5. Annealing in liquids produces an initial rapid increase in the period, followed by an asymptotic approach to a maximum limiting value.

An outstanding feature of the long meridional spacings of polymer fibers as well as single-crystal mats is their dependence on the crystallization or annealing temperature. Numerous significant studies of this phenomenon have been made [95–101], of which only a brief summary can be presented here. When polyethylene single-crystal mats are air annealed at temperatures about 20° below their melting point (137°C), the x-ray long periods display irreversible increases from initial values of the order of 100 Å to final values that are commonly as large as 400 Å [98, 100] (see Figure 5-31) and much larger under carefully controlled heating conditions [101]. The final period that is attained depends on both the time and temperature of anneal. The growth in the x-ray long period parallels a similar increase in the average thickness of the crystal platelets and is accompanied by the development of holes—two effects that are directly demonstrable by electron microscopy. The reader may compare the electron micrographs shown in Figures 1-12 and 5-32, which show respectively a polyethylene single crystal freshly grown from dilute solution and a similar crystal after being heated 35 min at 125.7°C, a few degrees below the melting point.

Morphological Interpretations; Chain Folding. To arrive at some understanding of the phenomena just described consider three possible models of chain folding within lamellar crystals, as sketched in Figure 5-33 [102]. On the basis of several lines of evidence Geil concludes that the most regularly folded model (*a*) is the one most likely to be approximated by an actual polymer, with possibly some admixture of type (*b*) [102]. Growth in thickness during annealing would then involve a self-diffusion of each molecule along its own "backbone." A possible mechanism permitting extension of the fold period by means of migration of point defects along a molecular chain under annealing conditions has been suggested by Reneker [103] and

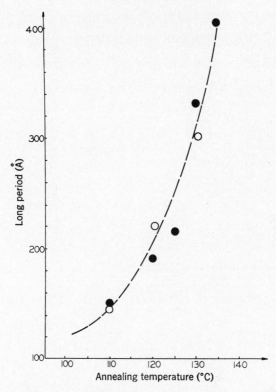

Figure 5-31 Dependence of the long period of polyethylene crystals on the annealing temperature. Closed circles – measured at room temperature by Statton and Geil [97] (original crystals 104 Å). Open circles – measured at annealing temperature by Schmidt and Fischer (unpublished data; original crystals 120 Å). (Fischer [99].)

is portrayed schematically in Figure 5-34. The overall effect of the self-diffusion of the folded molecules during annealing is to make individual regions of the platelike crystal taller but thinner, thus producing a distribution of void spaces, as plainly shown in Figure 5-32. In addition to growth by molecular self-diffusion, at temperatures very near the melting point it is possible that the greater mobility of the molecules will result in partial melting and subsequent recrystallization with a larger fold period [101].

Crystallization from the melt is of greater practical interest than crystallization from solution because of its numerous technological applications. Nevertheless, the concepts of chain folding and lamellar growth from solution crystallization have been helpful in understanding crystallization from the melt, which is inherently more

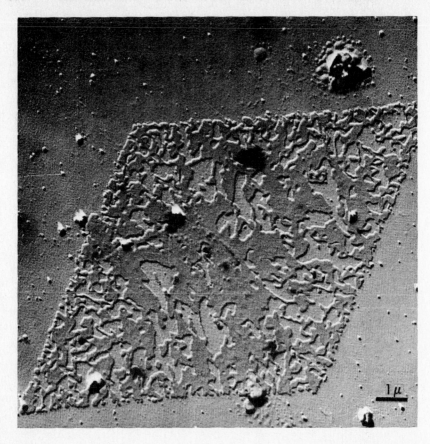

Figure 5-32   Electron micrograph of a polyethylene single crystal grown from dilute solution and annealed for 35 min at 125.7°C. Magnification 14,000×. (Courtesy of the E. I. du Pont de Nemours and Company, Inc.; micrograph by P. H. Geil [113].)

complex. On the strength of much recently accumulated experimental evidence it can hardly be doubted that there are important points of similarity between the two crystallization processes and between the resulting morphologies. In Section 1-2.4 a brief description was given of the most common mode of crystallization from the melt, *spherulitic* growth, as well as mention of intermediate structures called *axialites* and *hedrites*. Detailed electron-microscopic studies of these three morphologies clearly reveal lamellar substructures, from which it is natural to infer the presence of folded molecular chains. Actually, on the basis of x-ray and electron diffraction as well as the electron-microscopic evidence it is now generally believed that folded-chain

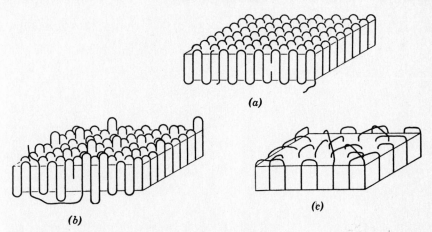

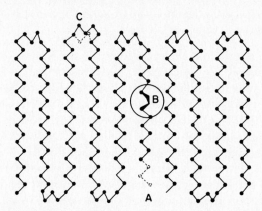

Figure 5-33 Three proposed models for the lamellar crystals that are observed in polymers crystallized from the melt: (a) *regular*, adjacent reentry folds similar to those postulated as present in pyramidal crystals that have been grown from solution; (b) *irregular*, adjacent reentry folds in which the extent or thickness of the irregular layer is suggested to be proportional to the temperature; (c) *switchboard*, or nonadjacent reentry, model in which an even more nonordered amorphous layer is present on both sides of the lamellae than in the irregular model. (Geil [102].)

lamellae comprise the basic structural units of all polymer morphologies crystallized from the melt. Figure 5-35 shows clearly the lamellae exposed on the interior fracture surface of a sintered polytetrafluoroethylene specimen. It has been already mentioned in Section 1-2.4 that the observation of banded spherulite extinction patterns under the polarizing microscope indicates twisting of the

Figure 5-34 Illustration of point defects originating at A, progressing through the lattice at B, and extending the fold period at C. (Reneker [103].)

Figure 5-35 Interior fracture surface of polytetrafluoroethylene showing lamellae. Electron micrograph, magnification 30,000×. (Courtesy of the E. I. du Pont de Nemours and Company, Inc.; micrograph by P. H. Geil.)

radially oriented lamellar crystallites. The electron micrograph of the central portion of a banded polyethylene spherulite shown in Figure 5-36*a* and the enlargement shown in Figure 5-36*b* unmistakably reveal the twisting of the lamellae.

In contrast to the abundant experimental evidence for the existence of folded-chain lamellae in solution- and melt-grown polymers, evidence that such lamellae are also present in drawn, or in drawn and annealed, fibers and films is comparatively meager, although currently on the increase. We may cite in particular the following lines of evidence that appear to constitute strong support for the presence of folded-chain lamellae in drawn fibers and other cold-worked polymers: (a) the close resemblance between the meridional

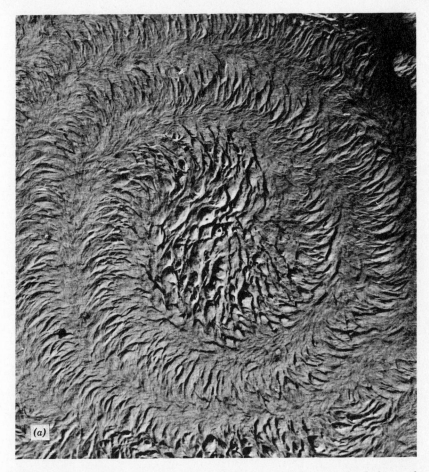

(a)

Figure 5-36 (a) Electron micrographs of banded spherulite of polyethylene. The twisted lamellae are seen alternately by their planar surfaces and edges. Magnification 9000×. [Part (b) on page 348]

interferences produced by drawn fibers and single-crystal mats (see again Figure 5-27), (b) the observation of a manyfold increase in intensity of the meridional interferences of drawn fibers on annealing[104, 105], and (c) direct observation of transverse lamellae in annealed fibers by means of a nitric-acid etching treatment[106]. Dismore and Statton[105] noted a fortyfold increase in intensity of the meridional reflection on annealing drawn fibers of nylon 66, and an increase of similar magnitude has been observed by Pollack and Magill[107] on annealing drawn fibers of polytetramethyl-p-silphenylenesiloxane at 133°C, about 10° below the melting point.

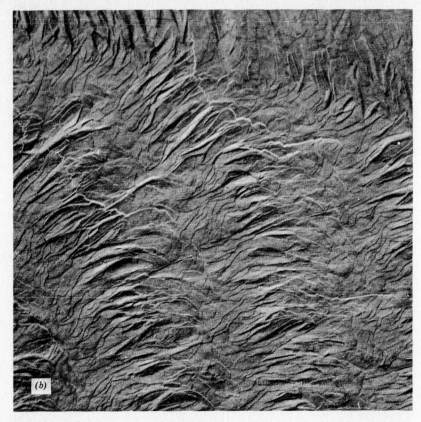

Figure 5-36 [Continued from page 347]   (b) Enlargement of a portion of (a) at 25,000×. (Courtesy of the E. I. du Pont de Nemours and Company, Inc.; micrograph by P. H. Geil.)

It is clearly impossible to account for such a dramatic heightening of the intensity on the basis of alternating crystalline and amorphous regions simply because of the lack of sufficient scattering power [the electron-density contrast, $\Delta\rho$, of (K5-23) is too small]. On the other hand, if drawing of the fiber destroys most of the folded-chain lamellae as suggested in Figure 5-37a and if annealing results in redevelopment of folded-chain lamellae interspersed with fissures or very sparsely occupied interstices (Figure 5-37b), the large contrast in electron density between lamellae and interstices abetted by the large increase in the number of folded chains can easily explain the pronounced rise in intensity.

Peterlin[108, 109] has been making an intensive study of the structural metamorphoses that occur during the deformation and

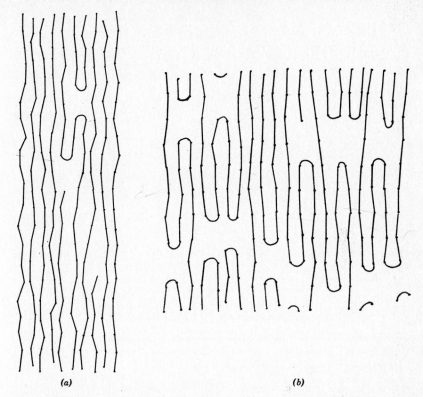

Figure 5-37 Schematic concept of molecular chains in (*a*) drawn and (*b*) drawn and annealed fibers. (Dismore and Statton [105].)

annealing of polymers. Figure 5-38 illustrates schematically Peterlin's ideas for fiber formation by mechanisms of chain tilting, slip, and breaking off of blocks of folded chains with subsequent re-formation of folded-chain domains constituting a fiber of different orientation. Geil [102] points out that, increasingly, ideas familiar in studying cold work and annealing of metals are being applied to polymers. Thus edge and screw dislocations as well as dislocation networks have been observed—and twinning, phase changes, segmental tilt, and slip have been found to be important features of the deformation process in polymers. For further information on these topics the reader must be directed to the literature [103, 110–117].

Huggins [118] has proposed that in many linear polymers the *intra*chain forces can lead to orientations of atoms or groups characterized by strong *inter*chain interactions that are favorable only over a limited region and that this results in a change of structure when the

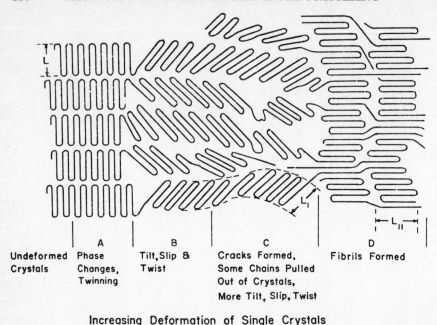

	A	B	C	D
	Undeformed Crystals	Phase Changes, Twinning	Tilt, Slip & Twist	Cracks Formed, Some Chains Pulled Out of Crystals, More Tilt, Slip, Twist	Fibrils Formed

Increasing Deformation of Single Crystals

Figure 5-38   A possible model for fiber formation by mechanisms of chain tilting, slip, and breaking off of blocks of folded chains. (Peterlin [109].)

limits of this region are reached. He suggests that this principle may underlie the generation of long x-ray spacings and the uniform folding of polymer chains.

Table 5-7   Symbols Used in This Chapter

Note: I and $\tilde{I}$ are intensities measured with collimators consisting of very small pinholes and very long narrow slits, respectively. Intensities designated by other symbols and quantities derived from intensities are differentiated in the same way; for example, i_n and $\tilde{i}_n$, Q and $\tilde{Q}$, R and $\tilde{R}$.

Symbol	Definition	Units of Measure
A_c	Cross-sectional area of rodlike particle	Å^2
A_i	Atomic weight of ith constituent element of solute	
a	Specimen-to-receiver distance	cm
c_e	Ratio of the number of electrons in the solute to the number in the solution	
c_1	Concentration of solute	g/cm^3 of solution
d	Thickness of specimen	cm

Symbol	Definition	Units of Measure
d'	Effective thickness of specimen	cm
f	Cluster fraction	
f_a	Anisotropy, or shape, factor	
$I(S), \tilde{I}(S)$	Experimental intensity scattered at angle S; measured in relative units	
$I_n, \tilde{I}_n$	Scattered intensity in absolute units, I/P_0 and $\tilde{I}/P_0$	
$I(0), \tilde{I}(0)$	Experimental intensity scattered at zero angle	
$i_n, \tilde{i}_n$	Scattered intensity in normalized absolute units	
i_e	Thomson scattering constant of a free electron 7.9×10^{-26}	cm^2
K_0	Modified Bessel function of zeroth order	
l	Length of a fully extended polymer chain	Å
$\bar{l}_c$	Correlation distance, or length	Å
l_0	Transversal, or inhomogeneity, length of dispersant phase	Å
l_1	Transversal, or inhomogeneity, length of disperse phase	Å
M	Molecular weight	
m	Distance between incident and scattered rays in the plane of registration	cm
m_1	Number of electrons in one solute particle	
N	Avogadro's number	
n	Number of straight elements comprising an idealized molecular chain	
n_i	Stoichiometric proportion of ith element of solute	
O	Total area of interface separating the disperse and dispersant phases	Å^2
O_s	Specific inner surface $= O/V_1$	Å^{-1}
P_0	Intensity of the primary beam	
$Q, \tilde{Q}$	Invariant (without specification of dimensionality)	
$Q_s, \tilde{Q}_s$	Invariant as a function of s	Å^{-3}, Å^{-2}
$Q_m, \tilde{Q}_m$	Invariant as a function of m	cm^3, cm^2
q	Homogeneous swelling factor	
q'	Swelling factor of the dispersed fraction	
q_m	Macroscopic swelling factor	
R	Electronic radius of gyration	Å
$\tilde{R}$	Apparent electronic radius of gyration	Å
R_c	Electronic radius of gyration of the cross section of a rodlike particle	Å
R_{sp}	Radius of gyration of a sphere	Å
r	Experimentally accessible fraction of the scattering power	
$\overline{r^2}$	Mean-square end-to-end distance of a coiled polymer chain	Å^2
r_{sp}	Radius of a sphere	Å
S	$4\pi(\sin\theta)/\lambda = 2\pi s$	Å^{-1}

Symbol	Definition	Units of Measure
s	$2(\sin\theta)\lambda$	Å^{-1}
t	Thickness of a lamellar particle	Å
V	Total volume of a scattering system	cm^3
V_1	Volume occupied by the disperse phase	cm^3
v_1	Volume of one (average) particle	Å^3
$\bar{v}_1$	Partial specific volume of disperse phase	cm^3/g
w_1	Volume fraction of disperse phase (solute)	
w_0	Volume fraction of dispersant phase (solvent)	
w_1'	Volume fraction of solute swollen with solvent	
y	Persistence length	Å
$Z(m)$	Residual intensity curve after subtraction of Gaussian function	
Z_i	Atomic number of ith constituent element of solute	
z	Length of hypothetical straight elements comprising a polymer chain	Å
z_1	Ratio of sum of atomic numbers of solute to sum of atomic weights	
α	Internal solvation ratio, the ratio of the number of electrons swelling one solute particle to the number of electrons in the unswollen particle	
$\gamma(r)$	Correlation function, or characteristic	
η	Electrons per square centimeter of surface of flat specimen, a measure of thickness	cm^{-2}
θ	Bragg angle	degrees of arc
λ	Wavelength of x-rays	Å
μ	Number of electrons per unit length of rodlike particle	Å^{-1}
$\boldsymbol{\mu}$	Number of molecular-weight units per unit length of rodlike particle	Å^{-1}
ν	$\lambda^2 i_e = \lambda^2 \times 7.9 \times 10^{-26}$	$\text{Å}^2 \times \text{cm}^2$
ρ	Electron density	$\text{e}/\text{Å}^3$
ρ_1	Electron density of solute	$\text{e}/\text{Å}^3$
ρ_0	Electron density of solvent	$\text{e}/\text{Å}^3$
$\bar{\rho}$	Average electron density	$\text{e}/\text{Å}^3$
$\boldsymbol{\rho}_1$	Electron density of solute	mole electrons per cm^3
$\boldsymbol{\rho}_0$	Electron density of solvent	mole electrons per cm^3
$\overline{\boldsymbol{\rho}}$	Average electron density	mole electrons per cm^3
Υ	Angle of inclination between two successive polymer-chain elements	degrees of arc
$\phi(s)$	Residual intensity curve after subtraction of Gaussian function	
ψ	Electronic partial specific volume of disperse phase	$\text{Å}^3/\text{electron}$

GENERAL REFERENCES

H. Brumberger (editor), *Small-Angle X-ray Scattering*, Gordon and Breach, New York, 1967, Parts I, II, and III.

P. H. Geil, "Polymer Morphology," *Chem. Eng. News*, August 16, 1965, p. 72.

P. H. Geil, *Polymer Single Crystals*, Interscience, New York, 1963.

A. Guinier and G. Fournet, *Small-Angle Scattering of X-rays*, Wiley, New York, 1955.

O. Kratky, "X-Ray Small-Angle Scattering with Substances of Biological Interest in Diluted Solutions," in *Progress in Biophysics*, Vol. 13, Pergamon Press, New York, 1963, p. 105.

V. Luzzati, "Small-Angle X-Ray Scattering on an Absolute Scale," in *X-Ray Optics and X-ray Microanalysis*, H. H. Pattee, V. E. Coslett, and A. Engström (editors), Academic Press, New York, 1963, p. 133.

G. Porod, "Anwendung und Ergebnisse der Röntgenkleinwinkelstreuung in festen Hochpolymeren," *Fortschr. Hochpolym. Forsch.*, 2, 363 (1961).

J. M. Schultz, W. H. Robinson, and G. M. Pound, "Temperature-Dependent X-Ray Small-Angle Scattering from Melt-Crystallized Polyethylene," *J. Polymer Sci.*, Part A-2, 5, 511 (1967).

W. O. Statton, "Small-Angle X-Ray Studies of Polymers," in *Newer Methods of Polymer Characterization*, B. Ke (editor), Interscience, New York, 1964.

SPECIFIC REFERENCES

[1] A. Guinier, *Compt. Rend.*, 204, 1115 (1937).

[2] A. Guinier, *Ann. Phys.* (Paris), 12, 161 (1939).

[3] A. Guinier, *J. Chim. physique*, 40, 133 (1943).

[4] W. O. Statton, *J. Polymer Sci.*, 58, 205 (1962).

[5] G. Porod, *Fortschr. Hochpolym. Forsch.*, 2, 363 (1961).

[6] R. Hosemann, *Z. Physik*, 113, 751 (1939); 114, 133 (1939).

[7] R. Hosemann, *Z. Electrochem.*, 46, 535 (1940).

[8] R. Hosemann, *Kolloid-Z.*, 117, 13 (1950); 119, 129 (1950).

[9] O. Kratky and G. Porod, *Z. physik. Chem.*, 7, 236 (1956).

[10] O. Kratky, in *Progress in Biophysics*, Vol. 13, Pergamon Press, New York, 1963, p. 105.

[11] V. Luzzati, *Acta Cryst.*, 13, 939 (1960).

[12] V. Luzzati in *X-Ray Optics and X-Ray Microanalysis*, H. H. Pattee, V. E. Coslett, and A. Engström (editors), Academic Press, New York, 1963, pp. 133–156.

[13] O. Kratky, *Angew. Chem.*, 72, 467 (1960).

[14] O. Kratky, *Makromol. Chem.*, 35A, 12 (1960).

[15] O. Kratky and H. Wawra, *Monatsh. Chem.*, 94, 981 (1963).

[16] P. H. Hermans, D. Heikens, and A. Weidinger, *J. Polymer Sci.*, 35, 145 (1959).

[17] O. Kratky, G. Porod, and L. Kahovec, *Z. Electrochem.*, 55, 53 (1951).

[18] O. Kratky, I. Pilz, and P. J. Schmitz, *J. Colloid Interface Sci.*, 21, 24 (1966).

[19] I. Pilz and O. Kratky, *J. Colloid Interface Sci.*, 24, 211 (1967).

[20] A. Guinier and G. Fournet, *Small-Angle Scattering of X-Rays*, Wiley, New York, 1955, pp. 116–120.

[21] A. Guinier and G. Fournet, *J. Phys. Radium*, 8, 345 (1947).

[22] J. W. M. DuMond, *Phys. Rev.*, 72, 83 (1947).

[23] O. Kratky, G. Porod, and Z. Skala, *Acta Phys. Austriaca*, 13, 76 (1960).

[24] P. W. Schmidt and R. Hight, Jr., *Acta Cryst.*, 13, 480 (1960).

[25] P. W. Schmidt, Twenty-Second Pittsburgh Diffraction Conference, Program and Abstracts, November 4–6, 1964, Abstract G–1.

[26] O. Kratky and G. Miholic, *J. Polymer Sci.*, Part C, **2**, 449 (1963).

[27] C. G. Shull and L. C. Roess, *J. Appl. Phys.*, **18**, 295 (1947).

[28] V. Gerold, *Acta Cryst.*, **10**, 287 (1957).

[29] S. Heine and J. Roppert, *Acta Phys. Austriaca*, **15**, 148 (1962).

[30] S. Heine, *Acta Phys. Austriaca*, **16**, 144 (1963).

[31] P. Kent and H. Brumberger, *Acta Phys. Austriaca*, **17**, 263 (1964).

[32] Reference 20, p. 74.

[33] G. Porod, *Kolloid-Z.*, **124**, 83 (1951).

[34] L. Kahovec, G. Porod, and H. Ruck, *Kolloid-Z.*, **133**, 16 (1953).

[35] Reference 20, p. 157.

[36] V. Luzzati, J. Witz, and A. Nicolaieff, *J. Molec. Biol.*, **3**, 367 (1961).

[37] O. Kratky, *Pure and Appl. Chem.*, **12**, 483 (1966).

[38] H. B. Stuhrmann and R. G. Kirste, *Z. physik. Chem.*, **46**, 247 (1965).

[39] P. Debye and A. M. Bueche, *J. Appl. Phys.*, **20**, 518 (1949).

[40] P. Debye, H. R. Anderson, and H. Brumberger, *J. Appl. Phys.*, **28**, 679 (1957).

[41] G. Porod, *Kolloid-Z.*, **125**, 51 (1952).

[42] Reference 20, pp. 24–28.

[43] Reference 20, p. 130.

[44] O. Kratky, *Z. Elektrochem.*, **60**, 245 (1956).

[45] G. N. Watson, *A Treatise on the Theory of Bessel Functions*, 2nd ed., Cambridge University Press, 1952, pp. 80 and 181.

[46] N. W. McLachlan, *Bessel Functions for Engineers*, 2nd ed., Clarendon Press, Oxford, 1955, p. 203.

[47] Reference 46, Fig. 22 on p. 116.

[48] For example, reference 45, Table IV on p. 737.

[49] O. Kratky and G. Porod in *Die Physik der Hochpolymeren*, Vol. 2, H. A. Stuart (editor), Springer-Verlag, Berlin, 1953, p. 515.

[50] W. Kuhn, *Kolloid-Z.*, **68**, 2 (1934); *Exp. Basel*, **1**, 1 (1945); W. Kuhn and H. Kuhn, *Helv. Chim. Acta*, **26**, 1394 (1943).

[51] O. Kratky and G. Porod, *Rec. trav. chim. Pays-Bas*, **68**, 1106 (1949).

[52] G. Porod, *J. Polymer Sci.*, **10**, 157 (1953).

[53] A. Peterlin, *J. Polymer Sci.*, **47**, 403 (1960).

[54] S. Heine, O. Kratky, and J. Roppert, *Makromol. Chem.*, **56**, 150 (1962).

[55] S. Heine, *Makromol. Chem.*, **71**, 86 (1964).

[56] J. J. Hermans and R. Ulmann, *Physica*, **18**, 951 (1952).

[57] S. Heine, O. Kratky, G. Porod, and P. J. Schmitz, *Makromol. Chem.*, **44–46**, 682 (1961).

[58] V. Luzzati and H. Benoit, *Acta Cryst.*, **14**, 297 (1961).

[59] R. Kirste and W. Wunderlich, *Makromol. Chem.*, **73**, 240 (1964).

[60] W. Wunderlich and R. G. Kirste, *Ber. Bunsenges. physik. Chem.*, **68**, 646 (1964).

[61] Reference 20, pp. 151–155.

[62] R. Hosemann, *Ergeb. exakt. Naturwiss.*, **24**, 142 (1951).

[63] L. C. Roess and C. G. Shull, *J. Appl. Phys.*, **18**, 308 (1947).

[64] M. Champagne, A. Luzzati, and A. Nicolaieff, *J. Chim. Phys.*, **58**, 657 (1961).

[65] C. Fromageot and M. Privat de Garilhe, *Biochim. Biophys. Acta*, **4**, 509 (1950).

[66] J. Jollés, P. Jollés, and J. Jaurequi-Adell, *Bull. Soc. Chim. Biol.*, Paris, **42**, 1319 (1960).

[67] M. H. Halver, G. Nutting, and B. A. Brice, *J. Amer. Chem. Soc.*, **73**, 2786 (1951).

[68] P. A. Charlwood, *J. Amer. Chem. Soc.*, **79**, 776 (1957).

[69] H. N. Ritland, P. Kaesberg, and W. W. Beeman, *J. Chem. Phys.*, **18**, 1237 (1950).

[70] P. H. Hermans and P. Platzek, *Z. physik. Chem.*, **A185**, 260 (1939).

[71] P. H. Hermans and A. J. de Leeuw, *Kolloid-Z.*, **83**, 58 (1938).

[72] O. Kratky, A. Sekora, and R. Treer, *Z. Electrochem*, **48**, 587 (1942); O. Kratky and A. Wurster, ibid., **50**, 249 (1944); O. Kratky, ibid., **57**, 42 (1953).

[73] A. N. J. Heyn, *Nature*, **172**, 1000 (1953).

[74] A. Frey-Wissling, *Submicroscopic Morphology of Protoplasm and Its Derivatives*, Elsevier, New York, 1948, pp. 69–81; *Experientia*, **9**, 181 (1953).

[75] W. O. Statton, *J. Polymer Sci.*, **22**, 385 (1956).

[76] D. Heikens, *J. Polymer Sci.*, **35**, 139 (1959).

[77] W. O. Statton in *Newer Methods of Polymer Characterization*, B. Ke (editor), Interscience, New York, 1964, Chapter 6.

[78] W. O. Statton and G. M. Godard, *J. Appl. Phys.*, **28**, 1111 (1957).

[79] K. Hess and H. Kiessig, *Z. physik. Chem.*, **A193**, 196 (1944).

[80] O. Kratky and A. Sekora, *Makromol. Chem.*, **1**, 113 (1943).

[81] R. S. Bear, *J. Amer. Chem. Soc.*, **65**, 1784 (1943); **66**, 2043 (1944).

[82] R. S. Bear and O. E. A. Bolduan, *Acta Cryst.*, **3**, 230, 236 (1950).

[83] R. S. Bear and O. E. A. Bolduan, *J. Appl. Phys.*, **22**, 191 (1951).

[84] K. Hess and H. Kiessig, *Naturwiss.*, **31**, 171 (1943).

[85] K. Hess and H. Kiessig, *Kolloid-Z.*, **130**, 10 (1953).

[86] H. Kiessig, *Physikal. Verhandl.*, **5**, 87 (1954).

[87] H. Kiessig, *Kolloid-Z.*, **152**, 62 (1957).

[88] For example: (a) R. Bonart and R. Hosemann, *Kolloid-Z., Z. Polymere*, **186**, 16 (1962); (b) R. Hosemann, *J. Appl. Phys.*, **34**, 25 (1963). (See also Chapter 7.)

[89] See Chapter 7 and H. P. Klug and L. E. Alexander, *X-Ray Diffraction Procedures*, Wiley, New York, 1954, Chapter 9.

[90] Reference 77, p. 273.

[91] L. M. Arnett, E. P. H. Meibohm, and A. F. Smith, *J. Polymer Sci.*, **5**, 737 (1950).

[92] E. P. H. Meibohm and A. F. Smith, *J. Polymer Sci.*, **7**, 449 (1951).

[93] B. Belbéoch and A. Guinier, *Makromol. Chem.*, **31**, 1 (1959).

[94] P. Predecki and W. O. Statton in *Small-Angle X-Ray Scattering*, H. Brumberger (editor), Gordon and Breach, New York, 1967, pp. 131–143.

[95] H. Zahn, *Melliand Textilber.*, **32**, 534 (1951).

[96] H. Zahn and U. Winter, *Kolloid-Z.*, **128**, 142 (1952).

[97] W. O. Statton and P. H. Geil, *J. Appl. Pol. Sci.*, **3**, 357 (1960).

[98] W. O. Statton, *J. Appl. Phys.*, **32**, 2332 (1961).

[99] E. W. Fischer, *Ann. N. Y. Acad. Sci.*, **89**, 620 (1961).

[100] E. W. Fischer and G. F. Schmidt, *Angew. Chem. (Int. Ed.)*, **1**, 488 (1962).

[101] L. Mandelkern, A. S. Posner, A. F. Diorio, and D. E. Roberts, *J. Appl. Phys.*, **32**, 1509 (1961).

[102] P. H. Geil, *Chem. Eng. News*, August 16, 1965, p. 72.

[103] D. H. Reneker, *J. Polymer Sci.*, **59**, S39 (1962).

[104] D. R. Beresford and H. Bevan, *Polymer*, **5**, 247 (1964).

[105] P. F. Dismore and W. O. Statton, *J. Polymer Sci.*, Part C, **13**, 133 (1966).

[106] I. L. Hay and A. Keller, *Nature*, **204**, 862 (1964).

[107] S. S. Pollack and J. H. Magill, *J. Polymer Sci.*, Part A, **7**, 551 (1969).

[108] A. Peterlin in *Small-Angle X-Ray Scattering*, H. Brumberger (editor), Gordon and Breach, New York, 1967, pp. 145-155.

[109] A. Peterlin, *Structure of Drawn Polymers*, Technical Report AFML-TR-67-6,

December 1966, U.S. Air Force Materials Laboratory, Wright-Patterson AFB, Ohio.

[110] R. Jaccodine, *Nature*, **176**, 305 (1955).

[111] F. C. Frank, A. Keller, and A. O'Connor, *Phil. Mag.*, **3**, 64 (1958).

[112] D. A. Zaukelies, *J. Appl. Phys.*, **33**, 2797 (1962).

[113] P. H. Geil, *Polymer Single Crystals*, Interscience, New York, 1963.

[114] V. F. Holland, *J. Appl. Phys.*, **35**, 3235 (1964).

[115] V. F. Holland and P. H. Lindenmeyer, *J. Appl. Phys.*, **36**, 3049 (1965).

[116] P. Predecki and W. O. Statton, *J. Appl. Phys.*, **37**, 4053 (1966).

[117] W. O. Statton, *J. Polymer Sci.*, Part C, **20**, 117 (1967).

[118] M. L. Huggins, *Makromol. Chem.*, **92**, 260 (1966); *Pure and Appl. Chem.*, **15**, 368 (1967).

6

Microstructure from Wide-Angle Diffraction

By microstructural studies we refer to investigations of (a) the stereochemistry of the individual chain molecules and (b) the manner in which these molecules are associated to form the solid polymer, which may in general be crystalline, semicrystalline, or noncrystalline (so-called totally disordered or amorphous). The first and larger portion of this chapter is concerned principally with the structure of the individual chain molecule, whereas the concluding section shifts the emphasis to the arrangement of the chains with respect to one another in the crystalline state. The solution of the molecular structures of technological polymers rests very largely on the analysis of their wide-angle x-ray diffraction patterns, whereas in the investigation of biological fibers the chain diameters and longitudinal periodicities are so large as to require the interpretation of x-ray diffraction effects in the small-angle region as well.

6-1 MOLECULAR STRUCTURE

The molecular structure is determined first of all by the *chemical constitution* of the monomeric units of which the chains are composed and secondly by the *modes of addition of successive monomeric units*, including the resulting degree of steric regularity of substituent groups. In the usual course of events, of course, the chemical constitution of the monomer units is known in advance of any needed x-ray studies, and commonly there is also information concerning the mode of addition; for example, whether the monomer units are linked head to tail, alternately head to head and tail to tail, or in a mixed mode.

The complete description of the structure of a chain molecule must include specification of both its *configuration* and its *conformation*. With Dunitz and Prelog[1] and Corradini[2] we shall define *configuration* as meaning the spatial arrangement of bonds in a molecule of a given constitution without regard to the multiplicity of spatial arrangements that may arise by rotation about single bonds. *Conformations* are the different spatial arrangements of the atoms in a molecule of a given constitution and configuration that may arise through rotation about single bonds; for example, the placement of side groups R relative to the chain axis in all right-hand, or all left-hand, or alternating right- and left-hand positions is a matter of configuration; whereas in a molecular chain that is characterized by all right-hand placements twisting of the chain (by means of rotations about single bonds) out of a planar zigzag form into a helical form, thereby minimizing steric interferences between successive R groups, is a matter of conformation. An important means for the description of the configuration of a molecular chain is the specification of its *tacticity*.

6-1.1 Configuration and Tacticity

The reader may wish to read again Sections 1-2.1 and 1-2.2. Following Huggins et al.[3] we shall now briefly describe the important tactic types, including modes that are more complex than the simple isotactic and syndiotactic types introduced in Chapter 1. For a fuller discussion the reader is referred to the original paper[3].

Tacticity can be defined very simply as the presence of steric order in the main chain. In a tactic polymer there is an ordered structure with respect to the configuration around at least one main-chain site of steric isomerism in the *conventional base unit* (usually identical with the monomer residue). In *holotactic* polymers *all* main-chain sites of steric isomerism are ordered, whereas in *monotactic*, *ditactic*, etc., polymers only *one*, *two*, etc., of such sites are ordered per base unit. In principle the term "atactic" is applicable only to a polymer in which complete randomness of configuration exists at all the main-chain sites of steric isomerism, although in actual practice it is often applied to polymers whose degree of order is so low that they differ very little in macroscopic properties from strictly atactic types.

We now define an *isotactic* polymer, more precisely than was done in Chapter 1, as a polymer composed of base units each of which possesses, as a component of the main chain, a carbon atom (or other sterically similar atom) with two different lateral substituents so arranged that a hypothetical observer, advancing along the bonds of the main chain, finds each of these chain atoms with all of its substituents in *the same steric order.* Contrariwise, in a *syndiotactic* polymer these two lateral substituents are so arranged that the observer finds *opposite steric configurations* around these chain atoms in successive conventional base units.

Monotactic examples portrayed as if based on hypothetically extended zigzag chains are the following:

Isotactic polyethylidene (R = CH₃)

$$\begin{array}{c}
\text{R}\quad\text{H}\quad\text{R}\quad\text{H}\quad\text{R}\quad\text{H}\quad\text{R}\quad\text{H}\\
\diagdown\diagup\ \diagdown\diagup\ \diagdown\diagup\ \diagdown\diagup\\
\text{C}\quad\text{C}\quad\text{C}\quad\text{C}\\
\diagup\diagdown\ \diagup\diagdown\ \diagup\diagdown\ \diagup\diagdown\\
\text{C}\quad\text{C}\quad\text{C}\quad\text{C}\\
\text{H}\quad\text{R}\quad\text{H}\quad\text{R}\quad\text{H}\quad\text{R}\quad\text{H}\quad\text{R}
\end{array}$$

Base unit
(true and conventional)

$$\begin{array}{c}\text{R}\quad\text{H}\\ \diagup\\ \text{C}\\ \diagdown\end{array}$$

Syndiotactic polyethylidene (R = CH₃)

$$\begin{array}{c}
\text{R}\quad\text{H}\quad\text{R}\quad\text{H}\quad\text{R}\quad\text{H}\quad\text{R}\quad\text{H}\\
\diagup\ \diagdown\diagup\ \diagdown\diagup\ \diagdown\diagup\ \diagdown\\
\text{C}\quad\text{C}\quad\text{C}\quad\text{C}\\
\diagdown\ \diagup\diagdown\ \diagup\diagdown\ \diagup\diagdown\\
\text{C}\quad\text{C}\quad\text{C}\quad\text{C}\\
\text{R}\quad\text{H}\quad\text{R}\quad\text{H}\quad\text{R}\quad\text{H}\quad\text{R}\quad\text{H}
\end{array}$$

True base
unit

$$\begin{array}{c}\text{R}\quad\text{H}\\ \diagup\\ \text{C}\\ \diagdown\\ \text{C}\\ \diagup\\ \text{R}\quad\text{H}\end{array}$$

Conventional
base unit

$$\begin{array}{c}\text{R}\quad\text{H}\\ \diagdown\diagup\\ \text{C}\\ \diagup\diagdown\end{array}$$

In a *di-isotactic* polymer the base unit possesses, as components of the main chain, two carbon atoms each with two different lateral substituents R and H or R' and H, the steric orientations in successive units being such as to make the molecule isotactic with respect to the configuration around corresponding chain atoms of either type considered separately. In an *erythro-di-isotactic* polymer the configurations at the two main-chain sites of steric isomerism in the base unit are *alike.* Thus in a hypothetically extended (zigzag) molecule (CHRCHR')ₙ the substituents R are all on one side of the plane of the chain atoms and the R' substituents are all on the other side. In a *threo-di-isotactic* polymer the configurations at the two main-chain sites of steric isomerism in each base unit are *opposite*. In a hypothetically extended (zigzag) molecule $(CHRCHR')_n$ the substituents R and R' are all on the same side of the plane of the chain atoms.

Erythro-di-isotactic poly-2-pentene $(R = CH_3, \quad R' = C_2H_5)$

Threo-di-isotactic poly-2-pentene $(R = CH_3, \quad R' = C_2H_5)$

For a more complete description of the nomenclature and the characterization of modes of stereoregularity in linear polymers the reader is referred to publications of Natta, Farina, and Peraldo [4], Miller and Nielsen [5], Newman [6], and Miller [7, 8].

6-1.2   Conformational Nomenclature

Since the molecular chains of so many natural and synthetic polymers assume helical conformations, at least in the crystalline state, an appropriate system of nomenclature is needed to describe them. Actually, two systems are presently in use, in both of which a helical net of points is generated from an origin point by repetitive rotation through an angle $\Delta\Phi$ accompanied by a translation Δz along the helix axis Z. The first system, which is to be preferred for the description of polymer helices, may be referred to as the *helical point-net* system, and the second is the conventional *crystallographic screw-axis* designation. For the description of these systems we define the following quantities (see Figure 6-1) [9]:

$P = Z$ translation equivalent to one turn of the helix (*pitch* of the helix);

$c =$ crystallographic identity period parallel to Z;

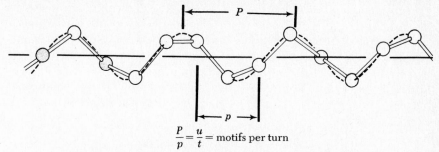

$$\frac{P}{p} = \frac{u}{t} = \text{motifs per turn}$$

Figure 6-1   Helical nomenclature. (Courtesy of E. S. Clark.)

p = Z-axis projection of the distance between consecutive equivalent points on the helix;

u = number of points, or motifs, on the helix corresponding to the period c (an integer);

t = number of turns of the helix in the identity period c (an integer);

$\Delta\Phi$ = projection on a plane perpendicular to Z of the central angle defined by two successive equivalent points on the helix;

γ = a positive integer > 0;

ϵ = a positive integer ≥ 0.

Helical Point Net H.   This system has been discussed in detail by Hughes and Lauer[10] and also treated by Nagai and Kobayashi [11]. The symmetry is expressed as u/t, and the operators $\Delta\Phi_H$ and Δz_H are defined by

$$\Delta\Phi_H = 2\pi\frac{p}{P} = 2\pi\frac{t}{u}, \tag{6-1}$$

$$\Delta z_H = p = \frac{c}{u}. \tag{6-2}$$

Screw Axis S.   The symmetry is expressed as u_γ, and the rotational and translational operators are defined by

$$\Delta\Phi_S = \frac{2\pi}{u}, \tag{6-3}$$

$$\Delta z_S = \gamma\left(\frac{c}{u}\right). \tag{6-4}$$

The two systems of notation are related by the expression

$$\gamma t = \epsilon u + 1. \tag{6-5}$$

For a given helix the numerical descriptions u/t and u_γ are equivalent ($u = u$ and $t = \gamma$) if a value of ϵ exists such that $\gamma = t$ and $\gamma^2 = t^2 = \epsilon u + 1$.

Figures 6-2 and 6-3 illustrate the application of these notations with the example of polyoxymethylene. The symmetry of the actual polymer is approximately, but not precisely, 9/5 in the H-notation; however, the idealized symmetry 9/5 will be assumed for purposes of illustration. Thus the period c contains nine $C-O$ motifs and five turns of the helix. Figure 6-3 depicts the action of the rotational operator through one identity period. By (6-1) the rotational operation involved is

$$\Delta \Phi_H = 2\pi \frac{t}{u} = \frac{5}{9} \times 360° = 200°.$$

If we commence at A (point 0), a rotation of 200° results in a rotation to C (point 1), accompanied by a translation of

$$p = \frac{c}{u} = \frac{c}{9}$$

[from (6-2)]. Successive operations then take us to points 2, 3, 4, $\cdots$, 9(0), accompanied by equal translational increments p.

In the S-system, however, the rotational operation is

$$\Delta \Phi_S = \frac{2\pi}{u} = 40°,$$

accompanied by a translation of.

$$\Delta z_S = \gamma \frac{c}{u} = \frac{2c}{9}.$$

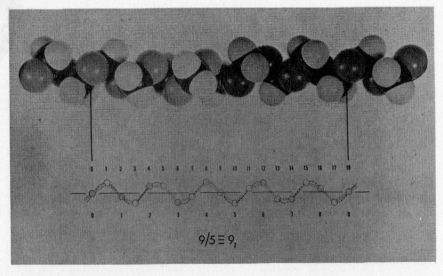

Figure 6-2   Helical conformation of polyoxymethylene. (Courtesy of E. S. Clark.)

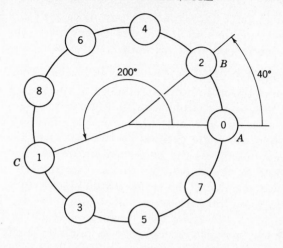

Figure 6-3 Helical conformation 9/5 ≡ 9_2, showing action of rotational operators.

Successive operations, then, take us from 0 to 2, 2 to 4, 4 to 6, etc., to 9(0). The net translation equivalent to the nine operations is in this case $9 \times 2c/9$, or $2c$, rather than c as in the case of the helical-point system of nomenclature. Since $\gamma = 2$, the screw-axis designation of this helix is 9_2. It follows from (6-5) that the numerical value of ϵ relating the two notations for the polyoxymethylene helix is

$$\epsilon = \frac{\gamma t - 1}{u} = \frac{2 \cdot 5 - 1}{9} = 1.$$

We may note that, if the carbon and oxygen atoms could be treated as completely equivalent, the helical designations in the two systems would be 18/5 and 18_{11}.

It is important to observe that in the above systems of nomenclature all the atoms that constitute one unit of structure (*true base unit* in tacticity parlance and usually also the *monomer unit*) are termed the *motif* and are represented by one point on the helix. The radial and azimuthal coordinates of the component atoms with respect to some arbitrary origin are not specified. A more complete description of helical structure and its diffraction effects is presented in Appendix 1 and in Section 6-1.6.

The helical-point-net nomenclature has been adopted by Miller and Nielsen[12, 13] and by Miller[14] in their valuable tabulations of crystallographic data for various polymers, a recent edition of which comprises Appendix 3 of the present monograph. In his more recent editions of the tables Miller prefixes the symbol u/t with the number n of *skeletal* atoms in the chain motif. At the same time he employs the format $n*u-t$ in order to permit direct photo-offset reproduction of the computer printout.

Internal-Rotation Angles[15]. A useful way of describing the conformation of the main polymer chain is to specify the sequence of internal-rotation angles. Referring to Figure 6-4, we consider any three successive chain bonds, L_1, L_2, and L_3. The internal-rotation angle σ is then the angle defined by the two planes L_1L_2 and L_2L_3, which is seen to express the conformation of the chain around the bond L_2. We adopt the convention that σ is positive if, looking along L_2 from L_1, we see that L_1 must be rotated clockwise by an angle σ, which is less than 180°, in order that it superpose L_3. This is completely equivalent to saying that if we look along L_2 from L_3, we see that L_3 must be rotated clockwise through the angle σ to superpose it on L_1. Counterclockwise rotations of less than 180° are designated negative.

Bunn's Notation[16, 17]. In the course of early studies of polymer chain structures, in particular vinyl polymers, Bunn suggested the following scheme of nomenclature for the most commonly encountered single-bonded carbon-chain conformations. The orientation of a given bond with respect to the *two preceding bonds* in the chain is designated A, B, or C as follows:

$A = trans$ $(\sigma = 180°)$,

$B = \text{left } gauche$ $(\sigma = +60°)$,

$C = \text{right } gauche$ $(\sigma = -60°)$.

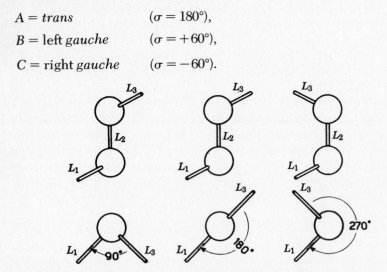

Figure 6-4 Conventions for the specification of internal-rotation angles σ. (Natta, Corradini, and Bassi[15].)

With Tadokoro[18] we now simplify the nomenclature by replacing A, B, and C with T, G, and $\overline{G}$, respectively. These three basic bond sequences in single-bonded carbon chains, represented in projection in Figure 6-5, are the only ones that are permitted by the principle of staggered bonds. When the chain motif, or geometrical repeating unit, comprises more than one bond sequence of the same kind, it is denoted by the appropriate numerical subscript; for example, T_2, $(TG)_3$, etc. Mixed modes such as $TGT\overline{G}$ or $(TG)_2(T\overline{G})_2$ may also be specified.

On the basis of idealized bond distances and angles we may expect characteristic identity periods to be observed for a number of single-bonded carbon-chain conformations as follows:

Mode	Identity Period (Å)	Mode	Identity Period (Å)
T_2	2.5	$(TG)_3$	6.2
T_4	5.0	$(T\overline{G})_3$	6.2
$TGT\overline{G}$	4.4	$T_3GT_3\overline{G}$	8.8
G_4	3.6	$(TG)_2(T\overline{G})_2$	8.5

Five of these conformations are depicted in Figure 6-6. If substituent groups of appreciable size are attached to alternate carbon atoms in the chain, the ideally tetrahedral angles in the chain tend to open up a little, thus increasing the identity periods somewhat; for example, the observation of periods between 6.2 and 6.7 Å in vinyl polymers indicates the chain form to be $(TG)_3$ or $(T\overline{G})_3$, which are left- and right-handed isotactic helices, respectively. On the other hand, periods slightly larger than 5.0 Å in vinyl polymers are indicative of simple syndiotactic placements based on the chain form T_4. An example of chain form T_4 with alternate left- and right-hand

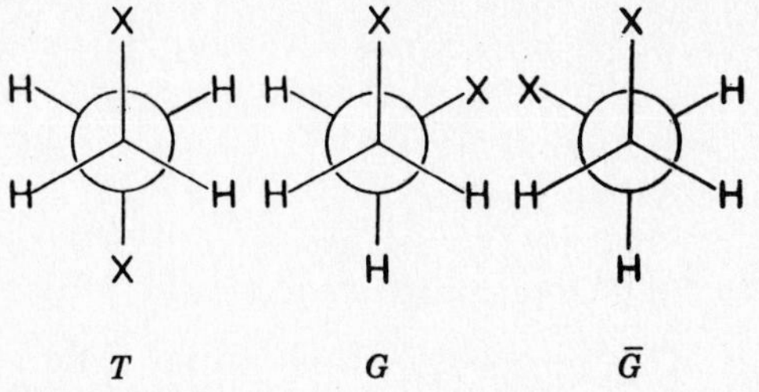

Figure 6-5 The three bond sequences in single-bonded carbon chains that conform to the principle of staggered bonds: T, *trans*; G, left *gauche*; $\overline{G}$, right *gauche*. (Tadokoro [18].)

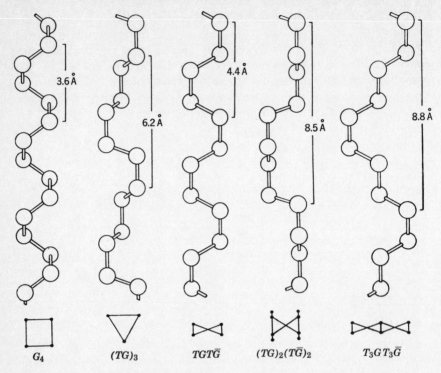

Figure 6-6 Five chain conformations based on the T, G, and $\overline{G}$ staggered bond orientations of Figure 6-5. (After Bunn [16].)

placement of substituent groups is supplied by syndiotactic poly-(vinyl chloride) (see Figure 1-11), which displays an axial identity period of 5.1 Å[19]. Figure 6-7 shows a good example of a $(TG)_3$ structure, that of isotactic polystyrene[17, 20], which shows an axial periodicity of 6.65 Å. We may note that in H and S helical notation $(TG)_3$ structures are designated 3/1 and 3_1, respectively. Some additional examples of $(TG)_3$ structures with their observed periodicities are: polypropene, 6.50 Å; poly-1-butene, 6.50 Å; poly-1-pentene, 6.49 Å; poly(N,N-dibutyl acrylamide), 6.3 Å; 1,2-poly-1,3-butadiene, 6.5 Å; and poly(butyl vinyl ether), 6.30 Å.

6-1.3 Factors That Determine Chain Conformation

As the result of numerous studies Natta, Corradini, and Ganis [21, 22] have concluded that in crystalline linear polymers (a) the monomeric units, or motifs, occupy geometrically equivalent positions in relation to the axis of the macromolecule (*equivalence*

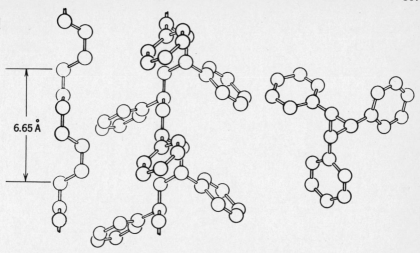

Figure 6-7 Left: the chain $(TG)_3$. Center and right: side and end views of the isotactic polystyrene molecule. (Bunn and Howells [17].)

postulate) and (b) the conformation actually assumed approaches the one of *minimum potential energy* that would be assumed by an *isolated* macromolecule conforming to the symmetry restriction (a). This means that chain-packing considerations ordinarily play only a secondary role in governing the conformations that are assumed by the molecular chains in the crystalline state. Important exceptions are polymers such as polyamides in which adjacent molecular chains are linked together by hydrogen bonds, which strongly influence the conformation adopted by the individual molecule.

So far as consistent with other stereochemical constraints a carbon-chain macromolecule "strives" (a) to assume normal bond lengths and angles, and (b) to adopt a conformation that satisfies the principle of staggered bonds (see again Figure 6-5) while at the same time permitting the molecules to pack together in a manner that permits comfortable intermolecular van der Waals contacts. It is evident that this is simply an alternative and somewhat more detailed way of expressing the minimum-potential-energy postulate. It should be emphasized that actual linear macromolecules, especially those characterized by bulky substituent groups, cannot simultaneously satisfy all of these geometrical criteria, with the result that appreciable deviations from ideality will be observed. The foregoing principles were first extensively invoked in the investigation and prediction of polymer structure by C. W. Bunn and co-workers [16, 17, 23].

Two classic examples of linear polymers in which the adoption of helical conformations minimizes overcrowding of substituent groups are polyisobutene and polytetrafluoroethylene, respectively; that is,

$$[-CH_2C(CH_3)_2-]_n \quad \text{and} \quad (-CF_2CF_2-)_n.$$

In stretched polyisobutene[23–25] the methyl groups on alternate carbon atoms would be only 2.54 Å apart if the chain possessed a fully extended, planar zigzag conformation with normal tetrahedral bond angles. The situation is relieved by (a) adoption of an 8/5 helical conformation and (b) opening and closing of alternate tetrahedral angles in the main chain such that

$$CH_2 \diagup_\diagdown = 126° \quad \text{and} \quad (CH_3)_2C \diagup_\diagdown = 107°,$$

which is equivalent to alternating internal-rotation angles of 102.5 and 51°. This chain conformation, depicted in Figure 6-8, is remarkably well defined by one of the best known polymer diffraction patterns. (See again Figure 1-3.)

In polytetrafluoroethylene[26–28] the ·relatively small fluorine-atom substituents, with a van der Waals radius of 1.4 Å, would nevertheless cause overcrowding in a planar zigzag conformation since they are substituents on all, rather than alternate, carbon atoms. Below the 19°C transition temperature there occurs a small twist of 20° about every chain bond, resulting in a 13/6 helix, which together with an opening up of the chain bond angles to 116° increases the shortest $F\cdots F$ distance to an acceptable value of 2.7 Å. The conformation of the polytetrafluoroethylene molecule is shown in Figure 6-9.

6-1.4   Randomly Oriented Systems

The importance of this section lies in its applicability to polymers possessing less than three-dimensional order and especially to noncrystalline specimens. Randomness of orientation, a state sometimes

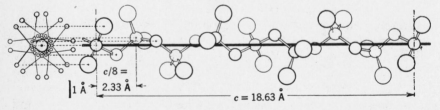

Figure 6-8   The 8/5 helix of the polyisobutene molecule. Small circles $= CH_2$; large circles $= CH_3$. (Liquori[25].)

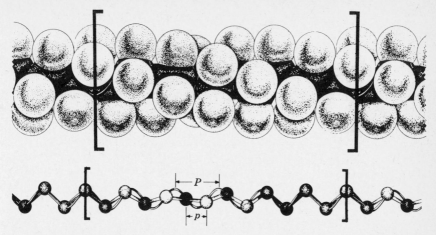

Figure 6-9 Molecular conformation of polytetrafluoroethylene: top—repeat unit of one molecule; bottom—carbon skeleton showing 13/6 helix. (Clark and Muus [27].)

referred to as *isotropic*, signifies that one orientation of a molecule or molecular element, such as a monomer residue, is as probable as any other and, furthermore, that the number of such orientations is very large from the point of view of statistics. The mathematical basis of x-ray scattering by such systems is presented in condensed form in Appendix 1, and more detailed treatments are given in the literature [29–32].

The x-ray pattern of a system characterized by orientational randomness is spherically symmetrical and can be mathematically transformed, or inverted, into a radial-distribution function, which specifies the atomic or electronic density as a function of the distance from every atom or electron in the system regarded successively as the origin. Thus the method is capable of giving information relative to the magnitudes but not the directions of the interatomic vectors, which constitutes such a severe limitation that only infrequent applications have been made to polymer structures.

The testing of alternative structure models by radial-distribution analysis can be performed in two ways, which in a formal sense are equivalent. First, the Fourier transform of the normalized experimental intensity curve $i(S)$ [see Appendix 1, (T-17)] may be compared with the radial-distribution curve calculated for the model in question or, second, the calculated and experimental intensity curves may be compared directly. On the basis of considerable practical experience Arndt and Riley [33] concluded that the second method is capable of revealing finer structural distinctions than the first. They attribute this to the fact that the effects of small structural variations tend to be concentrated in one portion of the $i(S)$ curve, particularly the region $1 < S < 3.5$, which is precisely the portion of the intensity curve that can be measured most accurately. On the other hand, Riley[34] considers the first method to be well suited to more rapid diagnosis, thereby providing a convenient means of narrowing down the field of choice preliminary to more detailed analysis based on the $i(S)$ curve. We shall now give condensed accounts of two investigations — the first illustrating the comparison of observed and calculated radial-distribution curves, and the second, and more precise, illustrating the comparison of $i(S)$ curves.

Conformation of Poly(hexamethylene adipamide) [35].   With the aim of devising procedures that would prove useful in cellulose analysis, Bjørnhaug, Ellefsen, and Tønnesen[35] compared the experimental radial-distribution curves of selected synthetic linear polymers with curves calculated for various models. Debye-Scherrer patterns were recorded with both CuKα and MoKα radiations monochromatized with a rock-salt crystal. The patterns were analyzed with a microphotometer and converted to intensity curves in electron units by standard procedures[29, 36]. These investigators made use of Finbak's electronic distribution function [36], which has the form of (T-17) (Appendix 1) but differs from it in that $i(S)$ is defined as $I(S) - Nf^2$ rather than $[I(S) - Nf^2]/Nf^2$. The removal of the factor $1/Nf^2$ has the beneficial effect of virtually eliminating the adverse effects of small errors in the intensity curve at large values of S. The electronic distribution may be expressed as

$$\sigma(r) = \frac{2r}{\pi} \int_0^\infty S\,[I(S) - Nf^2]\sin rS\,dS. \qquad (6\text{-}6)$$

Limitations of space permit us to describe only the results obtained by Bjørnhaug, Ellefsen, and Tønnesen for poly(hexamethylene adipamide). The solid line of Figure 6-10a is the experimental $\sigma(r)$ curve of a randomly oriented specimen, whereas the dotted line is the $\sigma(r)$ curve calculated for a fully extended planar zigzag conformation of poly(hexamethylene adipamide) with a repetition period of 17.4 Å [37]. It will be noticed that there are rather large-scale areas of disagreement even though the individual maxima in the two functions agree on a detailed scale. Figure 6-10b is the function obtained by smoothing out the difference curve obtained from the two curves of part a. Since the calculated curve of Figure 6-10a takes no account

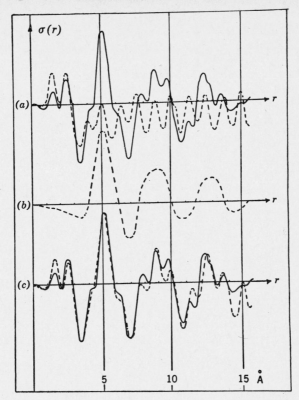

Figure 6-10 Radial-distribution analysis of poly(hexamethylene adipamide): (a) comparison of experimental $\sigma(r)$ curve (solid line) with curve calculated for an infinitely long planar zigzag conformation (dotted line); (b) smoothed-out difference of the two curves in (a), which should correspond to intermolecular distances; (c) comparison of experimental $\sigma(r)$ curve (solid line) with the sum of the two dotted curves of (a) and (b) (dotted line). (Bjørnhaug, Ellefsen, and Tønnesen [35].)

of interferences between molecular chains, curve b must represent the effect of such interchain interferences on the experimental intensities. This explanation is also supported by Figure 6-10c, in which the sum of the two dotted curves from parts a and b is seen to compare well with the experimental distribution function $\sigma(r)$. It should also be noted that the major peak in the experimental function $\sigma(r)$, which is situated at 5 Å, corresponds reasonably well to the expected interchain distance [38]. (See Section 6-1.5.)

Thus the results of the foregoing analysis can be interpreted as meaning that randomly oriented, semicrystallized poly(hexamethylene adipamide) consists principally of rather fully extended, planar zigzag chains with a periodicity of 17.4 Å, just as in the crystalline polymer[37]. By way of caution, however, we emphasize that identical scattering and radial-distribution curves would be generated by *homometric* structures[34], which Patterson[39] defines as two different structures possessing the same interatomic distances terminated by similar pairs of like or unlike atoms. Stated another way, there exists in radial-distribution analysis an ambiguity of scattering-equivalent systems similar to, but usually less severe than, that which characterizes small-angle-scattering analysis. (See Chapter 5, p. 281.) Therefore we may conclude only that the aforesaid structure model is not contradicted by these experimental results and so has a reasonable probability of being correct. In conclusion we may also note that the efforts of Bjørnhaug, Ellefsen, and Tønnesen[35] to differentiate isotactic and syndiotactic chain conformations from the evidence supplied by the radial-distribution curves were unsuccessful, although admittedly this constituted a severe test of the method.

Helical Characterization of Proteins[33, 34]. Arndt and Riley employed the spherical-transform method to obtain information concerning the types of helices present in 30 proteins and 2 synthetic polypeptides. We shall summarize only their results for α-proteins, which were the most numerous type studied. In order to heighten the power of the method they (a) made use of refined experimental techniques and (b) interpreted their structure models by comparison of experimental and calculated $i(S)$ curves.

The diffraction specimens were in the form of compressed plates of powder or chopped-up fibers. The patterns were recorded diffractometrically by the symmetrical-transmission technique with $CuK\alpha$, and in some cases also with $AgK\alpha$, radiation from a high-power rotating-anode x-ray tube. Monochromatization was effected by the use of a β-filter in conjunction with a proportional counter and pulse-height discrimination. The diffractometer that was employed has been described by Arndt, Coates, and Riley[40]. An unusual degree of stability was achieved with the aid of a monitor counter, which together with special attention to other aspects of the experimental technique permitted a 1% level of accuracy to be achieved within specific portions of the intensity curve.

The fundamental element of structure considered in the investigation of Arndt and Riley was the α-helix[41]. The stereochemistry around the asymmetric α carbon atom in a single amino-acid residue of a polypeptide chain may be represented as follows:

$$\vdots$$
$$C'{=}O$$
$$\cdots HN-\underset{\underset{R'}{|}}{\overset{\underset{|}{C_\beta}}{\underset{|}{C_\alpha}}}-H \qquad \text{(6-I)}$$

There are four possible varieties of this helix consisting of two enantiomorphic pairs. If L and D denote *laevo* and *dextro*, respectively, and if LH and RH refer to left- and right-handed helices, these enantiomorphic pairs α_1, α_1' and α_2, α_2' may be represented as follows:

$$\alpha_1 = \text{LH } C_{\beta,2} \text{ L-}C_\alpha, \qquad \alpha_1' = \text{RH } C_{\beta,2} \text{ D-}C_\alpha$$

and

$$\alpha_2 = \text{RH } C_{\beta,1} \text{ L-}C_\alpha, \qquad \alpha_2' = \text{LH } C_{\beta,1} \text{ D-}C_\alpha.$$

The members of an enantiomorphic pair, of course, give identical x-ray patterns. The fundamental polypeptide chain itself remains the same in all helices of the same category except for possible differences in handedness.

Intensity curves $i(S)$ were calculated for the following helical models of a folded polypeptide chain:

$$\alpha, \gamma, \pi, 4_{13}, 3_{10}, 3_8, 2_7 b,$$

expressed in the convention of Bragg, Kendrew, and Perutz[42]. In their notation S_R represents an S-fold helix containing R atoms in the hydrogen-bonded ring. The radial-distribution curves of these helices were first computed using a Gaussian smoothing function, $\exp(-9.5x^2)$, and then inverted by Fourier transformation to the $i(S)$ functions, which are shown in Figures 6-11 and 6-12. In the calculation of the transforms a temperature factor, $\exp(-0.013S^2)$, was applied.

The experimental $i(S)$ curves of all the α-proteins that were examined were found to be very similar. The composite $i(S)$ curve in the upper part of Figure 6-13 was formed by averaging together the curves of six of the α-proteins—namely, ribonuclease, ox hemoglobin, ox serum albumin, whale myoglobin, horse myoglobin, and ox insulin; the lower curve is that of serum albumin alone. The process of averaging together these $i(S)$ curves would be expected to enhance the interference effects due to the common polypeptide chain configuration, at the same time diminishing any effects arising from the side chain R′ (formula 6-I).

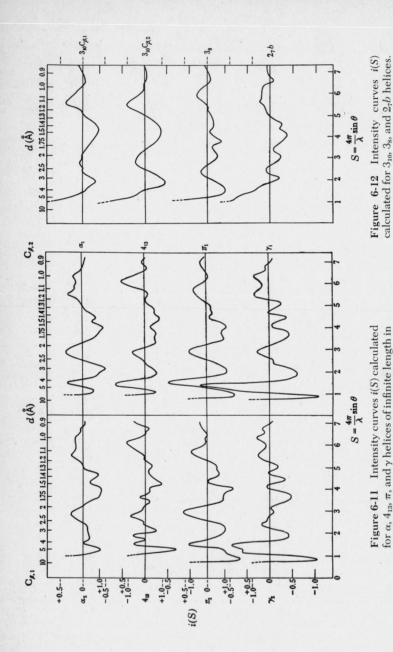

Figure 6-11 Intensity curves $i(S)$ calculated for α, 4_{13}, π, and γ helices of infinite length in random orientation. The left-hand set of curves refers to models in which the C_β atoms are in position 1, the right-hand set to C_β in position 2. (Arndt and Riley [33].)

Figure 6-12 Intensity curves $i(S)$ calculated for 3_{10}, 3_8, and 2_7b helices. Two alternative C_β positions are considered for the 3_{10} helix. The ordinate scales are the same as for Figure 6-11. (Arndt and Riley [33].)

374

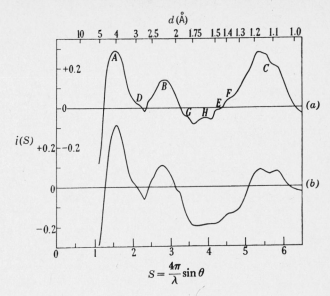

Figure 6-13 Experimental $i(S)$ curves for α-proteins: (a) composite curve obtained by averaging $i(S)$ curves of six α-proteins; (b) curve of serum albumin alone. (Arndt and Riley[33].)

Careful comparison of the experimental $i(S)$ functions with those calculated for the helical models under consideration, Figures 6-11 and 6-12, led to the following significant observations and attendant conclusions:

1. The principal experimental peak at $S = 1.5$ is absent from the curves of the 3_{10}, 3_8, and $2_7 b$ helices, thus eliminating them at once.

2. The $4_{13}C_{\beta,1}$ helix exhibits a marked peak at $S = 1.9$ that is missing from Figure 6-13, thus eliminating this model.

3. The γ_2-helix displays a shoulder at $S \simeq 1.1$ that is not visible in Figure 6-13. Furthermore, the broad peak B centered at $S = 2.8$ in Figure 6-13 is resolved in the γ_2-curve into three peaks at 2.4, 2.8, and 3.2. This evidence suffices to rule out the γ_2-helix.

4. The γ_1-helix produces an excessively high peak at $S = 1.55$ compared with experiment and can therefore be discarded.

The foregoing results leave only α_1, α_2, π_1, π_2, and $4_{13}C_{\beta,2}$ for further consideration. On the basis of additional more detailed comparisons Arndt and Riley showed that the configurational model in best agreement with the experimental data was the α_1-helix. They also found that remarkably close agreement existed between most of the features of the experimental $i(S)$ curve of the synthetic polypeptide poly-γ-

benzyl-L-glutamate and the $i(S)$ curve calculated for an α_1-helix of infinite length. (See Figure 6-14.)

In concluding this section we may quote from Riley [34]:

> In general, the essential prerequisite for this type of work is that the structure to be tested be expressible in mathematical form, however complex. The spherical-transform method ... is not capable of leading directly, without any assumptions, to a detailed picture of atomic positions in the way that single-crystal Fourier techniques can with small molecules. Some chemical knowledge must be used and rational guesses made concerning the molecular structure. It is then a relatively simple matter to put these suppositions to the test, and the crux of the matter is whether the method is intrinsically sufficiently sensitive to distinguish between the various possibilities.

It is also probably fair to conclude that the most useful function of spherical-transform analysis is to eliminate various models that have been proposed, thereby narrowing the field of choice.

6-1.5   Cylindrically Symmetrical Systems†

When some degree of axial (fiber) orientation exists, the x-ray diffraction pattern displays a corresponding degree of cylindrical symmetry, and the structure can be investigated with the aid of cylindrical transforms, which, being two-dimensional, are capable of disclosing more information than spherical transforms. Varying degrees of success have been had by Norman [43, 44], Heyn [45, 46], and Milberg [47–49] in applying the theory of cylindrical Patterson functions (also referred to as cylindrical distribution functions) to appropriate polymer specimens. However, the greatest yield of molecular-structure information has been realized in the analysis of *helical* molecules possessing a high degree of axial orientation. Strictly speaking, as a consequence of their rotational symmetry in diffraction analysis, helical structures comprise a particular case of cylindrically symmetrical systems, but because of their special importance they are discussed separately in the section that follows.

†In this section and in the one that follows, "**Helical Molecules**," we shall for the sake of clarity employ the conventions of Appendix 1 in specifying coordinates in direct and reciprocal space. This results in one major point of difference with the usage adhered to elsewhere in this monograph; namely, $|\mathbf{r}'|$ will be used in precisely the same sense as s to designate the magnitude of the radius vector in reciprocal space. Hence

$$|\mathbf{r}'| = s = \frac{2\sin\theta}{\lambda} = \frac{S}{2\pi}.$$

At the same time z' and R' will be used to denote the axial and radial reciprocal-space coordinates, respectively, rather than ζ and ξ as in Chapter 1.

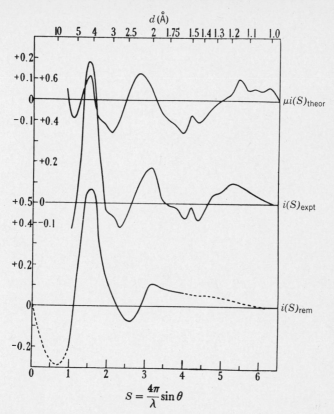

Figure 6-14   Comparison of the theoretical $i(S)$ curve of the α_1-helix with the experimental curve of poly-γ-benzyl-L-glutamate. The lower curve represents the effect of the "remainder" atoms not included in the calculations, principally the non-C_β atoms in the side chains. (Arndt and Riley[33].)

One-Dimensional (Radial) Analysis.   Heyn[45, 46] studied the transverse structure of cellulose fibers by treating them as assemblages of parallel, solid, rodlike fibrils of infinite length without radial or longitudinal structure. (See Appendix 1.) For this simplified cylindrical problem it was necessary for Heyn to measure only the intensity of scattering in the equatorial plane—and by employing only the scattering curve at small angles he eliminated the effects of the internal structure of the fibrils, which were assumed to be identical and of circular cross section. The intensities were recorded photographically using $CuK\alpha$ and $CrK\alpha$ radiation monochromatized with a double quartz-crystal monochromator, and the photographic densities were converted to intensities in the usual way with the aid of a densitometer. (See Section 2-3.2 and reference [29], pp. 368–376.)

Figure 6-15 shows the normalized small-angle intensity curves obtained from Fortisan fibers. The fiber swollen in water shows a discrete interference maximum, indicative of a considerable degree of regularity in the two-dimensional arrangement of the fibrils in the plane perpendicular to the fiber axis. Figure 6-16 shows the corresponding radial-distribution functions as determined with the Fourier-Bessel transform equation (T-33) (Appendix 1). The locations of the first maxima in these curves indicate the closest approach of the centers of adjoining fibrils to be about 45 Å in the dry fiber, 68 Å in the water-swollen fiber, and 80 Å in the highly swollen fiber. From independent measurements the fibril diameter is known to be close to 45 Å, which leads to the conclusion that in the dry state the fibrils are in actual contact with each other, leaving room for little intervening "amorphous" material. On the other hand, a similar study of jute fibers led to an interfibril distance of 35 Å in the dry fiber compared with a known mean value of the fibril diameter of 28 Å, which suggests that in dry jute fibers the constituent fibrils are separated somewhat by disordered matter (for example, lignin). Figure 6-17 is a schematic

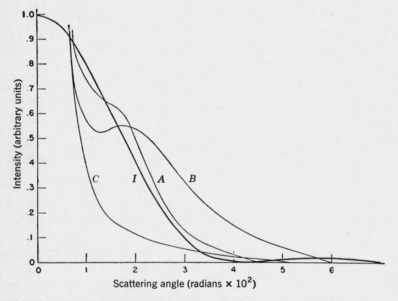

Figure 6-15 Normalized small-angle intensity curves of Fortisan rayon: *I* — theoretical curve for independent scattering; *A* — fiber swollen beyond the water-swollen state; *B* — fiber swollen in water; *C* — fiber in dry state. (Heyn [46].)

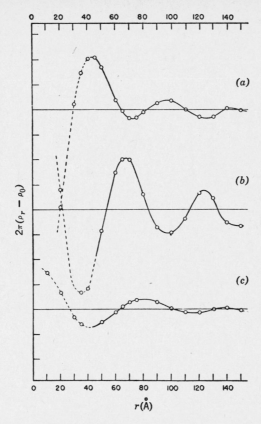

Figure 6-16 Radial-distribution functions of fibril centers for Fortisan: (*a*) dry fiber; (*b*) fiber swollen in water; (*c*) fiber swollen beyond the water-swollen state. Abscissae: distance from origin in Å; ordinates: each division equals 5 units if ρ is given in number of fibril centers per $(100 \text{ Å})^2$. (Heyn[46].)

representation of the cross-sectional sizes and distributions of the elementary fibrils in Fortisan rayon and jute.

Interchain Distance. If we assume that an amorphous fiber composed of molecular chains contains a large proportion of chains approximately parallel to the fiber axis, we may use the model of parallel rods or cylinders discussed above to derive an expression giving the interchain distance as a function of the position of the principal peak in the equatorial plane of the amorphous scattering pattern. For assemblages of parallel cylinders[50] or lines [51] with a most frequently occurring distance r between their centers in the equatorial plane, (T-31) for the intensity of scattering in the same

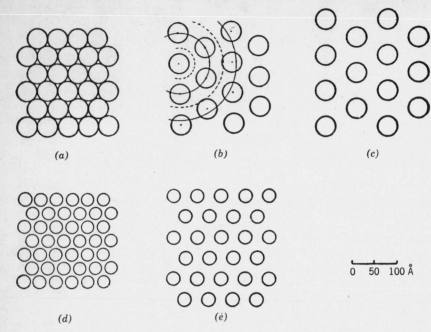

Figure 6-17 Schematic representation of the cross-sectional sizes and distributions of the elementary fibrils in Fortisan rayon (top) and jute (bottom). (a)—Fortisan dry; (b)—same swollen in water; (c)—same swollen beyond the water-swollen state; (d)—jute dry; (e)—jute swollen in water. (Heyn[46].)

plane simplifies to

$$I(S) = F^2[1 + J_0(Sr)], \tag{6-7}$$

where F is the scattering factor of an isolated cylinder or line and J_0 is a Bessel function of zero order. The first and principal maximum in $J_0(Sr)$ occurs at $Sr = 7$, from which

$$r = \frac{7}{2\pi} \times \left(\frac{\lambda}{2 \sin \theta}\right) = 1.11 \times d_{\text{Bragg}}. \tag{6-8}$$

This equation states that a *hypothetical* Bragg spacing calculated from the position of the principal amorphous peak should be multiplied by 1.11 to give the interchain distance.

The above result for aggregates of parallel rodlike scattering elements may be. compared with the result for randomly oriented noncrystalline systems, to which Debye's intensity equation is applicable. [See Appendix 1, (T-14).] It can be shown that for this model the most frequently occurring interatomic distance r is in principle related to the position of the first and most intense amorphous peak by [38, 52]

$$r = \tfrac{5}{4} \times \left(\frac{\lambda}{2 \sin \theta} \right), \tag{6-9}$$

although in practical calculations it is somewhat more realistic to use the expression [38]

$$r = 1.22 \times d_{\text{Bragg}}. \tag{6-10}$$

Since (6-8) and (6-10) apply to ideal limiting cases, it is difficult to say which one should be used in any given case. For highly oriented but amorphous linear polymers (6-8) is probably to be preferred, whereas for unoriented amorphous polymers a value of r that is intermediate between those given by (6-8) and (6-10) may be more realistic. It should also be emphasized that these equations are valid only insofar as the actual structures correspond to the models on which the equations are based. Thus (6-8) takes no account of details of structure within the molecular chains — for example, bulky side groups or substituent atoms of high scattering power — that may profoundly modify the calculated scattering pattern. Therefore it must be kept in mind that (6-8) and (6-10) possess only rather limited quantitative character and, indeed, may even lead to erroneous conclusions.

Two-Dimensional Patterson (Distribution) Function.† For the analysis of axially oriented structures as functions of the radial (R) and axial (z) coordinates, we may employ (T-29) or, in case there is periodicity in z, (T-30). (See Appendix 1.) The intensity and its transform, the distribution function D, will now for convenience be expressed in spherical coordinates r', α' and r, α, respectively. (See Figure T-1.) Since both $I(r')$ and $D(r)$ possess cylindrical symmetry and inversion centers, they may be expanded as series of Legendre polynomials P_{2n} of even order [43]:

$$I(\mathbf{r'}) = I(r', \alpha') = \sum_{n=0}^{\infty} I_{2n}(r') P_{2n}(\cos \alpha') \tag{6-11}$$

and

$$D(\mathbf{r}) = D(r, \alpha) = \sum_{n=0}^{\infty} D_{2n}(r) P_{2n}(\cos \alpha), \tag{6-12}$$

in which the coefficients $I_{2n}(r')$ and $D_{2n}(r)$ are respectively given by

---

†To preserve consistency with the related literature we shall designate the distribution function D rather than P, as in Appendix 1, and reserve the symbol P to denote the Legendre polynomial.

$$I_{2n}(r') = \frac{4n+1}{2} \int_0^\pi I(r', \alpha') P_{2n}(\cos \alpha') \sin \alpha' \, d\alpha' \qquad (6\text{-}13)$$

and

$$D_{2n}(r) = \frac{4n+1}{2} \int_0^\infty D(r, \alpha) P_{2n} (\cos \alpha) \sin \alpha \, d\alpha. \qquad (6\text{-}14)$$

Norman[43] has shown that (6-14) can be converted to the form

$$D_{2n}(r) = (-1)^n 4\pi \int_0^\infty (r')^2 I_{2n}(r') J_{2n} (2\pi r'r) \, dr', \qquad (6\text{-}15)$$

where J_{2n} is the spherical Bessel function of order $2n$.

The expansion of the distribution function as a series of Legendre polynomials will be of most value when the symmetry of the system, though actually cylindrical, does not differ greatly from spherical, for the reason that the series then converges rapidly. We may remark that the first term of (6-12), corresponding to $n = 0$, is identical with the radial-distribution function for spherically symmetrical systems. (See Appendix 1.)

Norman[43, 44] investigated the structure of cellulose specimens with the aid of their two-dimensional distribution functions. He selected Fortisan and ramie, respectively, as examples of fibers with and without appreciable z-axis periodicity as evidenced by the well-defined layer lines displayed by the former and the relatively diffuse pattern of the latter. The patterns were recorded photographically using crystal-monochromatized radiation and evacuated or hydrogen-filled cameras to eliminate air scatter, and the photographic densities were converted to intensities with a microdensitometer. The experimental intensities were normalized to electron units, and the incoherent scattering and structure-independent scattering, $\sum_m f_m^2$, were subtracted in the usual way (see Appendix 1, (T-24), and more particularly reference[29], pp. 588–592) prior to Fourier transformation to obtain the cylindrical-distribution functions.

Figure 6-18 compares these functions as derived from the intensity data of ramie in three ways. The function of part a was derived by using the complete two-dimensional distribution function as given by (T-29); part b was calculated by using only the first two terms of a series of Legendre polynomials [see (6-12)]; part c was calculated similarly but by using the first three terms. The packing of the cellulose chains parallel to z is evident in all three diagrams, but, of course, most clearly in part a. This feature can be seen, although less plainly, in part b with the use of only two polynomial terms,

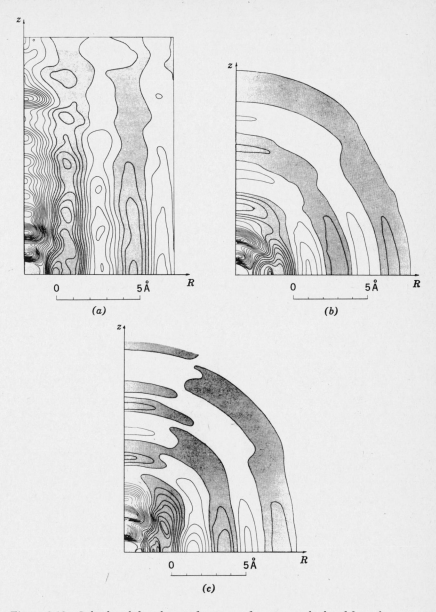

Figure 6-18 Cylindrical-distribution functions of ramie as calculated from the experimental intensity curve in three ways: (*a*) by using the complete transform (T-29), (*b*) by using the first two terms of a series of Legendre polynomials, and (*c*) by using the first three terms of the series. Contour levels on the same, but arbitrary, scale in (*a*), (*b*), and (*c*). Negative regions shaded. (Norman [43].)

whereas in part c it is evident that the use of three terms yields considerable improvement. These results demonstrate that Legendre polynomials consisting of only a few terms do not constitute a good approximation of a complete Fourier transformation when the cylindrical symmetry is pronounced.

Norman [43, 44] compared the two-dimensional contour plots of Fortisan and ramie with various cellulose-structure models that have been proposed but was unable to make a definite decision among them. However, he found that the model of Meyer and Misch [53] did not conform well to the experimental results, whereas the agreement was improved if alternate pyranose rings along the chain were rotated from the positions originally proposed. Furthermore, the contour maps suggest that successive glucose residues along the molecular chain are related by a twofold screw axis.

The Difference-Distribution Function. Milberg [47] pointed out that, if the term with $n = 0$ in (6-12) is omitted, the difference between the cylindrical- and radial-distribution functions is obtained and that this difference function must contain all the experimentally available information relative to the state of orientation of the system. If the difference between the observed intensity distribution $I(r', \alpha')$ and its average over α', which is $I_0(r')$, is designated $I'(r', \alpha')$, that is,

$$I'(r', \alpha') = I(r', \alpha') - I_0(r'), \qquad (6\text{-}16)$$

and if its transform is $D'(r, \alpha)$, then it can be shown mathematically that

$$D'(r, \alpha) = D(r, \alpha) - D_0(r). \qquad (6\text{-}17)$$

The transform to be evaluated is then (6-12) summed from $n = 1$ to ∞, or therefore

$$D'(r, \alpha) = \sum_{n=1}^{\infty} D'_{2n}(r) P_{2n}(\cos \alpha). \qquad (6\text{-}18)$$

It is important to realize that, since all spherically symmetrical scattering is included in $I_0(r')$ and therefore *excluded* from the transform $D'(r, \alpha)$, independent scattering of any kind (including incoherent) as well as air scattering may be neglected. Moreover, the difference-intensity function is not as sensitive to intensity errors at large values of r' as is the complete function $I(r', \alpha')$ [49], and in addition a high degree of monochromatization is not required inasmuch as extraneous wavelengths produce very largely isotropic scattering, so that normally a β-filter suffices. The difference-distribution method is particularly easy to apply to diffraction patterns consisting of one or more anisotropic halos superposed on an isotropic background, in which case (6-18) needs to be summed only over the m anisotropic halos. It may then be written

$$D'(r,\alpha) = \sum_{i=1}^{m} \sum_{n=1}^{\infty} D'_{2n,i}(r)P_{2n}(\cos \alpha), \qquad (6-19)$$

where i refers to the ith anisotropic halo. A further simplification that can be employed if the halos are well resolved is to assume that the difference-intensity distribution for each halo, $I'_i(r',\alpha')$, can be resolved into two factors — one a function of r' alone and the other of α' alone. Figure 6-19 shows some typical diffraction patterns of noncrystalline polymers containing diffuse halos that display increased intensity on the equator, accompanied by diminution of intensity at other azimuths, especially on the meridian [51].

The simplified procedure just described for deriving the difference-distribution function suffers from the limitations that the results are on an arbitrary scale and measured from an unknown base line. Nevertheless, the information that it is capable of supplying relative to the nature and degree of preferred orientation corresponding to particular interatomic distances can be of assistance in choosing among different molecular models.

We shall describe very briefly the investigation by Milberg and Daly [49] of the molecular conformation and orientation of noncrystalline sodium metaphosphate fibers. Figure 6-20 shows the experimental intensity distribution plotted as a function of the cylindrical coordinates R' and z' (see (T-2), Appendix 1), and Figure 6-21 is the corresponding electronic difference-distribution function, which was computed with (6-18) using 25 terms in the summation. The series proved to be more slowly convergent as a function of n than had been anticipated, an indication that the molecular chains are rather highly oriented. A high degree of preferential alignment of the chains of PO_4 tetrahedra parallel to the fiber axis is also suggested by the following features of the distribution function (Figure 6-21):

1. Negative values of the electron-density difference on the z-axis for $r > 1.2$ Å.
2. Positive values on the R-axis for $r > 1.2$ Å.
3. The well-resolved peaks at $r = 1.6$ Å, $\alpha = 20°$ and $r = 1.4$ Å, $\alpha = 58°$.
4. The alternating positive and negative bands parallel to the R-axis.

Best agreement between the experimental and calculated distribution functions was obtained for a model based on a puckered

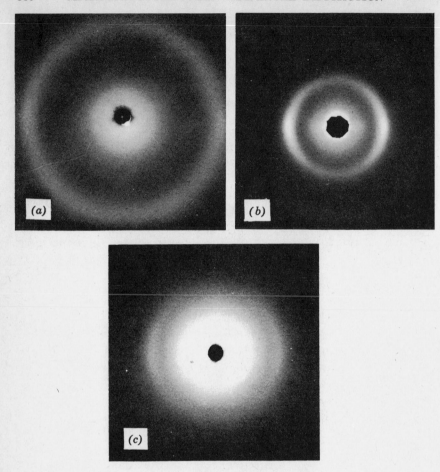

Figure 6-19 Diffraction patterns of noncrystalline polymers showing preferred molecular orientation (fiber axis vertical): (*a*) polybutadiene elongated 535%; (*b*) silicone elastomer elongated 540%; (*c*) sodium-metaphosphate glass fiber. (Alexander and Michalik [51].)

version of the type of planar chain that has been found to characterize sodium metasilicate. Each PO_4 tetrahedron in the chain is tilted relative to the chain axis through two mutually perpendicular angles of approximately 20 to 30°. Because of the complexity of Milberg and Daly's interpretative procedure the reader must be referred to the original paper for further details [49].

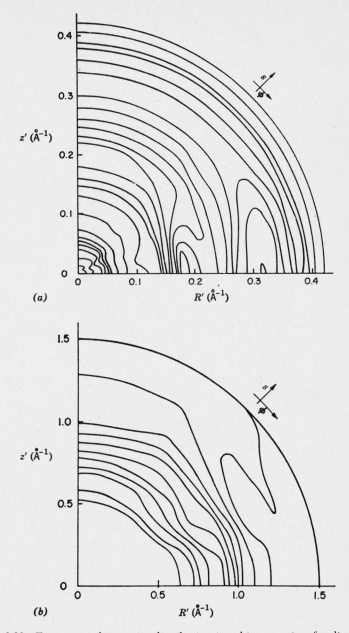

Figure 6-20 Experimental intensity distribution in arbitrary units of sodium-meta-phosphate glass fibers: (*a*) small *s* with contours at intervals of 2; (*b*) large *s* with contours at intervals of 1. (Milberg and Daly [49].)

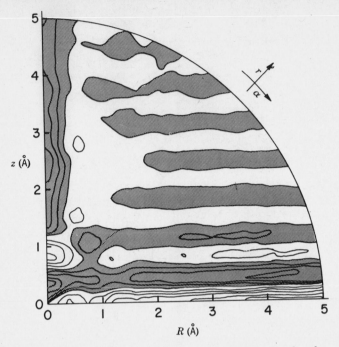

Figure 6-21 Electronic difference-distribution function (in cylindrical coordinates) of sodium-metaphosphate glass fiber. Negative regions shaded. Contours at intervals of $0.005\ e^2/\text{Å}^6$ except near the origin, where interval is doubled. (Milberg and Daly [49].)

6-1.6 Helical Molecules

The conformations of an impressive number of helical polymer molecules have been analyzed successfully with the aid of the helical-transform theory [54], which appeared in 1952. The most widely heralded and dramatic results have been achieved in the investigation of the structures of fibrous proteins, nucleic acids, and viruses; nevertheless, a number of simpler, technological-polymer molecules have also yielded to this analytical approach.

The complete helical transform for the lth layer line corresponding to one motif (asymmetric unit) of structure is given by (T-38) of Appendix 1:

$$F_c\!\left(R',\psi',\frac{l}{c}\right) = \sum_n \sum_j f_j J_n(2\pi R'R_j)\exp\left\{i\left[n\left(\psi-\psi_j+\tfrac{1}{2}\pi\right)+\frac{2\pi l z_j}{c}\right]\right\}\!.$$

$$(\text{T-38})$$

For a given atom j, the first summation is over the orders n of Bessel functions permitted by the selection rule (T-36), which in terms of $t = c/P$ and $u = c/p$ (see Section 6-1.2) may be written

$$tn + um = l. \tag{6-20}$$

The second summation in (T-38) is over all the atoms comprising the asymmetric unit of structure, the helical coordinates of the jth atom being given by R_j, ψ_j, z_j. (Refer to Figure T-2.)

Because of the rotational symmetry of helices in actual fibers, the observed distribution of intensity in the lth layer line as a function of R' is given by the square of the modulus of (T-38) as expressed by (T-39), (T-40), and (T-41). Although the ψ' average of the intensity distribution, $I_{\bar{\psi}'}[R', (l/c)]$, is of fundamental importance to the understanding of diffraction by helical molecules, of even more direct value in analysis is the selection rule [(T-36) and (6-20)], which is the most distinctive feature of helical diffraction. Two useful devices for the portrayal of diffraction by a helical molecule are its radial projection and the l, n net of its Fourier transform [55, 56]. We illustrate these items with the simple case of a 10/3 helix containing 10 single-atom motifs of structure in 3 turns. All 10 atoms lie on the surface of a common cylinder of radius R_0. If this cylindrical surface is slit lengthwise, unrolled, and flattened out, the locations of the atoms mark out a two-dimensional helical net, or *radial projection*, as shown in Figure 6-22. In a more complicated structure each net point rep-

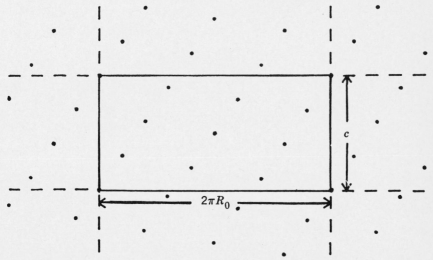

Figure 6-22 Helical net, or radial projection, of a 10/3 helix composed of single-atom motifs. (Klug, Crick, and Wyckoff [55].)

resents an asymmetric unit, or motif, consisting of a number of atoms all of which lie on helices of different radii but equal pitch. The Fourier transform of the helical net of Figure 6-22 is the reciprocal net shown in Figure 6-23, which can be conveniently numbered with the appropriate values of l and n in the ordinate and abscissa directions, respectively. By comparison of Figures 1-26 and 6-23 the reader will perceive that a parallel relationship between the direct and reciprocal lattices exists for both ideal crystals and helices.

By inspection of Figure T-4 of Appendix 1 we observe that (a) at $x = 0$ the Bessel function of zero order, $J_0(x)$, is large, whereas all higher order Bessel functions are zero, and (b) the first maximum in $J_n(x)$ retreats from the origin as n increases and its amplitude diminishes. Consideration of (T-38) in the light of these features indicates that *on the average* the overall intensity of scattering on a given layer line l will be strong only if Bessel terms of low order are

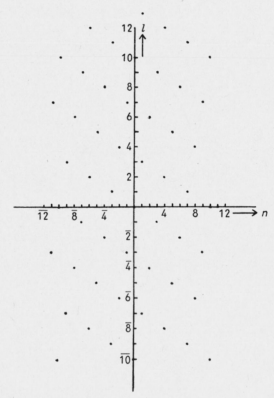

Figure 6-23   The l,n net reciprocal to the 10/3 helical net of Figure 6-22. (Klug, Crick, and Wyckoff [55].)

present. It also follows that the strongest maxima will lie close to the meridian† ($R' = 0$) as a general rule, with only zero-order Bessel terms producing reflections precisely on the meridian. In favorable cases these properties of helical diffraction afford a simple means of discriminating between alternative models purely on the basis of degree of correlation between the experimental layer-line intensities and the orders of Bessel terms permitted by (6-20).

Optical Transforms. Diffraction by helical structures can be studied very effectively with the aid of optical transforms. (See Appendix 1 and Section 2-5.) Figure 6-24 shows two-dimensional projections of three helices together with their optical diffraction patterns, which were prepared with the light beam perpendicular to the helical axis. Since the transform of a projection is a section containing the Z' and R' axes, these patterns correspond closely to what would be observed by x-ray diffraction. The pattern of a continuous helix (Figure 6-24a) shows a characteristic feature of helical patterns, the "cross" through the origin, which arises from the increase in the order of the Bessel function with increase in the layer-line number l. [See (T-35) and (T-36).] The additional weaker maxima of these Bessel functions are also visible on the layer lines at points further removed from the meridian. In this connection the reader should examine Figure T-4 of Appendix 1.

In discontinuous helices the introduction of the motif periodicity p along the helical axis results in a more or less pronounced modification of the simple cross pattern, as illustrated in Figures 6-24b, and c for helices composed of 10 and 5 points per turn, respectively. It can be shown that the transform of a discontinuous helix will be the convolution of the transform of a continuous helix and the transform of a line of points spaced at the interval p parallel to Z. The net effect is to introduce new "origins" for the cross as one proceeds along the Z' axis on the diffraction pattern, which produces diamond-shaped patterns about the meridian. As p increases and the number of motifs per turn decreases, the new origins draw closer to each other and to the true origin, causing increasing degrees of overlapping of the cross patterns. If P is not an integral multiple of p, as is the case in a substantial proportion of polymer helices, and if interhelical (crystalline) interference effects are considered, the typical cross- or diamond-shaped features of the helical diffraction pattern become

†The investigator is reminded of the necessity of preparing diffraction patterns with the fiber axis inclined toward the incident beam in the case of specimens with a high degree of axial orientation. This topic is discussed in Section 1-4.6.

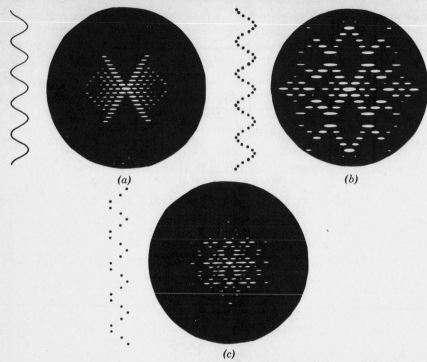

Figure 6-24 Two-dimensional projections of three helices together with their optical transforms: (*a*) continuous helix; (*b*) helix with 10 points per turn (10/1); (*c*) helix with 5 points per turn (5/1). (Holmes and Blow [56]; figure prepared by C. A. Taylor.)

further obscured but are, nevertheless, usually recognizable by the experienced observer.

Poly-γ-Methyl-L-Glutamate [54].   As a first example of the simplified approach to helical analysis on the basis of overall layer-line intensities we shall describe very briefly the classic example of the synthetic polypeptide poly-γ-methyl-L-glutamate (PMLG),

In this first application of their theory of helical diffraction Cochran, Crick, and Vand[54] were able to show a convincing degree of correlation between the observed distribution of layer-line intensities from a well-oriented specimen[57] and the incidence of low-order Bessel terms calculated for the α-helix model of Pauling, Corey, and Branson[41]. At the outset we must bear in mind that all atoms in the monomeric unit will in general lie on concentric helices of different radii although of identical pitch. If one is designated as the reference helix, the other helices will be rotated and translated with respect to it by individual increments $\Delta\psi_j$ and Δz_j, respectively, as indicated by (T-38). However, the simplified method of analysis makes use only of the selection rule (6-20) and takes no notice of the coordinates of the individual atoms.

The axial identity period of the α-helix model contains 18 monomer residues and 5 turns. In helical-point-net notation $u = 18$ and $t = 5$, and the descriptive symbol is 18/5. Thus the selection rule (6-20) becomes

$$5n + 18m = l. \tag{6-21}$$

Figure 6-25 is a plot of l versus n for $-18 \leq n \leq 18$, and Table 6-1 presents a comparison of the permitted smaller orders n of Bessel functions for each layer line corresponding to the α-helix model with the observed integral intensities of the same layer lines in the fiber pattern of PMLG. It is seen that appreciable intensity is observed only for layer lines characterized by near-zero values of n, which constitutes evidence that the α-helix conformation is present in PMLG.

Polyisobutene. A second example of the application of the selection rule to the determination of the helix type is that of polyisobutene. An early study by Fuller, Frosch, and Pape[58] showed the fiber period of the stretched elastomer to be $c = 18.63$ Å and also revealed a subperiodicity of 2.33 Å, or $c/8$. However, the helical conformation proposed by these investigators was not borne out by subsequent studies of Bunn and Holmes[23] and of Liquori[25] based on the helical-transform theory. They independently deduced an 8/5 conformation, which can be shown to proceed from the selection-rule analysis as follows[25].

Four possible conformations, each containing eight monomer units in the axial identity period of 18.63 Å were considered by Liquori. These possessed 1, 3, 5, and 7 helical turns in the identity period. It was possible to reject even numbers of turns at the outset since they

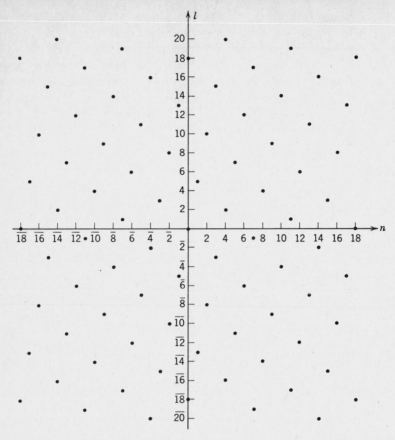

Figure 6-25 The l, n net of the α-helix for $-18 \leqslant n \leqslant 18$.

would have the effect of dividing the observed axial identity period by 2 or 4. It was also found that one or three turns were unacceptable because they would result in impossibly close approaches between consecutive methyl groups along the chain. Therefore it was necessary to analyze only the 8/5 and 8/7 conformations, for which the selection rule (6-20) assumes the respective forms

$$5n + 8m = l \qquad \text{(8/5 helix)}, \qquad (6\text{-}22)$$
$$7n + 8m = l \qquad \text{(8/7 helix)}. \qquad (6\text{-}23)$$

The l, n diagrams of the 8/5 and 8/7 helices are shown in Figure 6-26, and Table 6-2 compares the smaller Bessel orders n of each helix with the estimated experimental intensities of corresponding layer lines of the diffraction pattern. (See again Figure 1-3b.) The first

Table 6-1 Poly-γ-Methyl-L-Glutamate: Comparison of Experimental Layer-Line Intensities with Orders n of Bessel Functions for the α-Helix ($-18 \leq n \leq 18$)

Index of Layer Line l	Estimated Average Intensity of the Layer Line[a]	Orders n of Bessel Functions		
0	Strong	−18	0	18
1		−7	11	
2	Weak	−14	4	
3	Very weak	−3	15	
4		−10	8	
5	Medium	−17	1	
6		−6	12	
7		−13	5	
8	Weak	−2	16	
9		−9	9	
10	Weak	−16	2	
11		−5	13	
12		−12	6	
13	Very weak	−1	17	
14		−8	10	
15		−15	3	
16		−4	14	
17		−11	7	
18	Medium	−18	0	18
19		−7	11	
20		−14	4	

[a]Data from [57].

zeroth-order Bessel function occurs for both helices on the eighth layer line, in agreement with a subperiodicity of $c/8$ corresponding to the monomeric unit of the chain. However, in other respects the correlation between average intensity and the presence of small-order J_n terms is seen to be good for the 8/5 helix but noticeably poor for the 8/7 helix, thus tending to rule out the latter conformation. The idealized 8/5 helical conformation of stretched polyisobutene is shown in Figure 6-8 [25].

Complete Helical-Structure-Factor Calculations [59]. Employing the complete cylindrically averaged transform expressions (T-39), (T-40), and (T-41), Davies and Rich [59] used a digital computer to calculate the layer-line intensity profiles for the five helical polypeptide models described in Table 6-3, all of which were regarded

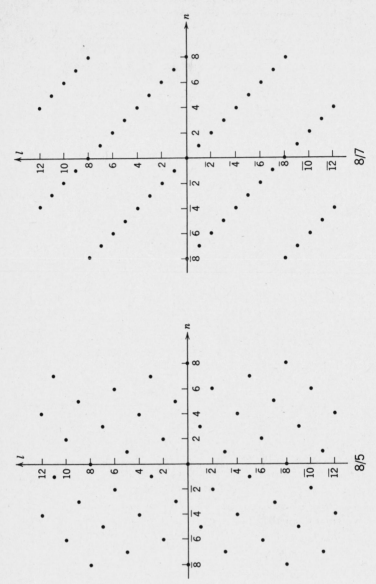

Figure 6-26 The *l, n* nets of the 8/5 and 8/7 helices.

Table 6-2 Polyisobutene: Comparison of Experimental Layer-Line Intensities with Orders n of Bessel Functions for the 8/5 and 8/7 Helices ($-8 \leq n \leq 8$)[a]

Index of Layer Line l	Estimated Average Intensity of the Layer Line	Orders n of Bessel Functions					
		8/5 Helix			8/7 Helix		
0	Very strong	−8	0	8	−8	0	8
1	Medium weak	−3	5		−1	7	
2	Medium	−6	2		−2	6	
3	Strong	−1	7		−3	5	
4	Medium	−4	4		−4	4	
5	Strong	−7	1		−5	3	
6	Medium	−2	6		−6	2	
7	Weak	−5	3		−7	1	
8	Strong	−8	0	8	−8	0	8
9	Very weak	−3	5		−1	7	
10	Weak	−6	2		−2	6	
11	Medium weak	−1	7		−3	5	

[a]Data from [25].

as possible models for poly-L-alanine. Since successive Bessel functions on a given layer line differ in order by the integer $u = c/p$ (column 4 of Table 6-3), except when u is small (as for the 3.0-residue helix) a sufficiently good approximation to the intensity profile can be realized by including in the calculation only the Bessel function of the lowest order that is permitted by the selection rule.

The results for the five models are shown in Figure 6-27. The general similarity of the patterns of the α- and π-helices can be

Table 6-3 Five Helical Polypeptide Models[a]

Model	Helix Type (u/t)	$c\,(\text{Å})$	u	N[b]
γ-Helix	36/7	35.28	36	1
3.0-Residue helix	3/1	6.00	3	4
π-Helix	22/5	25.3	22	1
α-Helix	18/5	26.64	18	1
Modified π-helix	23/5	26.64	23	1

[a]Data from [59].
[b]N = number of Bessel functions per layer line used in the calculation.

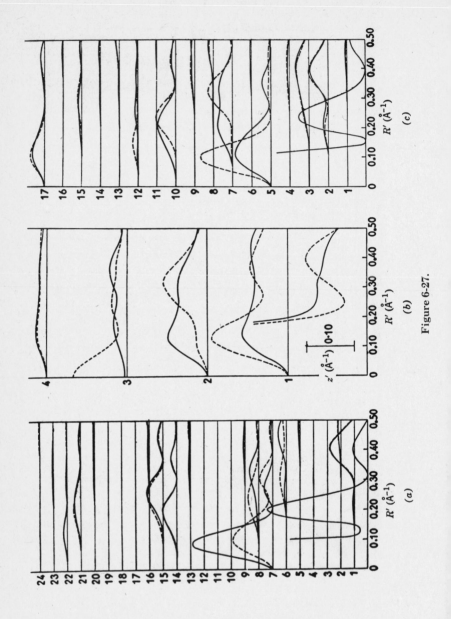

Figure 6-27.

398

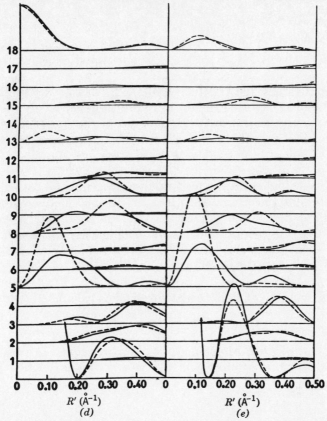

Figure 6-27 Cylindrically-averaged intensity distributions calculated for five helical polypeptide models: (*a*) γ-helix; (*b*) 3.0-residue helix; (*c*) π-helix; (*d*) α-helix; (*e*) modified π-helix. The z', or ζ, scale is shown in (*b*). The solid line represents the intensity of a right-handed helix of L-amino acids (or a left-handed helix of D-amino acids). The dashed line represents a left-handed helix of L-amino acids. (Davies and Rich [59].)

greatly increased by "twisting" the latter slightly to give 23 residues in 5 turns and increasing c to 26.64 Å. This helix, further modified slightly to improve the bond angles, is referred to as the modified π-helix in Table 6-3. The principal differences between the patterns of the α- and modified π-helices are seen to occur on the equatorial and 18th layer lines. Thus the 18th layer line of the α-helix contains a bona fide meridional maximum ($n = 0$) of medium intensity, whereas the modified π-helix displays an off-meridian reflection ($n = 1$) of much weaker intensity.

Well-oriented diffraction patterns of α-poly-L-alanine have been prepared by Brown and Trotter[60] (Figure 6-28) showing a 1.5-Å reflection that can only be indexed as the meridional reflection (0·0·0·47) in hexagonal Miller-index notation. This reflection, which is clearly visible in Figure 6-28*b*, corresponds to an axial period of 70.3 rather than 26.64 Å, which is calculated for the idealized α-helix. Nevertheless, it indicates 3.62 rather than 3.6 monomer residues per turn, which shows that it actually differs only very slightly from the ideal α-helix. Furthermore, the observed equatorial intensity distribution clearly favors the α as opposed to the modified π-model. It is also plain that the γ and 3.0-residue helices may be dropped from further consideration because of their obvious points of disagreement with the experimental intensities. Davies and Rich[59] emphasize that whereas the high-quality experimental data of Brown and Trotter[60] permit a clear choice to be made between the α and π helical models, such a clear distinction would not have been possible from inspection of a poorly oriented fiber pattern.

In the province of polymers possessing technological interest, as examples of successful solutions of molecular structures by comparison of the experimental layer-line intensity distributions with cylindrically averaged helical structure factors calculated for different molecular conformations, we may cite investigations of poly-*m*-methylstyrene[61] and poly(ethylene oxide)[62].

Complex Biological Molecules. We now present a brief survey of some of the significant biological-structure solutions in which helical-

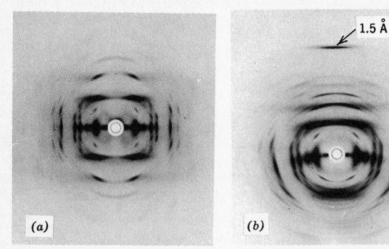

Figure 6-28 Fiber-diffraction patterns of α-poly-L-alanine: (*a*) fiber axis vertical, cylindrical-film axis vertical; (*b*) fiber tilted 31° to vertical, cylindrical-film axis horizontal. (Brown and Trotter [60].)

diffraction theory has played an important role. It is hoped that the literature reviews and other references cited will suffice to direct the interested reader to suitable source material for a study of this subject in greater depth and breadth. We shall first devote special attention to one biologically important molecule, *deoxyribonucleic acid* (DNA), and then much more succinctly refer to several other structural investigations.

Deoxyribonucleic acid is of unusual stereochemical interest as the carrier of the genetic specificity of the chromosomes. X-ray studies beginning in the early 1950s showed that the molecule assumes at least four similar, but not identical, helical conformations as functions of the relative humidity and the particular cation present [63–66]. Figure 6-29 shows the x-ray fiber patterns of sodium deoxyribonucleic acid at relative humidities of 75 (A form) and 92% (B form). The characteristic helical-diffraction cross is evident in the more diffuse pattern of the B form, which closely approximates the cylindrically averaged transform of parallel-oriented helices scattering independently. [See (T-39), (T-40), and (T-41) of Appendix 1.] In the pattern of the A form the characteristic helical cross or diamond pattern is concealed by the crystalline ordering of the individual helices, which causes a decomposition of the intensity profile of any given layer line into sharp crystalline reflections. The analysis of the A and B patterns shows that both forms have an integral number of nucleotides† per helical turn, 11 for A and 10 for B, so that the helix types are 11/1 and 10/1, respectively. The axial periodicities c are 28.1 and 34 Å for the A and B forms, respectively.

The presently accepted structure model of the B form is rather similar to the original proposal of Watson and Crick[67, 68], being composed of paired chains spiraling about a common axis. The great complexity of the DNA structure becomes evident from an inspection of the molecular model shown in Figure 6-30, which is compared with a similarly scaled model of ribonucleic acid (RNA). The phosphate groups lie on the outside of the paired helical chains, which are connected across the helical axis by hydrogen bonds linking a purine base of one helix to a pyrimidine base of its counterpart. Figure 6-31 depicts in two dimensions the relationship between the pair of chains comprising one double helix. Although a purine base is always hydrogen bonded to a pyrimidine, there is no necessity that purine and pyrimidine bases alternate in a regular manner along the helices.

The general structural features of the *tobacco mosaic virus* as

†*Nucleotide* refers to a monomeric residue consisting of a *phosphate group, sugar,* and *base* (purine or pyrimidine).

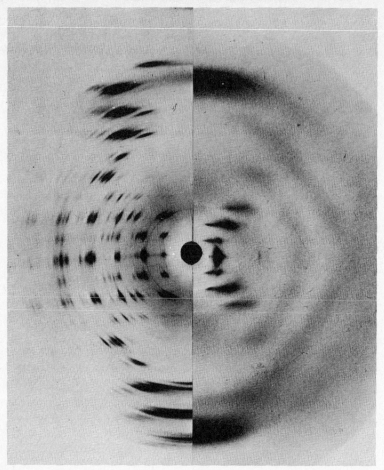

Figure 6-29 X-Ray fiber patterns of sodium deoxyribonucleic acid (DNA). Left: relative humidity 75%, A form. Right: relative humidity 92%, B form. (Holmes and Blow [56]; photograph prepared by W. Fuller.)

reported by Franklin, Klug, and Holmes [69] in 1957 have not been improved upon in any important respect [66, 70]. Because of the weakness of interparticle interferences except at very small angles, the fiber-diffraction pattern of the oriented virus in solution (Figure 6-32) corresponds to the cylindrically averaged transform of the single virus particle. X-ray and electron-microscopic studies show the particles to be rod shaped, about 150 Å in diameter and 3000 Å long. The x-ray diagrams show that the structure of an individual rod is based on a 49/3 helix of 23-Å pitch. As portrayed in Figure 6-33, an

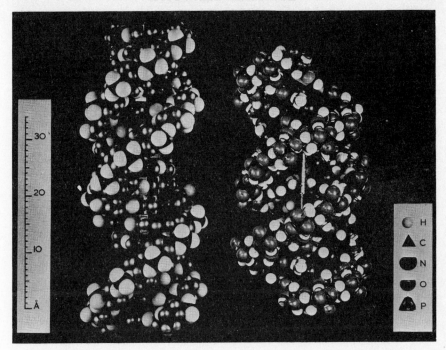

Figure 6-30 Molecular models of (left) part of a DNA molecule in the B configuration and (right) part of the helical region of an RNA molecule. (Holmes and Blow [56]; figure prepared by W. Fuller.)

inner RNA helix surrounds a hollow core 35 to 40 Å in diameter. Outside of and attached to the RNA nucleotides are the approximately 2130 roughly ellipsoidal protein subunits that comprise the great bulk of the volume of the particle. For further details the reader is

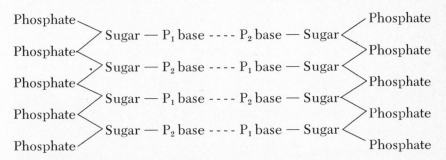

Figure 6-31 Two-dimensional schematic formula of one pair of DNA chains. $P_1 =$ purine base, $P_2 =$ pyrimidine base. Broken lines represent hydrogen bonds.

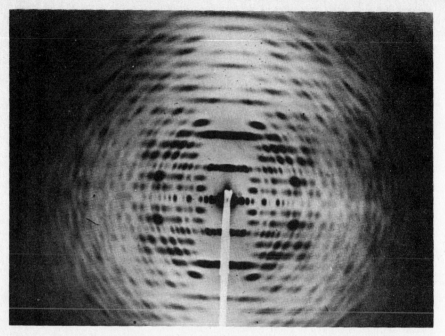

Figure 6-32 X-Ray fiber pattern of oriented tobacco mosaic virus in solution. Fiber axis vertical. Bent-quartz line-focusing monochromator, vacuum camera, specimen-to-film distance 12 cm. (Franklin and Holmes [73].)

Figure 6-33 Helical structure of tobacco mosaic virus. The elongated objects are the protein subunits, and the beads on the helical band represent nucleotides in the RNA. (Klug and Caspar [70].)

referred to papers by Franklin and associates [71–73] and to the excellent review of Klug and Caspar [70].

X-ray diffraction and electron-microscopic studies have shown that *collagen* consists of three coiled polypeptide strands wound slowly (with a relatively large pitch) about each other [66, 74]. Additional features of the structure are still somewhat unclear, an important unsolved problem being the origin of the well-known axial macroperiod of 640 Å. The reader can do much to familarize himself with the recent developments in our knowledge of collagen structure by consulting the valuable review paper by Crick and Kendrew [74] and reading the detailed account written by Rich and Crick [75] of their investigations as well as papers by Ramachandran et al.[76, 77]. Optical transforms were used to advantage by Rich and Crick in the early stages of their evaluation of the collagen I and II models.

There has been general agreement for some years that the α-helix is an important building unit of α-*keratin* [66], although the structure in its entirety is very complex and the detailed arrangement of the constituent α-helices has not been ascertained. There is a consensus of opinion, however, that the most acceptable structure model is one that is composed of several α-helices winding "slowly" around each other in a regular way so as to form a *coiled coil* without, at the same time, evoking significant distortions of the component helices [78, 79]. Crick [80] has derived an expression for the Fourier transform of such a coiled coil. In the area of investigation of muscle structure it has been found that the calculated transform of a coiled coil composed of two α-helices agrees in its major features with the observed diffraction pattern of *paramyosin*, a component of molluscan "catch" muscle [81, 82].

6-2 CRYSTAL STRUCTURE

The analysis of the crystal structure of a molecular compound necessarily includes both the determination of the relative positions of the atoms comprising one molecule (molecular structure) and the positions and orientations of the molecules with respect to one another in the crystal lattice. In general a three-dimensional-structure solution is only possible on the basis of three-dimensional x-ray diffraction data—or, more specifically, a complete three-dimensional elucidation of the identities and positions of the atoms in one unit cell can only proceed from a measurement of the intensities and positions of all the x-ray reflections from a single crystal. The conditions that govern diffraction by a single crystal and the relationship between the atomic coordinates and the diffraction pattern have been discussed in Sections 1-4.1, 1-4.2, and 1-4.3 and in the section "Three-Dimensional Transforms" of Appendix 1.

Because of the prevalence and severity of lattice imperfections (for example, atomic and molecular dislocations, and other distortions of the ideally perfect lattice rows and planes) in polymers, it has not been found possible to grow single crystals of suitable quality and sufficient size to permit application of the standard modern x-ray crystallographic procedures for the solution and refinement of three-dimensional crystal structures. Instead it is necessary to utilize diffraction patterns of highly oriented fibers or, in favorable cases, of specimens with higher degrees of orientation. (See Chapter 4.) Fiber patterns are analogous to rotation photographs of single crystals (Section 1-4.6) and suffer from the same failing in that the individual reflections on a given layer line are usually too closely spaced to permit their complete resolution according to indices (hkl) and corresponding intensities $I(hkl)$. Furthermore, in polymer patterns this problem is considerably accentuated by the greater diffuseness of, and areas occupied by, the reflections in comparison with those of rotating single crystals. This spreading and diffusing result from both lattice distortions and imperfect axial orientation. The latter factor causes the reflection to assume an arclike shape centered about the undeviated beam image.

Another adverse effect on the lattice pattern of lattice imperfections is a rapid diminution of the reflection intensities with increasing scattering angle 2θ, which manifests itself during the process of matching observed and calculated structure factors by the necessity of introducing anomalously large temperature-vibration coefficients B_n in the structure-factor expression (1-30). Thus numerical values in the range 15 to 20 are not uncommon, whereas the analysis of good single crystals of typical organic compounds calls for values between 5 and 10. The scantiness of diffraction data at large 2θ values has the net effect of imposing an ultimate limit on the accuracy with which the atoms can be located (so-called *resolution* of the structure). A further characteristic of polymer fiber patterns, which arises partly from the anomalously large B_n values and partly from the parallelism of molecular chains in the crystalline regions, is the tendency for a large proportion of the total intensity of diffraction to be lumped together into two or three reflections, usually equatorial, at small Bragg angles.

In attempting to solve the crystal structure of a polymer we must therefore recognize at the outset that the limitations affecting the intensity data at our disposal will in general deprive us of the use of the more conventional modern structure-solving tools, in particular the symbolic-addition procedure of phase determination and the three-dimensional Patterson synthesis. Moreover, we shall have at best only limited opportunities in the subsequent refinement process to make use of the usual powerful methods of Fourier difference synthesis or full-matrix least squares.† Instead we are compelled to revert historically, so to speak, to more purely trial-and-error methods of deducing promising models of the structure and to evaluate their relative merits as well as gage the degree of refinement accomplished by the quality of agreement between the observed and calculated structure factors [see (1-30)]. This state of affairs puts a premium on the ingenuity and general knowledge of the investigator, who must contrive the best possible trial structure on the basis of all the information at his disposal. First of all, of course, the distribution of intensities in the diffraction pattern is likely to afford some direct help; for example, (a) whether the main chain possesses a planar zigzag or helical conformation, (b) if helical, what the particular conformation is, (c) number of monomeric units in the unit cell, (d) probable orientation of any planar elements (amide groups, aromatic groups, etc.) on the basis of relative reflection intensities, (e) direct preliminary phasing of the stronger reflections on the basis of one or more heavy-atom substituents, and (f) indication of the presence of screw axes or glide planes by systematically extinguished classes of reflections. Maximum use should be made of independent sources of structural information; for example, (a) knowledge of the stereochemical formula on the basis of the method of chemical synthesis, (b) infrared and nuclear-magnetic-resonance spectra, (c) optical-rotatory-dispersion data, (d) comparison with crystal structures of similar polymers, and (e) compilations of standard bond lengths and angles, involving in particular carbon, nitrogen, and oxygen atoms and hydrogen bonds.

A notable advance toward bias-free and automatic refinement of polymer structures by least-squares-matrix methods has recently been made by Arnott and Wonacott[83]. Their method greatly reduces the number of parameters to be refined by allowing cooperative movements of large groups of atoms, at the same time permitting realistic stereochemical constraints to be maintained, such as standard bond lengths and angles, and meaningful modes of linking of successive monomer units. Important benefits gained as a result of the reduced number of parameters are (a) sharper averaging over un-

---

†See reference to the least-squares method of Arnott and Wonacott below, which constitutes a significant exception to this statement.

avoidable errors in poor-quality data such as are commonly obtained from polymers and (b) elimination of meaningless shifts in individual atomic parameters. The method is generally applicable to the refinement of any molecular structure but has particular advantages when applied to polymers because of the very limited amount of intensity data normally available. Arnott and Wonacott have adapted their method to a digital computer and have successfully applied it to refinement of the structures of poly(ethylene terephthalate) and α-poly-L-alanine.

In concluding this introduction we wish to emphasize that the crystal structures that can be derived from typical polymer diffraction patterns tend to represent *averaged* or somewhat *idealized* versions of the actual structures, which are always permeated by imperfections of various kinds and different degrees of severity. In fact not infrequently the disordering and/or lattice distortions are so great as to leave certain features of the structure in doubt. In particular, when purely trial-and-error methods of solution and refinement are employed, there is no way of knowing whether the *best fit obtained* between the observed and calculated structure factors is really the *best fit possible* or not. A direct consequence of the special limitations that affect the measurement of intensity data from polymers as well as the subsequent refinement of parameters is that the reliability index,

$$R = \frac{\Sigma \ |F_o - F_c|}{\Sigma \ F_o}, \qquad (6\text{-}24)$$

calculated from all the reflection data at the termination of refinement cannot be expected to fall much below 0.20, whereas for organic structures determined by photographic techniques from good-quality single crystals, R should normally diminish to between 0.10 and 0.15.

6-2.1   Crystal Structure of Poly(ethylene adipate)

As an illustration of the crystal-structure determination of a polymer from its fiber-diffraction pattern by "classical" methods we shall choose the investigation of the poly(ethylene adipate) structure by Turner-Jones and Bunn[84]. It is fitting that a paper from the researches of Bunn and associates be selected in recognition of Dr. Bunn's extensive pioneering work and early pre-eminence in the x-ray crystallography of polymers.

Specimens. This compound is a polyester of ethylene glycol and adipic acid with the formula

$$[-(CH_2)_2-O-\overset{\overset{\displaystyle O}{\|}}{C}-(CH_2)_4-\overset{\overset{\displaystyle O}{\|}}{C}-O-]_n.$$

The polymer, of moderate molecular weight, was linked with a few percent of hexamethylene diisocyanate, which had the effect of increasing the molecular-chain length and heightening the degree of orientation that could be achieved. It was demonstrated that this treatment improved the sharpness and resolution of the crystalline diffraction pattern without otherwise modifying it. The diffraction photographs were made of cold-drawn specimens or rubberlike specimens stretched above the melting point. Efforts to produce doubly oriented samples in order to further improve the resolution of the diffraction patterns were unsuccessful.

Experimental Measurements. Fiber patterns were recorded in a cylindrical camera with $CuK\alpha$ radiation (Figure 6-34). The unit cell was found to be monoclinic with the following dimensions:

$a = 5.47 \pm 0.03$ Å,
$b = 7.23 \pm 0.02$ Å,
$c = 11.72 \pm 0.04$ Å (fiber axis),
$\beta = 113°30'$,
Volume of cell, $V = abc \sin \beta = 425.38$ Å^3,
Molecular weight of monomer, $M = 172.184$.

From the unit-cell volume and molecular weight the density calculated for two monomer units in the cell ($Z = 2$) is

$$d_x = \frac{ZM}{VA} = \frac{2 \times 172.18}{425.38 \times 10^{-24} \times 6.023 \times 10^{23}} = 1.34 \text{ g/cm}^3,$$

which may be compared with the experimental density of the drawn fiber of 1.26 g/cm^3. The agreement is satisfactory since the density of the drawn fiber, which is partially amorphous, would be expected to be less than that of the crystalline regions.

Except for a weak reflection $(10\bar{1})$, class $(h0l)$ with l odd was missing, supporting the choice of $P2_1/a$ as the most likely space group[85]. Additional evidence showed that in all probability the weak $(10\bar{1})$ reflection was generated by a very minor amount of a second crystalline phase, designated β, with the same cell dimensions but lower symmetry $P2_1$. It was decided that the contributions of the β phase to the measured reflection intensities could be neglected. The intensities were determined by visual comparison with a calibration scale of spots of controlled relative exposures. Correc-

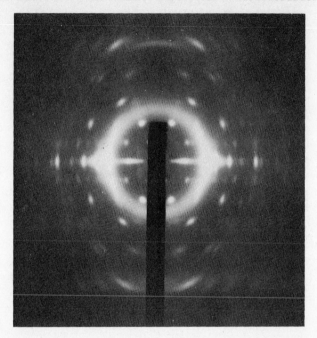

Figure 6-34   Fiber pattern of a rubberlike specimen of poly(ethylene adipate) linked with hexamethylene diisocyanate and stretched above the melting point. Fiber axis vertical. (Courtesy of A. Turner-Jones.)

tions were applied to the reflections to compensate for variations in the area occupied by each on the film, after which these normalized intensities were multiplied by $1/L$, $1/P$, and $1/j$ [see (1-20)] to yield the squared experimental structure factors F_o^2 listed in Table 6-4. The absorption correction $1/A$ was neglected because of the small numerical value of μt. (Refer to Sections 1-3.1 and 1-4.3, and Table 2-1.) An inspection of Table 6-4 shows that a large number of the data are observed as unresolved doublets and triplets.

Solution of the Structure.   On the assumption of standard bond lengths and angles involving carbon and oxygen the calculated identity period of one chemical monomer unit for a planar zigzag conformation is 12.2 Å, which may be compared with the experimental fiber identity period of $c = 11.72$ Å. The shortening of about 0.5 Å must be the result of deviations of some segment of the chain from a strictly planar zigzag conformation. There are two pieces of evidence indicating that the deformation occurs in the glycol rather than in the acid component of the chain. First, a similar shortening of 0.5 Å

Table 6-4 Poly(ethylene adipate): Observed and Calculated Squared Structure Factors[a,b,c]

hkl	F_o^2	F_c^2	hkl	F_o^2	F_c^2	hkl	F_o^2	F_c^2	hkl	F_o^2	F_c^2
110	608	480	14$\bar{1}$, 31$\bar{1}$	20	14	11$\bar{3}$, 013	52	73	32$\bar{4}$	7	8
020	608	474	231, 141	12	9	12$\bar{3}$, 20$\bar{3}$, 023	8	15	134	(4)	2
120	56	61	32$\bar{1}$	(2)	4				214, 14$\bar{4}$, 204	11	5
200	18	18	311, 24$\bar{1}$	4	2	21$\bar{3}$	16	22	044	(5)	2
210	1	1	002	41	29	113	17	19	224, 24$\bar{4}$, 334	(5)	4
130	71	48	012	—	1	22$\bar{3}$	8	11			
220	13	14	11$\bar{2}$	10	9	123	(1)	1	11$\bar{5}$	18	20
040	16	4	022	—	4	13$\bar{3}$, 033	10	14	005	19	35
230, 140	5	10	12$\bar{2}$, 112	73	81	31$\bar{3}$, 23$\bar{3}$, 203	20	20	20$\bar{5}$, 015	78	61
310	5	5	20$\bar{2}$	7	10	133	(2)	7	21$\bar{5}$	1	2
320	(1)	—	21$\bar{2}$	5	1				12$\bar{5}$	32	47
240	5	7	122	19	28	32$\bar{3}$, 14$\bar{3}$, 04$\bar{3}$	5	5	025	2	7
330, 150	2	2	032	(1)	4				22$\bar{5}$	36	13
001	4	4	22$\bar{2}$, 13$\bar{2}$	20	15	223	(> 2)	2	13$\bar{5}$, 115	14	21
011	10	14	202, 212, 132	29	36	24$\bar{3}$, 143	(> 2)	2	31$\bar{5}$, 035	33	31
(10$\bar{1}$)	(1.5)	—	23$\bar{2}$	(2)	2	004, 11$\bar{4}$	33	32	23$\bar{5}$, 125	19	40
11$\bar{1}$	18	25	31$\bar{2}$, 222	8	15	014	(1)	—	32$\bar{5}$	6	3
111	—	—	042, 14$\bar{2}$	12	3	20$\bar{4}$	58	67	14$\bar{5}$, 135	16	16
021	1	2	322, 142	5	5	21$\bar{4}$, 124	20	12			
12$\bar{1}$	4	—	232	(3)	—	024	(1)	1	205, 045, 33$\bar{5}$, 24$\bar{5}$, 215	16	12
20$\bar{1}$, 121	33	33	24$\bar{2}$	15	7	22$\bar{4}$, 114	9	7			
21$\bar{1}$	1	5	003	9	5	13$\bar{4}$	(2)	3			
031	9	5				124, 034	(2)	—			
201	22	22				31$\bar{4}$, 23$\bar{4}$	5	2			
13$\bar{1}$, 211, 22$\bar{1}$	18	16									
131	4	9									
221	(1)	5									
041, 23$\bar{1}$	(1)	2									

[a]Numerical values are in electrons$^2 \times 10^{-1}$.
[b]Figures in parentheses denote half the minimum observable F_o^2.
[c]Data from [84].

411

has been noted in polyesters of ethylene glycol with a number of dibasic acids. Second, in a homologous series of such polyesters with n varying by unitary steps in the acid residue, $-CO(CH_2)_nCO-$, the fiber identity period displays a concurrent variation in steps of precisely 2.5 Å, whereas the overall contraction of the monomer identity period below the calculated fully extended zigzag value remains constant at 0.5 Å. If the deformation occurred in the acid element of the chain it is highly probable that (a) the very constant increase of 2.5 Å in c per unitary increase in n would not be maintained and (b) the overall contraction of 0.5 Å would tend to increase with n.

Comparison of the dimensions of the cross sections of the unit cells of poly(ethylene adipate) and polyethylene perpendicular to the direction of the molecular chains shows them both to be rectangles of very similar dimensions (compare Figures 6-38 and 6-40):

Poly(ethylene adipate)	Polyethylene
$b = 7.23$ Å	$a = 7.40$ Å
$a \sin \beta = 5.02$ Å	$b = 4.93$ Å

This observation suggested that a trial structure might appropriately be based on an orthogonal packing of molecular chains in two dimensions similar to that of polyethylene[86]. Two ways of arranging the chains were found to be possible, as indicated in Figure 6-35. These two arrangements were then subjected to several modes of bond-angle distortion from strictly planar zigzag in order to effect the required contraction to $c = 11.72$ Å while at the same time preserving as far as possible acceptable intermolecular contact distances. The net results of these efforts were to demonstrate (a) that only models that incorporate bond-angle distortions in the glycol residue were acceptable and (b) that the chain arrangement of Figure 6-35a rather than 6-35b gave an encouraging degree of agreement between the F_o^2 and F_c^2 values.

At this point a limited refinement of the atomic coordinates was effected by trial and error. In order to confirm the essential correctness of the structure as developed to this point as well as to accomplish, if possible, some further improvement in the atomic positions, three-dimensional Fourier syntheses of electron density were now computed along lines parallel to a, b, and c passing through the x, y, z coordinates of each atom. In terms of the general equation T-11 (Appendix 1) this meant computing $\rho(x, y, z)$ in the following three ways:

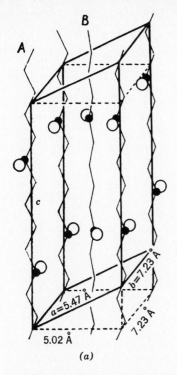

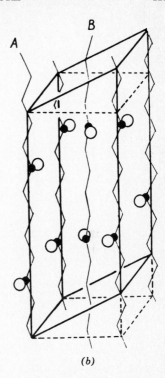

(a)   (b)

Figure 6-35 Unit cell of poly(ethylene adipate) and the two possible orientations of the molecular chain. (Turner-Jones and Bunn [84].)

1. With y and z fixed and x variable.
2. With x and z fixed and y variable.
3. With x and y fixed and z variable.

One of these line Fouriers, defined by $x = -0.073$, $y = 0.025$, z, is shown in Figure 6-36. Its significance in terms of the main-chain conformation can be appreciated by relating it to Figures 6-37, 6-38, and 6-39. If the conformation were truly planar zigzag, the line $(-0.073, 0.025, z)$ would pass close to the centers of atoms C_4, C_2, C_1, O_1, and C_3 successively. The fact that the C_1 peak is relatively weak confirms that the (glycol) carbon atoms are considerably displaced from the lines paralleling c on which the remaining chain atoms are situated.

Improved agreement with the experimental structure factors was now achieved by including in the calculated structure factors the contributions of the methylene hydrogen atoms placed in their expected tetrahedral positions relative to, and at a distance of 1.09 Å

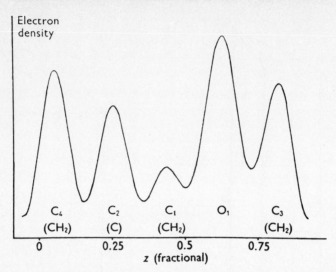

Figure 6-36   Fourier summation along the line $(x = -0.073, y = 0.025, z)$ parallel to the c-axis. (Turner-Jones and Bunn [84].)

from, their respective carbon atoms. Some further improvement in the agreement of the F_o and F_c values was also accomplished by trial-and-error adjustments of the atomic coordinates. In the calculation of the final F_c^2 values presented in Table 6-4 an overall isotropic temperature-vibration coefficient of $B = 17 \times 10^{-16}$ cm was employed, which may be compared with a value of 18×10^{-16} cm required in the refinement of the structure of poly(ethylene terephthalate)[87]. As explained above in the introduction to this section, lattice distortions in polymers undoubtedly contribute as strongly to the formal thermal-vibration coefficient B as do the thermal motions themselves. The F_o values were scaled to the F_c values on the basis of all the reflections except (110) and (020), which were omitted from this calculation because of their excessively large share of the total intensity in the diffraction pattern. On the basis of the F_o and F_c data of Table 6-4, values of the reliability index were determined both from the numerical structure factors themselves with (6-24) and from the F_o^2 and F_c^2 values with the formula

$$R' = \frac{\sum |F_o^2 - F_c^2|}{\sum F_o^2}. \tag{6-25}$$

The results are as follows:

	R	R'
Using all reflections	19	27
Excluding (110), (020)	21	32

Table 6-5 lists the coordinates of the atoms comprising one asymmetric unit of structure, in this case one-half a monomeric residue since each residue is situated on a crystallographic symmetry center. These atomic parameters were used in computing the F_c^2 values of Table 6-4. Since full three-dimensional Fourier or least-squares procedures were not employed in the refinement of the atomic coordinates, it was not possible to assign to each numerical parameter a statistically meaningful standard deviation.

Brief Description of the Structure. Figure 6-37 shows views of the chain configuration looking approximately (a) perpendicular and (b) parallel to the principal zigzag plane. Figures 6-38 and 6-39 are respectively c- and b-projections of the contents of the unit cell. Figures 6-38 and 6-40 show that the mode of packing of the molecular chains in poly(ethylene adipate) resembles very closely that of polyethylene, as was anticipated. The dihedral angle between the adipate zigzag plane and the ac plane is 40°, which may be compared with 41.2° between the zigzag plane and the bc plane in polyethylene.

Table 6-5 Poly(ethylene adipate):
Atomic Coordinates Expressed in
Fractions of the Unit-Cell Edges[a]

Atom	x	y	z
C_1	+0.059	+0.069	+0.468
O_1	+0.102	−0.035	+0.374
C_2	−0.060	+0.025	+0.258
C_3	+0.068	−0.050	+0.173
C_4	−0.063	+0.035	+0.044
H_1	+0.248	+0.125	+0.535
H_2	−0.081	+0.182	+0.426
$O_2(CO)$	−0.280	+0.105	+0.230
H_3	+0.281	−0.014	+0.214
H_4	+0.044	−0.202	+0.167
H_5	−0.043	+0.186	+0.051
H_6	−0.276	0.000	+0.004

[a]Data from [84].

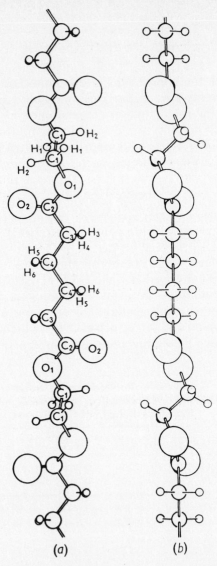

Figure 6-37 Chain configuration of poly(ethylene adipate) viewed (*a*) perpendicular and (*b*) parallel to the plane of the adipate chain. (Turner-Jones and Bunn [84].)

The pronounced displacement of the C_1 atoms from the molecular plane is clearly evident in Figures 6-37*b* and 6-38.

The nature of the deformation from the fully extended planar chain configuration, which affects only the glycol residue, can be quantita-

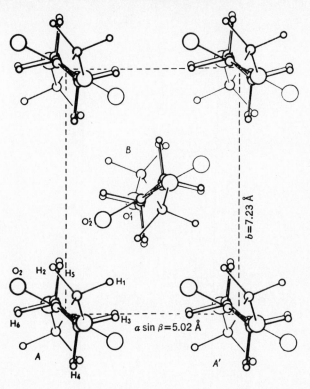

Figure 6-38 Arrangement of the molecules in the c-projection. (Turner-Jones and Bunn [84].)

tively defined by means of the following three internal-rotation angles (refer again to "Internal-Rotation Angles" in Section 6-1.2):

Plane L_1L_2	Plane L_2L_3	Rotation about	σ (degrees)
$C_4C_3C_2$	$C_3C_2O_1$	C_3C_2	170
$C_3C_2O_1$	$C_2O_1C_1$	C_2O_1	162
$C_2O_1C_1$	$O_1C_1C_1^*$	O_1C_1	-114

In this tabulation C_1^* is centrosymmetrically related to C_1. For the fully extended conformation σ is of course always 180°. In addition to the choice of these internal-rotation angles, a shortening of the O_1—C_2 bond length from the initial value of 1.43 to 1.37 Å was required in order to satisfy the monomeric identity period of 11.72 Å without a decided worsening of the agreement between the observed

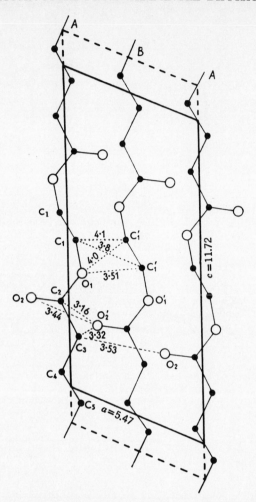

Figure 6-39   Arrangement of the molecules in the b-projection. All distances are in angstroms. (Turner-Jones and Bunn [84].)

and calculated structure factors. A shortening of the CO—O bond of similar amount has been observed in ester groups of other structures.

Within the expected limits of about ± 0.25 Å the shorter distances between carbon and oxygen atoms in adjacent molecules agree with the sums of the atomic van der Waals radii involved. The three shortest distances are $C_2 \cdots O_2' = 3.16$, $C_3 \cdots O_2' = 3.32$, and $C_4 \cdots O_2' = 3.35$ Å, which may be compared with the sum of Pauling's van der Waals radii [88]:

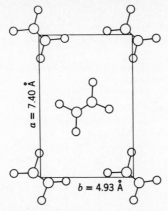

Figure 6-40 Arrangement of the molecules of polyethylene in the *c*-projection. (Courtesy of E. S. Clark.)

$$2.0(CH_2) + 1.4(O) = 3.4 \text{ Å}.$$

The closest methylene $\cdots$ methylene approaches are 3.80 and 4.10 Å, as compared with the van der Waals radius of 2.0 Å.

6-2.2 Further Illustrative Structures

In addition to the general references dealing with structure that are cited at the end of this chapter, from the available literature we may make a few specific selections as follows:

Structure	Investigator
Poly(ethylene terephthalate)	Daubeny, Bunn, and Brown[87]
Nylon 6	Holmes, Bunn, and Smith[89]
Nylon 77	Kinoshita[90]
Isotactic polypropene	Natta and Corradini[91]
Isotactic poly-1-butene	Natta, Corradini, and Bassi[92]

GENERAL REFERENCES

F. H. C. Crick and J. C. Kendrew, "X-Ray Analysis and Protein Structure," *Advances in Protein Chemistry*, **12**, 133 (1957).

D. R. Davies, "X-Ray Diffraction Studies on Polypeptide Conformations," *Progr. Biophys.*, **15**, 189 (1965).

K. C. Holmes and D. M. Blow, *The Use of X-Ray Diffraction in the Study of Protein and Nucleic Acid Structure*, Interscience, New York, 1966. [Reprinted from *Methods of Biochemical Analysis*, **13**, 113 (1965).]

M. I. Huggins, G. Natta, V. Desreux, and H. Mark, "Report on Nomenclature Dealing with Steric Regularity in High Polymers," *J. Polymer Sci.*, 56, 153 (1962).

A. Klug, F. H. C. Crick, and H. W. Wyckoff, "Diffraction by Helical Structures," *Acta Cryst.*, 11, 199 (1958).

H. P. Klug and L. E. Alexander, "Diffraction Studies of Noncrystalline Materials," Chapter 11 in *X-Ray Diffraction Procedures*, Wiley, New York, 1954.

G. Natta and P. Corradini, "General Considerations on the Structure of Crystalline Polyhydrocarbons," *Nuovo Cimento*, 15, Suppl. No. 1, 9 (1960).

N. Norman, *On the Cylindrically Symmetrical Distribution Method in X-Ray Analysis*, Ph.D. Thesis, University of Oslo, 1954.

S. C. Nyburg, *X-Ray Analysis of Organic Structures*, Academic Press, New York, 1961.

A. Rich and D. W. Green, "X-Ray Studies of Compounds of Biological Interest," *Ann. Rev. Biochem.*, 30, 93 (1961).

H. Tadokoro, "Structural Studies of Several Helical Polymers," *J. Polymer Sci.*, Part C, 15, 1 (1966).

B. K. Vainshtein, *Diffraction of X-Rays by Chain Molecules*, Elsevier, Amsterdam, New York, 1966.

SPECIFIC REFERENCES

[1] J. D. Dunitz and V. Prelog, Conference on Stereochemistry, Bürgenstock (1965).

[2] P. Corradini, *J. Macromol. Sci. (Chem.)*, Al(2), 301 (1967).

[3] M. L. Huggins, G. Natta, V. Desreux, and H. Mark, *J. Polymer Sci.*, 56, 153 (1962).

[4] G. Natta, M. Farina, and M. Peraldo, *Makromol. Chem.*, 38, 13 (1960).

[5] R. L. Miller and L. E. Nielsen, *J. Polymer Sci.*, 46, 303 (1960).

[6] S. Newman, *J. Polymer Sci.*, 47, 111 (1960).

[7] R. L. Miller, *J. Polymer Sci.*, 56, 375 (1962).

[8] R. L. Miller, *J. Polymer Sci.*, 57, 975 (1962).

[9] E. S. Clark, *Program and Abstracts*, American Crystallographic Association Meeting, Bozeman, Montana, July 26–31, 1964, Abstract F–12.

[10] R. E. Hughes and J. L. Lauer, *J. Chem. Phys.*, 30, 1165 (1959).

[11] K. Nagai and M. Kobayashi, *J. Chem. Phys.*, 36, 1268 (1961).

[12] R. L. Miller and L. E. Nielsen, *J. Polymer Sci.*, 44, 391 (1960).

[13] R. L. Miller and L. E. Nielsen, *J. Polymer Sci.*, 55, 643 (1961).

[14] R. L. Miller in *Polymer Handbook*, J. Brandrup and E. H. Immergut (editors), Interscience, New York, 1966. III–1.

[15] G. Natta, P. Corradini, and I. W. Bassi, *J. Polymer Sci.*, 51, 505 (1961).

[16] C. W. Bunn, *Proc. Roy. Soc. (London)*, A180, 67 (1942).

[17] C. W. Bunn and E. R. Howells, *J. Polymer Sci.*, 18, 307 (1955).

[18] H. Tadokoro, *J. Polymer Sci.*, Part C, 15, 1 (1966).

[19] G. Natta and P. Corradini, *J. Polymer Sci.*, 20, 251 (1956).

[20] G. Natta, P. Corradini, and I. W. Bassi, *Nuovo Cimento*, 15, Suppl. No. 1, 68 (1960).

[21] G. Natta and P. Corradini, *Nuovo Cimento*, 15, Suppl. No. 1, 9 (1960).

[22] G. Natta, P. Corradini, and P. Ganis, *J. Polymer Sci.*, 58, 1191 (1962).

[23] C. W. Bunn and D. R. Holmes, *Disc. Faraday Soc.*, No. 25, 95 (1958).

[24] C. W. Bunn, *J. Chem. Soc.*, 1947, 297.

[25] A. M. Liquori, *Acta Cryst.*, 8, 345 (1955).

[26] C. W. Bunn and E. R. Howells, *Nature*, 174, 549 (1954).

[27] E. S. Clark and L. T. Muus, *Z. Krist.*, 117, 108 (1962).

[28] E. S. Clark and L. T. Muus, *Z. Krist.*, 117, 119 (1962).

[29] H. P. Klug and L. E. Alexander, *X-Ray Diffraction Procedures*, Wiley, New York, 1954, Chapter 11.

[30] B. E. Warren and N. S. Gingrich, *Phys. Rev.*, **46**, 368 (1934).

[31] N. S. Gingrich, *Revs. Modern Phys.*, **15**, 90 (1943).

[32] R. W. James, *The Crystalline State*, Vol. II., *The Optical Principles of the Diffraction of X-Rays*, G. Bell, London, 1948, Chapter 9.

[33] U. W. Arndt and D. P. Riley, *Phil. Trans. Roy. Soc. (London)*, **A247**, 409 (1955).

[34] D. P. Riley in *Non-Crystalline Solids*, V. D. Frechette (editor), Wiley, New York, 1960, pp. 26–52.

[35] A. Bjørnhaug, O. Ellefsen, and B. A. Tønnesen, *J. Polymer Sci.*, **12**, 621 (1954).

[36] C. Finback, *Acta Chem. Scand.*, **3**, 1279, 1293 (1949).

[37] C. W. Bunn and E. V. Garner, *Proc. Roy. Soc. (London)*, **A189**, 39 (1947).

[38] Reference 29, pp. 631–633.

[39] A. L. Patterson, *Phys. Rev.*, **56**, 195 (1944).

[40] U. W. Arndt, W. A. Coates, and D. P. Riley, *Proc. Phys. Soc. (London)*, **B66**, 1009 (1953).

[41] L. Pauling, R. B. Corey, and H. R. Branson, *Proc. Nat. Acad. Sci. U.S.*, **37**, 205 (1951).

[42] W. L. Bragg, J. C. Kendrew, and M. F. Perutz, *Proc. Roy. Soc. (London)*, **A203**, 321 (1950).

[43] N. Norman, *On the Cylindrically Symmetrical Distribution Method in X-Ray Analysis*, Ph.D. Thesis, University of Oslo, 1954.

[44] N. Norman, *Acta Cryst.*, **7**, 462 (1954).

[45] A. N. J. Heyn, *J. Appl. Phys.*, **26**, 519 (1955).

[46] A. N. J. Heyn, *J. Appl. Phys.*, **26**, 1113 (1955).

[47] M. E. Milberg, *J. Appl. Phys.*, **33**, 1766 (1962).

[48] M. E. Milberg, *J. Appl. Phys.*, **34**, 722 (1963).

[49] M. E. Milberg and M. C. Daly, *J. Chem. Phys.*, **39**, 2966 (1963).

[50] G. Oster and D. P. Riley, *Acta Cryst.*, **5**, 272 (1952).

[51] L. E. Alexander and E. R. Michalik, *Acta Cryst.*, **12**, 105 (1959).

[52] W. O. Statton in *Handbook of X-Rays in Research and Analysis*, E. Kaelble (editor), McGraw-Hill, New York, 1967, Chapter 21, p. 14.

[53] K. H. Meyer and L. Misch, *Helv. chim. Acta*, **20**, 232 (1937).

[54] W. Cochran, F. H. C. Crick, and V. Vand, *Acta Cryst.*, **5**, 581 (1952).

[55] A. Klug, F. H. C. Crick, and H. W. Wyckoff, *Acta Cryst.*, **11**, 199 (1958).

[56] K. C. Holmes and D. M. Blow, *The Use of X-Ray Diffraction in the Study of Protein and Nucleic Acid Structure*, Interscience, New York, 1966. [Reprinted from *Methods of Biochemical Analysis*, **13**, 113 (1965).]

[57] C. H. Bamford, L. Brown, A. Elliott, W. E. Hanby, and I. F. Trotter, *Nature*, **169**, 357 (1952).

[58] C. S. Fuller, C. J. Frosch, and N. R. Pape, *J. Amer. Chem. Soc.*, **62**, 1905 (1940).

[59] D. R. Davies and A. Rich, *Acta Cryst.*, **12**, 97 (1959).

[60] L. Brown and I. F. Trotter, *Trans. Faraday Soc.*, **52**, 537 (1956).

[61] P. Corradini and P. Ganis, *J. Polymer Sci.*, **43**, 311 (1960).

[62] H. Tadokoro, Y. Chatani, T. Yoshihara, S. Tahara, and S. Murahashi, *Makromol. Chem.*, **73**, 109 (1964).

[63] M. H. F. Wilkins and J. T. Randall, *Biochim. Biophys. Acta*, **10**, 192 (1953).

[64] R. Langridge, H. R. Wilson, C. W. Hooper, M. H. F. Wilkins, and L. D. Hamilton, *J. Mol. Biol.*, **2**, 19 (1960).

[65] R. Langridge, D. A. Marvin, W. E. Seeds, H. R. Wilson, C. W. Hooper, M. H. F. Wilkins, and L. D. Hamilton, *J. Mol. Biol.*, **2**, 38 (1960).

[66] A. Rich and D. W. Green, *Ann. Rev. Biochem.*, **30**, 93 (1961).

[67] J. D. Watson and F. H. C. Crick, *Nature*, **171**, 737 (1953).

[68] J. D. Watson and F. H. C. Crick, *Nature*, **171**, 964 (1953).

[69] R. E. Franklin, A. Klug, and K. C. Holmes, *The Nature of Viruses*, Churchill, London, 1957, p. 39.

[70] A. Klug and D. L. D. Caspar, *Advances in Virus Research*, **7**, 225 (1960).

[71] R. E. Franklin, *Nature*, **175**, 379 (1955).

[72] R. E. Franklin and A. Klug, *Biochim. Biophys. Acta*, **19**, 403 (1956).

[73] R. E. Franklin and K. C. Holmes, *Acta Cryst.*, **11**, 213 (1958).

[74] F. H. C. Crick and J. C. Kendrew, *Advances in Protein Chemistry*, **12**, 134 (1957).

[75] A. Rich and F. H. C. Crick, *J. Mol. Biol.*, **3**, 483 (1961).

[76] G. N. Ramachandran and G. Kartha, *Nature*, **174**, 269 (1954); *Nature*, **176**, 593 (1955).

[77] G. N. Ramachandran and V. Sasisekharan, *Current Sci.* (India), **30**, 127 (1961).

[78] F. H. C. Crick, *Nature*, **170**, 882 (1952).

[79] L. Pauling and R. B. Corey, *Nature*, **171**, 59 (1953).

[80] F. H. C. Crick, *Acta Cryst.*, **6**, 685 (1953).

[81] C. Cohen and K. C. Holmes, *J. Mol. Biol.*, **6**, 423 (1963).

[82] Reference 56, pp. 231–232.

[83] S. Arnott and A. J. Wonacott, *Polymer*, **7**, 157 (1966).

[84] A. Turner-Jones and C. W. Bunn, *Acta Cryst.*, **15**, 105 (1962).

[85] For example, see *International Tables for X-Ray Crystallography*, Vol. I, Kynoch Press, Birmingham, 1952, 1965; M. J. Buerger, *X-Ray Crystallography*, Wiley, New York, 1942, Chapter 4; or reference 29, pp. 157–161.

[86] C. W. Bunn, *Trans. Faraday Soc.*, **35**, 482 (1939).

[87] R. de P. Daubeny, C. W. Bunn, and C. J. Brown, *Proc. Roy. Soc. (London)*, **A226**, 531 (1954).

[88] L. Pauling, *Nature of the Chemical Bond*, 3rd ed., Cornell University Press, Ithaca, N.Y., 1960, p. 260.

[89] D. R. Holmes, C. W. Bunn, and D. J. Smith, *J. Polymer Sci.*, **17**, 159 (1955).

[90] Y. Kinoshita, *Makromol. Chem.*, **33**, 21 (1959).

[91] G. Natta and P. Corradini, *Nuovo Cimento*, **15**, Suppl. No. 1, 40 (1960).

[92] G. Natta, P. Corradini, and I. W. Bassi, *Nuovo Cimento*, **15**, Suppl. No. 1, 52 (1960).

7

Lattice Distortions
and Crystallite Size

In relation to polymer systems the current state of the theory of lattice distortions and their diffraction effects is one of active flux including very appreciable elements of uncertainty and controversy. Under these circumstances it is out of the question to attempt a comprehensive or definitive treatment of the subject. We shall deal with it, instead, by focusing attention in turn on two of the most significant attacks that have been directed at clarifying the uncertainties and resolving the controversies. The first comprises the fundamental and largely theoretical investigations that have emanated from the school of Hosemann over the past two decades. The second refers to the important recent investigation of Buchanan and Miller that subjected the several theories of diffraction by distorted lattices to precise experimental testing and thereby defined some preliminary guide lines.

7-1 THE SCHERRER EQUATION

For a powder composed of relatively perfect crystalline particles the mean crystallite size can be determined with the familiar Scherrer equation [1, 2],

$$L_{hkl} = \frac{K\lambda}{\beta_0 \cos \theta},$$ \hfill (7-1)

where L_{hkl} is the mean dimension of the crystallites perpendicular to the planes (hkl), β_0 is the integral breadth or breadth at half-maximum intensity of the pure reflection profile in radians,† and K is a constant that is commonly assigned a value of unity. Equation 7-1 can be modified to express the reflection breadth in δs units $[s = 2(\sin \theta)/\lambda]$ as follows:

$$L_{hkl} = \frac{K}{[(2 \cos \theta)/\lambda]\delta\theta} \simeq \frac{1}{(\delta s)_0}. \qquad (7\text{-}2)$$

We may take note of the fact that L_{hkl} can be expressed as

$$L_{hkl} = \bar{N} \cdot d_{hkl}, \qquad (7\text{-}3)$$

in which $\bar{N}$ is the mean number of planes (hkl) corresponding to the dimension L_{hkl}.

Although the Scherrer equation, as expressed above, is sometimes used to obtain an estimate of the size of the crystalline regions in polymers, such numerical results are unreliable because of the prevalence of lattice distortions, which may also cause substantial broadening of the reflections. Thus it is necessary to replace the Scherrer equation with an expression that encompasses both size and distortion broadening. As a result of the relatively poor quality of most polymer diffraction patterns, this poses formidable experimental and interpretative difficulties.

7-2 THE TREATMENT OF HOSEMANN

7-2.1 Concept of the Paracrystal

The structure of an ideal crystal is characterized by the dimensions of the unit cell, the electron-density distribution within the cell, and the crystal size. We may introduce the concept of the paracrystal as a crystal possessing these same three characteristics except that now the dimensions of the unit cell and the electron-density distribution vary statistically from cell to cell. Figure 7-1 is a schematic representation of a two-dimensional paracrystalline lattice showing how successive translation vectors in any given lattice row line vary in both length and direction. These irregularities in the paracrystalline lattice cause the reflections in the diffraction pattern to be broader and more diffuse than would be expected on the basis of the classical crystallite-size effect. At the same time both the paracrystalline-lattice

---

†In Chapter 4 the symbol ω was used to denote the breadths of reflections or their component profiles in degrees of arc. Here, however, for the sake of clarity we have chosen to conform to the usage of the related literature in designating the angular breadth in radians β.

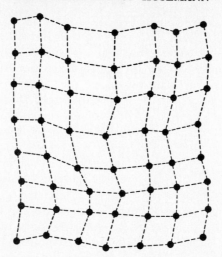

Figure 7-1 Schematic representation of a two-dimensional paracrystalline lattice. [Bonart and Hosemann, Z. *Elektrochem.*, **64**, 314 (1960).]

irregularities and the statistical fluctuations of the electron-density distribution within the unit cells give rise to a diffuse background scatter.

7-2.2 Distortions of the First and Second Kinds

Hosemann[3, 4] postulates that real crystal structures are subject to two kinds of distortions, referred to as of the *first* (I) and *second* (II) kinds. In lattices affected only by *distortions of the first kind* the long-range periodicity (order) is preserved, the distortions therefore being displacements of the structural elements (for example, atoms, monomer units, or motifs) from the equilibrium positions prescribed by the ideal lattice points. Figure 7-2*a* is an optical-diffraction model of a lattice characterized by distortions of the first kind in one dimension only, the horizontal direction. If the distribution of displacements from the ideal sites is assumed to be Gaussian, the net effect of type I distortions on the diffraction pattern is functionally indistinguishable from the effect of thermal vibrations and therefore may be expressed as

$$D(m) = \exp(-4\pi^2 g_I^2 m^2). \tag{7-4}$$

For this reason distortions of the first kind are often referred to as frozen-in thermal displacements. In (7-4) g_I is the relative linear displacement, $(\Delta d)_I/\bar{d}$, and m is the order of the reflection. Like thermal motion, distortions of type I cause a falling off of the intensity of a series of reflections with increasing order h but no broadening [5–7]. Figure 7-3a shows the one-dimensional optical-diffraction pattern of the model given in Figure 7-2a, and Figure 7-4a is the intensity profile of this pattern.

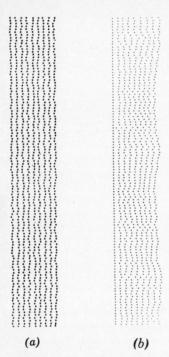

(a)   *(b)*

Figure 7-2 Optical-diffraction models of lattices with one-dimensional distortions *(a)* of the first kind with $g_I = 9.0\%$ and *(b)* of the second kind with $g_{II} = 13.7\%$. (Bonart, Hosemann, and McCullough[7].)

Distortions of the second kind were first characterized by Zernike and Prins[8] and by Kratky[9] in discussions of the structure of liquids. In a lattice possessing paracrystalline distortions (second kind) the long-range order is lost and each lattice point varies in position only in relation to its nearest neighbors rather than to the ideal lattice points. Thus the absolute magnitudes of the displacements from the positions corresponding to a strictly periodic arrangement increase with the distance from any arbitrarily chosen reference point, and in fact as the *square of the distance*[10]. When distortions of both the first and second kinds are simultaneously present, we may say that the displacements of the lattice points constitute the distortions of the second kind, whereas the displacements of the structural elements with respect to these (paracrystalline) lattice points constitute the distortions of the first kind. J. J. Hermans[11] has given a mathematical treatment of a linear lattice with distortions of the second kind; Porod[12, 13] and Hosemann[14] have treated special models of two and three dimensions. Figures 7-2b, 7-3b, and 7-4b illustrate distortions of the second kind by means of one-dimensional optical transforms. Distortions of the second kind result in both a diminution of intensity and an increase in reflection breadth with increasing angle 2θ (or with s or reflection order m).

The theory of paracrystalline diffraction developed by Hosemann and co-workers has contributed greatly to our understanding of x-ray scattering by semicrystalline systems of all kinds, including poly-

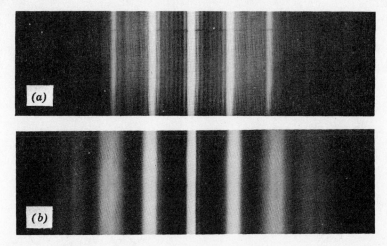

Figure 7-3 Optical-diffraction patterns of the models shown in Figure 7-2: (a) $g_I =$ 9.0%; (b) $g_{II} = 13.7\%$. (Bonart, Hosemann, and McCullough [7].)

mers. The theory may be applied to one-dimensional systems — for example, natural and synthetic linear polymers having macroperiodicities in the direction of the fiber axis — without invoking any questionable assumptions or supplementary hypotheses. However, when the theory is applied to two- and three-dimensional systems, it becomes necessary to introduce certain assumptions of which one in particular is open to question — namely, that statistical correlations between the various lattice fluctuation vectors may be neglected. It can be appreciated that this assumption may be unrealistic for linear polymers, possessing, as they do, strong primary valence bonds within the molecular chains but comparatively weak forces between them. We shall now describe Hosemann's method for separating crystallite-size and distortion parameters and follow it with applications to (a) macroperiodicities in a polymer fiber and (b) microstructure of the crystalline regions in drawn linear polymer fibers and of polymer single crystals. The reader who wishes to gain a better understanding of Hosemann's theory of paracrystalline diffraction is referred to the extensive available literature. A number of key references are cited among the general and specific references at the end of this chapter.

7-2.3 Separation of Size and Distortion Parameters

To a reasonable approximation the effect of lattice distortions of the second kind on the intensities of several orders m of reflection

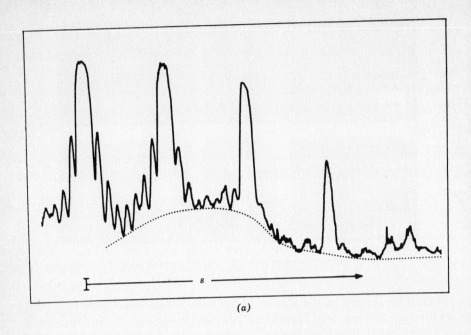

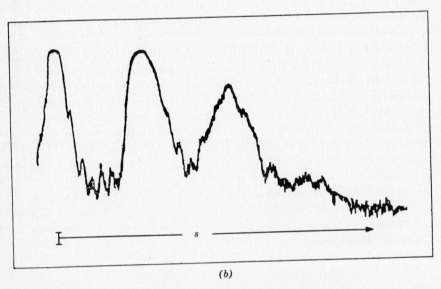

Figure 7-4 Intensity profiles of the optical-diffraction patterns shown in Figure 7-3: (a) $g_I = 9.0\%$; (b) $g_{II} = 13.7\%$. (Bonart, Hosemann, and McCullough[7].)

from a given family of planes may be expressed by the factor [7]

$$|F| = \exp(-2\pi^2 g_{II}^2 m^2), \dagger \qquad (7\text{-}5)$$

in which g_{II} is the relative distance fluctuation, $(\Delta d)_{II}/\bar{d}$, of the lattice vector perpendicular to the planes concerned. With reference to Figure 7-4b it is instructive to note that for g_{II} as small as 13.7% the diffraction effects have begun to assume the diffuse character of so-called amorphous scattering. When g_{II} is of the order of 25 to 35% only the first-order reflection is visible, and the higher orders flow together and become indistinguishable from the background [3, 4, 15, 16].

When distortions of the second kind cannot be disregarded as a possible source of broadening, (7-2) must be replaced by [7, 17]

$$(\delta s)_0^2 = (\delta s)_c^2 + (\delta s)_{II}^2 = \frac{1}{\bar{L}_{hkl}^2} + \frac{(\pi g_{II})^4 m^4}{\bar{d}_{hkl}^2}, \qquad (7\text{-}6)$$

which is valid if $2\pi^2 g_{II}^2 m^2$ is much smaller than unity and insofar as the broadening profiles $(\delta s)_c$ and $(\delta s)_{II}$ are of Gaussian shape, probably a reasonable assumption when the polydispersity is small. For low-order reflections the overall broadening $(\delta s)_0$ is determined mainly by $(\delta s)_c$, whereas for high orders it is influenced by both $(\delta s)_c$ and $(\delta s)_{II}$.

Equation 7-6 permits the separation of crystallite-size and lattice-distortion broadening provided at least two well-resolved orders of reflection from a given set of planes (hkl) are present in the diffraction pattern. After correction of the experimental reflection profiles for instrumental broadening, the squared integral breadths $(\delta s)_0^2$ are plotted against m^4, which yields a straight line with ordinate intercept $1/\bar{L}_{hkl}^2$ and slope $(\pi g_{II})^4/\bar{d}_{hkl}^2$, permitting $\bar{L}_{hkl}$ and g_{II} to be determined. It should be noted that this simplified approach does not lead to values of type I distortions (g_I) nor of *directional fluctuations* of the translational vectors comprising the lattice row lines. Besides $\bar{L}_{hkl}$, only g_{II}, the mean distance fluctuation between successive planes of the family (hkl), is obtained.

7-2.4   Applications to Linear Polyethylene

Macrostructure from Small-Angle Scattering [16]. Cold-drawn linear polyethylene gave the small-angle diagram shown in Figure 7-5, consisting of a horizontally elongated region of diffuse scatter on

---

$\dagger |F|$ is not to be confused with the conventional crystallographic structure factor (Section 1-4.3). Here it is the positive real value of $F(s)$, the Fourier transform of the probability function for the distortions of the second kind [7].

Figure 7-5 Small-angle diagram of cold-drawn linear polyethylene. Nickel-filtered CuKα radiation. Specimen-to-film distance 400 mm. Fiber axis vertical. (Bonart and Hosemann [16]; reprinted by permission of Dr. Dietrich Steinkopff, *Kolloid-Z.*, *Z. Polymere*.)

the equator and a single weaker meridional interference, also diffuse and horizontally elongated. From the absence of higher meridional orders of reflection it could be concluded that the mean paracrystalline fluctuation g_z of the large-scale period in the fiber direction z was approximately equal to or greater than 35%. Within experimental error the horizontal profile of the meridional reflection (radial direction R' of Figures T-2b and 7-6) showed no dependence on z', from which it was concluded that there were no appreciable statistical correlations between the positions of laterally adjacent crystalline regions, as illustrated schematically by Figure 1-21a. Because of this longitudinal randomness of adjacent fibrils, the horizontal (R') reflection profile is determined exclusively by the cross section of the fiber. Figure 7-7 shows a Guinier plot, $\log I(R')$ versus $(R')^2$, of the meridional interference, from which the assumption of a circular cross section leads to a fibril diameter of 96 Å.

Figure 7-7 also shows a Guinier plot of the horizontal intensity profile of the equatorial scattering, which is seen to be almost parallel to the plot of the meridional reflection except at the lower angles, where it curves sharply upward. The slope of the linear portion of the equatorial plot yielded a fibril diameter of 104 Å, in good agreement with the value of 96 Å from the meridional reflection. If this linear "monodisperse" component I' is subtracted from the Guinier plot of the equatorial scattering, as shown in Figure 7-8, it is possible to resolve most of the remaining visible intensity into straight-line

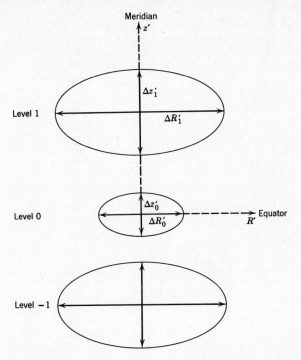

Figure 7-6 Schematic representation of the small-angle pattern of a synthetic linear polymer showing specification of the radial (R') and axial (z') dimensions.

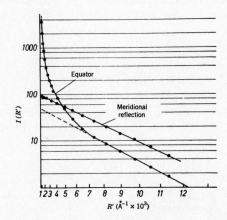

Figure 7-7 Cold-drawn linear polyethylene: Guinier plots of the R' intensity profiles of the equatorial scattering and meridional streak of the small-angle diagram shown in Figure 7-5. (Bonart and Hosemann [16]; reprinted by permission of Dr. Dietrich Steinkopff, *Kolloid-Z., Z. Polymere.*)

431

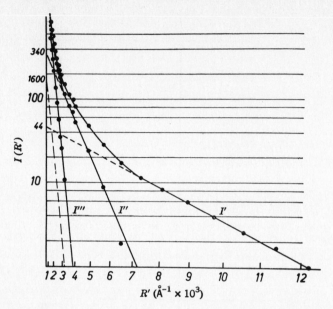

Figure 7-8   Cold-drawn linear polyethylene: resolution of the equatorial intensity profile into Gaussian components by means of Guinier plots. (Bonart and Hosemann [16]; reprinted by permission of Dr. Dietrich Steinkopff, *Kolloid-Z., Z. Polymere.*)

(Gaussian) components I'' and I'''. The slopes of these components lead to diameters of 211 and 437 Å, suggesting that substantial portions of the fibrils are grouped into clusters of 4 and 16 with respective diameters of approximately 200 and 400 Å. The portion of the intensity that is still not included in the analysis, shown by a broken line at very small angles in Figure 7-8, indicates the existence of still larger clusters that cannot be quantitatively evaluated since their small-angle-scattering contributions lie at angles inaccessible to the technique that was employed. On the basis of the ordinate intercepts the relative numerical abundances of clusters of 1, 4, and 16 fibrils are

$$N_1 = \frac{44}{1}, \qquad N_2 = \frac{340}{4^2}, \qquad N_3 = \frac{1600}{16^2},$$

from which the numerical percentage abundances are respectively 19, 37, and 44 if the larger clusters inaccessible to analysis are excluded.

In conclusion we may observe that the Bonart-Hosemann analysis [16] of the R' intensity profiles of the equatorial and first-order meridional scattering shows them to be in mutual agreement with a model consisting of clusters of parallel fibrils and some isolated fibrils, all without correlations among their longitudinal positions. This explanation must be viewed as alternative or supplementary to that of elongated microvoids as being the predominant source of diffuse equatorial scattering. (See Section 5-5.)

Microstructural Distortions in Drawn Fibers [17]. Hosemann and Wilke analyzed the wide-angle diffraction pattern of a highly extended hot-drawn specimen of linear polyethylene. Preliminary small-angle analysis similar to that described above revealed the presence of isolated fibrils 91 ($\pm$20) Å in diameter as well as clusters with mean diameters of 192 ($\pm$30) and 309 ($\pm$30) Å, indicating clustering in groups of four and nine. The wide-angle patterns were photographically recorded with a Guinier double camera of the Jagodzinski design [18] with crystal-monochromatized $CuK\alpha_1$ radiation. Paracrystalline-lattice distortions were determined by the method of (7-6) from the experimental profiles of the reflections (110), (220), and (330), which were well resolved and of adequate intensity.

With the reflections (110), (220), and (330) being designated $m = 1$, 2, and 3, respectively, a plot of log $|F|$ versus m^2 was prepared to ascertain whether or not (7-5) was satisfied. The plot was clearly nonlinear, so that it was necessary to postulate an inhomogeneous statistical fluctuation with terms in $(g_{II})_1$ and $(g_{II})_2$ as follows:

$$|F| = A \exp \left[-2\pi^2 (g_{II})_1^2 m^2 \right] + (1-A) \exp \left[-2\pi^2 (g_{II})_2^2 m^2 \right]. \quad (7\text{-}7)$$

The numerical solution was found to be

$$A = 0.929, \qquad (g_{II})_1 = 0.007,$$
$$1 - A = 0.071, \qquad (g_{II})_2 = 0.24.$$

These results were interpreted as a rough preliminary picture of the paracrystalline distortions in the absence of consideration of the crystallite-size effect. They indicate a very small fluctuation of 0.7% within the lattice and a much larger fluctuation of 24% that is encountered after the traversal of

$$N = \frac{1}{0.071} = 14.1 \, d_{110} \text{ spacings,}$$

or about 58 Å for polyethylene. This dimension is interpreted as a first approximation of the diameter of the ultimate, or "ultra," fibrils of which the polymer is composed.

In order to arrive at more accurate values of L and g_{II} according to the method of (7-6) it was necessary to begin by resolving each $(hh0)$ reflection profile into two Gaussian profiles, I_1 and I_2. This could be most conveniently accomplished by using plots of $\log I$ versus s^2 and deducing acceptable pairs of straight-line components, as illustrated in Figure 7-9. For the (330) reflection the I_1 term had become too broad for meaningful measurement, so that only the I_2 term was available. Figure 7-10 shows the squared integral breadths of the Gaussian-component profiles I_1 and I_2 plotted as a function of m^4. The numerical results are

$$(g_{\text{II}})_1 = 3.2\%, \qquad N_1 = 20.2, \qquad L_1 = N_1 d_{110} = 83 \text{ Å},$$

$$(g_{\text{II}})_2 = 2.2\%, \qquad N_2 = 42.7, \qquad L_2 = N_2 d_{220} = 175 \text{ Å}.$$

The agreement between the findings from small- and wide-angle scattering for the diameters of the ultrafibril and the quadruple cluster — 91 versus 83 and 192 versus 175 Å, respectively — is very good and is considered by Hosemann and Wilke[17] to support the essential correctness of these numerical results and the related concepts. These authors do not give a definite explanation for the absence of any information concerning the ninefold clusters from the wide-angle data.

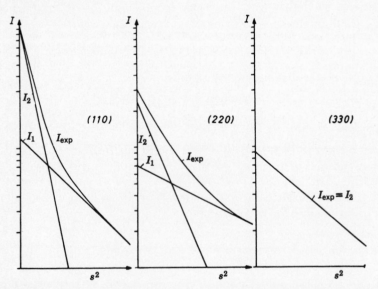

Figure 7-9 Hot-drawn linear polyethylene: resolution of the $(hh0)$ reflection profiles into Gaussian components by means of Guinier plots. (Hosemann and Wilke [17].)

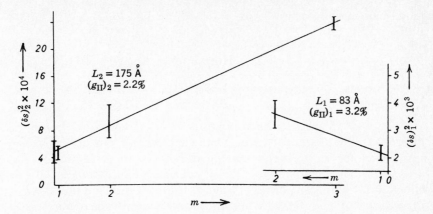

Figure 7-10 Hot-drawn linear polyethylene: plots of $(\delta s)_1^2$ and $(\delta s)_2^2$ versus m^4 for the (110), (220), and (330) reflections, with derived values of L and g_{II}. (Hosemann and Wilke [17].)

Microstructural Distortions in Single Crystals [19]. Hosemann, Čačković, and Wilke prepared single crystals of two commercial linear polyethylenes, Hostalen G and Lupolen 6011 H, by slow cooling of 0.05% xylene solutions. Mats of parallel crystal plates displayed small-angle x-ray periodicities of 108 ($\pm$2) and 99.7 ($\pm$2) Å. (In this connection the reader should refer to Section 5-5.2 and Figure 5-27a.) The individual crystals were lozenge shaped and from 100 to 200 Å thick. Diffraction patterns were obtained in a Guinier double camera of the Jagodzinski design [18] with CuKα_1 x-rays from a Johannson monochromator (see Section 2-3.4) incident on the polycrystalline mats parallel to the layering plane. In order to permit correction of the observed reflection profiles (110), (220), and (330) for instrumental broadening reference specimens of the same thickness, consisting of mixed powders of polyethylene and calcium tungstate, were prepared. When the instrumental broadening $(\delta s)_g$ is smaller than the breadth $(\delta s)_0$ of the pure diffraction profile, this correction is commonly made on the assumption of Gaussian profiles and consequent additivity of the squares of the breadths [see Section 4-3.1 and (4-3)]:

$$(\delta s)_{\text{exp}}^2 = (\delta s)_0^2 + (\delta s)_g^2.$$

In no case did this correction decrease the integral reflection breadth $(\delta s)_{\text{exp}}$ by more than 2%. The corrected δs values were multiplied by $\cos \theta$, squared, and plotted as a function of m^4 to give the excellent straight lines shown in Figure 7-11. From the slopes and ordinate intercepts the following numerical results were derived:

		Hostalen G	Lupolen 6011 H
Slopes,	g_{II} (percent)	1.88 (±0.07)	2.14 (±0.06)
Intercepts,	L_{110} (Å)	299 (±20)	332 (±15)

The g_{II} values are seen to be slightly smaller than those found for the drawn fibers, indicative of somewhat greater perfection in the single crystals. It is likely that the difference in perfection would have been more pronounced were it not that the drawn material was annealed for a longer period of time. It is also significant that the L_{110} dimensions of about 300 Å are in pronounced disagreement with the much larger crystal dimensions of 10,000 Å or more that were observed directly by electron microscopy. This striking difference shows that the single-crystal platelets are highly mosaic, consisting of mosaic blocks with a mean dimension parallel to the basal plane of about 300 Å.

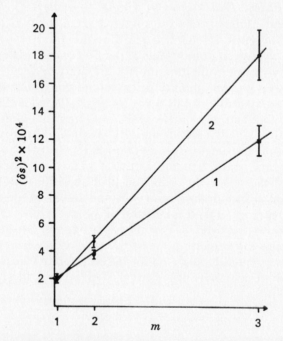

Figure 7-11   Polyethylene single crystals: plots of $(\delta s)^2$ versus m^4 for the (110), (220), and (330) reflections of two samples. Sample 1 is Hostalen G and sample 2 is Lupolen 6011 H. (Courtesy of R. Hosemann.)

7-3   THE ANALYSIS OF BUCHANAN AND MILLER

Because of the special significance that must be attached to Buchanan and Miller's[20] pioneering and critical evaluation of the several methods for separating size and distortion broadening that are potentially useful in the field of polymers, we shall present a rather extended account of their investigation here. The value of their studies is enhanced by reason of (a) the painstaking experimental tests employed and (b) the special emphasis accorded to the Fourier-transform method developed by Warren and Averbach[21, 22] for the analysis of local lattice distortions (microstrains) and crystallite size in metals. Warren and Averbach's method leads to the *number-average crystallite size* $\bar{N}$ and *mean-square local lattice distortions* $\langle g_t^2 \rangle$ in the direction t perpendicular to the diffracting planes concerned, which must afford at least three orders of reflections for the analysis.

7-3.1   Method of Fourier Transforms

After correction for instrumental broadening the experimental diffraction profile is the convolution of a crystallite-size profile (subscript c) and distortion profile (subscript D). According to (T-6) (Appendix 1), their transforms $A_c(t)$ and $A_D(t, s)$ are related to the transform $A(t, s)$ of the corrected experimental profile by

$$A(t, s) = A_c(t) \cdot A_D(t, s), \tag{7-8}$$

in which $s = 2(\sin \theta)/\lambda$ as usual.

With Bertaut[23] we now imagine that each crystallite consists of an assemblage of columns parallel to t, each column in turn being composed of a stack of unit cells sufficient in number to traverse the crystal from one surface to the other. If there are n_i columns composed of i cells and if $p\,(i)\,di$ expresses the number of columns with a length between i and $i + di$, $A_c(t)$ may be expressed as

$$A_c(t) = \frac{1}{N} \int_{i=|t|+1}^{\infty} (i - |t|)p(i)\,di. \tag{7-9}$$

In this equation N is the total number of unit cells,

$$N = \sum_{i=1}^{\infty} in_i.$$

The derivative of $A_c(t)$ with respect to t at $t = 0$ is

$$\left[\frac{dA_c(t)}{dt}\right]_{t=0} = -\frac{1}{N} \int\limits_{i=1}^{\infty} p(i)\, di = -\frac{N_c}{N} = \frac{1}{\bar{N}}, \qquad (7\text{-}10)$$

in which N_c is the total number of columns in the crystallite and $\bar{N}$ is the number-average crystallite size in the direction t.

If we now denote the mean-square lattice distortion over the domain t by $\langle g_t^2 \rangle$, it can be shown that for small values of g_t and low orders of reflection [21–23]

$$A_D(t, s) = \exp\left(-2\pi^2 s^2 t^2 \langle g_t^2 \rangle\right), \qquad (7\text{-}11)$$

where

$$L_t = g_t t = t\frac{\Delta d}{d}.$$

If we now substitute (7-11) into (7-8) and take logarithms, the result is

$$\log_e A(t,s) = \log_e A_c(t) - 2\pi^2 s^2 t^2 \langle g_t^2 \rangle. \qquad (7\text{-}12)$$

Provided several orders of reflection are available, plots of $\log_e A\,(t,s)$ versus s^2 for any desired values of t make it possible to obtain $A_c(t)$ and $\langle g_t^2 \rangle$ respectively from the ordinate intercepts and slopes at low angles. At larger angles the plots tend to deviate from linearity, so that no less than three orders of reflection are required to effect an acceptable extrapolation to $s = 0$. Finally the values of $A_c(t)$ corresponding to a range of values of t are plotted as a function of t, and from the slope of this function at low t the number-average crystallite size $\bar{N}$ is obtained as indicated by (7-10).

7-3.2 Method of Integral Breadths

The labor of evaluating the Fourier transforms in the method of Warren and Averbach [21, 22] can be avoided, although at the cost of reduced information and accuracy, by resorting to integral breadths in place of the complete profiles of the reflections. Wilson [24] has shown that

$$(\delta s)_D = 4e\frac{\sin\theta}{\lambda} = \frac{2e}{d_{hkl}}, \qquad (7\text{-}13)$$

in which $1/d_{hkl} = s$ and

$$e \simeq \left(\frac{\Delta d}{d}\right)_{hkl},$$

a somewhat vaguely defined upper limit of the lattice distortions. According to an analysis of Buchanan, McCullough, and Miller [25], the relationship between e and $\langle g_t^2 \rangle$ is

$$e \simeq 1.25 \langle g_t^2 \rangle^{1/2}. \tag{7-14}$$

If the reflection profiles are plotted on the 2θ scale expressed in radians, we have for the integral breadth of the distortion profile from (7-13)

$$\beta_D = 4e \tan \theta \quad \text{radians}, \tag{7-15}$$

and (7-1) gives the integral breadth of the crystallite-size profile,

$$\beta_c = \frac{K\lambda}{L \cos \theta}.$$

If at least two orders of reflection are available and if only one source of broadening is operative (distortion or size, but not both), it is in principle possible to distinguish which factor is present by determining whether the broadening of the pure diffraction profile conforms to $\tan \theta$ or $1/\cos \theta$.

When both size and distortion broadening are operative, the magnitude of the combined broadening effect depends on the shapes of the two contributing profiles. Most commonly these are assumed to be either Cauchy or Gaussian,

$$\frac{1}{(1 + k^2 x^2)} \quad \text{or} \quad \exp(-k^2 x^2),$$

respectively. If both profiles are Cauchy, the net integral breadth $(\delta s)_0$ of the pure diffraction profile will be the sum of the size and distortion breadths,

$$(\delta s)_0 = (\delta s)_c + (\delta s)_D \quad \text{(Cauchy)}, \tag{7-16}$$

whereas for two Gaussian profiles

$$(\delta s)_0^2 = (\delta s)_c^2 + (\delta s)_D^2 \quad \text{(Gaussian)}. \tag{7-17}$$

In (7-16) and (7-17) we have returned to the expression of the integral breadths in terms of the variable s and Å^{-1} units.

If the profiles are assumed to be Cauchy, substitution of (7-2) and (7-13) into (7-16) yields

$$(\delta s)_0 = \frac{1}{L} + 2es \quad \text{(Cauchy)}, \tag{7-18}$$

whereas on the assumption of Gaussian profiles, substitution of (7-2) and (7-13) into (7-17) gives the result

$$(\delta s)_0^2 = \left(\frac{1}{L}\right)^2 + (2es)^2 \quad \text{(Gaussian)}. \tag{7-19}$$

Thus on the basis of the Cauchy assumption we obtain from a plot of $(\delta s)_0$ versus s the mean crystallite dimension L from the ordinate intercept and the "maximum" lattice distortion e from the slope. For the Gaussian assumption, however, we utilize a plot of $(\delta s)_0^2$ versus s^2 in the same manner.

7-3.3 Application to Isotactic Polystyrene

The relatively high crystal symmetry (trigonal) of isotactic polystyrene[26] suits it better, perhaps, than most other linear polymers for a test of the Warren-Averbach[21, 22] method of analysis. Not only are the $(hk0)$ reflections numerous and rather well resolved, including three or more orders of $(hh0)$, but the sixfold diffraction symmetry tends to reduce disparities between $(hh0)$ and $(hk0)$ reflections from the standpoint of the size and distortion broadening effects, thereby providing a larger number of potentially useful reflections for the analysis.

Preparation of Samples. Polystyrene films 0.5 mm thick were molded in a press at 240°C and quenched rapidly in cold water. Three samples for analysis were prepared as follows from this film:

No. 0–165. Annealed 20 min at 165°C.
No. 4–165. Drawn to an elongation of 300% at 105°C, quenched and annealed at constant length for 5 min at 165°C.
No. 4–120. Drawn to an elongation of 300% at 105°C, quenched and annealed at constant length for 5 min at 120°C.

X-Ray Diffraction Measurements. Very stringent experimental conditions were imposed in order to provide intensity data of the best possible quality. *Wide-angle x-ray scattering* was measured with crystal-monochromatized $CuK\alpha_1$ radiation in conjunction with a scintillation counter and pulse-height discrimination. (The reader may refer again to Section 2-4.) The intensity curve was measured point by point on the 2θ scale, keeping counting statistics favorable by counting either a total of 10,000 counts or for a period of 30 min at each point, whichever came first. The symmetrical-reflection geometry (see Figure 2-8a) was used in measuring the $(hk0)$ reflections, and the symmetrical-transmission geometry (see Figure 2-8c) was also employed for measuring a few (hkl) reflections from sample 4–165. Figure 7-12 is a representative experimental intensity curve, that of sample 4–120. The small-angle scattering was also measured, but the results are not directly pertinent to the present discussion and will not be considered here.

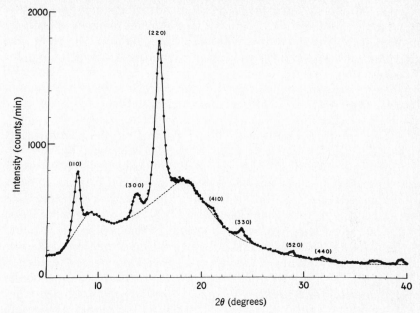

Figure 7-12 Uncorrected experimental intensity curve of isotactic-polystyrene sample 4-120. Broken line represents the "amorphous" background. (Buchanan and Miller[20].)

Reduction of Data. The numerical calculations were performed with a digital computer. The procedure was as follows:

1. In the case of the symmetrical-reflection data a *background correction* was established by comparison of the experimental intensity curve with the curve of a quenched, isotactic-polystyrene sample prepared under the same instrumental conditions. A degree of guesswork was required in plotting an acceptable background, such as is illustrated by the broken-line curve of Figure 7-12.

2. After subtraction of the background intensities any *overlapping reflections* in the corrected curve were resolved by means of an iterative trial-and-error procedure in which it was assumed that each component profile was symmetrical about its maximum point. Despite the somewhat dubious accuracy of this method, Buchanan and Miller [20] found that the precision and reproducibility were good. Furthermore, results agreeing to within 3 to 5% were obtained whether the resolution of overlapping reflections was carried out before or after the correction of the experimental profile for instrumental broadening by the Fourier deconvolution procedure. (See item 3 below.) The principal disadvantage of the foregoing resolution procedure is that

a large number of iterations are required for convergence. If the component profiles are assumed to conform to some particular mathematical function, the resolution process can be performed analytically. Noble and Hayes[27] have invented an electronic analog computer that permits complex line profiles to be rapidly resolved by direct visual inspection into the sums of two or more Gaussian or Cauchy profiles, or profiles of some intermediate shape. A commercial model of this curve analyzer is now on the market.†

3. The most accurate methods of correcting the experimental reflection profiles for *instrumental broadening* are Fourier deconvolution (unfolding) methods, such as that of Stokes[28], which are described in Appendix 2. A well-crystallized sample of hexamethylenetetramine in the form of a compacted disk 0.036 cm thick was used as the reference specimen, characterized by dimensions and x-ray absorption equivalent to those of the polymer specimens but producing reflection profiles that were free of size and distortion broadening. The Stokes deconvolution procedure yields the integral breadth β_0 of the pure diffraction profile in terms of the Fourier coefficients A_n of that profile:

$$\beta_0 = \frac{1}{\sum\limits_{-\infty}^{\infty} A_n} \cdot \qquad (7\text{-}20)$$

Alternatively β_0 may be determined directly from the pure diffraction profile.

If the instrumental and "pure" broadening profiles can with good reason be regarded as at least approximately Gaussian in shape, the correction can be speedily performed by subtracting the squared integral breadth of the instrumental profile b from that of the experimental profile B,

$$\beta_0^2 = B^2 - b^2 \qquad \text{(Gaussian)} \qquad (7\text{-}21)$$

For Cauchy profiles, on the other hand,

$$\beta_0 = B - b \qquad \text{(Cauchy).} \qquad (7\text{-}22)$$

In (7-21) and (7-22) the integral breadths are understood to be expressed in either radians or degrees on the 2θ scale. Table 7-1 lists the experimental and instrumental breadths, together with numerical values of the corrected breadths as determined in the three ways just

†The "310 Curve Resolver," available from E. I. du Pont de Nemours and Company, Inc., Instrument Products Division, Wilmington, Delaware 19898.

Table 7-1 Integral-Breadth Data for Reflections of Isotatic-Polystyrene Samples[a]

		Breadths ($^\circ 2\theta$)		Corrected Breadths (Å^{-1})[b]			
				From Fourier Deconvolution		From Breadths	
Sample	(hk0)	Observed (B)	Instrumental (b)	Directly from Corrected Profile	$\left(\sum\limits_{-\infty}^{\infty} A_n\right)^{-1}$	Gaussian Assumption	Cauchy Assumption
4–120	(110)	0.751	0.235	0.0079	0.0076	0.0080	0.0090
	(300)	0.946	0.235	0.0100	0.0091	0.0103	0.0124
	(220)	1.028	0.235	0.0111	0.0109	0.0112	0.0138
	(410)	0.806	0.227	0.0081	0.0078	0.0086	0.0101
	(330)	0.773	0.227	0.0078	0.0075	0.0082	0.0095
4–165	(110)	0.624	0.235	0.0061	0.0059	0.0065	0.0044
	(300)	0.726	0.235	0.0080	0.0078	0.0077	0.0055
	(220)	0.756	0.235	0.0070	0.0069	0.0080	0.0058
	(410)	0.804	0.227	0.0087	0.0085	0.0085	0.0064
	(330)	0.853	0.227	0.0084	0.0082	0.0091	0.0069
0–165	(110)	0.374	0.235	0.0035	—	0.0033	0.0024
	(300)	0.435	0.235	0.0053	0.0051	0.0041	0.0035
	(220)	0.436	0.235	0.0044	0.0043	0.0041	0.0035
	(330)	0.863	0.227	0.0055	0.0052	0.0057	0.0063

[a]Data from [20].
[b]On s scale.

discussed and represented by (7-20), (7-21), and (7-22). The reflections measured include (110), (220), (330), (300), and (410). Unlike B and b, the corrected breadths $(\delta s)_0$ are expressed in reciprocal angstrom units. It can be seen that the Gaussian approximation agrees in general with the deconvolution method to within 10%, whereas the Cauchy postulate results in much greater discrepancies. Consequently the numerical breadths that are obtained on the assumption of Cauchy profiles are not utilized in the evaluations of size and distortion broadening that follow.

Results. Table 7-2 lists numerical values of the crystallite-size and lattice-distortion parameters as derived for the three polystyrene samples by the five methods of separating size and distortion broadening specified in column 2. With the exception of the Fourier-transform method (item No. 1), a comparison is given of numerical parameters based on two procedures for correcting the experimental reflection profiles for instrumental broadening: (a) the deconvolution method of Stokes and (b) the Gaussian method, $\beta_0^2 = B^2 - b^2$.

Figures 7-13 and 7-14 illustrate the application of the Fourier-transform method to the experimental data. By means of the data for sample 4–165, Figure 7-13 illustrates the plotting of the logarithm of the distortion transform, $\log A(t, s)$, versus m^2 for a wide range of t values [see (7-12)]. The orders m of the reflections $(hh0)$ are proportional to s, which satisfies the functionality of the equation. The upward inflection of the curves at the lower values of m^2 is an indication that the strain distribution falls off more slowly than does a Gaussian function[29]. Figure 7-14 shows plots of $A_c(t)$ versus t for all three polystyrenes, from which the number-average crystallite sizes are determined from the slopes according to (7-10). Figure 7-15 shows the corrected integral breadths for sample 4–165 plotted as a function of $(\sin\theta)/\lambda$. The solid line is the best linear plot through

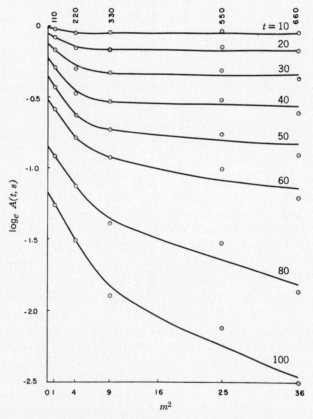

Figure 7-13   Isotactic-polystyrene sample 4–165: extrapolation curves for the Fourier-transform method of Warren and Averbach. (Buchanan and Miller [20].)

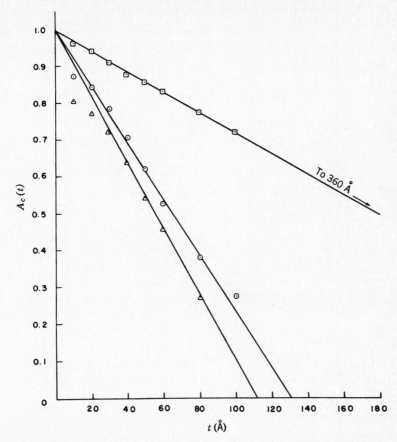

Figure 7-14   Isotactic-polystyrene samples: plots of $A_c(t)$ versus t. (Buchanan and Miller [20].)

the first three orders of $(hh0)$, and it can be seen that the $(hk0)$ points fit this line as well as the remaining $(hh0)$ orders, although the experimental scatter is considerable. This agrees with what was said at the beginning of this section about the near equivalence of the broadening effects of $(hh0)$ and $(hk0)$ reflections to be expected because of the sixfold diffraction symmetry in the equatorial plane. This equivalence can be an important consideration in the analysis of other polymers of relatively high symmetry that yield only one or two resolved orders of reflection from any given set of planes. Unfortunately the crystallographic symmetry of most polymers is much lower, which rather severely curtails the application of this equivalence principle. Its applicability to isotactic polystyrene,

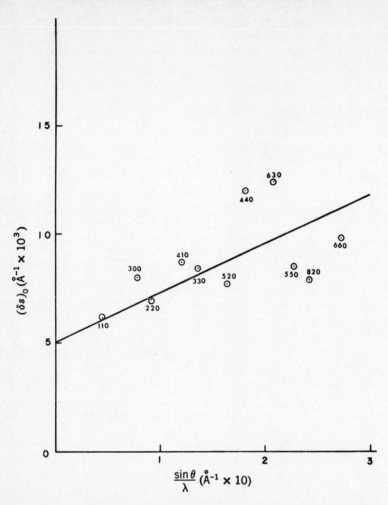

Figure 7-15 Isotactic-polystyrene sample 4–165: corrected integral breadths as a function of $(\sin\theta)/\lambda$. (Buchanan and Miller [20].)

however, was further confirmed by a least-squares fitting of the broadening data of Figure 7-15 to the Cauchy relationship (7-18) for the extraction of the size and distortion parameters. The resulting values of L and e as derived from the $(hh0)$ data taken alone and from the $(hh0)$ and $(hk0)$ data together are in good agreement:

From $(hh0)$: $L = 181\,(\pm 28)\,$Å, $e = 0.46\,(\pm 0.10)\%$.
From $(hh0)$ and $(hk0)$: $L = 172\,(\pm 25)\,$Å, $e = 0.47\,(\pm 0.10)\%$.

Weight-average values of L and "maximum" local lattice distortions e were next determined with the aid of (7-18) and (7-19) respectively for the Cauchy and Gaussian assumptions of broadening profiles. Prior corrections for instrumental broadening were carried out by deconvolution and by the Gaussian postulate, as shown in Table 7-2. The numerical results with L in angstroms and e in percent are given as items 2 and 3 in Table 7-2.

Item 4 of Table 7-2 presents comparative numerical results derived from the integral breadths by a Fourier-transform method due to Schoening[30], which postulates that the size broadening is Cauchy but the distortion broadening is Gaussian. Schoening's method yields a weight-average size and a so-called "maximum" distortion e, as defined earlier. Finally, item 5 of Table 7-2 gives a weight-average size L and a value of the paracrystalline distortion, $g_{II} = (\Delta d)_{II}/\bar{d}$, according to the principle of (7-6). The numerical results given in item 5 implicitly assume that any distortions present are of the second kind, which requires that the integral breadth due to distortions alone be proportional to the square of the order of reflection m. When both size and paracrystalline broadening are present, this means that a plot of $(\delta s)_0^2$ versus m^4 must also be linear [see (7-6)].

Interpretation of Results.   An evaluation of the numerical findings leads to a number of conclusions as follows:

1. The extent and accuracy of the experimental data do not suffice to characterize the distortions in these polystyrene samples unequivocally as being of the microstrain or paracrystalline types. For low levels of distortion these types should respectively broaden the orders of reflection according to the first and second power of the order m (or s). In Figure 7-16 the squared integral breadths of the (110), (220), and (330) reflections of samples 4–165 and 0–165 are plotted as functions of h^4 and h^2. It can be said that the linearity is somewhat better for the h^2 plot, thus favoring the microstrain over the paracrystalline model of distortion.

2. The numerical data of Table 7-2, with the exception of the Cauchy item 2, show a consistent increase in size and decrease in distortion with increasing·annealing (proceeding from left to right in the table), in accordance with expectations.

3. Another rather consistent feature of these numerical results is that the weight-average sizes (from integral breadths) are larger than the corresponding number-average sizes (from Fourier transforms). That there is a theoretical basis for this apparent disparity has been demonstrated by Buchanan, McCullough, and Miller[25],

Table 7-2   Crystallite-Size and Lattice-Distortion Parameters Parallel to the Basal Plane (001) for Isotactic-Polystyrene Samples[a]

Item No.	Method of Separation	Type of Parameter		Method of Instrumental Correction	Sample 4-120		Sample 4-165		Sample 0-165	
		Size	Distortion[b]		Size (Å)	Distortion (%)	Size (Å)	Distortion (%)	Size (Å)	Distortion (%)
1	Transform (7-12)	Number average	$\langle g_t^2 \rangle^{1/2}$	Deconvolution	111	0.66	130	0.51	360	0.41
2	Cauchy (7-18)	Weight average	e	Deconvolution	217	1.78	208	0.63	400	0.54
				Gaussian	213	1.78	192	0.72	400	0.45
3	Gaussian (7-19)	Weight average	e	Deconvolution	150	0.61	177	0.28	305	0.13
				Gaussian	150	0.61	160	0.31	338	0.11
4	Cauchy-Gaussian[c]	Weight average	e	Deconvolution	172	1.92	176	0.70	330	0.56
				Gaussian	162	1.95	176	1.02	355	0.55
5	Paracrystal (7-6)	Weight average	g_{II}	Deconvolution	131	4.95	162	2.80	275	2.21
				Gaussian	129	4.95	156	3.76	310	2.92

[a] Data from [20].

[b] Distortion parameters:

$\langle g_t^2 \rangle^{1/2}$ = root-mean-square local lattice distortions averaged over the distance t.

$e = 1.25 \langle g_t^2 \rangle^{1/2}$ = "maximum" lattice distortion [24].

g_{II} = lattice distortions of the second kind (paracrystalline) expressed as the mean relative distance fluctuation [7].

[c] Fourier-transform method of Schoening [30], which postulates that size broadening is Cauchy but distortion broadening is Gaussian.

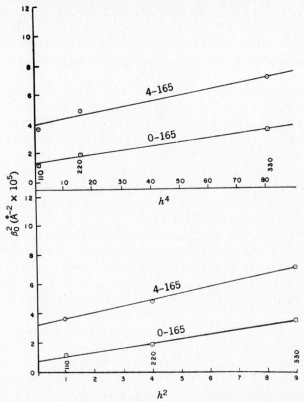

Figure 7-16 Isotactic-polystyrene samples 4–165 and 0–165: squared integral breadths (corrected for instrumental broadening) of $(hh0)$ reflections as a function of h^4 and h^2. Here h is the order of reflection. (Buchanan and Miller [20].)

so that these two types of size averages should be regarded as mutually complementary rather than contradictory. In this regard the apparently anomalous result for sample 0–165 can be attributed to the loss in efficacy of the deconvolution process when the breadth of the polymer reflection is but slightly greater than that of the reference sample. (See Appendix 2.)

4. Of the methods investigated, the Fourier-transform technique is capable of extracting the most information from the experimental data relative to size and distortions. It is very important to note, however, that very accurate intensities and no less than *three orders of reflection* are required for the determination of the curvature and the carrying out of a reliable extrapolation in accordance with (7-12) for the determination of $A_c(t)$ and $\langle g_t^2 \rangle$ and for the evaluation of the distribution of lattice distortions as a function of distance t. Thus it

is evident that the usefulness of the Fourier-transform method of Warren and Averbach[21, 22] is restricted to the very few polymer systems that possess exceptionally high crystallographic symmetry. When only two orders of reflection are available, linear extrapolation according to (7-12) leads to values of both the number-average size and local distortions that are too small[29]. This is evidently a failing of the results for sample 4–120. (See item 1 of Table 7-2.) It is indicative of the high demands imposed by the Fourier-transform method that, in spite of the painstaking experimental techniques employed by Buchanan and Miller[20] as well as the thoroughness of their data-processing and interpretative procedures, their best set of data (for sample 4–165) did not suffice to establish the form of the size transform $A_c(t)$ precisely enough to yield reliable information about the crystallite-size distribution.

5. In evaluating the size and distortion parameters from integral breadths the assumption of Gaussian profiles or of the mixed Cauchy and Gaussian mode of Schoening[30] leads to numerical results in reasonable accord with those derived by full Fourier-transform analysis. On the other hand, the assumption that both profiles are Cauchy is found to be unacceptable. If the same functionality is to be postulated for both the size and distortion profiles, there are reasons to suppose that the Gaussian form is more appropriate for broad profiles but the Cauchy form for narrow ones. In actual practice polymer reflection profiles are nearly always broad, so that the pure Gaussian assumption is very generally applicable.

6. Compared with the results obtained by the integral-breadth methods, the paracrystalline model (item 5 of Table 7-2) yields crystallite sizes that are somewhat smaller and lattice-distortion parameters that are somewhat larger. For g_{II} values below 5% Buchanan and Miller[20] found the paracrystalline approach as represented by (7-6) to be very insensitive and concluded that considerably larger distortions would be required to permit a categorical distinction to be made between the paracrystalline and microstrain models of lattice distortion.

7. The rapid falloff to a constant value of the magnitude of lattice distortions with t (see Figure 7-17), the size of the domain concerned, is compatible with a dislocation theory of lattice distortions. The approximate density of edge dislocations that would correspond to the observed values of lattice distortions e parallel to the basal plane [given by $(hk0)$ data] can be calculated from

$$D' = \frac{4e}{|\mathbf{b}|}, \tag{7-23}$$

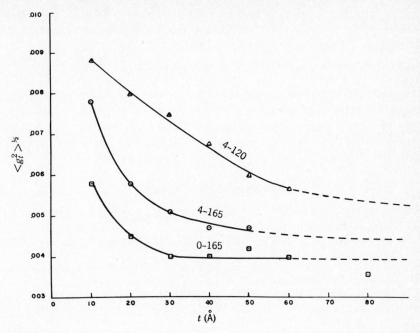

Figure 7-17 Isotactic-polystyrene samples: plots of $\langle g_t^2 \rangle$ versus t. (Buchanan and Miller [20].)

where $|\mathbf{b}|$ is the magnitude of the Burgers vector and D' is the number of edge dislocations per angstrom in the direction $\mathbf{b}$. If $|\mathbf{b}|$ is taken to be 10.95 Å, the interplanar spacing of the (110) planes, and if 0.2, 0.5, and 1.0% are considered to be representative of the range of distortion magnitudes in Table 7-2, the resulting values of D' are 5×10^9, 3×10^{10}, and 1×10^{11} dislocations per square centimeter, respectively. These values may be characterized as slightly less than those that have been observed in highly strained metals [31].

7-3.4 General Conclusions

Buchanan and Miller [20] have demonstrated convincingly that a meaningful separation of the size and distortion broadening effects in the x-ray patterns of polymers is possible under favorable conditions. For size broadening two quite different parameters, weight- and number-average size, may be derived by the method of integral breadths and by the method of Fourier transforms. For distortion broadening a minimum of three orders of reflection is required to permit a choice to be made between the microstrain and paracrystalline theories of distortion for any given set of data. When distortion is the only cause of broadening, results that are different and also not necessarily compatible will issue from the different methods of analysis.

Since both the integral-breadth and the Fourier-transform methods require the same minimum number of orders of reflection and since the quality of the data required for the use of integral breadths is but little inferior to that demanded by the transform method, in the interests of obtaining the maximum amount of information it is recommended that the full Fourier-transform method be utilized whenever possible.

GENERAL REFERENCES

D. R. Buchanan and R. L. Miller, "X-Ray Line Broadening in Isotactic Polystyrene," *J. Appl. Phys.*, 37, 4003 (1966).
R. Hosemann and S. N. Bagchi, *Direct Analysis of Diffraction by Matter*, North Holland, Amsterdam, 1962.
B. E. Warren, "X-Ray Studies of Deformed Metals," Chapter 3 in *Progress in Metal Physics*, Vol. 8, Pergamon Press, New York, 1959.
B. K. Vainshtein, *Diffraction of X-Rays by Chain Molecules*, Elsevier, Amsterdam, New York, 1966.

SPECIFIC REFERENCES

[1] P. Scherrer, *Göttinger Nachrichten*, 2, 98 (1918).
[2] H. P. Klug and L. E. Alexander, *X-Ray Diffraction Procedures*, Wiley, New York, 1954, Chapter 9.
[3] R. Hosemann, Z. *Physik*, 128, 1, 464 (1950).
[4] R. Hosemann and S. N. Bagchi, *Direct Analysis of Diffraction by Matter*, North Holland, Amsterdam, 1962, pp. 239–246, 654 ff.
[5] P. Debye, *Verh. dtsch. phys. Ges.*, 15, 678, 857 (1913).
[6] I. Waller, Z. *Physik*, 17, 398 (1923); *Ann. Physik*, Lpz., 79, 261 (1926).
[7] R. Bonart, R. Hosemann, and R. L. McCullough, *Polymer*, 4, 199 (1963).
[8] F. Zernike and J. A. Prins, Z. *Physik*, 41, 184 (1927).
[9] O. Kratky, *Physik. Z.*, 34, 482 (1933).
[10] G. Porod, *Fortschr. Hochpolym. Forsch.*, 2, 363 (1961).
[11] J. J. Hermans, *Rec. trav. chim. Pays-Bas*, 63, 5 (1944).
[12] G. Porod, *Kolloid-Z.*, 125, 51 (1952).
[13] G. Porod, *Acta Phys. Austriaca*, 3, 66 (1949).
[14] R. Hosemann, *Acta Cryst.*, 4, 520 (1951).
[15] R. Bonart, Z. *Krist.*, 109, 296 (1957).
[16] R. Bonart and R. Hosemann, *Kolloid-Z.*, Z. *Polymere*, 186, 16 (1962).
[17] R. Hosemann and W. Wilke, *Faserforsch. Textiltechnik*, 15, 521 (1964).
[18] R. Glocker, *Materialprüfung mit Röntgenstrahlen*, Springer Verlag, Berlin, 1958, p. 191.
[19] R. Hosemann, H. Čačković, and W. Wilke, *Naturwiss.*, 54, 278 (1967).
[20] D. R. Buchanan and R. L. Miller, *J. Appl. Phys.*, 37, 4003 (1966).
[21] B. E. Warren and B. L. Averbach, *J. Appl. Phys.*, 21, 595 (1950).
[22] B. E. Warren, *Acta Cryst.*, 8, 483 (1955).

[23] E. F. Bertaut, *Compt. Rend.*, **228**, 492 (1949).

[24] A. J. C. Wilson, *X-Ray Optics*, Methuen, London, 1949, p. 5.

[25] D. R. Buchanan, R. L. McCullough, and R. L. Miller, *Acta Cryst.*, **20**, 922 (1966).

[26] G. Natta, P. Corradini, and I. W. Bassi, *Nuovo Cimento*, Suppl. No. 15, 68 (1960).

[27] F. W. Noble and J. E. Hayes, Jr., *Ann. N.Y. Acad. Sci.*, **115**, 644 (1964).

[28] A. R. Stokes, *Proc. Phys. Soc. (London)*, **A61**, 382 (1948).

[29] B. E. Warren, *Acta Met.*, **11**, 995 (1963).

[30] F. R. L. Schoening, *Acta Cryst.*, **18**, 975 (1965).

[31] C. Kittel, *Introduction to Solid State Physics*, Wiley, New York, 1957, pp. 554–557.

Fourier and Fourier-Bessel Transforms

In what follows the principal concepts and formulations that are potentially useful in the analysis of polymers have been collected and succinctly presented for reference purposes. Necessarily this has resulted in a considerable oversimplification of the subject, including a conspicuous lack of both mathematical rigor and comprehensiveness. For excellent and more thorough treatments the reader is referred to the literature[1–4]. In order to make the presentation as clear and systematic as possible corresponding quantities in direct and reciprocal (or vector) space are denoted respectively with unprimed and primed, but otherwise identical, symbols; for example, x and x', r and r', R and R', z and z'. A consequence of this method of simplification is that the nomenclature of Appendix 1 disagrees in some points with that customarily employed in the x-ray diffraction literature, including the main body of this monograph. These differences of notation are summarized in Table T-1.

If a one-dimensional function $f(x)$ in direct coordinates is defined by

$$f(x) = \int_{-\infty}^{+\infty} T(x') \exp\left(2\pi i x x'\right) dx', \qquad \text{(T-1)}$$

according to Fourier's integral theorem[3, 4] a corresponding function $T(x')$ in reciprocal coordinates is given by

$$T(x') = \int_{-\infty}^{+\infty} f(x) \exp\left(-2\pi i x x'\right) dx. \qquad \text{(T-2)}$$

The function $T(x')$ is said to be the transform of $f(x)$, and vice versa. Fourier transforms are of fundamental importance in x-ray diffraction

455

Table T-1 Differences in Notation

Quantity	Body of Monograph, Especially Chapter 1	Appendix 1
Reciprocal-lattice vector	$\boldsymbol{\rho}$ or $\boldsymbol{\rho}_{hkl}$	$\mathbf{r}'$
Magnitude of reciprocal-lattice vector, $(2 \sin \theta)/\lambda$	s	$\|\mathbf{r}'\|$
Axial reciprocal-lattice coordinate	ζ	z'
Radial reciprocal-lattice coordinate	ξ	R'

because the electron-density distribution and the diffraction amplitude that corresponds to it are related to each other as $f(x)$ and $T(x')$. A simple example of a one-dimensional Fourier transform is shown in Figure T-1. Let $f(x) = 1$ for $-\frac{1}{2}a < x < +\frac{1}{2}a$ and let $f(x) = 0$ for $-\frac{1}{2}a \geqslant x \geqslant +\frac{1}{2}a$. Then

$$T(x') = \int_{-\frac{1}{2}a}^{+\frac{1}{2}a} 1 \exp(-2\pi i x x') \, dx = \frac{a \sin \pi x' a}{\pi x' a}. \tag{T-3}$$

Convolution, or Fold. Another relationship of great value in diffraction problems is the convolution (German *Faltung*), or fold, of two functions, defined by

$$f(x)\widehat{\,}g(x) = h(x) = \int_{-\infty}^{+\infty} g(\eta) f(x-\eta) \, d\eta. \tag{T-4}$$

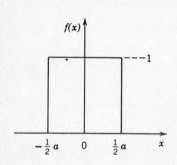
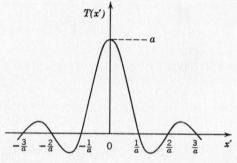

Figure T-1.

Let the Fourier transforms of $h(x)$, $f(x)$, and $g(x)$ be $H(x')$, $F(x')$, and $G(x')$, respectively. Then by application of Fourier's integral theorem it can be shown that

$$\int_{-\infty}^{+\infty} F(x')\,G(x')\exp(-ixx')\,dx' = \int_{-\infty}^{+\infty} g(\eta)\,f(x-\eta)\,d\eta$$

$$= \int_{-\infty}^{+\infty} H(x')\exp(-ixx')\,dx'. \qquad \text{(T-5)}$$

Evidently it also follows that

$$F(x')\,G(x') = H(x'). \qquad \text{(T-6)}$$

From (T-4), (T-5), and (T-6) we see that *convoluting* and *multiplying* are corresponding operations in direct and reciprocal space (or vice versa). Thus the convolution of two functions is equal to the transform of the product of their individual transforms.

Three-Dimensional Transforms. With reference to Figure T-2a, let a point in direct space with Cartesian coordinates x, y, z or cylindrical coordinates R, ψ, z be specified by a radius vector $\mathbf{r}$, and let equivalent primed coordinates denote a point in reciprocal space (Figure T-2b). Then the electron density at $\mathbf{r}$ and the transform at $\mathbf{r}'$ of the electron-density distribution may be written respectively

$$\rho(\mathbf{r}) = \int_{dr'} T(\mathbf{r}')\exp(-2\pi i\mathbf{r}\cdot\mathbf{r}')\,d\mathbf{r}' \qquad \text{(T-7)}$$

and

$$T(\mathbf{r}') = \int_{dr} \rho(\mathbf{r})\exp(2\pi i\mathbf{r}\cdot\mathbf{r}')\,d\mathbf{r}. \qquad \text{(T-8)}$$

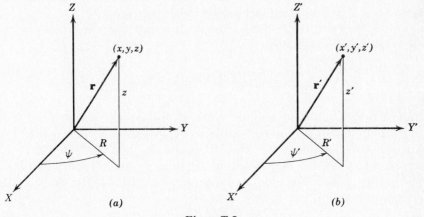

Figure T-2.

If the electron-density distribution is regarded as composed of discrete atoms with individual spherical scattering factors, $f_n(\mathbf{r}')$, (T-8) may be replaced by a summation over all the atoms in the system:

$$T(\mathbf{r}') = \sum_n f_n(r') \exp(2\pi i \mathbf{r}_n \cdot \mathbf{r}'). \tag{T-9}$$

If now the electron-density distribution has three-dimensional periodicity over large distances, as in ideal crystals, and if the summation in (T-9) extends over one motif, or unit cell, the continuous transform $T(\mathbf{r}')$ becomes a discontinuous function with finite values only at, or very close to, the points of a three-dimensional reciprocal lattice. If $\mathbf{a}'$, $\mathbf{b}'$, $\mathbf{c}'$ are the reciprocal-lattice translation vectors in the $\mathbf{x}'$, $\mathbf{y}'$, $\mathbf{z}'$ directions, respectively, and h, k, and l are integers,

$$\mathbf{r}_n \cdot \mathbf{r}' = (\mathbf{x}_n + \mathbf{y}_n + \mathbf{z}_n) \cdot (h\mathbf{a}' + k\mathbf{b}' + l\mathbf{c}').$$

On the continued simplifying assumption of orthogonal direct and reciprocal crystallographic axes, since the direct unit-cell translations are $a = 1/a'$, $b = 1/b'$, $c = 1/c'$, (T-9) becomes

$$T(\mathbf{r}') = F(hkl) = \sum_n f_n \exp\left[2\pi i\left(\frac{hx_n}{a} + \frac{ky_n}{b} + \frac{lz_n}{c}\right)\right]$$

$$= \sum_n f_n \exp\left[2\pi i(hx_n + ky_n + lz_n)\right], \tag{T-10}$$

in which the significance of x_n, y_n, z_n has been changed to denote fractions of the unit-cell edges in agreement with conventional x-ray crystallographic usage. [See (1-17), (1-18), and (1-27).] This is the structure factor of classical x-ray crystallography. A rigorous derivation of (T-10) can be given by way of the convolution of the unit-cell contents and a three-dimensional periodic delta function [1].

For a periodic three-dimensional structure the electron-density expression (T-7) becomes

$$\rho(x, y, z) = \frac{1}{V_c} \sum_h \sum_k \sum_l F(hkl) \exp\left[-2\pi i(hx + ky + lz)\right], \tag{T-11}$$

in which V_c is the volume of the unit cell. It should be pointed out that in (T-7) to (T-10) it is not possible to experimentally measure $T(\mathbf{r}')$ but only the product of $T(\mathbf{r}')$ and its complex conjugate, $T(\mathbf{r}') T^*(\mathbf{r}')$, which means that it is possible to measure its magnitude (modulus) but not its phase. The idealized intensity of scatter at the point $\mathbf{r}'$ in reciprocal space is

$$I(\mathbf{r}') = T(\mathbf{r}') T^*(\mathbf{r}') = |T(\mathbf{r}')|^2. \tag{T-12}$$

Patterson Function, $P(u)$ [5]. If we convolute the electron-density distribution $\rho(r)$ and the electron-density distribution inverted through the origin, $\rho^*(r)$, the result is

$$P(\mathbf{u}) = \widehat{\rho(\mathbf{r})\rho^*}(\mathbf{r}) = \frac{1}{V} \int T(\mathbf{r}')\, T^*(\mathbf{r}') \exp\left(-2\pi i \check{\mathbf{u}} \cdot \mathbf{r}'\right) d\tau_{\mathbf{r}'}$$

$$= \frac{1}{V} \int |T(\mathbf{r}')|^2 \exp\left(-2\pi i \mathbf{u} \cdot \mathbf{r}'\right) d\tau_{\mathbf{r}'}. \tag{T-13}$$

In this equation $\mathbf{u}$ is the vector distance to a point in Patterson, or vector, space; V is the total volume of the system; $d\tau_{\mathbf{r}'}$ is a reciprocal volume element at the point $\mathbf{r}'$ in reciprocal space; and the integral is to be taken over all of reciprocal (diffraction) space. Since the Patterson function is the transform of the intensity, it can be evaluated without a knowledge of the phases.

Spherically Symmetrical Systems. We here refer to gases, liquids amorphous solids, or randomly oriented microcrystallites, which of course encompass randomly oriented polymers. Debye [6] showed that the intensity of x-ray scatter by such a system of atoms as a function of S is given by

$$I(S) = \sum_m \sum_n f_m f_n \frac{\sin Sr_{mn}}{Sr_{mn}}, \tag{T-14}$$

where S is related to the scattering angle 2θ by $S = 4\pi (\sin \theta)/\lambda$, f_m and f_n are respectively the atomic scattering factors of the mth and nth atoms, and r_{mn} is the distance separating them. For a system of N atoms of one kind (T-14) becomes

$$I(S) = Nf^2 \sum_m \frac{\sin Sr}{Sr}. \tag{T-15}$$

Replacing the summation over all the interatomic vectors by an integral over the equivalent continuous distribution of atomic density, we obtain

$$\frac{I(S)}{Nf^2} - 1 = i(S) = \int_0^\infty 4\pi r^2 (\rho_r - \rho_0) \frac{\sin Sr}{Sr}\, dr, \tag{T-16}$$

which with the aid of the Fourier integral theorem can be converted to the sine transform,

$$4\pi r^2 (\rho_r - \rho_0) = \frac{2r}{\pi} \int_0^\infty S\, i(S) \sin rS \, dS. \tag{T-17}$$

In (T-16) and (T-17) ρ_r is the number of atoms per unit volume at a distance r from every other atom taken successively as the reference atom and ρ_0 is the average atomic density in the system.

Equations T-16 and T-17 can be modified so as to apply to compounds composed of several kinds of atoms by assuming that all the atomic scattering factors have the same dependence on S and differ only in magnitude. This approximation is reasonably good for carbon, nitrogen, and oxygen—and therefore appropriate for many polymers if hydrogen scattering is neglected. Then, if f_e is the mean scattering factor per electron, the scattering factor of the mth kind of atom may be written as

$$f_m = K_m f_e, \tag{T-18}$$

in which K_m is the effective number of electrons per atom of type m.

Now let $m, n, \cdots$ denote the kinds of atoms that comprise some representative unit of structure, such as a monomeric formula unit, and let N be the number of such units. Equation T-15 may be expressed as

$$I = N \sum_p f_p^2 + \sum_m \sum_n^{m \neq n} f_m f_n \frac{\sin Sr_{mn}}{Sr_{mn}}, \tag{T-19}$$

in which the m summation embraces all the atoms in one unit and the n summation includes every pair of atoms in the specimen regardless of which units they belong to. If a weighted atomic density function, $\rho_m(r)$, is defined by

$$4\pi r^2 \rho_m(r) \, dr = \sum_m a_m f_m, \tag{T-20}$$

with the summation taken over all the atoms in the unit of structure as defined above, (T-19) may be converted to

$$I = N\left[\sum_m f_m^2 + \sum_m f_m \int_0^\infty 4\pi r^2 \rho_m(r) \frac{\sin Sr}{Sr} \, dr \right]. \tag{T-21}$$

With the approximation of (T-18) $\rho_m(r)$ of (T-20) can be expressed as a function of f_e in the form

$$\rho_m(r) = f_e \, g_m(r), \tag{T-22}$$

in which $g_m(r)$ is the atomic density function appropriate to the new treatment. Substitution of (T-18) and (T-22) into (T-21) gives

$$I = N\left[\sum_m f_m^2 + 4\pi f_e^2 \int_0^\infty \left\{ \sum K_m g_m(r) \right\} r^2 \frac{\sin Sr}{Sr} \, dr \right], \tag{T-23}$$

and a transformation parallel to that employed in obtaining (T-16) yields

$$\frac{I}{N} - \sum_m f_m^2 = 4\pi f_e^2 \int_0^\infty \sum_m K_m \{g_m(r) - g_0\} r^2 \frac{\sin Sr}{Sr} dr. \tag{T-24}$$

The Fourier transform of (T-24) is then

$$4\pi r^2 \left\{ \sum_m K_m g_m(r) - g_0 \sum_m K_m \right\} = \frac{2r}{\pi} \int_0^\infty S \, i(S) \sin rS \, dS, \tag{T-25}$$

in which $i(S)$ now has the significance

$$i(S) = \frac{1}{f_e^2} \left(\frac{I}{N} - \sum_m f_m^2 \right) = \sum_m K_m^2 \left(\frac{I}{N \sum_m f_m^2} - 1 \right). \tag{T-26}$$

Cylindrically Symmetrical Systems. With reference again to Figure T-2, equations (T-7) and (T-8) expressed in cylindrical coordinates in direct and reciprocal space have the respective forms [9]

$$\rho(R, \psi, z) = \int_0^\infty \int_0^{2\pi} \int_{-\infty}^\infty T(R', \psi', z') \exp\{-2\pi i [RR' \cos(\psi - \psi') + zz']\}$$
$$\times R' \, dR' \, d\psi' \, dz' \quad \text{(T-27)}$$

and

$$T(R', \psi', z') = \int_0^\infty \int_0^{2\pi} \int_{-\infty}^\infty \rho(R, \psi, z) \exp\{2\pi i [RR' \cos(\psi - \psi') + zz']\}$$
$$\times R \, dR \, d\psi \, dz. \quad \text{(T-28)}$$

If $\rho(r)$ has cylindrical symmetry the Patterson function (T-13) assumes the form [2, 10–12]

$$P(R, z) = 4\pi \int_0^\infty \int_0^\infty R' |T(R', z')|^2 J_0(2\pi RR') \cos 2\pi zz' \, dR' \, dz', \tag{T-29}$$

in which $|T(R', z')|^2$ represents the intensity $I(R', z')$ at R', z' in reciprocal space and J_0 is a Bessel function of zero order. If $\rho(R, z)$ is periodic in z with interval c and if the number of periods, N_1, is large, (T-29) becomes [13]

$$P(R, z) = \frac{2\pi N_1}{c} \sum_{-\infty \atop l}^{+\infty} \cos 2\pi l \frac{z}{c} \int_0^\infty R' \left| T\left(R', \frac{l}{c}\right) \right|^2 J_0(2\pi RR') \, dR', \tag{T-30}$$

where l is an integer.

The cylindrical problem reduces to a one-dimensional radial case for an assemblage of parallel cylindrical (rodlike) scattering elements of infinite length that are assumed to be structureless along their lengths (z-dimension). If all rotational orientations of the system about the Z-axis are equally probable and if we consider only diffraction in the (equatorial) plane perpendicular to the axes of the rods, the intensity distribution $I(S)$ can be expressed in terms of the equatorial scattering factors F_m and F_n of the mth and nth cylinders and of the distance r_{mn} separating their centers. The result is[14]

$$I(S) = \sum_m \sum_n F_m F_n J_0(S r_{mn}), \qquad (\text{T-31})$$

which is the radial analog of Debye's spherical function applied to this cylindrical case. If we assume that the system consists of N identical cylinders, replace the summations by an integral, and proceed as in the spherical case, we obtain

$$\frac{I(S)}{NF^2} - 1 = i(S) = \int_0^\infty 2\pi r (\rho_r - \rho_0) J_0(Sr) \, dr. \qquad (\text{T-32})$$

As Oster and Riley[14] have shown, the transform of (T-32) can now be obtained with the Fourier-Bessel, or Hankel, inversion theorem [15], with the result

$$2\pi (\rho_r - \rho_0) = \int_0^\infty S \, i(S) J_0(rS) \, ds. \qquad (\text{T-33})$$

In (T-32) and (T-33) ρ_0 is the mean number of cylinder centers per unit area in the equatorial plane and ρ_r is the number per unit area at a distance r from every other cylinder center taken successively as origin. The form of the radial cylindrical scattering factor of a single cylinder, F, which is contained in $i(S)$ of (T-33), is a function of its radial structure. For a solid cylinder of uniform density and radius R it is[14]

$$F(SR) = \frac{2J_1(SR)}{SR}. \qquad (\text{T-34})$$

Helical Structures[16, 17].   As shown in Figure T-3, let Z be the axis of the helix, R its radius, and P its axial translation interval per turn (pitch). If the helix is continuous and of infinite length, its Fourier transform is[16]

$$T\left(R', \psi', \frac{n}{P}\right) = J_n(2\pi RR') \exp\left[in(\psi' + \tfrac{1}{2}\pi)\right], \qquad (\text{T-35})$$

where n is an integer and J_n is a Bessel function of the nth order. The transform has finite values only on layer lines at height n/P in reciprocal space. Figure T-4 shows how the amplitude of the Bessel function $J_n(x)$ diminishes rapidly for small or moderate values of the argument x as n increases.

In actual structures the helices are discontinuous with like atoms situated regularly along the helix and separated by a Δz component of p. For a single helix the transform will then have nonzero values only in planes at heights

$$z = \frac{n}{P} + \frac{m}{p} = \frac{l}{c} \qquad \text{(T-36)}$$

in reciprocal space. Here l is the number of the layer line and c is the structural translation period in the z-direction. If P/p can be expressed as a rational number, discrete layer lines are observed at heights l/c and only special values of n are permitted by the *selection rule* of (T-36). For a single discontinuous helix (T-35) becomes

$$F\left(R', \psi', \frac{l}{c}\right) = \sum_n T\left(R', \psi', \frac{n}{P}\right) = \sum_n J_n(2\pi RR') \exp\left[in(\psi' + \tfrac{1}{2}\pi)\right].$$
$$\text{(T-37)}$$

The summation is taken over all values of n permitted by the selection rule.

The foregoing mathematical model does not take into account (a) the presence of more than one kind of atom, (b) the presence of con-

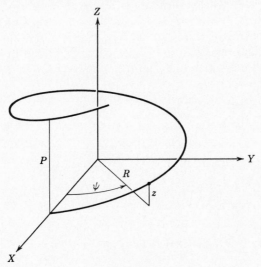

Figure T-3.

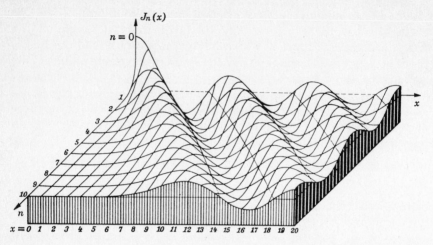

Figure T-4.   E. Jahnke and F. Emde, *Tables of Functions with Formulae and Curves,*
Dover, New York, 1945; reprinted with permission of the publisher.

centric helices of different radii, (c) the scattering factors of the
different kinds of atoms, and (d) the cylindrical randomness of
orientation presumed to exist in most actual fibers. Nevertheless,
the comparison of intensity distributions given by (T-37) with experimental data sometimes suffices to effect a choice between alternative
helical models. The complete helical transform, or structure factor,
can be expressed by extending (T-37) so as to allow for the four
additional factors just cited. Let f_j be the scattering factor of the jth
atom, R_j the radius of the helix on which it is situated, and ψ_j and z_j
the rotational and translational phase components of this atom. Then
the complete helical-structure factor becomes [16]

$$F_c\!\left(R', \psi', \frac{l}{c}\right) = \sum_n \sum_j f_j J_n(2\pi R' R_j)\ \exp\left\{i\left[n(\psi - \psi_j + \tfrac{1}{2}\pi) + \frac{2\pi l z_j}{c}\right]\right\}.$$
$$(T\text{-}38)$$

If rotational symmetry about the Z-axis is postulated, the square of
the F-modulus, or intensity, may be written as [18, 19]

$$F_{\psi'}^2\!\left(R', \frac{l}{c}\right) = \sum_n (A_n^2 + B_n^2),$$
$$(T\text{-}39)$$

where

$$A_n = \sum_j f_j J_n(2\pi R' R_j) \cos\left[n(\tfrac{1}{2}\pi - \psi_j) + \frac{2\pi l z_j}{c}\right]$$
$$(T\text{-}40)$$

and

$$B_n = \sum_j f_j J_n(2\pi R'R_j) \sin\left[n(\tfrac{1}{2}\pi - \psi_j) + \frac{2\pi l z_j}{c} \right]. \qquad \text{(T-41)}$$

In conclusion it must be emphasized that (T-38) gives the idealized *amplitude* of *scattering* by one helical molecule, or, by what is equivalent, a collection of identical helices, identically oriented, scattering independently. If now crystalline ordering of the assemblage occurs, the resulting intermolecular interferences result in a modulation, or decomposition, of the continuous layer-line intensity profiles into individual crystalline reflections. The intensities of these individual reflections can then be specified by the general structure-factor expression (T-10), and, furthermore, they will be found to coincide approximately with the envelope defined by (T-38) or (T-39) for a given layer line diffracted by one component helix. It may be said that the reflections comprising a particular layer line of a crystalline assemblage of helices represent samplings of the layer-line transform of the isolated helix.

Optical Transforms[1, 20, 21]. It can be shown that (T-9), which gives the total x-ray scattering from a system of n atoms, is perfectly general and applies to radiations other than x-rays. Thus it can be interpreted to give the distribution of scattering of light by a mask containing a particular arrangement of holes, which therefore means that $T(\mathbf{r}')$ is the Fourier transform of that assemblage of holes. It follows that a suitable analog apparatus, or optical diffractometer, will within certain limitations circumvent the complex and sometimes tedious mathematical procedures that have been described above for obtaining the transforms of proposed structure models. Clearly the optical diffractometer will be of prime importance in the early stages of a structural investigation, permitting the relationships between molecular models and their transforms to be studied rapidly by direct visual operation. It can also fairly be said that optical diffractometry complements computer methods since it can lead quickly to approximate structural solutions which it may be necessary to confirm subsequently by detailed numerical calculations.

For structure models characterized by three-dimensional (crystalline) order it is the usual practice to prepare optical transforms of the contents of one unit cell for comparison with the x-ray diffraction pattern—whereas for less orderly aggregates of atoms, such as polymers, it is necessary to include a considerably larger portion of the structure in order to achieve a representative sample. We may note that in interpreting (T-9) as an optical transform, $\mathbf{r}_n$ and $\mathbf{r}'$ will normally be confined to planes normal to the incident beam in direct and reciprocal space, respectively, which means that we must evaluate a *two-dimensional transform* of a *two-dimensional projection* of the actual *three-dimensional molecular model*. The optical transform is thus the equatorial section normal to the incident beam of the complete *three-dimensional transform* that in principle could have been obtained from the actual molecular model. This two-dimensional limitation that characterizes the usual mode of optical-transform analysis is not a significant handicap in the investigation of most polymer problems, in particular fiber structures, which because of rotational symmetry are themselves two-dimensional and definable in terms of coordinates R and z in direct space and R' and z' in transform space.

REFERENCES

[1] H. Lipson and C. A. Taylor, *Fourier Transforms and X-Ray Diffraction*, G. Bell, London, 1958.

[2] D. Wrinch, *Fourier Transforms and Structure Factors*, American Society for X-Ray and Electron Diffraction, Monograph No. 2, 1946.

[3] I. N. Sneddon, *Fourier Transforms*, McGraw-Hill, New York, 1954.

[4] E. C. Titchmarsh, *Introduction to the Theory of Fourier Integrals*, Oxford: Clarendon Press, 1948.

[5] A. L. Patterson, *Phys. Rev.*, **46**, 372 (1934).

[6] P. Debye, *Ann. Physik*, **46**, 809 (1915).

[7] F. Zernike and J. A. Prins, *Z. Physik*, **41**, 184 (1927).

[8] B. E. Warren and N. S. Gingrich, *Phys. Rev.*, **46**, 368 (1934).

[9] A. M. Cormack, *Acta Cryst.*, **10**, 354 (1957).

[10] G. H. Vineyard, *Acta Cryst.*, **4**, 281 (1951).

[11] N. Norman, Ph.D. Thesis, *On the Cylindrically Symmetrical Distribution Method in X-Ray Analysis*, University of Oslo, 1954.

[12] N. Norman, *Acta Cryst.*, **7**, 462 (1954).

[13] C. H. MacGillavry and E. M. Bruins, *Acta Cryst.*, **1**, 156 (1948).

[14] G. Oster and D. P. Riley, *Acta Cryst.*, **5**, 272 (1952).

[15] R. N. Watson, *Theory of Bessel Functions*, 4th ed., Cambridge University Press, 1948.

[16] W. Cochran, F. H. C. Crick, and V. Vand, *Acta Cryst.*, **5**, 581 (1952).

[17] A. Klug, F. H. C. Crick, and H. W. Wyckoff, *Acta Cryst.*, **11**, 199 (1958).

[18] R. E. Franklin and A. Klug, *Acta Cryst.*, **8**, 777 (1955).

[19] D. R. Davies and A. Rich, *Acta Cryst.*, **12**, 97 (1959).

[20] C. A. Taylor and H. Lipson, *Optical Transforms*, Cornell University Press, Ithaca, N.Y., 1965. (Also published by G. Bell, London, 1964.)

[21] C. A. Taylor, *J. Polymer Sci.*, Part C, **20**, 19 (1967).

Fourier Analysis of Reflection Profiles by the Method of Stokes[1,2]

In precise studies of the shapes and breadths of x-ray reflections it is necessary to know the pure diffraction profile free of instrumental effects. As experimentally observed the reflection profile $h(\epsilon)$ is the convolution (fold) of the instrumental profile $g(\epsilon)$ and the pure diffraction profile $f(\epsilon)$:

$$h(\epsilon) = \int_{-\infty}^{+\infty} \big| g(\eta) f(\epsilon - \eta) \, d\eta. \qquad (\text{S-1})$$

The variable ϵ and its auxiliary η denote any appropriate angular scale on which the reflection may be portrayed, such as 2θ or s. In the deconvolution method of Stokes[1] the pure diffraction profile f is calculated from the Fourier transforms of the observable functions h and g.

Let the Fourier transforms of h, g, and f be H, G, and F, respectively.
According to (T-6) these transforms are related by

$$F = \frac{H}{G}. \qquad (\text{S-2})$$

By Fourier's integral theorem[3]

$$f(\epsilon) = \frac{1}{\sqrt{2\pi}} \int_{-\infty}^{+\infty} F(\zeta) \exp\left(-2\pi i \epsilon \zeta\right) d\zeta, \qquad (\text{S-3})$$

and therefore

$$f(\epsilon) = \frac{1}{\sqrt{2\pi}} \int_{-\infty}^{+\infty} \frac{H(\zeta)}{G(\zeta)} \exp\left(-2\pi i \epsilon \zeta\right) d\zeta. \qquad (\text{S-4})$$

By employing the exponential form of the Fourier integral, generality is preserved, permitting the integrals to be applied to asymmetrical as well as symmetrical functions.

According to Stokes' method, the integrals are replaced by the corresponding summations and the limits of ϵ are changed from $\pm\infty$ to $\pm\frac{1}{2}\epsilon_m$, the minimum values of ϵ beyond which the intensity can be considered to have fallen to its background value. Equation S-4 may then be written as

$$f(\epsilon) = \frac{1}{\sqrt{2\pi}} \sum_{\zeta} \frac{H(\zeta)}{\epsilon_m\, G(\zeta)} \exp\left(\frac{-2\pi i \epsilon \zeta}{\epsilon_m}\right) \Delta\zeta. \tag{S-5}$$

In practice the constant factors $1/\sqrt{2\pi}$, $1/\epsilon_m$, and $\Delta\zeta$ can be eliminated from the computations since they affect only the height of a line profile and not its shape. This gives

$$f(\epsilon) = \sum_{\zeta} \frac{H(\zeta)}{G(\zeta)} \exp\left(\frac{-2\pi i \epsilon \zeta}{\epsilon_m}\right), \tag{S-6}$$

in which

$$H(\zeta) = \sum_{\epsilon} h(\epsilon)\, \exp\left(\frac{2\pi i \epsilon \zeta}{\epsilon_m}\right) \tag{S-7}$$

and

$$G(\zeta) = \sum_{\epsilon} g(\epsilon)\, \exp\left(\frac{2\pi i \epsilon \zeta}{\epsilon_m}\right). \tag{S-8}$$

Now, since

$$e^{iz} = \cos z + i \sin z, \tag{S-9}$$

it is evident that each of the series above consists of real and imaginary parts; for example,

$$G(\zeta) = \sum_{\epsilon} g(\epsilon) \cos \frac{2\pi\epsilon\zeta}{\epsilon_m} + i \sum_{\epsilon} g(\epsilon) \sin \frac{2\pi\epsilon\zeta}{\epsilon_m}, \tag{S-10}$$

which may be represented by

$$G(\zeta) = G_r(\zeta) + i\, G_i(\zeta). \tag{S-11}$$

Similarly, $H(\zeta)$ may be written

$$H(\zeta) = H_r(\zeta) + i\, H_i(\zeta). \tag{S-12}$$

From (S-11) and (S-12)

$$F(\zeta) = \frac{H(\zeta)}{G(\zeta)} = \frac{H_r + iH_i}{G_r + iG_i} = \frac{(H_r + iH_i)(G_r - iG_i)}{G_r^2 + G_i^2}.$$

Evidently then F has the real and imaginary components

$$F_r = \frac{H_r G_r + H_i G_i}{G_r^2 + G_i^2}$$ (S-13)

and

$$F_i = \frac{H_i G_r - H_r G_i}{G_r^2 + G_i^2}.$$ (S-14)

These equations provide a basis for the numerical calculation of F_r and F_i from the Fourier series G_r, G_i, H_r, and H_i, which can be evaluated from experimental data.

To return to (S-6), the desired pure diffraction profile $f(\epsilon)$ can be computed as follows:

$$f(\epsilon) = \sum_{\zeta} F(\zeta) \exp\left(\frac{-2\pi i \epsilon \zeta}{\epsilon_m}\right)$$

$$= \sum_{\zeta} \left[F_r(\zeta) + i F_i(\zeta)\right]\left(\cos\frac{2\pi\epsilon\zeta}{\epsilon_m} - i \sin\frac{2\pi\epsilon\zeta}{\epsilon_m}\right)$$

$$= \underbrace{\sum_{\zeta} F_r(\zeta) \cos\frac{2\pi\epsilon\zeta}{\epsilon_m}}_{(A)} + \underbrace{\sum_{\zeta} F_i(\zeta) \sin\frac{2\pi\epsilon\zeta}{\epsilon_m}}_{(B)}$$

$$\underbrace{- i \sum_{\zeta} F_r(\zeta) \sin\frac{2\pi\epsilon\zeta}{\epsilon_m}}_{(C)} + \underbrace{i \sum_{\zeta} F_i(\zeta) \cos\frac{2\pi\epsilon\zeta}{\epsilon_m}}_{(D)}.$$ (S-15)

But $F(-\zeta)$ is the complex conjugate of $F(\zeta)$ [see (S-2)], and therefore $F_r(-\zeta) = F_r(\zeta)$ and $F_i(-\zeta) = -F_i(\zeta)$. Since the summations extend over all positive and negative values of ζ, it follows that terms (C) and (D) in (S-15) reduce to zero, and $f(\epsilon)$ can be calculated from the equation

$$f(\epsilon) = \sum_{\zeta} F_r(\zeta) \cos\frac{2\pi\epsilon\zeta}{\epsilon_m} + \sum_{\zeta} F_i(\zeta) \sin\frac{2\pi\epsilon\zeta}{\epsilon_m}.$$ (S-16)

Obviously the application of Stokes' method is practicable only if equipment such as a digital computer is available for the rapid summation of Fourier series. This involves the measurement of the diffracted intensity over the experimental profiles at small intervals. It is convenient to divide the angular range ϵ_m within which a line has measurable intensity into 60 parts and to select an artificial unit of angle so small that $\epsilon_m = 60$ units and so that intensity measurements are made at intervals $\Delta\epsilon$ equal to unity. These experimental intensity data then become the Fourier coefficients for the series H_r, H_i, G_r, G_i, which assume the forms

$$H_r(\zeta) = \frac{1}{60} \sum_{-30}^{+30} h(\epsilon) \cos \frac{2\pi\epsilon\zeta}{60},$$

$$H_i(\zeta) = \frac{1}{60} \sum_{-30}^{+30} h(\epsilon) \sin \frac{2\pi\epsilon\zeta}{60},$$

$$G_r(\zeta) = \frac{1}{60} \sum_{-30}^{+30} g(\epsilon) \cos \frac{2\pi\epsilon\zeta}{60}, \quad \text{(S-17)}$$

$$G_i(\zeta) = \frac{1}{60} \sum_{-30}^{+30} g(\epsilon) \sin \frac{2\pi\epsilon\zeta}{60}.$$

It should be kept in mind that $h(\epsilon)$ refers to an experimental intensity profile broadened as a result of small crystallite size, or lattice distortions, or both, and $g(\epsilon)$ refers to the profile of a reference specimen that is free of size and distortion broadening. The values of H_r, H_i, G_r, and G_i computed with (S-17) over the required range of ζ values are next used to compute the corresponding values of $F_r(\zeta)$ and $F_i(\zeta)$ with (S-13) and (S-14). These F_r and F_i values then become the Fourier coefficients for evaluating the pure diffraction profile, $f(\epsilon)$, which can be expressed as a Fourier series in the "working" form [see (S-16)]:

$$f(\epsilon) = \frac{1}{60} \sum_{\zeta} F_r(\zeta) \cos \frac{2\pi\epsilon\zeta}{60} + \frac{1}{60} \sum_{\zeta} F_i(\zeta) \sin \frac{2\pi\epsilon\zeta}{60}. \quad \text{(S-18)}$$

Figure S-1 shows illustrative h, g, and f profiles from such a Fourier analysis of a diffraction line of cold-worked copper as reported by Stokes[1]. The broadened profile h of cold-worked copper filings is "unfolded" to give the pure profile f by reference to the corresponding line profile g of well-annealed copper. These results illustrate a valuable characteristic of the Fourier method—namely, that any degree of doubling or broadening of a line due to the $K\alpha_1 K\alpha_2$ doublet is automatically allowed for; the deduced profile f will be of the nature expected for the normal wavelength distribution prevailing in either the $K\alpha_1$ or $K\alpha_2$ radiation.

One condition for the success of such a Fourier analysis is that the measurement range $\pm\frac{1}{2}\epsilon_m$ be divided into a sufficiently large number of intervals; otherwise the summations (S-17) will not be accurate representations of the components of the Fourier coefficients F_r

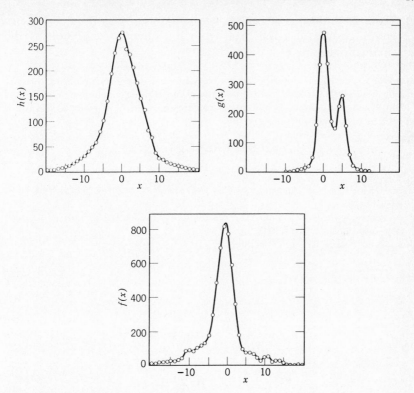

Figure S-1   Line profiles h, g, and f involved in a Fourier analysis of line shapes for a cold-worked copper specimen. (Stokes [1].)

and F_i required in the evaluation of $f(\epsilon)$. [See (S-18).] This condition is met if $H(\zeta)$ and $G(\zeta)$ fall to approximately zero for values of ζ not larger than the range $\pm\frac{1}{2}\epsilon_m$. If this is found not to be the case, it is necessary to use a larger number of divisions. In regard to the effect of experimental errors, Stokes points out as the result of a special analysis that $f(\epsilon)$ is deducible most accurately when the standardizing lines [$g(\epsilon)$ profiles] are as narrow as possible with respect to the broadened lines [$h(\epsilon)$ profiles] of the substance being studied. This result is not different from the familiar commentary on crystallite-size measurements from line broadening that the accuracy diminishes as the size increases, which is equivalent to saying as the width of the broadened line approaches that of the reference line. This observation points up the desirability of employing experimental techniques that contribute the smallest possible instrumental broadening. Despite its limitations the Fourier method is the most powerful technique currently available for deducing the pure diffraction profile of a line free from the effects introduced by the measuring apparatus.†

†Recently Ergun[4] has described a very effective procedure for unfolding the convolution product that is based on substitution of successive convolutions. The method is ideally suited to computers and gives accurate solutions if the function desired is integrable and everywhere differentiable.

REFERENCES

[1] A. R. Stokes, *Proc. Phys. Soc. (London)*, **A61**, 382 (1948).
[2] H. P. Klug and L. E. Alexander, *X-Ray Diffraction Procedures*, Wiley, New York, 1954, pp. 495–500.
[3] See, for example, E. C. Titchmarsh, *Introduction to the Theory of Fourier Integrals*, Clarendon Press, Oxford, 1948.
[4] S. Ergun, *J. Appl. Cryst.*, **1**, 19 (1968).

Crystallographic Data for Various Polymers[†]

ROBERT L. MILLER

Chemstrand Research Center
Durham, North Carolina

Contents

		Page
A.	Introduction	473
B.	Table of Crystallographic Data for Various Polymers	476
	1 – Poly-olefins	476
	2 – Poly-vinyls and Poly-vinylidenes	482
	3 – Poly-aromatics	485
	4 – Poly-dienes	487
	5 – Poly-amides	491
	6 – Poly-esters	495
	7 – Poly-urethanes and Poly-ureas	500
	8 – Poly-ethers	501
	9 – Poly-oxides	503
	10 – Poly-sulfides and Poly-sulfones	506
	11 – Cellulosics	508
	12 – Other Polymers	509
C.	References	510

A. INTRODUCTION

The following table presents crystallographic data for about 270 polymers. They have been grouped, according to the generic structure of the chain repeat unit, into poly-olefins, poly-vinyls and poly-

[†]Reprinted with revision by special permission from *Polymer Handbook*, J. Brandrup and E. H. Immergut (editors), Wiley-Interscience, New York, 1966.

vinylidenes, poly-aromatics, poly-dienes, poly-amides, poly-esters, poly-ethers, poly-oxides, poly-sulfides and poly-sulfones, poly-urethanes and poly-ureas, cellulosics, and others. Where a polymer might be included in more than one group, the following group definitions were used:

1. Poly-olefins: Those olefin polymers not containing aromatic rings; all substituted derivatives of these, excluding the vinyl and vinylidene polymers. For example, polyethylene, poly-tetrafluoro-ethylene, and polycyclopentene, but not polystyrene, or poly(vinyl chloride).

2. Poly-vinyls and Poly-vinylidenes: Those vinyl and vinylidene polymers containing atoms other than carbon and hydrogen, ex-cepting the polyethers. Thus, poly(vinyl chloride), but not poly-(methyl vinyl ether).

3. Poly-aromatics: Those hydrocarbon polymers containing aromatic rings. Thus, polystyrene and poly(p-xylylene).

4. The remaining specific categories contain polymers according to functional groups, as, for example, the poly-amides and the poly-esters. Poly-ethers and poly-oxides are differentiated according to whether the ether linkage is in the side-group or in the backbone, respectively. For example, poly(1-butoxy-2-chloroethylene) is a poly-ether and polyformaldehyde is a poly-oxide.

5. Those polymers not otherwise categorized, such as poly(phos-phonitrite chloride), are listed in the "other" category.

Within each category, polymers are listed alphabetically according to the basic structure ignoring substituents. Substituted polymers are listed alphabetically according to the substituent under the entry for the unsubstituted polymer. Thus, poly(tetrafluoroethylene) and poly(4-methyl-1-pentene) appear, respectively, under polyethylene and poly(1-pentene).

Included as part of the polymer name is the molecular weight of the chemical repeat unit in the chain. This appears below the name and is bracketed by asterisks, *. Unless otherwise indicated, the reference cited in the second column applies to the entire line of the table. Where an entry has been taken from a source other than that listed in column 2, a slash (/) separates the value of the entry from the reference. For example, one value of the heat of fusion of poly-ethylene is listed as 1.88/85 which is to be read as 1.88 kcal/mole according to reference 85. The crystal system, where known, is given according to the abbreviations given below. Following this is the space group symbol (the Schoenflies notation is used because of the limitations of electronic data processing symbolisms). In the space group indication, the subscript appears before the superscript. Thus, according to reference 14, the polyethylene crystal is orthorhombic (ORTHO) with space group D_{2h}^{16} ($D2H$-16). For a fuller discussion of space group symbols, see *International Tables for X-Ray Crystallography*, Vol. I, Kynoch Press, Birmingham, England, 1952.

The dimensions of the unit cell (a, b, and c) are given in angstroms. Unless otherwise indicated (by an *) c is the fiber axis. Unless otherwise noted, angles are 90°. Where required, unit cell angles are given in the order: α, β, and γ. Where only one angle is needed it is identified according to the abbreviations given below. The next entry is the number of chemical repeat units ("monomers") in the unit cell of dimensions listed. The densities of the crystal and of the completely amorphous polymer are those appropriate (normally) for room temperature. For uniformity, crystallographic densities have been recalculated. In certain cases the calculated density disagreed with that in the citation. Both are therefore given, with the recalculated value placed within parentheses. The melting point is in °C and the heat of fusion, in kcal/mole, corresponds to the unit of molecular weight listed under the polymer name.

The last column of the table indicates the conformation of the polymer chain in the crystal. The notation, $n * p$-q, specifies the number (n) of skeletal atoms in the asymmetric unit of the chain and the number of such units (p) per q turns of the helix in the crystallographic repeat. Thus, polyethylene has two carbon atoms in the backbone with one such unit per turn in the repeat ($2 * 1$-1). Note that polyethylene considered to be polymethylene would be designated $1 * 2$-1, an entirely equivalent description of the conformation. Isotactic polypropene, then, has two atoms in the backbone unit and three units per turn ($2 * 3$-1). On the other hand, syndiotactic polypropene has four atoms in the backbone unit (two chemical monomers) and two such units per turn ($4 * 2$-1). A fuller discussion of this notation is given in Hughes and Lauer, *J. Chem. Phys.*, **30**, 1165 (1959) and Nagai and Kobayashi, *J. Chem. Phys.*, **36**, 1268 (1961).

This list of polymers is not considered to be exhaustive. The compiler expands it as rapidly as new information is unearthed. All of the data in this table cannot be considered to have the same validity — the number of polymers for which detailed crystal structure analyses have been conducted is quite limited.

Abbreviations used in this table: TRI — triclinic, MONO — monoclinic, ORTHO — orthorhombic, TET — tetragonal, RHO — rhombohedral, HEX — hexagonal, P — pseudo, A — α, B — β, and G — γ.

B. TABLE OF CRYSTALLOGRAPHIC DATA FOR VARIOUS POLYMERS

POLYMER	REF	CRYST SYST.	SPACE GROUP	UNIT CELL PARAMETERS A	B	C	ANGLES	MON/ UNIT CELL	DENSITY (G./CC.) CRYSTAL	AMORPH.	MELT. POINT	HEAT OF FUSION KCAL/MON	CHAIN CONF.
POLY-OLEFINS -------- *HCH-CHR*													
POLY-ACETYLENE *26.04*	69	PHEX		4.2	4.2	2.43		1	1.16				2*1*1
POLY-ALLENE *40.06*	1 428	ORTHO	D2H-6	8.20	7.81	3.88		4	1.071				2*2*1
	II 497	MONO	C2-2 OR C2H-6	6.37	3.88*	5.12	B=96.6	2	1.058				2*2*1
	III 497					3.88							2*2-1
---, TETRAFLUORO* *112.03*	38	TET	C4-2 OR D4-3	6.88	6.88	15.4		8	2.041		126		2*4-1
POLY-1-BUTENE *56.10*	I. 35 252 391	RHO	D3D-6	17.7	17.7	6.50		18	0.951 0.950 0.95/12	0.87/45 0.868 0.860	126/12 132/250 136/277 135/345 142/313 135/320 138/380	3.33/82 1.45/277 1.71/345 1.68/320	2*3*1
	II. 207 350 442	TET TET	S4-1	14.89 14.85	14.89 14.85	20.87 20.6 76		44 44	0.866 0.902		122/345 126/313 124/277 130/380	1.50/277 0.97/320	2*11-3 2*11-3 2*40- 11
	III. 207 461	ORTHO		12.49	8.96	7.6		8	(0.876)		106/277	1.55/277	
**, 5-AMINO= *, N,N-DIISO- BUTYL- *197.35*	191					6.85					112		
---, **, N,N-DIISO- PROPYL- *169.30*	191					7.32					130		
***, 5-HYDROXY- *86.13*						6.5							

POLYMER	REF	CRYST SYST.	SPACE GROUP	CELL A	CELL B	CELL C	PARAMETERS ANGLES	MON/UNIT CELL	DENSITY (G./CC.) CRYSTAL	DENSITY AMORPH.	MELT. POINT	HEAT OF FUSION KCAL/MON	CHAIN CONF.
---, 3-METHYL- *70.13*	9	MONO		9.55	8.54	6.84	G=116.5	4	0.933		300/48	4.13/82	2*4*1
	231	MONO		19.1	17.8	6.85	G=116	16	0.890		310/90		2*4*1
	355	PORT		19.25	17.20	6.85	G=116.5	16	0.918		300/282		2*4*1
	431	TET	C4v-12	34.3	34.3			64	0.925				
	280	MONO		19.25	17.20	6.63	G=116.5	16	0.948				
	498	MONO		19.14	17.80	6.6	G=116.5	16	0.93				
---, 3-METHYL- (VIA HYDRIDE SHIFT) *70.13*	332			5.4		7.8					55 66/319		
---, 4-PHENYL- *132.20*	483	PORT	CS=2	10.4	18.0	6.61		6	1.064	0.962	158 160/90 168/187 159/282	1.1	2*3*1
	67					6.55							2*3*1
POLY-CYCLODECENE TRANS- III. 477 *138.24*	III. 477	TRI	CI=1	4.40	5.39	12.30	6610.4118	1	0.985				10*1*1
	IV. 228	MONO	C2H=5	7.42	5.00	12.41	B=85.8	2	1.000				10*1*1
POLY-CYCLODDECENE TRANS- III. 477 *166.30*	III. 477	TRI	CI=1	4.40	5.39	14.78	6610.4118	1	0.986				12*1*1
	IV. 126	MONO	C2H=5	7.43	5.00	14.85	B=86.5	2	1.003		80/439		12*1*1
POLY-CYCLOHEPTENE TRANS- 1. 97 *96.17*	1. 97	ORTHO	D2H=16	7.40	5.00	17.10		4	1.010		51/439		7*2*1
POLY-CYCLOOCTENE TRANS- III. 126 *110.19*	III. 126	TRI	CI=1	4.34	5.41	9.78	6410.5119	1	1.008		67/439 73/284	4.8/284	8*1*1
	476					9.78							
	IV. 228	MONO	C2H=5	7.43	5.00	9.91	B=84.8	2	0.998		62/439		8*1*1
POLY-CYCLOPENTENE TRANS- 1. 147 *68.11*	1. 147	ORTHO OR	C2v-9 or D2H-16	7.28	4.97	11.9		4	1.051		23		5*2*1
POLY-1-DECENE *140.26*	1. 150					13.2					34/250		
POLY-1-DODECENE *160.31*	1. 150					13.2					45/48 49/250		

POLYMER	REF	CRYST SYST	SPACE GROUP	UNIT CELL PARAMETERS A	B	C	ANGLES	MON/UNIT CELL	DENSITY (G./CC.) CRYSTAL	AMORPH.	MELT. POINT	HEAT OF FUSION KCAL/MON	CHAIN CONF.
POLYETHYLENE *28.05* I.	14	ORTHO	D2H=16	7.40	4.93	2.554		2	1.008	0.852/57	110/46	1.88/85	2*1=1
	15	ORTHO		7.36	4.94	2.554		2	1.011		137/85	1.84/86	
	363								0.9988		141/206	2.00/260	
	311	ORTHO		7.406	4.939	2.547		2	0.9998	0.8866	141/286	1.86/156	
	406								0.811	0.811	142/315	1.94/262	
	138								0.991	0.855	146/362	1.82/343	
											145/351		
SINGLE CRYSTAL	72	TRI		7.84	5.56	120	63,71,82	96	1.004				2*1=1
II.	58	PMONO	C2H=3	4.05	4.85	2.54	G=105	1	0.966				
	233	MONO		8.09	2.53*	4.79	B=107.9	2	0.988				
	254	TRI		4.285	4.820	2.54	9010108	1	1.002				2*1=1
***, CHLOROTRIFLUORO *116.48*	208	HEX		6.34	6.34	35		14	2.222	2.08/49	220/49	1.20/88	2*14*1
	21	HEX		6.5	6.5	35		14	2.11		210/46		2*17*
	296	HEX		6.385	6.385	42		17	2.217				
	116												
	205								2.192	2.032	215		2*16*1
	139					43		16	2.19/49	1.925	222/341		
***, ETHYLSILYL-ISOTACTIC *86.21*	489	HEX	C3I=2	21.60	21.60	6.50		18	0.981				2*3*1
***, TETRAFLUORO- *50.01*	11	PHEX		5.54	5.54	16.8	G=119.5	13	2.405				1*13=1
	266	TRI		4.882	4.875	5.105	90.87.87	4	2.741				1*13.6
	209	TRI		5.59	5.59	16.88	G=119.3	13	2.347		327/46		
ABOVE 20C	11	HEX		5.61	5.61	16.8		13	2.358		330	1.37/91	1*13=1
	209	HEX		5.66	5.66	19.50		15	2.302		330/101		1*15*7
									2.304/66				
***, TRIFLUORO* *82.03*	389	HEX		5.59	5.59	2.50		1	2.013				
POLY-ETHYLENE-CO-2*BUTENE *84.16*	324	MONO	C2H=5	10.92	7.73	9.10	A=130	4	0.950	0.87	135		4*2=1
POLY-ETHYLENE-CO-CYCLOHEPTENE *124.22*	259					9.00					74		
POLY-ETHYLENE-CO-CYCLOPENTENE *96.17*	247	ORTHO	D2H=17	8.76	7.83	9.02		4	1.032		185		4*2=1
	412	ORTHO OR	D2H=14	8.76	8.05	9.00		4	1.006				

POLYMER	REF	CRYST SYST.	SPACE GROUP	UNIT CELL PARAMETERS A	B	C	ANGLES	MON/UNIT CELL	DENSITY (G./CC.) CRYSTAL	AMORPH.	MELT. POINT	HEAT OF FUSION KCAL/MON	CHAIN CONF.
POLY-ETHYLIDENE *28.05*	178	ORTHO		12.38	6.28	2.5		4	0.958				
POLY-1-HEPTENE *98.18*	I. 150 359					6.45					17/250 *40/48		2*3*1 2*3*1
..., 5-METHYL- *112.21*	67					6.40					52/129		2*3*1
POLY-1-HEXADECENE *224.42*	II. 150	ORTHO		7.5	63.2	6.6		8	0.95		68/250		2*4*1
POLY-1-HEXENE *84.16*	150 OR 150	MONO ORTHO		22.2 11.7	8.89 26.9	13.7 13.7	G=94.5	14 28	0.726 0.907		-55/48		2*7*2 2*7*2
..., 4-METHYL- *98.18*	67	TET		19.64	19.64	14.00		28	0.845		200/9		2*7*2
..., 5-METHYL- *98.18*	9	HEX		10.2	10.2	6.50		3	0.835		130 110/282		2*3*1
POLY-ISOBUTENE *56.10*	34 115 139	ORTHO ORTHO	D2*4	6.94 6.94	11.96 11.96	18.63 18.63		16 16	0.964 0.964	0.915 0.912/53 0.842	44/46	2.87/82	2*8*5
POLY-1-OCTADECENE *252.47*	II. 150 III 493	ORTHO HEX		7.5 4.24	70.4 4.24	6.6		8	0.96		80/6 100/90		2*4*1
POLY-1-PENTENE *70.13*	IA. 355 OR	MONO		11.35 21.15	20.85 11.20	6.49	B=99.6	12 12	0.923 0.922		130/276 130/380		2*3*1
	IB. 355					6.5			0.96/408	0.85/408			2*3*1
	IIA. 152 355	PORT		19.30	16.90	6.60 7.08	G=116	16	0.898		75/48 80/9		2*4*1
	IIB. 355	PORT		19.60	16.75	7.08	G=115.3	16	0.887		78/282		2*4*1
..., 4,4-DIMETHYL- *98.18*	431 442	TET TET	S4-1	20.3 20.35	20.3 20.35	13.8 7.01		28 16	0.803 0.898		231/282 350/90 380/441		2*7*2 2*4*1

POLYMER	REF	CRYST SYST.	SPACE GROUP	UNIT CELL PARAMETERS A	B	C	ANGLES	MON/UNIT CELL	DENSITY (G./CC.) CRYSTAL	AMORPH.	MELT. POINT	HEAT OF FUSION KCAL/MON	CHAIN CONF.
---, 4-METHYL- *84.16*													
I.	55	TET	S4*1	18.66	18.66	13.80		28	0.814		235/48	4.71/82	
	452	TET		18.54	18.54	13.84		28	0.822		250/94		
	333	TET		18.50	18.50	13.76		28	0.831		228/282	2.85/370	IREG. 2*7*2
	67	TET		18.60	18.60	13.84		28	0.817				
									0.828/94	0.838/94			
II.	8			19.16	19.16	7.12		16	0.855		125		2*4*1
III.	8			19.36	19.36	7.05		16	0.846		75		2*4*1
---, 5-TRIMETHYL-SILYL- *142.31*	67					6.55					133/71		2*3*1
POLY-PROPENE ISOTACTIC *42.08*													
I.	127	MONO	C2H*6	6.65	20.96	6.50	B=99.3	12	0.938	0.85/45	176/10	2.37/82	2*3*1
	330	MONO	C2H*3					12	0.938		180/250	1.89/251	
	6	MONO		6.666	20.87	6.488	B=98.2	12	0.939		165/282	2.60/83	
	131	MONO		6.64	20.88	6.51	B=98.7	12	0.932	0.8535	178/349	2.40/358	
	136	MONO		6.69	20.98	6.504	B=99.5	12	0.932		183/380	1.47/182	
	170	TRI	CI*1	13.36	6.50*	10.99	87.10899	12	0.934		186/394	2.1/394	
	308	MONO		6.66	20.78	6.495	B=99.62	12	0.946		189/392	1.90/478 2.5/474	
II.	166	HEX	D3*4	12.74	12.74	6.35		12	0.939	.907/112			2*3*1
	196	RHO	OR*6	6.38	6.38	6.33		3	0.939				2*3*1
									0.88/167				2*3*1
	308	ORTHO	OR	19.08	11.01	6.490		18	0.922				2*3*1
	308	HEX		22.03	22.03	6.490		36	0.922				2*3*1
	485	RHO		19.08	19.08	6.49		27	0.922				2*3*1
III.	308	TRI		6.47	10.71		G=99.07						2*3*1
SYNDIOTACTIC I.	169	ORTHO	D2*2	14.5	5.8	7.4		8	0.898	.858/393	161/415		4*2*1
	430	ORTHO	D2*5	14.50	5.60	7.40		8	0.930				4*2*1
	409								0.898				
II.	306	PHEX				5.05				0.858			4*1*1
---, 3-CYCLOHEXYL- *124.22*	431	TET OR	C4*5 S4*2	21.06	21.06	20.09		40	0.926		230/90 215/282 214/328		2*10*3
II.	431	RHO OR	C4*3 C3I*2	19.12	19.12	6.33		9	0.926				2*3*1

POLYMER	REF	CRYST SYST.	SPACE GROUP	A	B	C	ANGLES	MON/UNIT CELL	DENSITY (G./CC.) CRYSTAL	AMORPH.	MELT. POINT	HEAT OF FUSION KCAL/MON	CHAIN CONF.
•••, 3-CYCLOPENTYL• •110.19•	431	TET OR	C4*5 OR S4*2	20.34	20.34	47.49		96	0.894				2*24*7
•••, HEXAFLUORO- •150.03•	268										160		2*4*1
•••, 3-PHENYL- (ALLYLBENZENE) •118.17•	67					6.40					230/90 208/187 185/282		2*3*1
•••, 3-SILYL• •72.18•	67					6.45					128/71		2*3*1
•••, TRIMETHYLSILYL- •114.26•	67					6.50					360/71		2*3*1
POLY-1-TETRADECENE •196.36•	II. 150	ORTHO		7.5	56.0	6.6		8	0.94		57/250		2*4*1
POLY-VINYLCYCLO-BUTANE •82.14•	431	TET		34.12	34.12	6.6		64	1.14		228/437		2*4*1
POLY-VINYLCYCLO-HEPTANE •124.22•	431	TET	C4H*6	23.4	23.4	6.5		16	0.927		300/437		2*4*1
POLY-VINYLCYCLO-HEXANE •110.19•	1. 431	TET	C4H*6	21.99	21.99	6.43		16	0.942		305/90		2*4*1
	67	TET		21.76	21.76	6.50		16	0.951		300/282		2*4*1
	95	TRI		11.6	7.8	6.6	92,10898	3	0.98		383/328		
	II. 431	TET	C4*5 OR S4*2	20.48	20.48	44.58		96	0.939		372/441		2*24*7
POLY-VINYLCYCLO-PENTANE •96.17•	1. 95	TRI		10.5	7.4	6.6	92,10899	3	1.00		292/328		2*4*1
	431	TET		20.14	20.14	6.50		16	0.969				2*4*1
	II. 431	TET	C4*5 OR S4*2	20.14	20.14	19.5		48	0.969				2*12*3
POLY-VINYLCYCLO-PROPANE •68.11•	III. 431	TET	D4H*1 OR D4H*7	37.3	37.3	19.8			0.927		230		2*10*3
	1. 175	TRI				6.5							
	95												

POLYMER	REF	CRYST SYST.	SPACE GROUP	UNIT CELL A	CELL B	PARAMETERS C	ANGLES	MON/UNIT CELL	DENSITY (G./CC.) CRYSTAL	AMORPH.	MELT. POINT	HEAT OF FUSION KCAL/MON	CHAIN CONF.
POLY-VINYLS AND POLY-VINYLIDENES													
-==-------==-------													
-HCH-CHR-													
POLY-ACRYLAMIDE													
-==, N, N=DIBUTYL- *183.29*	93	HEX		26.3	26.3	6.3		12	1.06 (0.97)				2=3=1
-==, N-ISOPROPYL- *113.16*	102								1.118	1.070	200		
POLY-ACRYLATES													
-=, ALLYL= *112.12*	278					6.5					90		2=3-1
-==, SEC=BUTYL- *128.17*	401	MONO	CS=2	17.92	10.34	6.5		6	1.062		130		2=3=1
	202	ORTHO				6.49							2=3=1
-==, TERT-BUTYL- *128.17*	67	RHO	C3=2			6.45							2=3-1
	401			17.92	10.50	6.5		6	1.047		193/16		2=3-1
	202	ORTHO				6.48					200		2=3-1
-==, ISOBUTYL- *128.17*	401	ORTHO		17.92	17.92	6.5		12	1.239				2=3=1
	202	ORTHO				6.42					81		2=3=1
-==, ISOPROPYL- ISOTACTIC *114.14*	67					6.5			1.08		162/120		2=3=1
	149					6.32					162		
SYNDIOTACTIC	113					5.18			1.18		115		
POLY-ACRYLONITRILE SYNDIOTACTIC *53.06*	76	HEX	C2v=16	5.99	5.99			4	1.110		317/77	1.16/77	4=1-1
	133	ORTHO		10.20	6.10	5.10		8	1.272				
	162	ORTHO		18.1	6.12	5.00		4	1.134				
	210	ORTHO		10.55	5.80	5.08		2	1.07				
	322	PHEX		6.1	6.11	(5.1)	G=120	16	1.125				
	486	ORTHO		21.18	11.60	(5.1)							
ISOTACTIC	133	TET		4.74	4.74	2.55		1	1.546				
POLY-ISOPROPENYL- METHYL KETONE *84.11*	347	TET		15.08	15.08	8.54		16	1.151		240		2=4-1
											240/255		
											200/379		

POLYMER	REF	CRYST SYST.	SPACE GROUP	UNIT CELL PARAMETERS A	B	C	ANGLES	MON/UNIT CELL	DENSITY (G./CC.) CRYSTAL	AMORPH.	MELT. POINT	HEAT OF FUSION KCAL/MON	CHAIN CONF.
POLY-METHACRYLATES													
***, METHYL- *100,11*													
ISOTACTIC	30	PORT		21.08	12.17	10.55		20	1.228	1.22/31	160/31		2.5.2
SYNDIOTACTIC	30									1.19/31	200/31		4.5.2
POLY-METHACRYLO- NITRILE I. *67.09*	427	PHEX		9.03	9.03	6.87		4	0.918		250/436		2.4.1
II.	427	MONO		13.5	7.71	7.62	B=97.7	8	1.134				
POLY-THIOLACRYLATES													
***, SEC-BUTYL- *144.23*	201	RHO	C3.2			6.35							2.3.1
	401	RHO	C3.2			6.5							
***, ISOBUTYL- *144.23*	201	MONO	CS.2			6.42							
	401	MONO	CS.2										
***, ISOPROPYL- *130.20*	200	RHO	C3.2			6.42							
	401												
***, PROPYL- *130.20*	200	MONO	CS.2			6.4							2.3.1
	401	MONO	CS.2			6.5							
POLY-VINYL ALCOHOL *44.05*	29	MONO	C2H.2	7.81	2.52*	5.51	B=91.7	2	1.350	1.29/1	232/310		2.1.1
	132	MONO		7.805	2.533	5.465	B=92.2	2	1.349	1.26/53	265/323		
	184	MONO	C2H.2	7.81	2.52*	5.50	B=92	2	1.345	1.269	228/433	1.64/433	
	261	MONO		7.81		5.43	B=91.5		1.352			267/450		
	316								1.34/410	1.27/410	243/480	1.67/480	
QUENCHED	261	ORTHO	C2V.20	7.42	2.52*	5.25		2	1.490				2.1.1
SINGLE CRYSTAL	151	HEX		5.45	5.45	2.51		1	1.133				
POLY-VINYL CHLORIDE *62.50*	7	ORTHO	D2H.11	10.6	5.4	5.1		4	1.42	1.41/413	273/372	2.7/372	4.1.1
	64	ORTHO		10.11	5.27	5.12		4	1.522		212/143	0.66/145	
	140	MONO		10.65	5.15*	5.00	B=90	4	1.455				
	235	ORTHO	D2H.11	10.40	5.30	5.10		4	1.477				
POLY-VINYL FLUORIDE *46.04*	62	HEX	C3V.6	4.93	4.93	2.53		1	1.436		200/79	1.80/79	2.1.1
	236	ORTHO	C2V.14	8.57	4.95	2.52		2	1.430		230/434		2.1.1
POLY-VINYL FORMATE *72.06* ISOTACTIC	232	RHO or	C3V.6 or D3D.6	15.9	15.9	6.55		18	1.502				

POLYMER	REF	CRYST. SYST.	SPACE GROUP	A	B	C	ANGLES	MON/UNIT CELL	CRYSTAL	AMORPH.	MELT. POINT	HEAT OF FUSION KCAL/MON	CHAIN CONF.
SYNDIOTACTIC 232						5.0							
POLY-VINYL METHYL KETONE *70.09*	164	TET	S4=1	14.52	14.52	14.40		28	1.073		170		2•7=2
	379	TET		14.56	14.56	14.10		28	1.090				2•7=2
POLY-VINYLIDENE BROMIDE *185.87*	33	MONO		25.88	4.77*	13.87	B=70.2	16	3.065				
POLY-VINYLIDENE CHLORIDE *96.95*	33	MONO		22.54	4.68*	12.53	B=84.2	16	1.959				4•1=1
	245	MONO	C2=2	6.73	4.68*	12.54	B=123.6	4	1.957				
	32	MONO		13.69	4.67*	6.296	B=55.2	4	1.948	1.66/43	190/46		2•2=1
	231									1.7754			
POLY-VINYLIDENE FLUORIDE *64.04*	I. 171	MONO		5.02	25.4*	4.62	B=107	10	1.888				4•1=1
	417	MONO	C2-2	17.78	4.57*	11.68	B=87.3	16	1.795				2•2=1
	418	ORTHO	C2v=9	9.66	4.96	4.64		4	1.913				6
	168	MONO	C2=2,1	9.64	4.64*	5.02	B=91.1	4	1.894				
	II. 417	ORTHO	C2v=14	8.45	4.88	2.55		2	2.022		171/373		2•1=1
	418	ORTHO	C2v=14	8.47	4.90	2.56		2	2.002		185/434		2•1=1
											220/443		

POLYMER	REF	CRYST SYST.	SPACE GROUP	UNIT CELL PARAMETERS A B C ANGLES	MON/ UNIT CELL	DENSITY (G./CC.) CRYSTAL AMORPH.	MELT. POINT	HEAT OF FUSION KCAL/MON	CHAIN CONF.
POLY-AROMATICS									
-HCH•CHR•									
POLY-P•PHENYLENE •76.09•	475	ORTHO		7.81 5.53 4.20	2	1.393			
POLY-STYRENE •104.14•	6 128 411 139	RHO RHO	D3D•6 D3D•6	22.08 22.08 6.626 21.9 21.9 6.65	18 18	1.112 1.127 1.114 1.12/45	1.04 TO 1.065/17 1.052 1.024	240/10 250/90 235/282	2.15/89 2.00/174	2•3•1
•••, P•TERT•BUTYL- •160.25•	198						0.950	300	
•••, O•FLUORO- •122.14•	123	RHO	C3V•6	22.15 22.15 6.63	18	1.296		270/75	2•3•1
•••, P•FLUORO- •122.14•	67			8.30				265/75	2•4•1
•••, A•METHYL- •118.17•	357	RHO		6.6					2•4•1
•••, O•METHYL- •118.17•	125	TET	C4V•12	19.01 19.01 8.10	16	1.072		360/74	2•4•1
•••, •••, P•FLUORO- •136.16•	67			8.05				360/75	2•4•1
•••, M•METHYL- •118.17•	80 103 163 430	TET TET OR	S4•1 C4•5 S4•2	19.81 19.81 21.74 57.0 19.9 19.9 57.1 78.9	44	1.012 1.005		215 215/74	2•11•3 2•29-8 2•40• 11
•••, P•METHYL- •118.17•	103			12.9			1.04/67		
•••, TRIMETHYLSILYL- •176.32•	103			60.4				284	
POLY-1•VINYL NAPHTHALENE •154.20•	122	TET	C4V•12	21.20 21.20 8.10	16	1.125		360/75	2•4•1

POLYMER	REF	CRYST SYST.	SPACE GROUP	UNIT CELL PARAMETERS A B C ANGLES	MON/ UNIT CELL	DENSITY (G./CC.) CRYSTAL AMORPH.	MELT. POINT	HEAT OF FUSION KCAL/MON	CHAIN CONF.
POLY-2-VINYLPYRIDINE *105.13*	185			6.7			212		
POLY-P-XYLYLENE									
ALPHA *104.14*	223	MONO		11.68 6.10 9.16 B=102.5	4	1.086	375/142		
	426	ORTHO		21.3 33.6 6.5A	31	1.138	420/219	7.20/82	
							375/82		
							435/323		
							400/395		
BETA	223	MONO		8.10 5.25 6.53 B=95	2	1.250	412		
	224			6.55			420		
	426	HEX		20.52 20.52 6.58	16	1.153			

POLYMER	REF	CRYST SYST.	SPACE GROUP	UNIT CELL PARAMETERS A	B	C	ANGLES	MON/ UNIT CELL	DENSITY (G./CC.) CRYSTAL	AMORPH.	MELT. POINT	HEAT OF FUSION KCAL/MON	CHAIN CONF.
POLY-DIENES													
1,2-POLY-1,3-BUTADIENE *54.09*													
SYNDIOTACTIC	7	ORTHO	D2H=11	10.98	6.60	5.14		4	0.964		154		4*1*1
ISOTACTIC	12	RHO	D3D=6	17.3	17.3	6.5		18	0.96		120 / 125/44		2*3=1
---, 4,4-DIMETHYL- ISOTACTIC II *82.14*	488	TET	D2D=10	17.80	17.80	36.50		72	0.849				2*18*5
1,4-POLY-1,3-BUTADIENE *54.09*													
TRANS= I.	37	PHEX		4.54	4.54	4.9		1	1.03		100/352 / 96/451	2.4/352 / 3.30/451	4*1*1 / 4*1*1
	496	MONO	C2H=5	8.63	9.11	4.83	B=114	4	1.036				
(ABOVE 65) II.	154	PHEX		4.88	4.88	4.68		1	0.930		141/352 / 148/44 / 145/451	1.10/352 / 1.43/342 / 1.10/451	
CIS=	124	MONO	CS=4	4.60	9.50	8.60	B=109	4	1.011		1/287	2.2/353	8*1*1
	60	MONO OR MONO	C2H=6 / C2H=5	8.53	8.16	12.66	B=83.33	8	0.821		6.3/335		
---, 2-TERT-BUTYL- CIS= *110.19*	367	TRI		13.95	20.78	15.3		22	0.907		106/299		4*11-3
---, 2-CHLORO- TRANS= (CHLOROPRENE) *88.54*	40	ORTHO	D2=4	8.84	10.24	4.79		4	1.356		80/81	2.00/81	
	109	ORTHO	ORTHO	9.0	8.23	4.79		4	1.657				
---, 1-CYANO- TRANS= *79.10*	275					4.8							
---, 2,3-DICHLORO- TRANS= *122.99*	221					4.86							

POLYMER	REF	CRYST SYST.	SPACE GROUP	UNIT CELL PARAMETERS A	B	C	ANGLES	MON/ UNIT CELL	DENSITY (G/CC.) CRYSTAL	AMORPH.	MELT. POINT	HEAT OF FUSION KCAL/MON	CHAIN CONF.
***, 2,3-DIMETHYL- TRANS* *82.14*	104					4.35					260		
CIS*	220					7.0					192		
***, 2-METHYL- (ISOPRENE) *68.11* TRANS*, ALPHA	288					8.70							4*2*1
BETA	23	ORTHO	D2*4	7.78	11.78	4.72		4	1.046		65/46	3.04/81	4*1*1
	288	ORTHO		11.9	4.8*	7.85		4	1.009		74/81		4*1*1
											68/287		
DELTA	492			7.84	5.99								
EPSILON	492			7.80	6.29								
CIS*	40	MONO	C2H*5	12.46	8.89	8.10	B=92	8	1.009	0.906/53	28/81	1.05/81	8*1*1
	288	ORTHO		8.97	8.20*	25.2		16	0.976	0.910/47	14/287		8*1*1
	124	ORTHO	D2H*15								36/50		8*1*1
***, 3-CHLORO- TRANS* *102.56*	165					4.9							
***, HYDROCHLORINATED *104.58*	229	PORTH	C2H*5	5.83	10.38	8.95	B=90	4	1.282		115		
	70	ORTHO		11.9	9.1*	10.4		8	1.23		110		
***, 2-METHYLACETOXY TRANS* *140.18*	165	ORTHO		16.2	9.3	4.75		4	1.30		135		
***, 2-PROPYL- TRANS* *96.17*	368	ORTHO		10.95	6.65	9.2		4	0.95		42		8*1*1
1,4-POLY*1,3- HEPTADIENE *96.17* TRANS* ISOTACTIC	375	MONO	C2*2	8.62	7.95	4.85	B=99	2	0.973		85/371		4*1*1
***, 6-METHYL- *110.19* TRANS* ISOTACTIC	371					4.85					119		

POLYMER	REF	CRYST SYST.	SPACE GROUP	UNIT CELL PARAMETERS A	B	C	ANGLES	MON/ UNIT CELL	DENSITY (G./CC.) CRYSTAL	AMORPH.	MELT. POINT	HEAT OF FUSION KCAL/MON	CHAIN CONF.
1,2•POLY•1,5• HEXADIENE *82•14*	267	ORTHO		7.69	6.21	4.80		2	1.190		146		4•1•1
	340	ORTHO		13.30	15.52	4.80		8	1.1101		128		
1,4•POLY•1,3• HEXADIENE *82•14* TRANS• ISOTACTIC	274	ORTHO	D2•4	14.02	8.02	4.85		4	1.000		82/371		
•••, 5•METHYL• *96.17* TRANS• ISOTACTIC	371					4.85					88		
2,5•POLY•2,4• HEXADIENE •••, 2,5•DIMETHYL• *110.19* TRANS•	183					4.8					265 265/376		
2,5•POLY•2,4• HEXADIENOATES TRANS•ERYTHRO• ISOTACTIC •••, BUTYL• *168.23*	384	ORTHO		11.36	9.70	4.80		2	1.056				8•1•1
•••, DIISOPROPYL• *226.26*	420	ORTHO		14.16	10.28	9.70		4	1.064				
•••, ET•YL• *140.18*	384					4.80							
•••, ISOAMYL• *182.25*	384					4.80							
•••, ISOBUTYL• *168.23*	384					4.80							
•••, ISOPROPYL• *154.20*	384					4.80							
•••, METHYL• *126.15*	384					4.80							
POLY•ISOPRENE POLY•BUTADIENE, •••, 2•METHYL•	SEE												

POLYMER	REF	CRYST SYST.	SPACE GROUP	UNIT CELL PARAMETERS A	B	C	ANGLES	MON/ UNIT CELL	DENSITY (G./CC.) CRYSTAL	AMORPH.	MELT. POINT	HEAT OF FUSION KCAL/MON	CHAIN CONF.
1,4-POLY-1,3- OCTADIENE *110.19*													
TRANS= ISOTACTIC	371					4.85					87		
1,4-POLY-1,3- PENTADIENE *68.11*													
TRANS- ISOTACTIC	189	ORTHO		19.73	4.85	4.8		4	0.98		95		4•1•1
CIS- SYNDIOTACTIC	234					8.50					53		8•1•1
ISOTACTIC	263					8.15					44		4•2•1
2,5=POLY-5=PHENYL- 2,4=PENTADIENOATES TRANS=ERYTHRO- ISOTACTIC													
···, BUTYL= *230.29*	384					4.80							
···, METHYL- *188.22*	384					4.80							

POLYMER	REF	CRYST SYST.	SPACE GROUP	UNIT CELL A	B	C	PARAMETERS ANGLES	MON/ UNIT CELL	DENSITY (G./CC.) CRYSTAL	AMORPH.	MELT. POINT	HEAT OF FUSION KCAL/MON	CHAIN CONF.
POLY-AMIDES -NH-X-NH-CO-Y-CO- OR -NH-X-CO-													
POLY-2-AMINOACETIC ACID *57.05* (GLYCINE)	273	HEX	C3-2	4.8	4.8	9.3		3	1.53				3*3*1
POLY-4-AMINOBUTYRIC ACID ALPHA *85.10*	272	MONO	C2-2	9.44	12.1*	8.22	B=64	8	1.340				
	457	MONO	C2-2	9.29	12.2*	7.97	B=114.5	8	1.375		260/337		
BETA	458 65	MONO			12.2*			2 2					
DELTA	458	HEX		4.65	4.65	4.65							
POLY-10-AMINOCAPRIC ACID *169.26*	65 61	PHEX HEX	C2-2	4.9	4.9	26.5		2	1.02		192/177 177/146 188/337		
POLY-6-AMINOCAPROIC ACID ALPHA *113.16* (CAPROLACTAM)	3	MONO	C2-2	9.56	17.2*	8.01	B=67.5	8	1.235		215	4.96/155	
	26	MONO	C2-2	4.81	17.10	7.61	B=79.5	4	1.221		223/153	4.32/216	
	135	MONO	ORTHO	9.65	17.2*	8.11	B=66.3	8	1.220		233/244	5.57/405	
	212	MONO	C2-2	9.45	8.02	17.08	B=68	8	1.252		226/293	5.15/343	
	334	MONO		9.66	8.32	17.0	G=65	8	1.214 1.23/213 1.25/449	1.10/213 1.11/449	214/304 228/346	4.26/346	
BETA	222	HEX		4.8	4.8	8.6		1	1.09				
GAMMA	336	MONO	C2H-5	9.35	16.6*	4.81	B=120	4	1.162				
	272	HEX		4.79	4.79	16.7		2	1.132				
	279	ORTHO		4.82	7.82	16.70		4	1.194				
	246	MONO	C2H-5	9.33	16.9*	4.78	B=121	4	1.163				
ABOVE 150	26	RHO		4.90	16.28	8.22							
*-, D-(-)-3-METHYL- *127.18*	440	MONO	C2-2	9.15	16.96	4.84	B=90	4	1.125		225/19 226/36		
POLY-8-AMINOCAPRYLIC ACID ALPHA *141.21*	272 65	MONO PHEX	C2-2	9.8	22.4*	8.3	B=65	8	1.14 1.18/153		185/146 202/153 209/244		

POLYMER	REF	CRYST SYST	SPACE GROUP	UNIT CELL PARAMETERS A	B	C	ANGLES	MON/UNIT CELL	DENSITY (G./CC.) CRYSTAL AMORPH.	MELT. POINT	HEAT OF FUSION KCAL/MON	CHAIN CONF.
BETA	121	MONO										
GAMMA	272	HEX		4.79	4.79	21.7		2	1.088			
	61	HEX		4.9	4.9	21.7		2	1.04			
POLY-7-AMINOENANTHIC ACID												
7 *127,18*	61	TRI	C1=1	4.9	5.4	9.85	49,77,63	1	1.21	225/146		
	65		OR C1=1					1	1.20/153	233/215		
										233/291		
										217/153		
*=0, (R)=3=METHYL- *141,21*	479					9.24				179		
*=0, (S)=4=METHYL- *141,21*	479					9.43				176		
*=0, (R)=5=METHYL- *141,21*	479					9.11				182		
*=0, (R)=6=METHYL- *141,21*	479					9.57				188		
POLY-9-AMINO- PELARGONIC ACID	421	TRI	C1=1	4.9	5.4	12.5	49,77,64	1	1.15	210		
9 *155,23*	65		OR C1=1							194/146		
										198/292		
										209/177		
POLY-3-AMINO- PROPIONIC ACID												
ALPHA *71,08*	491	MONO	C2=1	9.33	4.78*	8.73	B=60	4	1.400	340		
										260/244		
										330/337		
=, 2,2-DIMETHYL- *99,13* I.										189/73		
II.	73					8.4				273	3.1	
										270/402		
POLY-11-AMINO- UNDECANOIC ACID	61	TRI		4.9	5.4	14.9	49,77,63	1	1.15	182/146		
11 *183,29*	59	TRI		9.6	4.2	15.0	72,90,64	2	1.19	194/101	0.9/343	
	243	TRI		4.78	4.13	14.0	82,75,66	1	1.173	186		
	365	TRI		4.78	4.13	13.1	90,75,66	1	1.344	220		
	65		C1=1							188/444		
			OR C1=1							183/292		

POLYMER	REF	CRYST SYST.	SPACE GROUP	UNIT CELL PARAMETERS A	B	C	ANGLES	MON/ UNIT CELL	DENSITY (G./CC.) CRYSTAL	AMORPH.	MELT. POINT	HEAT OF FUSION KCAL/MON	CHAIN CONF.
POLY-DECAMETHYLENE ADIPAMIDE 10.6 *282.42*	110					20.0					230 236/244		
POLY-DECAMETHYLENE AZELAAMIDE 10.9 *324.49* GAMMA	65	PHEX						2			214/137	16.3/405 8.76/159	
POLY-DECAMETHYLENE SEBACAMIDE 10.10 *338.52*	65 110	TRI				25.6		1			203/244 196 198/454 216/137	12.2/118 8.29/159 17.2/405 7.82/160	
POLY-HEPTAMETHYLENE PIMELAMIDE 7.7 *254.36*	63	PHEX	CS=1	4.82	19.0	4.82	B=60	1	1.105		214/244 205/2 196/339		
GAMMA	65	PHEX		18.95				1					
POLY-HEXAMETHYLENE ADIPAMIDE ALPHA 6.6 *226.31*	25 402 407	TRI MONO TRI	CI=1	4.9 15.7 5.00	5.4 10.5 4.17	17.2 17.3 17.3	48,77,63 B=73 81,76.63	1 9 1	1.24 1.240 1.203 1.220/54	1.09/52 1.069/54	265/2 270/289	11.10/82 9.7/155 8.79/216 11.2/405	
BETA	25	TRI	CI=1	4.9	8.0	17.2	90,77,67	2	1.25		226/244 185/339		
POLY-HEXAMETHYLENE AZELAMIDE 6.9 *268.39	402	MONO		7.8	40.15	5.3	B=87	4	1.08				
POLY-HEXAMETHYLENE SEBACAMIDE ALPHA 6.10 *282.42*	25	TRI	CI=1	4.95	5.4	22.4	49,76.63	1	1.16 1.152/54 1.17/153 1.189/52	1.041/54	228/51 233/244 216/153 225/454 215/291	7.32/160 14.0/405 13.5/137	
BETA	25	TRI	CI=1	4.9	8.0	22.4	90,77,67	2	1.20				
POLY-NONAMETHYLENE AZELAAMIDE 9.9 *310.47*	110					24.0					177 189/2 165/339		

POLYMER	REF	CRYST SYST.	SPACE GROUP	UNIT A	CELL B	PARAMETERS C	ANGLES	MON/ UNIT CELL	DENSITY (G./CC.) CRYSTAL AMORPH.	MELT. POINT	HEAT OF FUSION KCAL/MON	CHAIN CONF.
GAMMA	65	PHEX						1				
POLY-PENTAMETHYLENE AZELAAMIDE *254.36* 5,9	2					19.5				179/291 178/339		
POLY-P-PHENYLENE PHTHALAMIDE *238.24*	396	ORTHO	D2H14	22.8	5.5	8.1		4	1.56			
POLY-PIPERAZINE ADIPAMIDE *196.24*	386					9.2				355		
POLY-PIPERAZINE SEBACAMIDE *252.35*										180/81	6,20/81	
POLY-M-XYLYLENE ADIPAMIDE *246.30*	84 459	MONO TRI	C1-1	5.10 12.01	4.70 4.83	15.2 29.8	G=69.6 75,26.65	1 2	1.198 1.250	246/326		
POLY-P-XYLYLENE SEBACAMIDE *302.40*	204	TRI		5.74	4.87	20.6	76,55.65	1	1.168	300/290 268/291 281/385 291/454		

POLYMER	REF	CRYST SYST.	SPACE GROUP	UNIT CELL A	B	C	PARAMETERS ANGLES	MON/UNIT CELL	DENSITY (G./CC.) CRYSTAL AMORPH.	MELT. POINT	HEAT OF FUSION KCAL/MON	CHAIN CONF.
POLY-ESTERS ----------------- -O-X-O-CO-Y-CO- OR -O-X-CO-												
POLY-1,3-CYCLO- BUTYLENE CARBONATE -", 2,2,4,4- TETRAMETHYL- TRANS* *170,20*	364	TRI		9.25	8.28		G=96,5		1.08	360		
CIS*	364	ORTHO		9.16	8.22	12.9		4	1.164	253		
POLY-1,4-CYCLO- HEXYLENEDIMETHYLENE TEREPHTHALATE *274,30* TRANS*	199	TRI		6.37	6.63	14.2	89,47114	1	1.266			
CIS*	199	TRI		6.02	6.01	13.7	89,53112	1	1.319			
POLY-DECAMETHYLENE ADIPATE 10.6 *284,38*	106	MONO		5.0	7.4	22.1		2	1.15	80/81 77/300	10.2/81 10.9/405	
POLY-DECAMETHYLENE AZELATE 10.9 *326,46*	106	MONO		5.0	7.4	51.7		4	1.13	69/137	10.0/81 12.1/405	
POLY-DECAMETHYLENE GLUTARATE 10.5 *270,36*	106	MONO		5.0	7.4	41.6		4	1.17			
POLY-DECAMETHYLENE OXALATE 10.2 *228,28*	106	MONO		5.28	7.00	17.0		2	1.206	79/305		
POLY-DECAMETHYLENE SEBACATE 10.10 *340,49*	106	MONO		5.0	7.4	27.1		2	1.13	80/137 73/291	12.0/81 12.3/160 13.5/405 7.2/301	
POLY-DECAMETHYLENE SUBERATE 10.8 *312,44*	106	MONO		5.0	7.4	24.6		2	1.14			

POLYMER	REF	CRYST SYST.	SPACE GROUP	UNIT CELL A	PARAMETERS B	C	ANGLES	MON/ UNIT CELL	DENSITY (G./CC.) CRYSTAL	AMORPH.	MELT. POINT	HEAT OF FUSION KCAL/MON	CHAIN CONF.
POLY-DECAMETHYLENE SUCCINATE 10.4 *256.33*	106	MONO		5.0	7.4	19.6		2	1.17		68/290		
POLY-DECAMETHYLENE TEREPHTHALATE 10.T *304.37*	68	TRI		4.62	6.30	20.10	1079 6113	1	1.022		138/81 129/99 131/453	11.00/81 10.5/405	
POLY-DIKETENE *84.07*	217	MONO		5.50	7.78	9.06	B=92	4	1.441		115/190 115		
POLY-ETHYLENE ADIPATE 2.6 *172.18*	211 203 105	MONO MONO MONO	C2H-5	7.26 5.47 25.7	5.40 7.23 30.7	10.85 11.72 11.71	A=67.7 B=113.5 B=103.8	2 2 40	1.453 1.345 1.274		47/265 54/291 52/27 50/46	3.80/265	
POLY-ETHYLENE AZELATE 2.9 *214.25*	108 109 105	ORTHO ORTHO MONO		7.45 5.0 25.7	4.97 7.4 30.7	31.5 31.2 31.2	B=103.8	4 4 80	1.220 1.23 1.190				
POLY-ETHYLENE ISOPHTHALATE I. 2.1 *192.16* II.	398 111 398					14.8 21.0			1.358	1.346	143 240		
POLY-ETHYLENE-P-OXYBENZOATE ALPHA *164.15*	494 356	ORTHO		10.52	4.75	15.68 15.7		4	1.391		220/354 203/327		
POLY-ETHYLENE 1.4-PIPERAZINE DICARBOXYLATE *200.19*	386					10.4					245		
POLY-ETHYLENE SEBACATE 2.10 *228.28*	100 109 105	MONO MONO MONO		5.5 5.52 25.7	15.0 7.4 30.7	16.9 16.9 16.67	B=65 B=65 B=103.8	4 2 40	1.20 1.21 1.187		72/265 76/137 78/302 79/46	3.30/265 6.95/137 6.11/158 8.3/405	
POLY-ETHYLENE SUBERATE 2.8 *200.23*	203 108	MONO MONO	C2H-5	5.51 5.0	7.25 7.4	14.28 14.1	B=114.5	2 2	1.261 1.27		55/27		

POLYMER	REF	CRYST SYST.	SPACE GROUP	UNIT CELL PARAMETERS A	B	C	ANGLES	MON/ UNIT CELL	DENSITY (G./CC.) CRYSTAL	AMORPH.	MELT. POINT	HEAT OF FUSION KCAL/MON	CHAIN CONF.
POLY-ETHYLENE SUCCINATE 2,4 *144,12*	108	MONO		5.0	7.4	8.32	B=102.8	2	1.55		108/290		
	105	MONO		9.05	11.09	8.32		4	1.176		103/302		
POLY-ETHYLENE TEREPHTHALATE 2,T *192,16*	27	TRI	CI=1	4.56	5.94	10.75	98 118 112	1	1.457		265	5.76/87	
	195	TRI		5.54	4.14	10.86	107 112 92	1	1.472		284/264	5.40/157	
	400	TRI		4.52	5.98	10.77	101 118 11	1	1.477		267/265	2.2/265	
											270/374	5.9/405	
											265/290	3.98/155	
											245/472	6.56/472	
POLY-GLYCOLIDE 2 *58,04*	312	ORTHO	D2=4	6.36	5.13	7.04		4	1.678		223/305		
	465								1.707	1.50	230	2.8/464	
	490	ORTHO	D2H=16	5.22	6.19	7.02		4	1.700		233		3*2=1
POLY-HEXAMETHYLENE SEBACATE 6,10 *284,38*											67/290	5.2/301	
POLY-HEXAMETHYLENE TEREPHTHALATE 6,T *248,27*	68	TRI		4.57	6.10	15.40	105 98 114	1	1.146		160/81	8.3/81	
											161/405	8.0/405	
											154/99	8.44/159	4*2=1
POLY-3-HYDROXY-BUTYRIC ACID *86,09*	419	ORTHO	D2=4	5.76	13.15	5.96		4	1.266		176/422		4*2=1
POLY-10-HYDROXY-DECANOIC ACID 10 *170,24*	108	ORTHO		7.45	4.96	27.1		4	1.129		80/309		
POLY-4,4'-ISOPROPYL-IDENEDIPHENYLENE CARBONATE *254,27*	5	ORTHO	D2=2=3	11.9	10.1	21.5		8	1.307	1.20	267/57		6*2=1
	435	MONO		12.3	10.1	20.8	G=84	8	1.314		240		6*2=1
									1.30/283	1.20/283	263/370		
											230/438		
POLY-4,4'-METHYLENE-DIPHENYLENE CARBONATE *226,22*	435	ORTHO	C2V=9	5.0	10.5	22.0		4	1.301	1.26	230		6*2=1
											300/438		
POLY-NONAMETHYLENE AZELATE 9,9 *312,44*											65/405	10.3/159	
												11.7/405	

POLYMER	REF	CRYST SYST.	SPACE GROUP	UNIT CELL PARAMETERS A	B	C	ANGLES	MON/UNIT CELL	DENSITY (G./CC.) CRYSTAL	AMORPH.	MELT. POINT	HEAT OF FUSION KCAL/MON	CHAIN CONF.
POLY-OXYDIETHYLENE SEBACATE *272.33*	105	TET		17.6	17.6	38.0		32	1.229		44/325		
POLY-PIVALOLACTONE *100.13*	321	MONO	C2H=5	9.02	11.64	6.02	B=121.5	4	1.234				4*2*1
POLY-B-PROPIOLACTONE 3 *72.06* ALPHA BETA	314 314					7.02 4.82					122/317		4*2*1 4*1*1
POLY-TETRAMETHYLENE ISOPHTHALATE 4,I *220.21*	111								1.309	1.268	152.5	10.1	
POLY-TETRAMETHYLENE SEBACATE 4,10 *256.33*											60/325 62/425	4.0/301 0.48/425	
POLY-TETRAMETHYLENE TEREPHTHALATE 4,T *220.22*									1.08/11		232/99 221/453	7.6/111 7.5/405	
POLY-4,4'-THIODI-PHENYLENE CARBONATE *244.25*	435	ORTHO	C2V-9	5.6	8.7	22.2		4	1.50	1.35	220		6*2*1
POLY-TRIMETHYLENE ADIPATE 3,6 *186.20*	107	MONO		5.0	7.4	21.5		4	1.55		38 45/300		
POLY-TRIMETHYLENE AZELAATE 3,9 *228.28*	107	MONO		5.0	7.4	27.7		4	1.48		50		
POLY-TRIMETHYLENE DODECANEDIOATE 3,12 *270.36*	107	MONO		5.0	7.4	35.8		4	1.36		61		
POLY-TRIMETHYLENE GLUTARATE 3,5 *172.18*	107	MONO		5.0	7.4	15.4		2	1.00		39		
POLY-TRIMETHYLENE OCTADECANEDIOATE 3,18 *354.51*	107	MONO		5.0	7.4	51.6		4	1.23		76 76/307		

POLYMER	REF	CRYST SYST.	SPACE GROUP	UNIT CELL PARAMETERS				MON/UNIT CELL	DENSITY (G./CC.)		MELT. POINT	HEAT OF FUSION KCAL/MON	CHAIN CONF.
				A	B	C	ANGLES		CRYSTAL	AMORPH.			
POLY-TRIMETHYLENE PIMELATE 3.7 *200.23*	107	MONO		5.0	7.4	23.6		4	1.52		37		
POLY-TRIMETHYLENE SEBACATE 3.10 *242.31*	107 105	MONO PTET		5.0 31.2	7.4 31.2	31.3 33.5	G=90	4 96	1.39 1.184		53 56/305		
POLY-TRIMETHYLENE SUBERATE 3.8 *214.25*	107	MONO		5.0	7.4	26.1		4	1.47		41		
POLY-TRIMETHYLENE SUCCINATE 3.4 *158.15*	107	MONO		5.0	7.4	15.2		4	1.87		47 52/300		
POLY-TRIMETHYLENE UNDECANEDIOATE 3.11 *256.33*	107	ORTHO		5.0	7.4	32.4		4	1.42		59		

POLYMER	REF	CRYST SYST.	SPACE GROUP	UNIT CELL A	B	C	PARAMETERS ANGLES	MON/ UNIT CELL	DENSITY (G./CC.) CRYSTAL AMORPH.	MELT. POINT	HEAT OF FUSION KCAL/MON	CHAIN CONF.
POLY-URETHANS AND POLY-UREAS												
--#--#--#--#--#--#--#--#--												
NH=X=NH=CO-O-Y-O-CO												
OR =NH-X=O=CO- ANU												
NHXNH-CO-NHYNH-CO												
OR =NH-X=NH-CO=												
POLY-ETHYLENE DECAMETHYLENE DIURETHAN *286,36* ALPHA	414					21.8				174		
BETA	414					18.9						
POLY-ETHYLENE 4,4'-ETHYLENE DIPHENYLENEDIURETHAN *326,34*	414	TRI				19.7				312		
POLY-ETHYLENE 4,4'-METHYLENE DIPHENYLENEDIURETHAN *312,31*	414	HEX				15.7				239		
POLY-ETHYLENE NONAMETHYLENE DIURETHAN *272,34*	414					36.2				168		
POLY-HEPTAMETHYLENE-UREA *154,22*											2.54/404	
POLY-HEXAMETHYLENE UREA *140,20*											3.31/404	
POLY-TETRAMETHYLENE HEXAMETHYLENE DIURETHAN *258,31*	402	TRI		4.95	8.69	19.17	90 10 60	2	1.248	180/291		
	334	TRI		9.05	19.1	8.35	90 63 65	4	1.510	173/101 184/402 182/473		

POLYMER	REF	CRYST SYST.	SPACE GROUP	UNIT CELL PARAMETERS A	B	C	ANGLES	MON/ UNIT CELL	DENSITY (G/CC) CRYSTAL	AMORPH.	MELT. POINT	HEAT OF FUSION KCAL/MON	CHAIN CONF.
POLY-ETHERS													

-HCH-CHOR-													
POLY-ETHYLENE													
---, 1-BUTOXY-2-													
CHLORO-													
-134.60-													
TRANS-	242					6.5							2-3-1
CIS-	242					8.6							2-4-1
---, 1-CHLORO-2-													
ISOBUTOXY-													
TRANS-	242					20.8							2-10-3
-134.60-													
---, 1-ISOBUTOXY-2-													
METHYL-													
TRANS-	67					13.77					226		2-7-2
-114.18-	141					13.8							
POLY-A-METHYLVINYL													
METHYL ETHER													
SYNDIOTACTIC	383	TET		15.2	15.2	16.4		32	1.011				2-8-3
-72.10-													
POLY-VINYL ETHERS													
B---, BENZYL-	253					6.30					162/114		2-3-1
-134.17-													
---, BUTYL-	360	RHO	C31-2	23.7	23.7	6.50		18	0.947	0.92	64/114		2-3-1
-100.16-													
---, SEC-BUTYL-	424	TET		18.25	18.25	35.5		68	0.956		170/382		2-17-5
-100.16-													
---, TERT-BUTYL-	269	TET	C4H-6	18.84	18.84	7.64		16	0.981		160 260/114 238/281		2-4-1
-100.16-													
---, ISOBUTYL-	161	ORTHO		16.8	9.70	6.50		6	0.942	0.94/269	115 117/269 165/114 170/281 115/46		2-3-1
-100.16-													
	152												2-3-1

POLYMER	REF	CRYST SYST.	SPACE GROUP	UNIT CELL PARAMETERS A	B	C	ANGLES	MON/ UNIT CELL	DENSITY (G./CC.) CRYSTAL	AMORPH.	MELT. POINT	HEAT OF FUSION KCAL/MON	CHAIN CONF.
***, ISOPROPYL- *86.13*	161	TET		17.2	17.2	35.5		68	0.926	0.93/269	191/281 98/269 190/114		2*17.5
***, METHYL- *58.08*	176	RHO	D3D.6	16.20	16.20	6.50		18	1.175		144/114		2*3-1
	429	RHO	D3D.6	16.25	16.25	6.50		18	1.168				2*3-1
***, 2-METHYLBUTYL- *114.18*	382					6.50					140		2*3-1
***, NEOPENTYL- *114.18*	161	ORTHO		18.2	10.51	6.50		6	0.915	0.91/269	216/281 155/269 216/114		2*3*1

POLYMER	REF	CRYST SYST.	SPACE GROUP	UNIT CELL PARAMETERS A B C ANGLES			MON/ UNIT CELL	DENSITY (G./CC.) CRYSTAL AMORPH.	MELT. POINT	HEAT OF FUSION KCAL/MON	CHAIN CONF.
POLY-OXIDES											
-X-O-Y-O₀ OR ₀R-O-											
POLY-ACETALDEHYDE *44.05*	92	TET	C4H₀6	14.60	14.60	4.79	16	1.146	165/329		2.4.1
2-CHLORO- *78.50*	387	TET	C4H₀6			4.80					2.4.1
2,2-DICHLORO- *112.95*	387	TET	C4H₀6			5.22					2.4.1
2,2,2-TRICHLORO *147.40*	387	TET	C4H₀6			6.45			220/388		2.4.1
POLY-ACETONE *58.08*	214	TET	S4₀1	14.65	14.65	10.22	28	1.231	60		2.7.2
POLY-2-BUTENE OXIDE *72.10*											
TRANS-	397	ORTHO	D2₀4	13.72	4.60	6.90	4	1.010	114/399		3.2.1
CIS-	397	ORTHO		11.20	10.44	7.01	8	1.168	162/399		3.2.1
POLY-BUTYRALDEHYDE *72.10*	92	TET	C4H₀6	20.01	20.01	4.78	16	1.001	225/329		2.4.1
POLY-1,3-CYCLO, BUTYLENE FORMAL 2,2,4,4-TETRA, METHYL- *156.21*											
TRANS- ALPHA	361					11.5			260 260/369		
BETA	361					5.75					
CIS- ALPHA	361					11.5			285 285/369		
BETA	361					5.75					
POLY-DECAMETHYLENE OXIDE *156.26*	482	ORTHO		7.40	4.94	27.49	4	1.033	79 79/181 72/180 60/291		11.2.1

POLYMER	REF	CRYST SYST.	SPACE GROUP	A	B	C	ANGLES	MON/UNIT CELL	DENSITY CRYSTAL	DENSITY AMORPH	MELT. POINT	HEAT OF FUSION KCAL/MON	CHAIN CONF.
POLY-DODECAMETHYLENE OXIDE *184.31*	482	ORTHO		7.40	4.94	32.53		4	1.029				13*2*1
POLY-EPICHLOROHYDRIN *92.53*	119	ORTHO	D2*4 OR C2V*9	12.14	4.90	7.05		4	1.461		117		
	194					7.07					121		
											135/318		
POLY-ETHYLENE OXIDE *44.05*	109	MONO		9.5	19.5*	12.0	B=101	36	1.207	1.13/366	66/81	1.98/81	3*7*2
	188	MONO	CS*2	8.03	13.09	19.52	B=125.1	28	1.220		62/180	2.8/466	3*7*2
	277	MONO		7.95	13.11	19.39	B=124.6	28	1.231		72/318		
	348	MONO		8.02	13.4	19.25	B=126.9	30	1.326				3*7*2
	303	ORTHO		12.83	12.83	19.3		56	1.289		66		3*7*2
	460	MONO						4	1.234				3*7*2
	390			8.1	12.99	19.30	B=126.1	28	1.239		66	1.75	
	190								1.235	1.124			
									1.220	1.123			
POLY-FORMALDEHYDE I. (OXYMETHYLENE) *30.03*	42	HEX	C3*2 OR C3*3	4.46	4.46	17.30		9	1.506	1.25	181	1.78/91	2*9*5
	134	HEX		4.43	4.43	17.25		9	1.531		198/217	1.59/186	2*29*
	258	HEX		4.470	4.470	56.00		29	1.492		178/329	1.76/343	16
	270										200/455		2*9*5
	362												2*2*1
II.	270	ORTHO	D2*4	4.767	7.660	3.563		4	1.533				
	249	ORTHO		7.75	4.46	17.30		18	1.501				2*4*1
***, CYANOMETHYL- *83.09*	240	PMON		9.44	5.32	4.95	G=102	2	1.135		176		
POLY-HEXAMETHYLENE OXIDE *100.16*	482	MONO	C2H*6	5.65	9.01	17.28	B=134.5	4	1.007		58		6*2*1
											58/181		
POLY- ISOBUTYRALDEHYDE *72.10*	96	TET				5.2					260/329		2*4*1
POLY- ISOVALERALDEHYDE *86.13*	96	TET		20.6	20.6	5.2		16	1.04				
POLY-NOVAMETHYLENE OXIDE *142.23*	482	ORTHO		7.36	4.94	12.45		2	1.043		73		10*1*1
POLY-OCTAMETHYLENE OXIDE I. *128.21*	482	MONO	C2H*6	5.67	9.04	22.45	B=134.5	4	1.038		67		9*2*1

POLYMER	REF	CRYST SYST.	SPACE GROUP	UNIT CELL PARAMETERS A	B	C	ANGLES	MON/UNIT CELL	DENSITY (G/CC) CRYSTAL	AMORPH.	MELT. POINT	HEAT OF FUSION KCAL/MON	CHAIN CONF.
	II. 482	ORTHO		7.36	4.93	22.43		4	1.046				9*2*1
POLY-OXACYCLOBUTANE (TRIMETHYLENE OXIDE) *58.08*	I. 446	MONO	C2H₉3	12.3	7.27	4.80	B=91	16	1.178		36/180 35/447 34/181		4*1*1
	II. 446	RHO	C3v*6	14.13	14.13	8.41		18	1.194				8*1*1
	III. 446	ORTHO	D2*5	9.23	4.82	7.27		4	1.193				8*1*1
-**, 3,3-BISCHLORO-METHYL- *155.03* ALPHA	344	ORTHO		17.85	8.16	4.8		4	1.47		190/173	7.69/271	3*1*1
	378	ORTHO	D2H*2	8.16	17.85	4.82		4	1.467 1.47/343	1.39/343	180/148	5.49/343	3*1*1
BETA	172	MONO	CS*1	6.85	11.42	4.75	B=109.8	2	1.472				3*1*1
	378	OR MONO	CS*2 CS*3	11.42	7.06	4.82	G=114.5	2	1.456				3*1*1
POLY-P-PHENYLENE OXIDE *92.09*	487	ORTHO		5.54	8.07	9.72		4	1.408	1.27	298		7*2*1
-**, 2,6-DIMETHYL- *120.14*	226			8.45	6.02		G=91				261 272/471 262/484	0.9/471 1.2/484	2*4*1
POLY-PROPIONALDEHYDE *58.08*	92	TET	C4H*6	17.52	17.52	4.78		16	1.051		185/329		2*4*1
POLY-PROPYLENE OXIDE *58.08*	13	ORTHO OR	C2V*9 D2*4	10.52	4.67	7.16		4	1.097	.998/139	75/18	1.03/18	3*2*1
	41	ORTHO	D2*4	10.52	4.68	7.10		4	1.104		75/377	2.0/377	
	76	ORTHO	D2*4	10.40	4.64	6.92		4	1.155		73/285		
	448	ORTHO	D2*4	10.46	4.66	7.03		4	1.126				
	495			10.52	4.68								
-**, 3-PHENOXY- *150.17*	236	ORTHO		17.0	8.2	5.48		4	1.30	1.27	215 210/297 208/318		3*2*1
-**, 8-**, O=CHLORO- *184.62*	445	ORTHO		12.6	9.90	6.93		4	1.418		200		
POLY-TETRAHYDROFURAN (TETRAMETHYLENE OXIDE) *72.10*	346	MONO	C2H*6	5.48	8.73	12.07	B=134.2	4	1.157		35	3.0/425	5*2*1
	403	MONO	C2H*6	5.61	8.92	12.25	B=134.5	4	1.095		60/366		5*2*1
	460	MONO	C2H*6	5.59	8.90	12.07	B=134.2	4	1.112		43 37/180		5*2*1

POLYMER	REF	CRYST. SYST.	SPACE GROUP	UNIT CELL A	B	PARAMETERS C	ANGLES	MON/ UNIT CELL	DENSITY (G./CC.) CRYSTAL AMORPH.	MELT. POINT	HEAT OF FUSION KCAL/MON	CHAIN CONF.
POLY-SULFIDES AND POLY-SULFONES												
POLYMER OF SULFUR *32.06*	288	MONO	C2H-2	26.4	9.26*	12.32	B=79.25		2.34			INDET. 1*10*3
	469	ORTHO	PORTH	8.11	9.20	9.25			2.059			
	192		C2-1	13.8*	32.4							
POLY-DIMETHYLENE SULFIDE *60.11*	241	HEX		4.92	4.92	6.74		2	1.413	210 190/181		3*2*1
POLY-ETHYLENE DISULFIDE *92.17*	134					8.8				130/294		
	468					8.8				113/295		4*3*1
POLY-ETHYLENE TETRASULFIDE *156.29*	256	ORTHO		8.57	5.0	4.27		1	1.42			
	257	MONO		8.68	5.03	4.32	G=87	1	1.378			4*1*1
	134					4.32						
POLY-HEXAMETHYLENE PENTAMETHYLENE SULFONE *282.41*	39	MONO		9.88	9.26	34.00	B=121.7	8	1.417			
POLY-HEXAMETHYLENE SULFONE *148.22*	39	MONO		9.88	9.26	18.24	B=121.7	8	1.387			
POLY-HEXAMETHYLENE TETRAMETHYLENE SULFONE *268.38*	39	MONO		9.88	9.26	15.68	B=121.7	4	1.460			
POLY-METHYLENE DISULFIDE *78.15*	467					4.18						3*2*1
POLY-METHYLENE SELENIDE *92.99*	I. 416	HEX		5.22	5.22	46.25		21	2.971	190		2*21*, 11
	423	HEX		5.22	5.22	46.25		21	2.971	178		2*21*, 11
	II. 463	ORTHO	D2*4	5.37	9.03	4.27		4	2.983	170		2*2*1

POLYMER	REF	CRYST SYST.	SPACE GROUP	UNIT CELL A	B	C	PARAMETERS ANGLES	MON/ UNIT CELL	DENSITY (G./CC.) CRYSTAL AMORPH.	MELT. POINT	HEAT OF FUSION KCAL/MON	CHAIN CONF.
POLY-METHYLENE SULFIDE *46.09*	237	ORTHO		12.7	12.0	5.10		16	1.575	260		
	331	HEX		5.07	5.07	36.52		17	1.600	245/181		2*17.9
										260/298		
POLY-PENTAMETHYLENE SULFONE *134.19*	39	MONO		9.88	9.26	.76	B=121.7	4	1.476			
POLY-PENTAMETHYLENE TETRAMETHYLENE SULFONE *254.35*	39	MONO		9.88	9.26	28.33	B=121.7	8	1.532			

POLYMER	REP	CRYST SYST.	SPACE GROUP	UNIT CELL A	PARAMETERS B	C	ANGLES	MON/UNIT CELL	DENSITY (G./CC.) CRYSTAL AMORPH.	MELT. POINT	HEAT OF FUSION KCAL/MON	CHAIN CONF.
CELLULOSICS												
==========												
CELLULOSE *162.14*												
I.	22	MONO		8.35	10.3*	7.9	B=84	4	1.594			
	98	MONO	C2=2	8.20	10.3*	7.90	B=83.3	4	1.625			
	20				10.34							
	28											
II.	22	MONO		8.14	10.3*	9.14	B=62	4	1.592			
	98	MONO		8.02	10.3*	9.03	B=62.8	4	1.623			
	28				10.34							
III.									1.62/28			
IV.	98	MONO		8.12	10.3*	7.99	B=90	4	1.612			
	239			7.9	10.3*							
X.	98	MONO		8.10	10.3*	8.16	B=78.3	4	1.615			
	218	MONO		8.12	10.3*	7.99	B=90.0	4	1.612			
***, TRIACETATE *288.25*	4	PORTH	C2=2	24.5	11.6*	10.43		8	1.292	306/144		
***, TRIBUTYRATE *372.40*										206/46 207/118	3.0/118	
***, TRICAPRYLATE *540.72*										116/117	3.1/117	
***, TRINITRATE *297.14*	24	ORTHO		12.25	25.4*	9.0		8	1.409	697/77 700/81	0.9 - 1.5/77	
***, 2,4-TRINITRATE										617/81	1.35/81	

OTHER POLYMERS

POLYMER	REF	CRYST SYST.	SPACE GROUP	A	B	C	ANGLES	MON/ UNIT CELL	DENSITY (G./CC.) CRYSTAL	AMORPH.	MELT. POINT	HEAT OF FUSION KCAL/MON	CHAIN CONF.
POLY-DIMETHYL KETENE (KETONE) I. *70.09* ALPHA	432 179	ORTHO	C2V-9	12.85	6.53	8.80 8.8		8	1.261		255/130 250		8*1*1
BETA	456					4.40							2*2*1
(ESTER) II. *140.18*													
POLY-(2,6-DIOXO-1,4-PIPERIDINEDIYL) TRIMETHYLENE *53.18*	481	TRI		9.64	11.32	15.80	9896114	8	1.326		170/179 180/248		
POLY-HYDROXY-METHYLENE *30.03*	338					2.5					280		
POLY-KETONE (ETHYLENE-CO-CARBON-MONOXIDE) *56.06*	193	ORTHO	D2H-16	7.97	4.76	7.57		4	1.296				3*2*1
POLY-OCTAMETHYLENE 5,5'-DIBENZIMIDAZOLE *344.44*	470					21							
POLY-PHOSPHONITRILE CHLORIDE *115.90*	230	ORTHO		11.07	4.92*	12.72		8	2.222				2*2*1
	56	ORTHO	C2V-9	11.07	4.92*	12.72		8	2.222	1.91			
POLY-SILOXANE ***, DIMETHYL* *74.15*	197	MONO		13.0	8.3*	7.75	B=60	6	1.02	0.98			
POLY-TETRAMETHYL-P-SILPHENYLENE SILOXANE *208.40*											148/225	4.35/225	

C. REFERENCES

[1] H. Tadokoro, K. Kozai, S. Seki, and I. Nitta, *J. Polymer Sci.*, **26**, 379 (1957).

[2] W. P. Slichter, *J. Polymer Sci.*, **35**, 77 (1959).

[3] D. R. Holmes, C. W. Bunn, and D. J. Smith, *J. Polymer Sci.*, **17**, 159 (1955).

[4] W. J. Dulmage, *J. Polymer Sci.*, **26**, 277 (1957).

[5] A. Prietzschk, *Kolloid-Z.*, **156**, 8 (1958).

[6] R. L. Miller, unpublished results.

[7] G. Natta and P. Corradini, *J. Polymer Sci.*, **20**, 251 (1956); *Atti Accad. Nazl. Lincei, Rend., Classe Sci. Fis., Mat. Nat.*, **19**, 229 (1955).

[8] Y. Tanada, K. Imada, and M. Takayanagi, *Kogyo Kagaku Zasshi*, **69**, 1971 (1966).

[9] G. Natta, P. Corradini, and I. W. Bassi, *Atti Accad. Nazl. Lincei, Rend., Classe Sci. Fis., Mat. Nat.*, **19**, 404 (1955).

[10] G. Natta, *SPE* (Soc. Plastics Engrs.) *J.*, **15**, 373 (1959).

[11] C. W. Bunn and E. R. Howells, *Nature*, **174**, 549 (1954).

[12] G. Natta, L. Porri, P. Corradini, and D. Morero, *Atti Accad. Nazl. Lincei, Rend., Classe Sci. Fis., Mat. Nat.*, **20**, 560 (1956).

[13] G. Natta, P. Corradini, and G. Dall Asta, *Atti Accad. Nazl. Lincei, Rend., Classe Sci. Fis., Mat. Nat.*, **20**, 408 (1956).

[14] C. W. Bunn, *Trans. Faraday Soc.*, **35**, 482 (1939).

[15] E. R. Walter and F. P. Reding, *J. Polymer Sci.*, **21**, 561 (1956).

[16] M. L. Miller and C. E. Rauhut, *J. Polymer Sci.*, **38**, 63 (1959).

[17] G. Natta, *J. Polymer Sci.*, **16**, 143 (1955).

[18] C. C. Price, M. Osgan, R. E. Hughes, and C. Shambelan, *J. Am. Chem. Soc.*, **78**, 690 (1956).

[19] G. C. Overberger and H. Jabloner, *J. Am. Chem. Soc.*, **85**, 3431 (1963).

[20] C. Legrand, *Acta Cryst.*, **5**, 800 (1952).

[21] H. S. Kaufman, *J. Am. Chem. Soc.*, **75**, 1477 (1953).

[22] P. H. Hermans, *Physics and Chemistry of Cellulose Fibres*, Elsevier, New York, 1949.

[23] C. W. Bunn, *Chemical Crystallography*, Clarendon Press, Oxford, 1946.

[24] H. S. Peiser, H. P. Rooksby, and A. J. C. Wilson, *X-Ray Diffraction by Polycrystalline Materials*, Chapman, London, 1955.

[25] C. W. Bunn and E. V. Garner, *Proc. Roy. Soc.* (London), **A189**, 39 (1947).

[26] A. Okada, *Kobunshi Kagaku*, **7**, 122 (1950).

[27] R. de P. Daubeny, C. W. Bunn, and C. J. Brown, *Proc. Roy. Soc.* (London), **A226**, 531 (1954).

[28] H. J. Wellard, *J. Polymer Sci.*, **13**, 471 (1954).

[29] C. W. Bunn, *Nature*, **161**, 929 (1948).

[30] J. D. Stroupe and R. E. Hughes, *J. Am. Chem. Soc.*, **80**, 1768 (1958).

[31] T. G Fox, B. S. Garrett, W. E. Goode, S. Gratch, J. F. Kincaid, A. Spell, and J. D. Stroupe, *J. Am. Chem. Soc.*, **80**, 1768 (1958).

[32] R. C. Reinhardt, *Ind. Eng. Chem.*, **35**, 422 (1943).

[33] S. Narita and K. Okuda, *J. Polymer Sci.*, **38**, 270 (1959).

[34] C. S. Fuller, S. J. Frosch, and N. R. Pape, *J. Am. Chem. Soc.*, **62**, 1905 (1940).

[35] G. Natta, P. Corradini, and I. W. Bassi, *Makromol. Chem.*, **21**, 240 (1956); *Nuovo Cimento*, Suppl. **15**, 52 (1960).

[36] V. M. Coiro, P. De Santis, L. Mazzarella, and L. Picozzi, *Chim. Ind.* (Milan), **47**, 1236 (1965).

[37] G. Natta, P. Corradini, and L. Porri, *Atti Accad. Nazl. Lincei, Rend., Classe Sci. Fis., Mat. Nat.*, **20**, 728 (1956).

REFERENCES

[38] J. D. McCullough, R. S. Bauer, and T. L. Jacobs, *Chem. Ind.* (London) **1957**, 706.

[39] H. D. Noether, *J. Polymer Sci.*, **25**, 217 (1957).

[40] C. W. Bunn, *Proc. Roy. Soc.* (London), **A180**, 40 (1942).

[41] C. Shambelan, Ph.D. Thesis, University of Pennsylvania, 1959 – *Dissertation Abstr.*, **20**, 120 (1959).

[42] C. F. Hammer, T. A. Koch, and J. F. Whitney, *J. Appl. Polymer Sci.*, **1**, 169 (1959).

[43] W. Goggin and R. Lowry, *Ind. Eng. Chem.*, **34**, 327 (1942).

[44] G. Natta, *Chem. Ind.* (London), **1957**, 1520.

[45] G. Natta, P. Pino, P. Corradini, F. Danusso, E. Mantica, G. Mazzanti, and G. Moraglio, *J. Am. Chem. Soc.*, **77**, 1708 (1955).

[46] R. Boyer, *Compt. rend. de la 2ᵉ reunion de chimie physique*, June 2–7, 1952, Paris, 383.

[47] D. E. Roberts and L. Mandelkern, *J. Am. Chem. Soc.*, **80**, 1289 (1958).

[48] F. P. Reding, *J. Polymer Sci.*, **21**, 547 (1956).

[49] J. D. Hoffman and J. J. Weeks, *J. Res. Natl. Bur. Std.*, **60**, 465 (1958).

[50] L. Wood, N. Bekkedahl, and R. E. Gibson, *J. Chem. Phys.*, **13**, 475 (1945).

[51] R. Beaman and F. Cramer, *J. Polymer Sci.*, **21**, 223 (1956).

[52] H. Starkweather, Jr., G. Moore, J. Hansen, T. Roder, and R. Brooks, *J. Polymer Sci.*, **21**, 189 (1956).

[53] R. Wiley, *Ind. Eng. Chem.*, **38**, 959 (1946).

[54] H. W. Starkweather, Jr. and R. E. Moynihan, *J. Polymer Sci.*, **22**, 363 (1956).

[55] F. C. Frank, A. Keller, and A. O'Connor, *Phil. Mag.*, **4**, 200 (1959).

[56] K. H. Meyer, W. Lotmar, and G. W. Pankow, *Helv. Chim. Acta*, **19**, 930 (1936).

[57] L. E. Nielsen, unpublished results.

[58] P. W. Teare and D. R. Holmes, *J. Polymer Sci.*, **24**, 496 (1957).

[59] K. Little, *Brit. J. Appl. Phys.*, **10**, 225 (1959).

[60] C. J. B. Clews, *Proc. Rubber Technology Conf.*, Heffer, Cambridge, England, 1938, p. 955.

[61] W. P. Slichter, *J. Polymer Sci.*, **36**, 259 (1959).

[62] R. C. Golike, *J. Polymer Sci.*, **42**, 583 (1960).

[63] Y. Kinoshita, *Makromol. Chem.*, **33**, 21 (1959).

[64] P. H. Burleigh, *J. Am. Chem. Soc.*, **82**, 749 (1960).

[65] Y. Kinoshita, *Makromol. Chem.*, **33**, 1 (1959).

[66] R. E. Moynihan, *J. Am. Chem. Soc.*, **81**, 1045 (1959).

[67] G. Natta, *Makromol. Chem.*, **35**, 93 (1960).

[68] J. Bateman, R. E. Richards, G. Farrow, and I. M. Ward, *Polymer*, **1**, 63 (1960).

[69] P. Corradini, *Atti Accad. Nazl. Lincei, Rend., Classe Sci. Fis., Mat. Nat.*, **25**, 517 (1958).

[70] S. D. Gehman, J. E. Field, and R. P. Dinsmore, *Proc. Rubber Technology Conf.*, Heffer, Cambridge, England, 1938, p. 961.

[71] G. Natta, G. Mazzanti, P. Longi, and F. Bernardini, *Chim. Ind.* (Milan), **40**, 813 (1958).

[72] W. D. Niegisch and P. R. Swan, *J. Appl. Phys.*, **31**, 1906 (1960).

[73] E. Martuscelli, R. Gallo, and G. Paiaro, *Makromol. Chem.*, **103**, 295 (1967).

[74] D. Sianesi, G. Natta, and P. Corradini, *Gazz. Chim. Ital.*, **89**, 775 (1959).

[75] G. Natta, F. Danusso, and D. Sianesi, *Makromol. Chem.*, **28**, 253 (1958); D. Sianesi, M. Rampichini, and F. Danusso, *Chim. Ind.* (Milan), **41**, 287 (1959).

[76] G. Natta, G. Mazzanti, and P. Corradini, *Atti Accad. Nazl. Lincei. Rend., Classe Sci. Fis., Mat. Nat.*, **25**, 3 (1958).

[77] W. R. Krigbaum and N. Tokita, *J. Polymer Sci.*, **43**, 467 (1960).

[78] E. Stanley and M. Litt, *J. Polymer Sci.*, **43**, 453 (1960).

[79] D. I. Sapper, *J. Polymer Sci.*, **43**, 383 (1960).

[80] P. Corradini and P. Ganis, *J. Polymer Sci.*, **43**, 311 (1960).

[81] L. Mandelkern, *Chem. Revs.*, **56**, 903 (1956).

[82] J. R. Schaefgen, *J. Polymer Sci.*, **38**, 549 (1959).

[83] F. Danusso, G. Moraglio, and E. Flores, *Atti Accad. Nazl. Lincei, Rend., Classe Sci. Fis., Mat. Nat.*, **25**, 520 (1958).

[84] N. Yoda and I. Matsubara, *J. Polymer Sci.*, Part A, **2**, 253 (1964).

[85] F. A. Quinn, Jr. and L. Mandelkern, *J. Am. Chem. Soc.*, **80**, 3178 (1958).

[86] F. W. Billmeyer, Jr., *J. Appl. Phys.*, **28**, 1114 (1957).

[87] M. Dole, *J. Polymer Sci.*, **19**, 347 (1956).

[88] A. M. Bueche, *J. Am. Chem. Soc.*, **74**, 65 (1952).

[89] F. Danusso and G. Moraglio, *Atti Accad. Nazl. Lincei, Rend., Classe Sci. Fis., Mat. Nat.*, **27**, 381 (1959).

[90] T. W. Campbell and A. C. Haven, Jr., *J. Appl. Polymer Sci.*, **1**, 73 (1959).

[91] H. W. Starkweather, Jr. and R. H. Boyd, *J. Phys. Chem.*, **64**, 410 (1960).

[92] G. Natta, G. Mazzanti, P. Corradini, and I. W. Bassi, *Makromol. Chem.*, **37**, 156 (1960); G. Natta, G. Mazzanti, P. Corradini, P. Chini, and I. W. Bassi, *Atti Accad. Nazl. Lincei, Rend., Classe Sci. Fis., Mat. Nat.*, **28**, 8 (1960); I. W. Bassi, *Rend. Ist. Lombardo Sci. Lettere*, **A94**, 579 (1960).

[93] D. V. Badami, *Polymer*, **1**, 273 (1960).

[94] J. H. Griffith and B. G. Ranby, *J. Polymer Sci.*, **44**, 369 (1960).

[95] C. G. Overberger, A. E. Borchert, and A. Katchman, *J. Polymer Sci.*, **44**, 491 (1960).

[96] G. Natta, G. Mazzanti, P. Corradini, A. Valvassori, and I. W. Bassi, *Atti Accad. Nazl. Lincei, Rend., Classe Sci. Fis., Mat. Nat.*, **28**, 18 (1960).

[97] G. Natta and I. W. Bassi, *European Polymer J.*, **3**, 33 (1967).

[98] O. Ellefsen, *Norelco Reporter*, **7**, 104 (1960).

[99] G. Farrow, J. McIntosh, and I. M. Ward, *Makromol. Chem.*, **38**, 147 (1960).

[100] N. G. Esipova, L. Pan-Tun, N. S. Andreeva, and P. V. Kozlov, *Vysokomolekul. Soedin.*, **2**, 1109 (1960).

[101] A. G. M. Last, *J. Polymer Sci.*, **39**, 543 (1959).

[102] D. J. Shields and H. W. Coover, Jr., *J. Polymer Sci.*, **39**, 532 (1959).

[103] S. Murahashi, S. Nozakura, and H. Tadokoro, *Bull. Chem. Soc. Japan*, **32**, 534 (1959).

[104] T. F. Yen, *J. Polymer Sci.*, **38**, 272 (1959).

[105] C. S. Fuller and C. L. Erickson, *J. Am. Chem. Soc.*, **59**, 344 (1937).

[106] C. S. Fuller and C. J. Frosch, *J. Am. Chem. Soc.*, **61**, 2575 (1939).

[107] C. S. Fuller, C. J. Frosch, and N. R. Pape, *J. Am. Chem. Soc.*, **64**, 154 (1942).

[108] C. S. Fuller and C. J. Frosch, *J. Phys. Chem.*, **43**, 323 (1939).

[109] C. S. Fuller, *Chem. Revs.*, **26**, 143 (1940).

[110] W. O. Baker and C. S. Fuller, *J. Am. Chem. Soc.*, **64**, 2399 (1942).

[111] A. Conix and R. Van Kerpel, *J. Polymer Sci.*, **40**, 521 (1959).

[112] J. A. Gailey and R. H. Ralston, *SPE* (Soc. Plastics Engrs.) *Trans.*, **4**, 29 (1964).

[113] H. S. Yanai, quoted in C. F. Ryan and J. J. Gormley, *Macromol. Synth.*, **1**, 30 (1963).

[114] E. J. Vandenberg, R. F. Heck, and D. S. Breslow, *J. Polymer Sci.*, **41**, 519 (1960).

[115] A. M. Liquori, *Acta Cryst.*, **8**, 345 (1955).

REFERENCES

[116] S. Furuya and M. Honda, *J. Polymer Sci.*, **28**, 232 (1958).
[117] P. Goodman, *J. Polymer Sci.*, **24**, 307 (1960).
[118] L. Mandelkern and P. J. Flory, *J. Am. Chem. Soc.*, **73**, 3206 (1951).
[119] S. Ishida and S. Murahashi, *J. Polymer Sci.*, **40**, 571 (1959).
[120] B. S. Garrett, W. E. Goode, S. Gratch, J. F. Kincaid, C. L. Levesque, A. Spell, J. D. Stroupe, and W. H. Watanabe, *J. Am. Chem. Soc.*, **81**, 1007 (1959).
[121] D. C. Vogelsong and E. M. Pearce, *J. Polymer Sci.*, **45**, 546 (1960).
[122] P. Corradini and P. Ganis, *Nuovo Cimento*, Suppl. **15**, 104 (1960).
[123] G. Natta, P. Corradini, and I. W. Bassi, *Nuovo Cimento*, Suppl. **15**, 83 (1960).
[124] G. Natta and P. Corradini, *Nuovo Cimento*, Suppl. **15**, 111 (1960); *Angew. Chem.*, **68**, 615 (1956).
[125] P. Corradini and P. Ganis, *Nuovo Cimento*, Suppl. **15**, 96 (1960).
[126] G. Natta and I. W. Bassi, *European Polymer J.*, **3**, 43 (1967).
[127] G. Natta and P. Corradini, *Nuovo Cimento*, Suppl. **15**, 40 (1960).
[128] G. Natta, P. Corradini, and I. W. Bassi, *Nuovo Cimento*, Suppl. **15**, 68 (1960).
[129] P. Pino and G. P. Lorenzi, *J. Am. Chem. Soc.*, **82**, 4745 (1960).
[130] G. Natta, G. Mazzanti, G. Pregaglia, M. Binaghi, and M. Peraldo, *J. Am. Chem. Soc.*, **82**, 4742 (1960).
[131] Z. W. Wilchinsky, *J. Appl. Phys.*, **31**, 1969 (1960).
[132] T. Mochizuki, *Nippon Kagaku Zasshi*, **81**, 15 (1960).
[133] R. Stefani, M. Chevreton, M. Garnier, and C. Eyraud, *Compt. Rend.*, **251**, 2174 (1960).
[134] M. L. Huggins, *J. Chem. Phys.*, **13**, 37 (1945).
[135] C. Ruscher and H. J. Schroder, *Faserforsch. Textiltech.*, **11**, 165 (1960).
[136] Z. Mencik, *Chem. Prumysl*, **10**, 377 (1960).
[137] M. Dole and B. Wunderlich, *Makromol. Chem.*, **34**, 29 (1959).
[138] G. Allen, G. Gee, and G. J. Wilson, *Polymer*, **1**, 456 (1960).
[139] G. Allen, G. Gee, D. Mangaraj, D. Sims, and G. J. Wilson, *Polymer*, **1**, 466 (1960).
[140] M. Asahina and K. Ohuda, *Kobunshi Kagaku*, **17**, 607 (1960).
[141] G. Natta, M. Farino, M. Peraldo, P. Corradini, G. Bressan, and P. Ganis, *Atti Accad. Nazl. Lincei, Rend., Classe Sci. Fis., Mat. Nat.*, **28**, 442 (1960).
[142] L. A. Auspos, C. W. Burnam, L. Hall, J. K. Hubbard, W. Kirk, J. R. Schaefgen, and S. B. Speck, *J. Polymer Sci.*, **15**, 19 (1955).
[143] A. T. Walter, *J. Polymer Sci.*, **13**, 207 (1954).
[144] C. J. Malm, J. W. Mench, D. L. Kendall, and G. D. Hiatt, *Ind. Eng. Chem.*, **43**, 688 (1951).
[145] C. E. Anagnostopoulos, A. Y. Coran, and H. R. Gamrath, *J. Appl. Polymer Sci.*, **4**, 181 (1960).
[146] C. F. Horn, B. T. Freure, H. Vineyard, and H. J. Decker, *Angew. Chem.*, **74**, 531 (1962).
[147] G. Natta, G. Dall Asta, and G. Mazzanti, *Angew. Chem.*, **76**, 765 (1964); *Angew. Chem.*, International Ed., **3**, 723 (1964); G. Natta and I. W. Bassi, *Atti Accad. Nazl. Lincei, Rend., Classe Sci. Fis., Mat. Nat.*, **38**, 315 (1965); *J. Polymer Sci.*, Part C, **16**, 2551 (1967).
[148] A. C. Farthing and W. J. Reynolds, *J. Polymer Sci.*, **12**, 503 (1954).
[149] H. S. Yanai, quoted in W. E. Goode, R. P. Fellman, and F. H. Owens, *Macromol. Synth.*, **1**, 25 (1963).
[150] A. Turner-Jones, *Makromol. Chem.*, **71**, 1 (1964).
[151] M. Niinomi, T. Fukuda, and M. Takayanagi, *J. Polymer Sci.* (in press).

[152] G. Natta, *Angew. Chem.*, **68**, 393 (1956).

[153] G. F. Schmidt and H. A. Stuart, *Z. Naturforschung*, **13A**, 222 (1958).

[154] G. Natta and P. Corradini, *Nuovo Cimento*, Suppl. **15**, 9 (1960).

[155] F. Rybnikar, *Chem. Listy*, **52**, 1042 (1958).

[156] B. Wunderlich and M. Dole, *J. Polymer Sci.*, **24**, 201 (1957).

[157] C. W. Smith and M. Dole, *J. Polymer Sci.*, **20**, 37 (1956).

[158] B. Wunderlich and M. Dole, *J. Polymer Sci.*, **32**, 125 (1958).

[159] P. J. Flory, H. D. Bedon, and E. H. Keefer, *J. Polymer Sci.*, **28**, 151 (1958).

[160] R. D. Evans, M. R. Mighton, and P. J. Flory, *J. Am. Chem. Soc.*, **72**, 2018 (1950).

[161] G. Dall Asta and N. Oddo, *Chim. Ind.* (Milan), **42**, 1234 (1960).

[162] Z. Mencik, *Vysokomolekul. Soedin.*, **2**, 1635 (1960).

[163] Y. Chatani, *J. Polymer Sci.*, **47**, 491 (1960).

[164] G. Wasai, T. Tsuruta, and J. Furukawa, *Kogyo Kagaku Zasshi*, **66**, 1339 (1963).

[165] M. Cesari, Private Communication.

[166] H. D. Keith, F. J. Padden, Jr., N. M. Walter, and H. W. Wyckoff, *J. Appl. Phys.*, **30**, 1485 (1959).

[167] G. Natta, M. Peraldo, and P. Corradini, *Atti Accad. Nazl. Lincei, Rend., Classe Sci. Fis., Mat. Nat.*, **26**, 14 (1959).

[168] K. Okuda, T. Yoshida, M. Sugita, and M. Asahina, *J. Polymer Sci.*, Part B, **5**, 465 (1967).

[169] G. Natta, I. Pasquon, P. Corradini, M. Peraldo, M. Pegoraro, and A. Zambelli, *Atti Accad. Nazl. Lincei, Rend., Classe Sci. Fis., Mat. Nat.*, **28**, 539 (1960).

[170] N. M. Walter, quoted in C. Y. Liang and F. G. Pearson, *J. Mol. Spectry.*, **5**, 290 (1960).

[171] S. S. Leshchenko, V. L. Karpov, and V. A. Kargin, *Vysokomolekul. Soedin.*, **1**, 1538 (1959); *Vysokomolekul. Soedin.*, **5**, 953 (1963); *Polymer Sci.* (USSR) (English transl.), **5**, 1 (1964).

[172] D. J. H. Sandiford, *J. Appl. Chem.*, **8**, 188 (1958).

[173] M. Hatano and S. Kambara, *Polymer*, **2**, 1 (1961).

[174] R. Dedeurwaerder and J. F. M. Oth, *Bull. Soc. Chim. Belges*, **70**, 37 (1961).

[175] G. Natta, D. Sianesi, D. Morero, I. W. Bassi, and G. Caporiccio, *Atti Accad. Nazl. Lincei, Rend., Classe Sci. Fis., Mat. Nat.*, **28**, 551 (1960).

[176] I. W. Bassi, *Atti Accad. Nazl. Lincei, Rend., Classe Sci. Fis., Mat. Nat.*, **29**, 193 (1960).

[177] G. Champetier, M. Laualov, and J. P. Pied, *Bull. Soc. Chim. France*, **1958**, 708.

[178] A. G. Nasini, L. Trossarelli, and G. Saini, *Makromol. Chem.*, **44–46**, 550 (1961).

[179] G. Natta, G. Mazzanti, G. F. Pregaglia, and M. Binaghi, *Makromol. Chem.*, **44–46**, 537 (1961).

[180] J. C. Swallow, *Proc. Roy. Soc.* (London), **A238**, 1 (1956).

[181] J. Lal and G. S. Trick, *J. Polymer Sci.*, **50**, 13 (1961).

[182] R. J. Wilkinson and M. Dole, *J. Polymer Sci.*, **58**, 1089 (1962).

[183] F. B. Moody, 140th National American Chemical Society Meeting, Chicago, September 1961—*ACS Polymer Preprints*, **2**, 285 (1961).

[184] I. Sakurada, K. Nukushina, and Y. Sone, *Kobunshi Kagaku*, **12**, 506 (1955).

[185] G. Natta, G. Mazzanti, P. Longi, G. Dall Asta, and F. Bernardini, *J. Polymer Sci.*, **51**, 487 (1961).

[186] M. Inoue, *J. Polymer Sci.*, **51**, S18 (1961).

[187] J. A. Price, M. R. Lytton, and B. G. Ranby, *J. Polymer Sci.*, **51**, 541 (1961).

[188] F. P. Price and R. W. Kilb, *J. Polymer Sci.*, **57**, 395 (1962).

[189] G. Natta, L. Porri, P. Corradini, G. Zanini, and F. Ciampelli, *J. Polymer Sci.*, **51**, 463 (1961).

REFERENCES

[190] P. Arlie and A. Skoulios, *Compt. Rend.*, **258**, 2570 (1964).

[191] U. Giannini, G. Bruckner, E. Pellino, and A. Cassata, *J. Polymer Sci.*, Part B, **5**, 527 (1967).

[192] S. Geller, *Science*, **152**, 644 (1966); M. D. Lind and S. Geller, *American Crystallographic Association Abstracts*, Summer Meeting, August 20–25, 1967, University of Minnesota, p. 108.

[193] Y. Chatani, T. Takizawa, S. Murahashi, Y. Sakata, and Y. Nishimura, *J. Polymer Sci.*, **55**, 811 (1961).

[194] J. R. Richards, Ph.D. Thesis, University of Pennsylvania, 1961, *Dissertation Abstr.*, **22**, 1029 (1961).

[195] H. G. Kilian, H. Haboth, and E. Jenckel, *Kolloid-Z.*, **172**, 166 (1960).

[196] E. J. Addink and J. Beintema, *Polymer*, **2**, 185 (1961).

[197] G. Damaschun, *Kolloid-Z.*, **180**, 65 (1962).

[198] F. L. Saunders, *J. Polymer Sci.*, Part A-1, **5**, 2187 (1967).

[199] C. A. Boye, *J. Polymer Sci.*, **55**, 275 (1961).

[200] A. Kawasaki, J. Furukawa, T. Tsuruta, Y. Nakayama, and G. Wasai, *Makromol. Chem.*, **49**, 112 (1961).

[201] A. Kawasaki, J. Furukawa, T. Tsuruta, Y. Nakayama, and G. Wasai, *Makromol. Chem.*, **49**, 136 (1961).

[202] T. Makimoto, T. Tsuruta, and J. Furukawa, *Makromol. Chem.*, **50**, 116 (1961).

[203] A. Turner-Jones and C. W. Bunn, *Acta Cryst.*, **15**, 105 (1962).

[204] D. C. Vogelsong, *J. Polymer Sci.*, **57**, 895 (1962).

[205] C. Y. Liang and S. Krimm, *J. Chem. Phys.*, **25**, 563 (1956).

[206] M. G. Broadhurst, *J. Res. Natl. Bur. Std.*, **66A**, 241 (1962); *J. Chem. Phys.*, **36**, 2578 (1962).

[207] R. L. Miller and V. F. Holland, *J. Polymer Sci.*, Part B, **2**, 519 (1964); V. F. Holland and R. L. Miller, *J. Appl. Phys.*, **35**, 3241 (1964).

[208] A. V. Ermolina, G. S. Markova, and V. A. Kargin, *Kristallografiya*, **2**, 623 (1957).

[209] E. S. Clark and L. T. Muus, *Z. Krist.*, **117**, 119 (1962).

[210] V. F. Holland, S. B. Mitchell, W. L. Hunter, and P. H. Lindenmeyer, *J. Polymer Sci.*, **62**, 145 (1962).

[211] J. J. Point, *Bull. Classe Sci. Acad. Roy. Belg.*, **30**, 435 (1953).

[212] L. G. Wallner, *Monatsh. Chem.*, **79**, 279 (1948).

[213] H. Hendus, K. Schmieder, G. Schnell, and K. A. Wolf, *Festschrift Carl Wurster der BASF vom 2/12/1960*.

[214] J. Furukawa, T. Saegusa, T. Tsuruta, S. Ohta, and G. Wasai, *Makromol. Chem.*, **52**, 230 (1962).

[215] G. Champetier and J. P. Pied, *Makromol. Chem.*, **44–46**, 64 (1961).

[216] F. Rybnikar, *Collection Czech. Chem. Commun.*, **24**, 2861 (1959).

[217] S. Okamura, K. Hayashi, and Y. Kitanishi, *J. Polymer Sci.*, **58**, 925 (1962).

[218] O. Ellefsen and N. Norman, *J. Polymer Sci.*, **58**, 769 (1962).

[219] L. A. Errede and R. S. Gregorian, *J. Polymer Sci.*, **60**, 21 (1962).

[220] T. F. Yen, *J. Polymer Sci.*, **35**, 533 (1959).

[221] J. F. Brown, Jr., and D. M. White, *J. Am. Chem. Soc.*, **82**, 5671 (1960).

[222] A. Ziabicki, *Kolloid-Z.*, **167**, 132 (1959).

[223] C. J. Brown and A. C. Farthing, *J. Chem. Soc.*, **1953**, 3270.

[224] M. H. Kaufman, H. F. Mark, and R. R. Mesrobian, *J. Polymer Sci.*, **13**, 3 (1954).

[225] R. L. Merker and M. J. Scott, *J. Polymer Sci.*, Part A, **2**, 15 (1964).

[226] W. A. Butte, C. C. Price, and R. E. Hughes, *J. Polymer Sci.*, **61**, S28 (1962).

[227] E. R. Walter and F. P. Reding, 133rd National American Chemical Society Meeting, San Francisco, April 1958.

[228] G. Natta, I. W. Bassi, and G. Fagherazzi, *European Polymer J.*, **3**, 339 (1967).

[229] C. W. Bunn and E. V. Garner, *J. Chem. Soc.*, **1942**, 654.

[230] E. Giglio, F. Pompa, and A. Ripamonti, *J. Polymer Sci.*, **59**, 293 (1960).

[231] F. Sakaguchi, R. Kitamaru, and W. Tsuji, *Bull. Inst. Chem.-Res., Kyoto Univ.*, **44**, 155 (1966).

[232] K. Fujii, T. Mochizuki, S. Imoto, J. Ukida, and M. Matsumoto, *Makromol. Chem.*, **51**, 225 (1962).

[233] K. Tanaka, T. Seto, and T. Hara, *J. Phys. Soc. Japan*, **17**, 873 (1962); T. Seto, T. Hara, and K. Tanaka, *Japan. J. Appl. Phys.*, **7**, 31 (1968).

[234] G. Natta, L. Porri, A. Carbonaro, F. Ciampelli, and G. Allegra, *Makromol. Chem.*, **51**, 229 (1962).

[235] G. Natta, I. W. Bassi, and P. Corradini, *Atti Accad. Nazl. Lincei, Rend.*, Classe *Sci. Fis., Mat. Nat.*, **31**, 17 (1961).

[236] G. Natta, I. W. Bassi, and G. Allegra, *Atti Accad. Nazl. Lincei, Rend.*, Classe *Sci. Fis., Mat. Nat.*, **31**, 350 (1961).

[237] J. B. Lando and V. Stannett, *J. Polymer Sci.*, Part B, **2**, 375 (1964); 148th National American Chemical Society Meeting, Chicago, September 1964–*ACS Polymer Preprints*, **5**, 969 (1964).

[238] A. Takahashi and S. Kambara, *Makromol. Chem.*, **72**, 92 (1964).

[239] T. Petitpas and J. Mering, *Compt. Rend.*, **254**, 2611 (1962).

[240] H. Sumitomo and K. Kobayashi, *J. Polymer Sci.*, Part A-1, **5**, 2247 (1967).

[241] S. Boileau, J. Coste, J.-M. Raynal, and P. Sigwalt, *Compt. Rend.*, **254**, 2774 (1962).

[242] G. Natta, M. Peraldo, M. Farina, and G. Bressan, *Makromol. Chem.*, **55**, 139 (1962).

[243] R. Aelion, *Ann. Chim.* (Paris), **3**, 5 (1948).

[244] K. Dachs and E. Schwartz, *Angew. Chem.*, **74**, 540 (1962).

[245] K. Okuda, *J. Polymer Sci.*, Part A, **2**, 1749 (1964).

[246] H. Arimoto, *J. Polymer Sci.*, Part A, **2**, 2283 (1964).

[247] G. Natta, G. Allegra, I. W. Bassi, P. Corradini, and P. Ganis, *Makromol. Chem.*, **58**, 242 (1962); G. Natta, P. Corradini, P. Ganis, I. W. Bassi, and G. Allegra, *Chim. Ind.* (Milan), **44**, 532 (1962).

[248] Y. Yamashita and S. Nunomoto, *Makromol. Chem.*, **58**, 244 (1962).

[249] L. Becker, *Wiss. Z. Karl-Marx-Univ. Leipzig, Math.-Naturw. Reihe*, **11**, 3 (1962).

[250] K. J. Clark, A. Turner-Jones, and D. J. H. Sandiford, *Chem. Ind.* (London), **1962**, 2010.

[251] E. Passaglia and H. K. Kevorkian, *J. Appl. Phys.*, **34**, 90 (1963).

[252] A. Nishioka and K. Yanagisawa, *Kobunshi Kagaku*, **19**, 667 (1962).

[253] S. Murahashi, H. Yuki, T. Sano, U. Yonemura, H. Tadokoro, and Y. Chatani, *J. Polymer Sci.*, **62**, S77 (1962).

[254] A. Turner-Jones, *J. Polymer Sci.*, **62**, S53 (1962).

[255] H. Watanabe, R. Koyama, H. Nagai, and A. Nishioka, *J. Polymer Sci.*, **62**, S74 (1962).

[256] J.-J. Trillat and R. Tertian, *Compt. Rend.*, **219**, 395 (1944).

[257] L. Ulicky, *Chem. Zvesti*, **16**, 818 (1962).

[258] E. Sauter, *Z. Physik. Chem.*, **B18**, 417 (1932).

[259] G. Natta, G. Dall Asta, and G. Mazzanti, *Chim. Ind.* (Milan), **44**, 1212 (1962).

[260] M. G. Broadhurst, *J. Res. Natl. Bur. Std.*, **67A**, 233 (1963).

[261] L. Becker, *Plaste Kautschuk*, **8**, 557 (1961).

REFERENCES

[262] B. Wunderlich and C. M. Cormier, *J. Polymer Sci.*, Part A-2, **5**, 987 (1967).
[263] G. Natta, L. Porri, G. Stoppa, G. Allegra, and F. Ciampelli, *J. Polymer Sci.*, Part B, **1**, 67 (1963).
[264] G. W. Taylor, *Polymer*, **3**, 543 (1962).
[265] O. B. Edgar and E. Ellery, *J. Chem. Soc.*, **1952**, 2633.
[266] H. G. Kilian, *Kolloid-Z.*, **185**, 13 (1962).
[267] H. S. Makowski, K. C. Shim, and Z. W. Wilchinsky, *J. Polymer Sci.*, Part A, **2**, 1549 (1964); *ACS Polymer Preprints*, **4**, 43 (1963).
[268] D. Sianesi and G. Caporiccio, *Makromol. Chem.*, **60**, 213 (1963).
[269] I. W. Bassi, G. Dall Asta, U. Campigli, and E. Strepparola, *Makromol. Chem.*, **60**, 202 (1963).
[270] G. Carazzolo, S. Leghissa, and M. Mammi, *Makromol. Chem.*, **60**, 171 (1963); G. Carazzolo and M. Mammi, *J. Polymer Sci.*, Part A, **1**, 965 (1963); G. Carazzolo, *Gazz. Chim. Ital.*, **92**, 1345 (1962); *J. Polymer Sci.*, Part A, **1**, 1573 (1963); G. Carazzolo and G. Putti, *Chim. Ind.* (Milan), **45**, 771 (1963).
[271] E. Baer, J. R. Collier, and D. R. Carter, *SPE* (Soc. Plastics Eng.) *Trans.*, **5**, 22 (1965).
[272] D. C. Vogelsong, *J. Polymer Sci.*, Part A, **1**, 1055 (1963).
[273] F. H. C. Crick and A. Rich, *Nature*, **176**, 780 (1955).
[274] G. Perego and I. W. Bassi, *Makromol. Chem.*, **61**, 198 (1963).
[275] U. Giannini, M. Cambini, and A. Cassata, *Makromol. Chem.*, **61**, 246 (1963).
[276] F. Danusso and G. Gianotti, *Makromol. Chem.*, **61**, 164 (1963).
[277] F. Danusso and G. Gianotti, *Makromol. Chem.*, **61**, 139 (1963).
[278] M. Donati and M. Farina, *Makromol. Chem.*, **60**, 233 (1963).
[279] E. M. Bradbury and A. Elliott, *Polymer*, **4**, 47 (1963).
[280] H. Utsunomiya, N. Kawasaki, M. Niinomi, and M. Takayanagi, *J. Polymer Sci.*, Part B, **5**, 907 (1967).
[281] E. J. Vandenberg, *J. Polymer Sci.*, Part C, **1**, 207 (1963).
[282] K. R. Dunham, J. Vandenberghe, J. W. H. Faber, and L. E. Contois, *J. Polymer Sci.*, Part A, **1**, 751 (1963).
[283] M. Tomika, *Kobunshi Kagaku*, **20**, 145 (1963).
[284] N. Calderon and M. C. Morris, *J. Polymer Sci.*, Part A-2, **5**, 1283 (1967).
[285] N. S. Chu and C. C. Price, *J. Polymer Sci.*, Part A, **1**, 1105 (1963).
[286] K. H. Meyer and A. van der Wyk, *Helv. Chim. Acta*, **20**, 1313 (1937).
[287] W. Cooper and R. K. Smith, *J. Polymer Sci.*, Part A, **1**, 159 (1963).
[288] K. H. Meyer, *Natural and Synthetic High Polymers*, Interscience, New York, 1950.
[289] J. R. Whinfield, *Nature*, **158**, 930 (1946).
[290] E. F. Izard, *J. Polymer Sci.*, **8**, 503 (1952).
[291] R. Hill and E. E. Walker, *J. Polymer Sci.*, **3**, 609 (1948).
[292] D. D. Coffman, M. L. Cox, E. L. Martin, W. E. Mochel, and F. J. van Natta, *J. Polymer Sci.*, **3**, 85 (1948).
[293] J. R. Schaefgen and P. J. Flory, *J. Am. Chem. Soc.*, **70**, 2709 (1948).
[294] J. C. Patrick, *Trans. Faraday Soc.*, **32**, 347 (1936).
[295] C. W. Bunn, *J. Polymer Sci.*, **16**, 323 (1955).
[296] E. L. Gal Perin, S. S. Dubov, E. V. Volkova, and M. P. Mlenik, *Kristallografiya*, **9**, 102 (1964); *Soviet Phys.-Cryst.*, **9**, 81 (1964).
[297] A. Noshay and C. C. Price, *J. Polymer Sci.*, **34**, 165 (1959).
[298] E. Gipstein, E. Wellisch, and O. J. Sweeting, *J. Polymer Sci.*, Part B, **1**, 237 (1963).

[299] W. Marconi, A. Mazzei, S. Cucinella, and M. Cesari, *J. Polymer Sci.*, Part A, **2**, 4261 (1964).

[300] W. H. Carothers and J. A. Arvin, *J. Am. Chem. Soc.*, **51**, 2560 (1929).

[301] K. Ueberreiter and N. Steiner, *Makromol. Chem.*, **74**, 158 (1964).

[302] W. H. Carothers and G. L. Dorough, *J. Am. Chem. Soc.*, **52**, 711 (1930).

[303] H. Tadokoro, Y. Chatani, T. Yoshihara, S. Tahara, and S. Murahashi, *Makromol. Chem.*, **73**, 109 (1964).

[304] W. H. Carothers and G. J. Berchet, *J. Am. Chem. Soc.*, **52**, 5289 (1930).

[305] W. H. Carothers, *Chem. Revs.*, **8**, 353 (1931).

[306] G. Natta, M. Peraldo, and G. Allegra, *Makromol. Chem.*, **75**, 215 (1964).

[307] W. H. Carothers and J. W. Hill, *J. Am. Chem. Soc.*, **54**, 1559 (1932).

[308] A. Turner-Jones, J. M. Aizlewood, and D. R. Beckett, *Makromol. Chem.*, **75**, 134 (1964).

[309] W. H. Carothers, *J. Am. Chem. Soc.*, **55**, 4714 (1933).

[310] M. I. Bessonov and A. P. Rudakov, *Fiz. Tverd. Tela*, **6**, 1333 (1964).

[311] P. R. Swan, *J. Polymer Sci.*, **56**, 403 (1962).

[312] K. Hirono, G. Wasai, T. Saegusa, and J. Furukawa, *Kogyo Kagaku Zasshi*, **67**, 604 (1964).

[313] J. Boor, Jr., and J. C. Mitchell, *J. Polymer Sci.*, Part A, **1**, 59 (1963).

[314] G. Wasai, T. Saegusa, and J. Furukawa, *Kogyo Kagaku Zasshi*, **67**, 601 (1964).

[315] B. Wunderlich and T. Arakawa, *J. Polymer Sci.*, Part A, **2**, 3697 (1964).

[316] K. Tsuboi and T. Mochizuki, *J. Polymer Sci.*, Part B, **1**, 531 (1963); *Kobunshi Kagaku*, **23**, 645 (1966).

[317] K. Hayashi, Y. Kitanishi, M. Nishii, and S. Okamura, *Makromol. Chem.*, **47**, 237 (1961).

[318] S. Kambara and A. Takahashi, *Makromol. Chem.*, **63**, 89 (1963).

[319] J. P. Kennedy, J. J. Elliott, and B. Groten, *Makromol. Chem.*, **77**, 26 (1964).

[320] H. Wilski and T. Grewer, *J. Polymer Sci.*, Part C, **6**, 33 (1964).

[321] G. Carazzolo, *Chim. Ind.* (Milan), **46**, 525 (1964).

[322] G. W. Urbanczyk, *Zeszyty Nauk Politech. Lodz, Wlokiennictwo*, **9**, 79 (1962) through *Chem. Abstr.*, **61**, 5836B (1964).

[323] K. Fujii, T. Mochizuki, S. Imoto, J. Ukida, and M. Matsumoto, *J. Polymer Sci.*, Part A, **2**, 2327 (1964).

[324] P. Corradini and P. Ganis, *Makromol. Chem.*, **62**, 97 (1963); G. Natta, G. Dall Asta, G. Mazzanti, I. Pasquon, A. Valvassori, and A. Zambelli, *J. Am. Chem. Soc.*, **83**, 3343 (1961).

[325] E. N. Zilberman, A. E. Kulikova, and N. M. Teplyakov, *J. Polymer Sci.*, **56**, 417 (1962).

[326] G. Allegra, A. Ponoglio, and I. Pasquon, *Rend. Ist. Lombardo Sci. Lettere*, **A95**, 335 (1961).

[327] M. Ishibashi, *J. Polymer Sci.*, Part A, **2**, 4361 (1964).

[328] A. D. Ketley and R. J. Ehrig, *J. Polymer Sci.*, Part A, **2**, 4461 (1964).

[329] O. Vogl, *J. Polymer Sci.*, Part A, **2**, 4621 (1964).

[330] A. Chiba, H. Futama, and J. Furuichi, *Reports on Progress in Polymer Phys. in Japan*, **7**, 51 (1964).

[331] G. Carazzolo and M. Mammi, *J. Polymer Sci.*, Part B, **2**, 1057 (1964).

[332] J. P. Kennedy and R. M. Thomas, *Makromol. Chem.*, **64**, 1 (1963).

[333] M. Litt, *J. Polymer Sci.*, Part A, **1**, 2219 (1963).

[334] R. Brill, *Z. Physik. Chem.*, **B53**, 61 (1943).

[335] J. C. Mitchell, *J. Polymer Sci.*, Part B, **1**, 285 (1963).

REFERENCES

[336] T. Ota, O. Yoshizaki, and E. Nagai, *Kobunshi Kagaku*, 20, 225 (1963); M. Ogawa, T. Ota, O. Yoshizaki, and E. Nagai, *J. Polymer Sci.*, Part B, 1, 57 (1963).
[337] E. Mueller, *Melliand Textilber.*, 44, 484 (1963).
[338] J. R. Schaefgen and R. Zbinden, *J. Polymer Sci.*, Part A, 2, 4865 (1964).
[339] D. D. Coffman, G. J. Berchet, U. R. Peterson, and E. W. Spanagel, *J. Polymer Sci.*, 2, 306 (1947).
[340] H. S. Makowsky, B. K. C. Shim, and Z. W. Wilchinsky, *J. Polymer Sci.*, Part A, 2, 4973 (1964).
[341] F. Rybnikar, *Collection Czech. Chem. Commun.*, 27, 2864 (1962).
[342] L. Mandelkern, M. Tryon, and F. A. Quinn, Jr., *J. Polymer Sci.*, 19, 77 (1956).
[343] M. Inoue, *J. Polymer Sci.*, Part A, 1, 2697 (1963).
[344] I. Heber, *Kolloid-Z.*, 189, 110 (1963).
[345] H. Wilski and T. Grewer, 145th National American Chemical Society Meeting, New York, September 1963 — *ACS Polymer Preprints*, 4, 464 (1963).
[346] G. B. Gechele and L. Crescentini, *J. Appl. Polymer Sci.*, 7, 1349 (1963).
[347] A. Nishioka, H. Watanabe, R. Koyama, and H. Nagai, *Reports on Progress in Polymer Phys. in Japan*, 6, 311 (1963).
[348] H. Tadokoro, Y. Chatani, M. Kobayashi, T. Yoshihara, and S. Murahashi, *Reports on Progress in Polymer Phys. in Japan*, 6, 303 (1963); K. Imada, T. Miyake, Y. Chatani, H. Tadokoro, and S. Murahashi, *Makromol. Chem.*, 83, 113 (1965).
[349] H. W. Wyckoff, *J. Polymer Sci.*, 62, 83 (1962).
[350] A. Turner-Jones, *J. Polymer Sci.*, Part B, 1, 455 (1963).
[351] M. G. Broadhurst, *J. Res. Natl. Bur. Std.*, 70A, 481 (1966).
[352] G. Natta and G. Moraglio, *Rubber Plastics Age*, 44, 42 (1963).
[353] G. Natta and G. Moraglio, *Makromol. Chem.*, 66, 218 (1963).
[354] M. Ishibashi, *Polymer*, 5, 305 (1964).
[355] A. Turner-Jones and J. M. Aizlewood, *J. Polymer Sci.*, Part B, 1, 471 (1963).
[356] M. Ishibashi, *J. Polymer Sci.*, Part B, 1, 629 (1963).
[357] Y. Sakurada, M. Matsumoto, K. Imai, A. Nishioka, and Y. Kato, *J. Polymer Sci.*, Part B, 1, 633 (1963).
[358] I. Kirshenbaum, Z. W. Wilchinsky, and B. Groten, *J. Appl. Polymer Sci.*, 8, 2723 (1964).
[359] N. P. Borisova and T. M. Birshtein, *Vysokomolekul. Soedin.*, 5, 279 (1963).
[360] G. Dall Asta and I. W. Bassi, *Chim. Ind.* (Milan), 43, 999 (1961).
[361] C. A. Boye, *Bull. Am. Phys. Soc.*, 8, 266 (1963).
[362] P. J. Flory and A. Vrij, *J. Am. Chem. Soc.*, 85, 3548 (1963).
[363] P. R. Swan, *J. Polymer Sci.*, 42, 525 (1960).
[364] A. Turner-Jones and R. P. Palmer, *Polymer*, 4, 525 (1963).
[365] M. Genas, *Angew. Chem.*, 74, 535 (1962).
[366] N. G. Gaylord (editor) *Polyethers*, Part I, Interscience, New York, 1963.
[367] M. Cesari, *J. Polymer Sci.*, Part B, 2, 453 (1964).
[368] W. Marconi, A. Mazzei, S. Cucinella, M. Cesari, and E. Pauluzzi, *J. Polymer Sci.*, Part A, 3, 123 (1965).
[369] W. J. Jackson, Jr. and J. R. Caldwell, *J. Appl. Polymer Sci.*, 7, 1975 (1963).
[370] R. B. Isaacson, I. Kirshenbaum, and W. C. Feist, *J. Appl. Polymer Sci.*, 8, 2789 (1964).
[371] G. Natta, L. Porri, and M. C. Gallazzi, *Chim. Ind.* (Milan), 46, 1158 (1964).
[372] D. C. Kockott, *Kolloid-Z.*, 198, 17 (1964).

[373] F. S. Ingraham and D. F. Wooley, Jr., *Ind. Eng. Chem.*, **56**, No. 9, 53 (1964).

[374] R. Janssen, H. Ruysschaert, and R. Vroom, *Makromol. Chem.*, **77**, 153 (1964).

[375] G. Natta, I. W. Bassi, and G. Perego, *Atti Accad. Nazl. Lincei, Rend., Classe Sci. Fis., Mat. Nat.*, **36**, 291 (1964).

[376] F. B. Moody, *Macromol. Synth.*, **1**, 67 (1963).

[377] G. Allen, C. Booth, M. N. Jones, D. J. Marks, and W. D. Taylor, *Polymer*, **5**, 547 (1964).

[378] G. Wasai, T. Saegusa, and J. Furukawa, *Kogyo Kagaku Zasshi*, **67**, 1428 (1964).

[379] T. Tsuruta, R. Fujio, and J. Furukawa, *Makromol. Chem.*, **80**, 172 (1964).

[380] F. Danusso and G. Gianotti, *Makromol. Chem.*, **80**, 1 (1964).

[381] F. Danusso, G. Gianotti, and G. Polizzotti, *Makromol. Chem.*, **80**, 13 (1964).

[382] G. P. Lorenzi, E. Benedetti, and E. Chiellini, *Chim. Ind.* (Milan), **46**, 1474 (1964).

[383] M. Goodman and Y.-L. Fan, *J. Am. Chem. Soc.*, **86**, 4922 (1964).

[384] G. Natta, P. Corradini, and P. Ganis, *J. Polymer Sci.*, Part A, **3**, 11 (1965).

[385] A. Bell, J. G. Smith, and C. J. Kibler, *J. Polymer Sci.*, Part A, **3**, 19 (1965).

[386] E. L. Wittbecker, W. S. Spliethoff, and G. R. Stine, *J. Appl. Polymer Sci.*, **9**, 213 (1965).

[387] G. Wasai, T. Iwata, K. Hirono, M. Kuragano, T. Saegusa, and J. Furukawa, *Kogyo Kagaku Zasshi*, **67**, 1920 (1964).

[388] D. E. Ilyina, B. A. Krentsel, and G. E. Semenido, *J. Polymer Sci.*, Part C, **4**, 999 (1964).

[389] E. L. Galperin and Yu. V. Strogalin, *Vysokomolekul. Soedin.*, **7**, 16 (1965); *Polymer Sci.* (USSR) (English transl.), **7**, 15 (1965).

[390] F. T. Simon and J. M. Rutherford, Jr., *J. Appl. Phys.*, **35**, 82 (1964).

[391] J. Powers, J. D. Hoffman, J. J. Weeks, and F. A. Quinn, Jr., *J. Res. Natl. Bur. Std.*, **69A**, 335 (1965).

[392] G. Farrow, *Polymer*, **4**, 191 (1963).

[393] F. Danusso, G. Moraglio, W. Chiglia, L. Motto, and G. Talamini, *Chim. Ind.* (Milan), **41**, 748 (1959).

[394] W. R. Krigbaum and I. Uematsu, *J. Polymer Sci.*, Part A, **3**, 767 (1965).

[395] W. F. Gorham, *ACS Polymer Preprints*, **6**, 73 (1965).

[396] H. Morawetz, S. Z. Zakabhazy, J. B. Lando, and B. Post, *Proc. Natl. Acad. Sci. U.S.*, **49**, 789 (1963).

[397] M. Barlow, *J. Polymer Sci.*, Part A–2, **4**, 121 (1966).

[398] R. Yamadera and C. Sonoda, *J. Polymer Sci.*, Part B, **3**, 411 (1965).

[399] E. J. Vandenberg, *J. Polymer Sci.*, **47**, 489 (1960).

[400] Yu. Ya. Tomashpolskii and G. S. Markova, *Vysokomolekul. Soedin.*, **6**, 27 (1964); *Polymer Sci.* (USSR) (English transl.), **6**, 316 (1964).

[401] G. Wasai, J. Furukawa, and A. Kawasaki, *Kogyo Kagaku Zasshi*, **68**, 210 (1965).

[402] V. V. Korshak and T. M. Frunze, *Synthetic Hetero-Chain Polyamides*, trans. by N. Kaner, Davey, New York, 1964, Tables I, II, XXI, XXVIII, and XXX.

[403] M. Cesari, G. Perego, and A. Mazzei, *Makromol. Chem.*, **83**, 196 (1965).

[404] H. Iiyama, M. Asakura, and K. Kimoto, *Kogyo Kagaku Zasshi*, **68**, 243 (1965).

[405] I. Kirshenbaum, *J. Polymer Sci.*, Part A, **3**, 1869 (1965).

[406] H. Kojima and K. Yamaguchi, *Kobunshi Kagaku*, **19**, 715 (1962).

[407] F. Echochard, *J. Chim. Phys.*, **43**, 113 (1946), through ref. 402, p. 363.

[408] F. A. Quinn, Jr. and J. Powers, *J. Polymer Sci.*, Part B, **1**, 341 (1963).

[409] G. Natta and M. Pegoraro, *Atti. Accad. Nazl. Lincei, Rend., Classe Sci. Fis., Mat. Nat.*, **34**, 110 (1963).

[410] H. Tadokoro, S. Seki, and I. Nitta, *Bull. Chem. Soc. Japan*, **28**, 559 (1955).

[411] G. Natta, F. Danusso, and G. Moraglio, *Makromol. Chem.*, **28**, 166 (1958).

REFERENCES

[412] G. Natta, G. Allegra, I. W. Bassi, P. Corradini, and P. Ganis, *Atti Accad. Nazl. Lincei, Rend., Classe Sci. Fis., Mat. Nat.*, 36, 433 (1964); G. Natta, G. Dall Asta, G. Mazzanti, I. Pasquon, A. Valvassori, and A. Zambelli, *Makromol. Chem.*, 54, 95 (1962).

[413] V. P. Lebedev, N. A. Okladnov, K. S. Minsker, and B. P. Shtarkman, *Vysokomolekul. Soedin.*, 7, 655 (1965).

[414] D. J. Lyman, J. Heller, and M. Barlow, *Makromol. Chem.*, 84, 64 (1965).

[415] J. Boor, Jr. and E. A. Youngman, *J. Polymer Sci.*, Part B, 3, 577 (1965).

[416] L. Mortillaro, L. Credali, M. Russo, and C. De Cheechi, *J. Polymer Sci.*, Part B, 3, 581 (1965).

[417] E. L. Galperin, Yu. V. Strogalin, and M. P. Mlenik, *Vysokomolekul. Soedin.*, 7, 933 (1965); *Polymer Sci.* (USSR) (English transl.), 7, 1031 (1965).

[418] J. B. Lando, H. G. Olf, and A. Peterlin, *J. Polymer Sci.*, Part A-1, 4, 941 (1966).

[419] K. Okamura, MS Thesis, Syracuse University, June 1965.

[420] M. Donati, G. Perego, and M. Farina, *Makromol. Chem.*, 85, 301 (1965).

[421] H. Komoto and K. Saotome, *Kobunshi Kagaku*, 22, 337 (1965).

[422] W. G. C. Forsyth, A. C. Hayward, and J. B. Roberts, *Nature*, 182, 800 (1958).

[423] G. Carazzolo, L. Mortillaro, L. Credali, and S. Bezzi, *J. Polymer Sci.*, Part B, 2, 997 (1965); G. Carazzolo and G. Valle, *J. Polymer Sci.*, Part A, 3, 4013 (1965).

[424] G. Natta, I. W. Bassi, and G. Allegra, *Makromol. Chem.*, 89, 81 (1965).

[425] K. Miki and R. Nakatsuka, *Reports on Progress in Polymer Phys. in Japan*, 6, 303 (1963).

[426] W. D. Niegisch, *J. Appl. Phys.*, 37, 4041 (1966); *J. Polymer Sci.*, Part B, 4, 531 (1966).

[427] Y. Joh, T. Yoshihara, Y. Kotake, F. Ide, and K. Nakatsuka, *J. Polymer Sci.*, Part B, 3, 933 (1965); T. Yoshihara, Y. Kotake, and Y. Joh, *J. Polymer Sci.*, Part B, 5, 459 (1967).

[428] H. Tadokoro, Y. Takahashi S. Otsuka, K. Mori and F. Imaizumi, *J. Polymer Sci.*, Part B, 3, 697 (1965).

[429] P. Corradini and I. W. Bassi, Preprint P421, Prague IUPAC High Polymer Symposium, September 1965.

[430] G. Natta, P. Corradini, P. Ganis, and P. A. Temussi, *J. Polymer Sci.*, Part C, 16, 2477 (1967).

[431] H. D. Noether, *J. Polymer Sci.*, Part C, 16, 725 (1967); M. G. Huguet, *Makromol. Chem.*, 94, 205 (1966).

[432] I. W. Bassi, P. Ganis, and P. A. Temussi, *J. Polymer Sci.*, Part C, 16, 2867 (1967).

[433] R. K. Tubbs, *J. Polymer Sci.*, Part A, 3, 4181 (1965).

[434] G. Natta, G. Allegra, I. W. Bassi, D. Sianesi, G. Caporiccio, and E. Torti, *J. Polymer Sci.*, Part A, 3, 4263 (1965).

[435] R. Bonart, *Makromol. Chem.*, 92, 149 (1966).

[436] G. Natta and G. Dall Asta, *Chim. Ind.* (Milan), 46, 1429 (1964).

[437] C. G. Overberger, H. Kaye, and G. Walsh, *J. Polymer Sci.*, Part A, 2, 755 (1964).

[438] H. Schnell, *Angew. Chem.*, 68, 633 (1956).

[439] G. Natta, G. Dall Asta, I. W. Bassi, and G. Carella, *Makromol. Chem.*, 91, 87 (1966).

[440] V. M. Coiro, P. De Santis, L. Mazzarella, and L. Picozzi, *J. Polymer Sci.*, Part A, 3, 4001 (1965).

[441] J. A. Faucher and F. P. Reding, *Crystalline Olefin Polymers*, R. A. V. Raff and K. W. Doak, editors, Interscience, New York, 1965, Part I, p. 677.

[442] A. Turner-Jones, *Polymer*, 7, 23 (1966).

[443] N. I. Makarevich, *Zh. Prikl. Spektroskopii, Akad. Nauk Belorussk. SSR*, 2, 341 (1965), through *Chem. Abstrs.*, 63, 8509A (1965).

[444] J. R. Schaefgen, F. H. Koontz, and R. F. Tietz, *J. Polymer Sci.*, 40, 377 (1959).

[445] T. B. Gibb, Jr., R. A. Clendinning, and W. D. Niegisch, *J. Polymer Sci.*, Part A-1, 4, 917 (1966).

[446] H. Tadokoro, Y. Takahashi, Y. Chatani, and H. Kakida, *Makromol. Chem.*, 109, 96 (1967).

[447] J. B. Rose, *J. Chem. Soc.*, 1956, 542, 546.

[448] M. Cesari, G. Perego, and W. Marconi, *Makromol. Chem.*, 94, 194 (1966).

[449] I. I. Novak and V. I. Vettegren, *Vysokomolekul. Soedin.*, 7, 1027 (1965); *Polymer Sci.* (USSR) (English transl.), 7, 1136 (1965).

[450] F. Hamada and A. Nakajima, *Kobunshi Kagaku*, 23, 395 (1966).

[451] G. Moraglio, G. Polizzotti, and F. Danusso, *European Polymer J.*, 1, 183 (1965).

[452] K-S. Chan, Ph.D. Thesis, State University College of Forestry at Syracuse University, *Dissertation Abstr.*, 26, 4260 (1966).

[453] J. G. Smith, C. J. Kibler, and B. J. Sublett, *J. Polymer Sci.*, Part A-1, 4, 1851 (1966).

[454] K. Saotome and H. Komoto, *J. Polymer Sci.*, Part A-1, 4, 1463 (1966).

[455] K. F. Wissbrun, *J. Polymer Sci.*, Part A-2, 4, 827 (1966).

[456] P. Ganis and P. A. Temussi, *European Polymer J.*, 2, 401 (1966).

[457] R. J. Fredericks, T. H. Doyne, and R. S. Sprague, *J. Polymer Sci.*, Part A-2, 4, 899 (1966).

[458] R. J. Fredericks, T. H. Doyne, and R. S. Sprague, *J. Polymer Sci.*, Part A-2, 4, 913 (1966).

[459] T. Ota, M. Yamashita, O. Yoshizaki, and E. Nagai, *J. Polymer Sci.*, Part A-2, 4, 959 (1966).

[460] H. Tadokoro, *J. Polymer Sci.*, Part C, 15, 1 (1966).

[461] H. Yasuda, Y. Tanada, and M. Takayanagi, *Kogyo Kagaku Zasshi*, 69, 304 (1966).

[462] R. W. Eykamp, A. M. Schneider, and E. W. Merrill, *J. Polymer Sci.*, Part A-2, 4, 1025 (1966).

[463] G. Carazzolo and M. Mammi, *Makromol. Chem.*, 100, 28 (1967).

[464] K. Chujo, H. Kobayashi, J. Suzuki, S. Tokuhara, and M. Tanabe, *Makromol. Chem.*, 100, 262 (1967).

[465] K. Chujo, H. Kobayashi, J. Suzuki, and S. Tokuhara, *Makromol. Chem.*, 100, 267 (1967).

[466] W. Braun, K.-H. Hellwege, and W. Knape, *Kolloid-Z.*, 215, 10 (1967).

[467] M. Hayashi, Y. Shiro, and H. Murata, *Bull. Chem. Soc. Japan*, 39, 1857 (1966).

[468] M. Hayashi, Y. Shiro, and H. Murata, *Bull. Chem. Soc. Japan*, 39, 1861 (1966).

[469] F. Tuinstra, *Acta Cryst.*, 20, 341 (1966).

[470] B. M. Ginzburg, L. N. Korzhavin, S. Ya. Frenkel, L. A. Laius, and M. B. Adrova, *Vysokomolekul. Soedin.*, 8, 278 (1966); *Polymer Sci.* (USSR) (English transl.), 8, 302 (1966).

[471] F. E. Karasz, J. M. O'Reilly, H. E. Bair, and R. A. Kluge, *ACS Polymer Preprints*, 9, 822 (1968).

[472] P. E. Slade and T. A. Orofino, *ACS Polymer Preprints*, 9, 825 (1968).

[473] W. J. MacKnight, M. Yang, and T. Kajima, *ACS Polymer Preprints*, 9, 860 (1968).

[474] V. G. Baranov, Bu. Zhu-Chan, T. I. Volkov, and S. Ya. Frenkel, *Vysokomol. Soedin.*, Ser. A, 9, 81 (1967); *Polymer Sci.* (USSR) (English transl.), Ser. A, 9, 87 (1967).

[475] P. Kovacic, M. B. Feldman, J. P. Kovacic, and J. B. Lando, *J. Appl. Polymer Sci.*, 12, 1735 (1968).

REFERENCES

[476] I. W. Bassi and G. Fagherazzi, *European Polymer J.*, 4, 123 (1968).

[477] G. Fagherazzi and I. W. Bassi, *European Polymer J.*, 4, 151 (1968).

[478] F. Danusso and G. Gianotti, *European Polymer J.*, 4, 165 (1968).

[479] C. G. Overberger and T. Takekoshi, *Macromolecules*, 1, 7 (1968).

[480] K. Kikukawa, S. Nozakura, and S. Murahashi, *Kobunshi Kagaku*, 25, 19 (1968).

[481] H. K. Reimschuessel, L. G. Roldan, and J. P. Sibilia, *J. Polymer Sci.*, Part A-2, 6, 559 (1968).

[482] S. Kobayashi, H. Tadokoro, and Y. Chatani, *Makromol. Chem.*, 112, 225 (1968).

[483] F. J. Golemba, J. E. Guillet, and S. C. Nyburg, *J. Polymer Sci.*, Part A-1, 6, 1341 (1968).

[484] F. E. Karasz, H. E. Bair, and J. M. O'Reilly, *J. Polymer Sci.*, Part A-2, 6, 1141 (1968).

[485] A. Turner-Jones and A. J. Cobbold, *J. Polymer Sci.*, Part B, 6, 539 (1968).

[486] J. J. Klement and P. H. Geil, *J. Polymer Sci.*, Part A-2, 6, 1381 (1968).

[487] H. M. van Dort, C. A. M. Hoefs, E. P. Magré, A. J. Schopf, and K. Yntema, *European Polymer J.*, 4, 275 (1968).

[488] G. Natta, P. Corradini, I. W. Bassi, and G. Fagherazzi, *European Polymer J.*, 4, 297 (1968).

[489] A. Carbonaro, A. Greco, and I. W. Bassi, *European Polymer J.*, 4, 445 (1968).

[490] Y. Chatani, K. Suehiro, Y. Okita, H. Tadokoro, and K. Chujo, *Makromol. Chem.*, 113, 215 (1968).

[491] J. Masamoto, K. Sasaguri, C. Ohizumi, and H. Kobayashi, *Reports on Progress in Polymer Phys. in Japan*, 11, 131 (1968).

[492] H. Utsunomiya, T. Mori, K. Imada, and M. Takayanagi, *ibid.*, p. 153.

[493] S. Minami, S. Manabe, and M. Takayanagi, *ibid.*, p. 155.

[494] S. Takamuku, K. Imada, and M. Takayanagi, *ibid.*, p. 159.

[495] M. Fujisaka, K. Imada, and M. Takayanagi, *ibid.*, p. 169.

[496] S. Iwayanagi, I. Sakurai, T. Sakurai, and T. Seto, *ibid.*, 10, 167 (1967).

[497] H. Tadokoro, M. Kobayashi, K. Mori, Y. Takahashi, and S. Taniyan, *ibid.*, p. 181.

[498] H. Utsunomiya, N. Kawasaki, M. Niinomi, and M. Takayanagi, *ibid.*, p. 197.

Mass Absorption Coefficients

The following table is reprinted by special permission from *International Tables for X-Ray Crystallography*,† Volume III, pp. 162–165. The numerical values are expressed in units of square centimeters per gram. The values given in italics are of low accuracy. A bar separating two successive numerical values denotes an intervening absorption edge.

†Kynoch Press, Birmingham, England, 1962.

Mass Absorption Coefficients μ/ρ of the Elements (Z = 1 to 83) for a Selection of Wavelengths

Target, Radiation, and Wavelength, λ (Å)

Absorber	Z	Ag Kα 0.5608	Ag Kβ₁ 0.4970	Mo Kα 0.7107	Mo Kβ₁ 0.6323	Cu Kα 1.5418	Cu Kβ₁ 1.3922	Co Kα 1.7902	Co Kβ₁ 1.6208	Fe Kα 1.9373	Fe Kβ₁ 1.7565	Cr Kα 2.2909	Cr Kβ₁ 2.0848
H	1	0.371	0.366	0.380	0.376	0.435	0.421	0.464	0.443	0.483	0.459	0.545	0.507
He	2	0.195	0.190	0.207	0.200	0.383	0.333	0.491	0.414	0.569	0.474	0.813	0.661
Li	3	0.187	0.177	0.217	0.200	0.716	0.571	1.03	0.804	1.25	0.978	1.96	1.52
Be	4	0.229	0.208	0.298	0.258	1.50	1.15	2.25	1.71	2.80	2.13	4.50	3.44
B	5	0.279	0.244	0.392	0.327	2.39	1.81	3.63	2.74	4.55	3.44	7.38	5.61
C	6	0.400	0.333	0.625	0.495	4.60	3.44	7.07	5.31	8.90	6.69	14.5	11.0
N	7	0.544	0.433	0.916	0.700	7.52	5.60	11.6	8.70	14.6	11.0	23.9	18.2
O	8	0.740	0.570	1.31	0.981	11.5	8.52	17.8	13.3	22.4	16.8	36.6	27.8
F	9	0.976	0.732	1.80	1.32	16.4	12.2	25.4	19.0	32.1	24.0	52.4	39.8
Ne	10	1.31	0.969	2.47	1.80	22.9	17.0	35.4	26.5	44.6	33.5	72.8	55.3
Na	11	1.67	1.22	3.21	2.32	30.1	22.3	46.5	34.8	58.6	44.0	95.3	72.5
Mg	12	2.12	1.54	4.11	2.96	38.6	28.7	59.5	44.6	74.8	56.3	121	92.4
Al	13	2.65	1.90	5.16	3.71	48.6	36.2	74.8	56.2	93.9	70.9	152	116
Si	14	3.28	2.35	6.44	4.61	60.6	45.1	93.3	70.1	117	88.3	189	144
P	15	4.01	2.85	7.89	5.64	74.1	55.2	114	85.5	142	108	229	175
S	16	4.84	3.44	9.55	6.82	89.1	66.5	136	103	170	129	272	209
Cl	17	5.77	4.09	11.4	8.14	106	79.0	161	122	200	152	318	246
A	18	6.81	4.82	13.5	9.62	123	92.4	187	142	232	177	366	284
K	19	8.00	5.66	15.8	11.3	143	107	215	164	266	204	417	325
Ca	20	9.28	6.57	18.3	13.1	162	122	243	186	299	231	463	363

Mass Absorption Coefficients μ/ρ of the Elements (Z = 1 to 83) for a Selection of Wavelengths – Continued

	Z												
Sc	21	10.7	7.57	21.1	15.1	184	139	273	210	336	260	513	405
Ti	22	12.3	8.70	24.2	17.3	208	158	308	237	377	293	571	453
V	23	14.0	9.91	27.5	19.7	233	178	343	266	419	327	68.4	502
Cr	24	15.8	11.2	31.1	22.3	260	199	381	296	463	363	79.8	60.7
Mn	25	17.7	12.6	34.7	24.9	285	219	414	323	57.2	395	93.0	70.8
Fe	26	19.7	14.0	38.5	27.7	308	238	52.8	349	66.4	50.0	108	82.2
Co	27	21.8	15.5	42.5	30.6	313	257	61.1	45.8	76.8	57.8	125	95.0
Ni	28	24.1	17.1	46.6	33.7	45.7	275	70.5	52.8	88.6	66.7	144	109
Cu	29	26.4	18.8	50.9	36.9	52.9	39.3	81.6	61.2	103	77.3	166	127
Zn	30	28.8	20.6	55.4	40.2	60.3	44.8	93.0	69.7	117	88.0	189	144
Ga	31	31.4	22.4	60.1	43.7	67.9	50.5	105	78.4	131	98.9	212	162
Ge	32	34.1	24.4	64.8	47.3	75.6	56.2	116	87.3	146	110	235	180
As	33	36.9	26.5	69.7	51.1	83.4	62.1	128	96.2	160	121	258	198
Se	34	39.8	28.6	74.7	54.9	91.4	68.1	140	105	175	133	281	216
Br	35	42.7	30.8	79.8	58.8	99.6	74.4	152	115	190	144	305	234
Kr	36	45.8	33.1	84.9	62.8	108	80.7	165	124	206	156	327	252
Rb	37	48.9	35.4	90.0	66.9	117	87.3	177	134	221	168	351	271
Sr	38	52.1	37.8	95.0	70.9	125	94.0	190	144	236	180	373	289
Y	39	55.3	40.3	100	75.0	134	101	203	154	252	193	396	308
Zr	40	58.5	42.8	15.9	79.0	143	108	216	165	268	205	419	326
Nb	41	61.7	45.3	17.1	82.9	153	115	230	175	284	218	441	345
Mo	42	64.8	47.8	18.4	13.1	162	123	243	186	300	231	463	363
Tc	43	67.9	50.3	19.7	14.1	172	131	257	197	316	244	485	382
Ru	44	10.7	52.8	21.1	15.1	183	139	272	209	334	259	509	403
Rh	45	11.5	55.2	22.6	16.2	194	148	288	222	352	274	534	424

Element	Z												
Pd	46	12.3	57.5	24.1	17.3	206	157	304	235	371	289	559	446
Ag	47	13.1	9.29	25.8	18.5	218	166	321	248	391	305	586	468
Cd	48	14.0	9.91	27.5	19.7	231	176	338	262	412	322	613	492
In	49	14.9	10.6	29.3	21.0	243	186	356	277	432	339	638	514
Sn	50	15.9	11.3	31.1	22.3	256	197	373	291	451	356	662	536
Sb	51	16.9	12.0	33.1	23.8	270	207	391	306	472	373	688	559
Te	52	17.9	12.7	35.0	25.2	282	218	407	320	490	389	707	578
I	53	19.0	13.5	37.1	26.7	294	228	422	333	506	404	722	594
Xe	54	20.1	14.3	39.2	28.2	306	238	436	346	521	418	763	609
Cs	55	21.3	15.1	41.3	29.8	318	248	450	358	534	431	793	621
Ba	56	22.5	16.0	43.5	31.4	330	258	463	370	546	444	461	661
La	57	23.7	16.9	45.8	33.2	341	268	475	382	557	456	202	681
Ce	58	25.0	17.8	48.2	34.9	352	278	486	394	601	468	219	409
Pr	59	26.3	18.8	50.7	36.7	363	288	497	405	359	479	236	183
Nd	60	27.7	19.8	53.2	38.6	374	298	543	416	379	519	252	196
Pm	61	29.1	20.8	55.9	40.6	386	308	327	428	172	538	268	209
Sm	62	30.6	21.9	58.6	42.6	397	319	344	461	182	328	284	222
Eu	63	32.2	23.0	61.5	44.8	425	329	156	478	193	344	299	234
Gd	64	33.8	24.2	64.4	47.0	439	340	165	295	203	157	314	247
Tb	65	35.5	25.4	67.5	49.2	273	352	173	309	214	165	329	259
Dy	66	37.2	26.6	70.6	51.6	286	369	182	140	224	173	344	271
Ho	67	39.0	27.9	73.9	54.0	128	231	191	146	234	181	359	283
Er	68	40.8	29.3	77.3	56.6	134	242	199	153	245	190	373	295
Tm	69	42.8	30.7	80.8	59.2	140	252	208	160	255	198	387	307
Yb	70	44.8	32.2	84.5	61.9	146	111	217	167	265	206	401	319
Lu	71	46.8	33.6	88.2	64.7	153	116	226	174	276	215	416	331
Hf	72	48.8	35.1	91.7	67.4	159	121	235	181	286	223	430	343
Ta	73	50.9	36.7	95.4	70.2	166	126	244	189	297	232	444	355

Mass Absorption Coefficients μ/ρ of the Elements ($Z = 1$ to 83) for a Selection of Wavelengths — Concluded

	Z												
W	74	53.0	38.2	99.1	73.1	172	132	253	196	308	241	458	368
Re	75	55.2	39.8	103	75.9	179	137	262	204	319	250	473	380
Os	76	57.3	41.4	106	78.7	186	143	272	212	330	259	487	393
Ir	77	59.4	42.9	110	81.4	193	148	282	219	341	269	502	406
Pt	78	61.4	44.5	113	83.9	200	154	291	228	353	278	517	419
Au	79	63.1	45.8	115	86.0	208	160	302	236	365	288	532	432
Hg	80	64.7	47.1	117	87.9	216	166	312	245	377	298	547	446
Tl	81	66.2	48.4	119	89.5	224	172	323	253	389	309	563	460
Pb	82	67.7	49.8	120	91.0	232	179	334	262	402	319	579	474
Bi	83	69.1	51.1	120	92.0	240	185	346	272	415	330	596	489

X-Ray Wavelengths

The following data are reprinted by special permission from *International Tables for X-Ray Crystallography*,[†] Volume III, pp. 60–62. The units are kX/1.00202. They may be taken as in angstroms to an accuracy of 1 in 25,000 only.[‡] Values given in parentheses have been calculated rather than measured spectroscopically. For additional information see the original reference above.

[†]Kynoch Press, Birmingham, England, 1962.
[‡]By definition the kX unit is based on the assumption that $d_{(200)}$ of sodium chloride at 18°C is equal to 2.814 units. For a fuller explanation see H. P. Klug and L. E. Alexander, *X-Ray Diffraction Procedures*, Wiley, New York, 1954, p. 90.

Wavelengths of K-Emission Series and K-Absorption Edges

Element	Z	$K\alpha_2$	$K\alpha_1$	$K\beta_1$	$K\beta_2$	K-Absorption Edge
H	1					
He	2					
Li	3	240 (D)[b]				226.5
Be	4	113 (D)				
B	5	67 (D)				
C	6	44 (D)				$43._{68}$
N	7	31.60 (D)				$30._{99}$
O	8	23.71 (D)				$23._{32}$
F	9	18.31 (D)				
Ne	10	14.616 (D)		14.464		
Na	11	11.909 (D)		11.617		
Mg	12	9.8889 (D)		9.558		9.5117
Al	13	8.33916	8.33669	7.981		7.9511
Si	14	7.12773	7.12528	6.7681		6.7446
P	15	6.1549 (D)		5.8038		5.7866

Element	Z	$K\alpha_2$	$K\alpha_1$	$K\beta_1$	$K\beta_2$	K-Absorption Edge
S	16	5.37471	5.37196	5.03169		5.0182
Cl	17	4.73050	4.72760	4.4031		4.3969
A	18	4.19456	4.19162	3.8848		3.8707
K	19	3.74462	3.74122	3.4538		3.43645
Ca	20	3.36159	3.35825	3.0896		3.07016
Sc	21	3.03452	3.03114	2.7795		2.757_3
Ti	22	2.75207	2.74841	2.51381		2.497_{30}
V	23	2.50729	2.50348	2.28434		2.269_{02}
Cr	24	2.29351	2.28962	2.08480		2.070_{12}
Mn	25	2.10568	2.10175	1.91015		1.896_{36}
Fe	26	1.93991	1.93597	1.75653		1.743_{34}
Co	27	1.79278	1.78892	1.62075		1.608_{11}
Ni	28	1.66169	1.65784	1.50010	1.48861	1.488_{02}
Cu	29	1.54433	1.54051	1.39217[a]	1.38102	1.380_{43}
Zn	30	1.43894	1.43511	1.29522	1.28366	1.283_3
Ga	31	1.34394	1.34003	1.20784	1.19595	1.195_{67}
Ge	32	1.25797	1.25401	1.12890	1.11682	1.116_{52}
As	33	1.17981	1.17581	1.05726	1.04498	1.044_{97}
Se	34	1.10875	1.10471	0.99212	0.97986	0.979_{78}
Br	35	1.04376	1.03969	0.93273	0.92064	0.91995
Kr	36	0.9841	0.9801	0.87845	0.86609	0.86547
Rb	37	0.92963	0.92551	0.82863	0.81641	0.81549
Sr	38	0.87938	0.875214	0.78288	0.77076	0.76969
Y	39	0.83300	0.82879	0.74068	0.72874	0.72762
Zr	40	0.79010	0.78588	0.70170	0.68989	0.68877
Nb	41	0.75040	0.74615	0.66572	0.65412	0.65291
Mo	42	0.713543	0.70926	0.632253	0.62099 (D)[b]	0.61977
Tc	43	0.67927	0.67493	0.60141	0.59018	(0.5891)
Ru	44	0.64736	0.64304	0.57246	0.56164	0.560_{47}
Rh	45	0.617610	0.613245	0.54559	0.53509 (D)	0.533_{78}
Pd	46	0.589801	0.585415	0.52052	0.51021	0.509_{15}
Ag	47	0.563775	0.559363	0.49701	0.48701	0.4858_2
Cd	48	0.53941	0.53498	0.475078	0.46531	0.46409
In	49	0.51652	0.51209	0.454514	0.444963	0.44388
Sn	50	0.49502	0.49056	0.435216	0.425900	0.42468

[a] Mean of β_1 and β_3.
[b] D signifies mean of doublet.

Element	Z	$K\alpha_2$	$K\alpha_1$	$K\beta_1$	$K\beta_2$	K-Absorption Edge
Sb	51	0.47479	0.470322	0.417060	0.407950	0.40663
Te	52	0.455751	0.451263	0.399972	0.391080	0.38972
I	53	0.437805	0.433293	0.383884	0.37547	0.373_{79}
Xe	54	0.42043	0.41596	0.36846	0.35989	0.35849
Cs	55	0.404812	0.400268	0.354347	0.346084	0.34474
Ba	56	0.389646	0.385089	0.340789	0.332745	0.33137
La	57	0.375279	0.370709	0.327959	0.32024 (D)	0.31842
Ce	58	0.361665	0.357075	0.315792	0.30826 (D)	0.30647
Pr	59	0.348728	0.344122	0.304238	0.29690 (D)	0.29516
Nd	60	0.356487	0.331822	0.293274	0.28631	0.28451
Pm	61	0.3249	0.3207	0.28209	(0.2761)	(0.2743)
Sm	62	0.31365	0.30895	0.27305	0.26629	0.26462
Eu	63	0.30326	0.29850	0.26360	0.25697	0.25552
Gd	64	0.29320	0.28840	0.25445	0.24812	0.24680
Tb	65	0.28343	0.27876	0.24601	0.23960	0.23840
Dy	66	0.27430	0.26957	0.23758	0.23175	0.23046
Ho	67	0.26552	0.26083	(0.2302)	(0.2244)	0.22290
Er	68	0.25716	0.25248	0.22260	0.21715	0.21566
Tu	69	0.24911	0.24436	0.21530	(0.2101)	0.2089
Yb	70	0.24147	0.23676	0.20876	0.20363	0.20223
Lu	71	0.23405	0.22928	0.20212	0.19689	0.19584
Hf	72	0.22699	0.22218	0.19554	0.19081	0.18981
Ta	73	0.220290	0.215484	0.190076	0.18508 (D)	0.18393
W	74	0.213813	0.208992	0.184363	0.17950 (D)	0.17837
Re	75	0.207598	0.202778	0.178870	0.17415 (D)	0.17311
Os	76	0.201626	0.196783	0.173607	0.16899 (D)	0.16780
Ir	77	0.195889	0.191033	0.168533	0.16404 (D)	0.16286
Pt	78	0.190372	0.185504	0.163664	0.15928 (D)	0.15816
Au	79	0.185064	0.180185	0.158971	0.15471 (D)	0.15344
Hg	80	0.17992	0.17504	0.15439	0.15020 (D)	0.14923
Tl	81	0.175028	0.170131	0.150133	(0.1461)	0.14470
Pb	82	0.170285	0.165364	0.145980	0.14201 (D)	0.14077
Bi	83	0.165704	0.160777	0.141941	0.13807 (D)	0.13706
Po	84	0.1608	0.1559	0.1382	0.1333	(0.1332)
At	85	(0.1570)	(0.1521)	(0.1343)	(0.1307)	(0.1295)
Rn	86	(0.1529)	(0.1479)	(0.1307)	(0.1271)	(0.1260)
Fr	87	(0.1489)	(0.1440)	(0.1272)	(0.1236)	(0.1225)
Ra	88	(0.1450)	(0.1401)	(0.1237)	(0.1203)	(0.1192)
Ac	89	(0.1414)	(0.1364)	(0.1205)	(0.1172)	(0.1161)
˙Th	90	0.137820	0.132806	0.117389	0.11416 (D)	0.11293

Element	Z	$K\alpha_2$	$K\alpha_1$	$K\beta_1$	$K\beta_2$	K-Absorption Edge
Pa	91	(0.1344)	(0.1294)	(0.1143)	(0.1112)	(0.1101)
U	92	0.130962	0.125940	0.111386	0.10864	0.10680
Np	93	(0.1278)	(0.1226)	(0.1085)	(0.1055)	(0.1045)
Pu	94	(0.1246)	(0.1195)	(0.1058)	(0.1029)	(0.1018)
Am	95	(0.1215)	(0.1165)	(0.1031)	(0.1003)	(0.0992)
Cm	96	(0.1186)	(0.1135)	(0.1005)	(0.0978)	(0.0967)
Bk	97	(0.1157)	(0.1107)	(0.0980)	(0.0953)	(0.0943)
Cf	98	(0.1130)	(0.1079)	(0.0956)	(0.0930)	(0.0920)
Es	99	(0.1103)	(0.1052)	(0.0933)	(0.0907)	(0.0897)
Fm	100	(0.1077)	(0.1026)	(0.0910)	(0.0885)	(0.0875)

Atomic Scattering Factors

The following table is reprinted by special permission with abridgement from *International Tables for X-Ray Crystallography*,† Volume III, pp. 202–207. The number of elements and range of $(\sin \theta)/\lambda$ have been limited so as to exclude much numerical data of little or no interest in polymer science. The numerical values have been calculated from self-consistent or variational wave functions. The values in parentheses have been obtained from the original calculated values by interpolation.

†Kynoch Press, Birmingham, England, 1962.

APPENDIX 6

Atomic Scattering Factors in Electron Units

Element and Ionic Charge	Z	$(\sin\theta)/\lambda$ (Å^{-1})						
		0.00	0.05	0.10	0.15	0.20	0.25	0.30
H	1	1.000	0.947	0.811	0.641	0.481	0.350	0.251
H^{-1}	1	2.000	1.566	1.064	0.742	0.519	0.364	0.255
He	2	2.000	1.955	1.832	1.654	1.452	1.249	1.058
He^{-1}	2	3.000						
Li	3	3.000	2.710	2.215	1.904	1.741	1.627	1.512
Li^{+1}	3	2.000	1.983	1.935	1.859	1.760	1.645	1.521
Li^{-1}	3	4.000	2.881	2.176	1.888	1.743	1.629	1.514
Be	4	4.000	3.706	3.067	2.469	2.067	1.838	1.705
Be^{+1}	4	3.000	2.877	2.583	2.267	2.017	1.843	1.721
Be^{+2}	4	2.000	1.991	1.966	1.925	1.869	1.802	1.724
B	5	5.000	4.726	4.066	3.325	2.711	2.276	1.993
B^{+1}	5	4.000	3.852	3.471	2.994	2.551	2.205	1.962
B^{+2}	5	3.000	2.933	2.757	2.524	2.290	2.088	1.928
B^{+3}	5	2.000	1.995	1.979	1.954	1.919	1.875	1.824
C	6	6.000	(5.760)	5.126	(4.358)	3.581	(2.976)	2.502
C (valence)	6	6.000	(5.750)	(5.093)	(4.313)	(3.561)	(2.956)	(2.506)
C^{+2}	6	4.000		3.686		2.992		2.338
C^{+3}	6	3.000	2.958	2.842	2.676	2.487	2.300	2.133
C^{+4}	6	2.000	1.996	1.986	1.969	1.945	1.914	1.880
N	7	7.000	6.781	6.203	5.420	4.600	3.856	3.241
N^{+3}	7	4.000	3.941	3.772	3.523	3.227	2.922	2.635
N^{+4}	7	3.000	2.971	2.890	2.768	2.619	2.461	2.306
N^{-1}	7	8.000		6.688		4.631		3.186
O	8	8.000	7.796	7.250	6.482	5.634	4.814	4.094
O^{+1}	8	7.000		6.493		5.298		4.017
O^{+2}	8	6.000		5.647		4.776		3.771
O^{+3}	8	5.000		4.760		4.151		3.410
O^{-1}	8	9.000		7.836		5.756		4.068
F	9	9.000		8.293		6.691		5.044
F^{-1}	9	10.000	9.763	9.108	8.174	7.126	6.103	5.188
Ne	10	10.000	9.834	9.363	8.661	7.824	6.942	6.087
Na	11	11.00	10.56	9.76	9.02	8.34	7.6	6.89
Na^{+1}	11	10.000	9.884	9.551	9.035	8.390	7.67	6.925
Mg	12	12.00	11.52	10.50	9.53	8.75	8.09	7.46
Mg^{+2}	12	10.00	9.91	9.66	9.26	8.75	8.15	7.51
Al	13	13.00	12.44	11.23	10.06	9.16	8.47	7.88
Al^{+1}	13	12.00		10.94		9.22		7.90

0.35	0.40	0.50	0.60	0.70	0.80	0.90	1.00	1.10
0.180	0.130	0.071	0.040	0.024	0.015	0.010	0.007	0.005
0.181	0.130	0.070	0.040	0.024	0.015	0.010	0.007	0.005
0.889	0.742	0.515	0.358	0.251	0.179	0.129	0.095	0.071
1.394	1.269	1.032	0.823	0.650	0.513	0.404	0.320	0.255
1.392	1.265	1.025	0.818	0.647	0.510	0.403	0.319	0.254
1.392	1.269	1.033	0.826	0.654	0.516	0.408	0.323	0.257
1.613	1.531	1.367	1.201	1.031	0.878	0.738	0.620	0.519
1.624	1.535	1.362	1.188	1.022	0.870	0.735	0.618	0.520
1.640	1.550	1.363	1.180	1.009	0.855	0.721	0.606	0.508
1.813	1.692	1.534	1.406	1.276	1.147	1.016	0.895	0.783
1.799	1.688	1.536	1.410	1.283	1.154	1.028	0.908	0.798
1.804	1.707	1.552	1.414	1.278	1.144	1.016	0.896	0.786
1.766	1.703	1.566	1.420	1.274	1.132	0.999	0.877	0.767
(2.165)	1.950	1.685	1.536	1.426	1.322	1.218	1.114	1.012
(2.182)	(1.975)	(1.712)	(1.553)	(1.434)	(1.322)	(1.207)	(1.096)	(0.993)
	1.910	1.672	1.533	1.429	1.332	1.233	1.131	1.030
1.991	1.874	1.697	1.564	1.447	1.335	1.225	1.116	1.012
1.838	1.794	1.692	1.579	1.459	1.338	1.219	1.104	0.994
2.760	2.397	1.944	1.698	1.550	1.444	1.350	1.263	1.175
2.382	2.172	1.869	1.682	1.558	1.461	1.373	1.287	1.199
2.164	2.038	1.837	1.690	1.573	1.472	1.375	1.281	1.188
	2.364	1.929	1.694	1.551	1.446	1.352	1.263	1.170
3.492	3.010	2.338	1.944	1.714	1.566	1.462	1.374	1.296
	3.016	2.356	1.956	1.717	1.567	1.461	1.374	1.296
	2.924	2.327	1.948	1.716	1.568	1.463	1.378	1.301
	2.745	2.246	1.913	1.701	1.562	1.463	1.382	1.308
	2.968	2.313	1.934	1.710	1.566	1.462	1.373	1.294
	3.760	2.878	2.312	1.958	1.735	1.587	1.481	1.396
4.416	3.786	2.885	2.323	1.972	1.747	1.596	1.486	1.399
5.305	4.617	3.536	2.794	2.300	1.976	1.760	1.612	1.504
6.16	5.47	4.29	3.40	2.76	2.31	2.00	1.78	1.63
6.196	5.510	4.328	3.424	2.771	2.314	2.001	1.785	1.634
6.83	6.20	5.01	4.06	3.30	2.72	2.30	2.01	1.81
6.85	6.20	4.99	4.03	3.28	2.71	2.30	2.01	1.81
7.32	6.77	5.69	4.71	3.88	3.21	2.71	2.32	2.05
	6.77	5.70	4.71	3.88	3.22	2.70	2.32	2.04

Element and Ionic Charge	Z	$(\sin\theta)/\lambda$ (Å^{-1})						
		0.00	0.05	0.10	0.15	0.20	0.25	0.30
Al^{+2}	13	11.00	10.84	10.40	9.81	9.17	8.55	7.95
Al^{+3}	13	10.00	9.93	9.74	9.43	9.01	8.52	7.98
Si	14	14.00	13.45	12.16	10.79	9.67	8.85	8.22
Si^{+3}	14	11.00		10.53		9.48		8.34
Si^{+4}	14	10.00	9.95	9.79	9.54	9.20	8.79	8.33
P	15	15.00	14.47	13.17	11.66	10.34	9.33	8.59
S	16	16.00	15.54	14.33	12.75	11.21	9.93	8.99
S^{-1}	16	17.00	16.45	15.00	13.14	11.36	9.95	8.95
S^{-2}	16	18.00	(17.19)	(15.16)	(12.73)	(10.74)	(9.45)	(8.66)
Cl	17	17.00	16.55	15.33	13.68	12.00	10.55	9.44
Cl^{-1}	17	18.00	17.46	16.02	14.12	12.20	10.60	9.40
A	18	18.00	17.54	16.30	14.65	12.93	11.42	10.20
K	19	19.00		16.73		13.73		10.97
K^{+1}	19	18.00	17.65	16.68	15.30	13.76	12.27	10.96
Ca	20	20.00	19.09	17.33	15.73	14.32	12.98	11.71
Ca^{+1}	20	19.00		17.21		14.35		11.70
Ca^{+2}	20	18.00		16.93		14.40		11.70
Sc	21	21.00	20.28	18.72	17.04	15.39	13.82	12.39
Sc^{+1}	21	20.00	19.59	18.50	17.03	15.43	13.86	12.43
Sc^{+2}	21	19.00	18.71	17.88	16.68	15.27	13.82	12.44
Sc^{+3}	21	18.00	17.77	17.11	16.12	14.92	13.63	12.38
Ti	22	22.00	21.17	19.41	17.65	16.07	14.58	13.20
Ti^{+1}	22	21.00	20.60	19.52	18.03	16.39	14.76	13.25
Ti^{+2}	22	20.00	19.70	18.86	17.63	16.19	14.68	13.25
Ti^{+3}	22	19.00	18.76	18.09	17.06	15.82	14.48	13.16
V	23	23.00	22.21	20.47	18.68	17.03	15.49	14.03
V^{+1}	23	22.00	21.61	20.54	19.05	17.37	15.69	14.11
V^{+2}	23	21.00	20.70	19.86	18.62	17.14	15.60	14.10
V^{+3}	23	20.00	19.76	19.07	18.03	16.76	15.37	13.99
Cr	24	24.00	23.39	21.93	20.17	18.37	16.64	15.01
Cr^{+1}	24	23.00	22.62	21.58	20.10	18.40	16.68	15.03
Cr^{+2}	24	22.00	21.70	20.87	19.62	18.13	16.55	15.00
Cr^{+3}	24	21.00	20.76	20.07	19.02	17.72	16.30	14.87
Mn	25	25.00	24.26	22.61	20.79	19.06	17.41	15.84
Mn^{+1}	25	24.00	23.63	22.60	21.12	19.42	17.65	15.96
Mn^{+2}	25	23.00	22.71	21.89	20.66	19.16	17.55	15.94
Mn^{+3}	25	22.00	21.76	21.07	20.02	18.71	17.26	15.78
Mn^{+4}	25	21.00	20.80	20.22	19.32	18.18	16.90	15.55

0.35	0.40	0.50	0.60	0.70	0.80	0.90	1.00	1.10
7.37	6.79	5.70	4.71	3.88	3.22	2.71	2.33	2.05
7.40	6.82	5.69	4.69	3.86	3.20	2.70	2.32	2.04
7.70	7.20	6.24	5.31	4.47	3.75	3.16	2.69	2.35
	7.27	6.25	5.30	4.44	3.73	3.14	2.67	2.34
7.83	7.31	6.26	5.28	4.42	3.71	3.13	2.68	2.33
8.02	7.54	6.67	5.83	5.02	4.28	3.64	3.11	2.69
8.32	7.83	7.05	6.31	5.56	4.82	4.15	3.56	3.07
8.27	7.79	7.05	6.32	5.57	4.83	4.16	3.57	3.08
(8.21)	(7.89)	(7.22)	(6.47)	(5.69)	(4.93)	(4.23)	(3.62)	(3.13)
8.64	8.07	7.29	6.64	5.96	5.27	4.60	4.00	3.47
8.59	8.03	7.28	6.64	5.97	5.27	4.61	4.00	3.47
9.25	8.54	7.56	6.86	6.23	5.61	5.01	4.43	3.90
	9.05	7.87	7.11	6.51	5.95	5.39	4.84	4.32
9.89	9.04	7.86	7.11	6.51	5.94			
10.59	9.64	8.26	7.38	6.75	6.21	5.70	5.19	4.69
	9.63	8.26	7.38	6.75	6.21	5.70	5.19	4.68
	9.61	8.25	7.38	6.75	6.22	5.70	5.18	4.68
11.15	10.12	8.60	7.64	6.98	6.45	5.96	5.48	5.00
11.18	10.13	8.61	7.64	6.98	6.45	5.96	5.48	5.00
11.22	10.18	8.64	7.65	6.98	6.45	5.96	5.48	5.01
11.22	10.22	8.68	7.67	6.98	6.44	5.96	5.49	5.02
	10.83	9.12	7.98	7.22	6.65	6.19	5.72	5.29
11.91	10.77	9.06	7.95	7.21	6.66	6.18	5.73	5.28
11.94	10.82	9.10	7.96	7.21	6.66	6.18	5.73	5.28
11.93	10.84	9.14	7.99	7.22	6.65	6.18	5.73	5.29
	11.51	9.63	8.34	7.48	6.86	6.39	5.94	5.53
12.69	11.46	9.57	8.31	7.47	6.87	6.39	5.95	5.52
12.72	11.51	9.61	8.32	7.47	6.86	6.38	5.95	5.52
12.69	11.52	9.65	8.36	7.48	6.87	6.38	5.95	5.53
	12.22	10.14	8.72	7.75	7.09	6.58	6.14	5.74
13.53	12.21	10.13	8.71	7.75	7.09	6.58	6.14	5.74
13.55	12.26	10.18	8.74	7.76	7.09	6.58	6.14	5.72
13.50	12.26	10.22	8.77	7.78	7.09	6.58	6.14	5.74
	13.02	10.80	9.20	8.09	7.32	6.77	6.32	5.93
14.39	13.00	10.75	9.17	8.08	7.33	6.78	6.33	5.93
14.42	13.05	10.80	9.19	8.09	7.33	6.77	6.33	5.93
14.35	13.04	10.84	9.23	8.12	7.34	6.78	6.32	5.93
14.23	12.98	10.84	9.25	8.13	7.35	6.79	6.34	5.94

Element and Ionic Charge	Z	$(\sin\theta)/\lambda$ ($\mathring{A}^{-1}$)						
		0.00	0.05	0.10	0.15	0.20	0.25	0.30
Fe	26	26.00	25.30	23.68	21.85	20.09	18.40	16.77
Fe^{+1}	26	25.00	24.64	23.63	22.16	20.45	18.66	16.92
Fe^{+2}	26	24.00	23.71	22.89	21.66	20.15	18.51	16.87
Fe^{+3}	26	23.00	22.76	22.09	21.04	19.72	18.25	16.74
Fe^{+4}	26	22.00	21.80	21.22	20.31	19.15	17.84	16.46
Co	27	27.00	26.33	24.74	22.92	21.13	19.41	17.74
Co^{+1}	27	26.00	25.65	24.66	23.20	21.49	19.67	17.89
Co^{+2}	27	25.00	24.72	23.91	22.68	21.17	19.52	17.84
Co^{+3}	27	24.00	23.77	23.09	22.04	20.71	19.23	17.68
Ni	28	28.00	27.35	25.80	23.99	22.19	20.44	18.73
Ni^{+1}	28	27.00	26.66	25.69	24.26	22.55	20.72	18.90
Ni^{+2}	28	26.00	25.72	24.93	23.71	22.21	20.54	18.83
Ni^{+3}	28	25.00	24.77	24.10	23.05	21.72	20.22	18.65
Cu	29	29.00	28.49	27.19	25.49	23.63	21.75	19.90
Cu^{+1}	29	28.00	27.67	26.71	25.30	23.59	21.76	19.92
Cu^{+2}	29	27.00	26.73	25.95	24.75	23.24	21.57	19.84
Cu^{+3}	29	26.00	25.77	25.11	24.07	22.75	21.24	19.65
Zn	30	30.00	29.39	27.92	26.14	24.33	22.54	20.77
Zn^{+2}	30	28.00	27.73	26.96	25.77	24.27	22.60	20.86
Ga	31	31.00	30.30	28.65	26.76	24.92	23.16	21.47
Ga^{+1}	31	30.00	29.55	28.35	26.74	24.98	23.22	21.50
Ga^{+3}	31	28.00	27.78	27.12	26.10	24.78	23.27	21.65
Ge	32	32.00	31.28	29.52	27.48	25.53	23.76	22.11
Ge^{+2}	32	30.00	29.64	28.64	27.21	25.58	23.89	22.22
Ge^{+4}	32	28.00	27.81	27.25	26.35	25.19	23.82	22.32
As	33	33.00	32.27	30.47	28.29	26.20	24.34	22.69
As^{+1}	33	32.00		29.79		25.85		22.38
As^{+2}	33	31.00		29.33		25.86		22.43
As^{+3}	33	30.00		28.74		25.79		22.47
As^{+5}	33	28.00	27.83	27.34	26.55	25.51	24.27	22.89
Se	34	34.00	33.27	31.43	29.13	26.91	24.95	23.24
Se^{+6}	34	28.00	27.85	27.42	26.71	25.78	24.65	23.38
Br	35	35.00	34.29	32.43	30.06	27.70	25.61	23.82
Br^{+7}	35	28.00	27.87	27.48	26.85	26.00	24.97	23.80
Br^{-1}	35	36.00	35.08	32.81	30.13	27.65	25.54	23.76

0.35	0.40	0.50	0.60	0.70	0.80	0.90	1.00	1.10
	13.84	11.47	9.71	8.47	7.60	6.99	6.51	6.12
15.29	13.82	11.41	9.67	8.45	7.60	6.99	6.52	6.11
15.30	13.86	11.46	9.69	8.46	7.60	6.99	6.51	6.11
15.26	13.87	11.50	9.73	8.48	7.61	6.99	6.52	6.12
15.08	13.78	11.51	9.77	8.52	7.64	7.00	6.52	6.11
	14.68	12.17	10.26	8.88	7.91	7.22	6.70	6.29
16.21	14.67	12.11	10.21	8.85	7.91	7.22	6.71	6.29
16.22	14.72	12.17	10.25	8.87	7.91	7.22	6.71	6.29
16.15	14.71	12.21	10.29	8.90	7.92	7.22	6.70	6.28
	15.56	12.91	10.85	9.33	8.25	7.48	6.90	6.47
17.17	15.57	12.86	10.80	9.31	8.24	7.48	6.91	6.46
17.17	15.61	12.91	10.84	9.32	8.25	7.48	6.91	6.46
17.08	15.58	12.95	10.88	9.36	8.26	7.48	6.90	6.46
	16.48	13.65	11.44	9.80	8.61	7.76	7.13	6.65
18.14	16.50	13.66	11.45	9.80	8.61	7.75	7.12	6.64
18.14	16.52	13.70	11.47	9.82	8.62	7.76	7.13	6.65
18.05	16.50	13.74	11.53	9.86	8.64	7.77	7.13	6.64
	17.42	14.51	12.16	10.37	9.04	8.08	7.37	6.84
19.13	17.48	14.54	12.18	10.37	9.04	8.07	7.36	6.83
19.84	18.26	15.38	12.95	11.02	9.54	8.46	7.64	7.05
19.84	18.26	15.37	12.94	11.02	9.54	8.45	7.65	7.05
20.00	18.38	15.41	12.94	11.00	9.53	8.44	7.64	7.05
20.54	19.02	16.19	13.72	11.68	10.08	8.87	7.96	7.29
20.60	19.05	16.18	13.71	11.68	10.08	8.87	7.97	7.29
20.76	19.21	16.26	13.72	11.66	10.06	8.85	7.96	7.29
21.15	19.69	16.95	14.48	12.37	10.67	9.34	8.32	7.57
	19.33	16.56	14.11	12.07	10.44	9.18	8.23	7.52
	19.33	16.54	14.09	12.06	10.43	9.17	8.23	7.52
	19.34	16.52	14.07	12.04	10.42	9.17	8.23	7.52
21.44	19.95	17.07	14.49	12.34	10.63	9.31	8.31	7.56
21.71	20.28	17.63	15.20	13.06	11.27	9.83	8.71	7.86
22.02	20.62	17.82	15.25	13.04	11.23	9.80	8.70	7.87
22.25	20.84	18.27	15.91	13.78	11.93	10.41	9.19	8.24
22.54	21.21	18.53	15.98	13.74	11.85	10.33	9.13	8.21
22.22	20.82	18.27	15.91	13.77	11.92	10.40	9.18	8.24

Filters for X-Rays

The following material is reprinted by special permission from *International Tables for X-Ray Crystallography*,[†] Volume III, pp. 75 and 79.

†Kynoch Press, Birmingham, England, 1962.

β-Filters for Seven Common Target Elements To Give Two Different Integrated-Intensity Ratios for $K\beta_1/K\alpha_1$

Target Element	β-Filter	$K\beta_1/K\alpha_1 = 1/100$			Percent Loss $K\alpha_1$	$K\beta_1/K\alpha_1 = 1/500$			Percent Loss $K\alpha_1$
		mm	mils[a]	g/cm²		mm	mils[a]	g/cm²	
Ag	Pd	0.062	2.4	0.074	60	0.092	3.6	0.110	74
	Rh	0.062	2.4	0.077	59	0.092	3.6	0.114	73
Mo	Zr	0.081	3.2	0.053	57	0.120	4.7	0.078	71
Cu	Ni	0.015	0.6	0.013	45	0.023	0.9	0.020	60
Ni	Co	0.013	0.5	0.011	42	0.020	0.8	0.017	57
Co	Fe	0.012	0.5	0.009	39	0.019	0.7	0.015	54
Fe	Mn	0.011	0.4	0.008	38	0.018	0.7	0.013	53
	Mn_2O_3	0.027	1.1	0.012	43	0.042	1.7	0.019	59
	MnO_2	0.026	1.0	0.013	45	0.042	1.6	0.021	61
Cr	V	0.011	0.4	0.007	37	0.017	0.7	0.010	51
	V_2O_5	0.036	1.4	0.012	48	0.056	2.2	0.019	64

[a]1 mil = 0.001 in.

Calculated Thickness of Ross-Filter Components for Commonly Used Radiations

Target Element	Filter Pair (A)	Filter Pair (B)	(A) Thickness mm	(A) Thickness mils[a]	(A) Thickness g/cm²	(B) Thickness mm	(B) Thickness mils[a]	(B) Thickness g/cm²
Ag	Pd	Mo	0.0275	1.08	0.033	0.039	1.53	0.040
Mo	Zr	Sr	0.0392	1.54	0.026	0.104	4.09	0.027
Mo	Zr	Y	0.0392	1.54	0.026	0.063	2.49	0.028
Cu	Ni	Co	0.0100	0.38	0.0089	0.0108	0.42	0.0095
Ni	Co	Fe	0.0094	0.37	0.0083	0.0113	0.45	0.0089
Co	Fe	Mn	0.0098	0.38	0.0077	0.0111	0.44	0.0083
Fe	Mn	Cr	0.0095	0.37	0.0071	0.0107	0.42	0.0077
Cr	V	Ti	0.0097	0.38	0.0059	0.0146	0.58	0.0066

[a] 1 mil = 0.001 in.

Formulas for Calculating Interplanar Spacings, d_{hkl}

Symbols:
h, k, l = Miller indices,
a, b, c = unit-cell edges (Å),
α, β, γ = interaxial angles.

Cubic system:

$$d = \frac{a}{\sqrt{h^2 + k^2 + l^2}}.$$

Tetragonal system:

$$d = \left(\frac{h^2}{a^2} + \frac{k^2}{a^2} + \frac{l^2}{c^2}\right)^{-1/2}.$$

Orthorhombic system:

$$d = \left(\frac{h^2}{a^2} + \frac{k^2}{b^2} + \frac{l^2}{c^2}\right)^{-1/2}.$$

Hexagonal system, hexagonal indexing:

$$d = \left[\frac{4}{3a^2}(h^2 + k^2 + hk) + \frac{l^2}{c^2}\right]^{-1/2}.$$

Hexagonal system, rhombohedral indexing:

$$d = a\left[\frac{(h^2 + k^2 + l^2)\sin^2\alpha + 2(hk + hl + kl)(\cos^2\alpha - \cos\alpha)}{1 + 2\cos^3\alpha - 3\cos^2\alpha}\right]^{-1/2}.$$

Monoclinic system:

$$d = \left[\frac{(h^2/a^2) + (l^2/c^2) - (2hl/ac)\cos\beta}{\sin^2\beta} + \frac{k^2}{b^2}\right]^{-1/2}.$$

Triclinic system:

$$d = \left(\frac{\dfrac{h}{a}\begin{vmatrix} \dfrac{h}{a} & \cos\gamma & \cos\beta \\ \dfrac{k}{b} & 1 & \cos\alpha \\ \dfrac{l}{c}\cos\alpha & 1 \end{vmatrix} + \dfrac{k}{b}\begin{vmatrix} 1 & \dfrac{h}{a} & \cos\beta \\ \cos\gamma & \dfrac{k}{b} & \cos\alpha \\ \cos\beta & \dfrac{l}{c} & 1 \end{vmatrix} + \dfrac{l}{c}\begin{vmatrix} 1 & \cos\gamma & \dfrac{h}{a} \\ \cos\gamma & 1 & \dfrac{k}{b} \\ \cos\beta & \cos\alpha & \dfrac{l}{c} \end{vmatrix}}{\begin{vmatrix} 1 & \cos\gamma & \cos\beta \\ \cos\gamma & 1 & \cos\alpha \\ \cos\beta & \cos\alpha & 1 \end{vmatrix}} \right)^{-1/2}$$

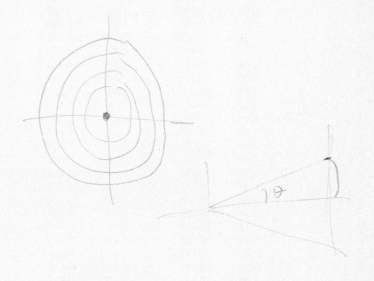

$\lambda = 2d \sin\theta$

$$d = \frac{\lambda}{2\sin\theta}$$

λ

Condensed Tables
of Lorentz and
Polarization Factors

The Polarization Factor, $(1+\cos^2 2\theta)/2$, as a Function of $\sin \theta$

$\sin \theta$	$\dfrac{1+\cos^2 2\theta}{2}$	$\sin \theta$	$\dfrac{1+\cos^2 2\theta}{2}$
0.00	1.000	0.50	0.625
0.02	0.999	0.52	0.605
0.04	0.997	0.54	0.587
0.06	0.993	0.56	0.570
0.08	0.987	0.58	0.553
0.10	0.980	0.60	0.539
0.12	0.972	0.62	0.527
0.14	0.962	0.64	0.516
0.16	0.950	0.66	0.508
0.18	0.937	0.68	0.503
0.20	0.923	0.70	0.500
0.22	0.908	0.72	0.501
0.24	0.891	0.74	0.504
0.26	0.874	0.76	0.512
0.28	0.856	0.78	0.524
0.30	0.836	0.80	0.539
0.32	0.816	0.82	0.559
0.34	0.795	0.84	0.584
0.36	0.774	0.86	0.615
0.38	0.753	0.88	0.651
0.40	0.731	0.90	0.692
0.42	0.709	0.92	0.740
0.44	0.688	0.94	0.794
0.46	0.666	0.96	0.856
0.48	0.645	0.98	0.924

The Combined Lorentz and Polarization Factor as a Function of sin θ

sin θ	Debye- Scherrer Method $\left(\dfrac{1+\cos^2 2\theta}{\sin^2 \theta \cos \theta}\right)$	Oscillating- or Rotating- Crystal Method $\left(\dfrac{1+\cos^2 2\theta}{\sin 2\theta}\right)$
0.00	∞	∞
0.025	3197	39.962
0.050	797.0	19.925
0.075	352.6	13.221
0.100	197.0	9.851
0.125	125.0	7.815
0.150	85.95	6.446
0.20	47.11	4.711
0.25	29.17	3.647
0.30	19.48	2.922
0.35	13.68	2.394
0.40	9.973	1.995
0.45	7.487	1.685
0.50	5.774	1.443
0.55	4.576	1.258
0.60	3.744	1.123
0.65	3.189	1.037
0.70	2.869	1.004
0.75	2.730	1.024
0.80	2.808	1.123
0.85	3.148	1.338
0.90	3.921	1.764
0.95	5.848	2.778
1.00	∞	∞

The Combined Lorentz and Polarization Factor as a Function of θ

θ (degrees)	Debye-Scherrer Method $\left(\dfrac{1+\cos^2 2\theta}{\sin^2 \theta \cos \theta}\right)$	Oscillating- or Rotating- Crystal Method $\left(\dfrac{1+\cos^2 2\theta}{\sin 2\theta}\right)$
0.0	∞	∞
1.0	6563	57.272
1.5	2916	38.162
2.0	1639	28.601
2.5	1048	22.860
3.0	727.2	19.029

θ (degrees)	Debye-Scherrer Method $\left(\dfrac{1+\cos^2 2\theta}{\sin^2\theta \cos\theta}\right)$	Oscillating- or Rotating-Crystal Method $\left(\dfrac{1+\cos^2 2\theta}{\sin 2\theta}\right)$
3.5	533.6	16.289
4.0	408.0	14.231
4.5	321.9	12.628
5.0	260.3	11.344
6.0	180.06	9.411
7.0	131.70	8.025
8.0	100.31	6.980
9.0	78.80	6.163
10.0	63.41	5.506
12.0	43.39	4.510
14.0	31.34	3.791
16.0	23.54	3.244
18.0	18.22	2.815
20.0	14.44	2.469
22.5	11.086	2.121
25.0	8.730	1.845
27.5	7.027	1.622
30.0	5.774	1.443
32.5	4.841	1.300
35.0	4.123	1.189
37.5	3.629	1.105
40.0	3.255	1.046
42.5	2.994	1.0115
45.0	2.828	1.000
47.5	2.744	1.0115
50.0	2.731	1.046
52.5	2.785	1.105
55.0	2.902	1.189
57.5	3.084	1.300
60.0	3.333	1.443
62.5	3.658	1.622
65.0	4.071	1.845
67.5	4.592	2.121
70.0	5.255	2.469
72.0	5.920	2.815
74.0	6.749	3.244
76.0	7.814	3.791
78.0	9.221	4.510
80.0	11.182	5.506
81.0	12.480	6.163
82.0	14.097	6.980
83.0	16.17	8.025

θ (degrees)	Debye-Scherrer Method $\left(\dfrac{1+\cos^2 2\theta}{\sin^2\theta \cos\theta}\right)$	Oscillating- or Rotating-Crystal Method $\left(\dfrac{1+\cos^2 2\theta}{\sin 2\theta}\right)$
84.0	18.93	9.411
85.0	22.78	11.344
85.5	25.34	12.628
86.0	28.53	14.231
86.5	32.64	16.289
87.0	38.11	19.029
87.5	45.76	22.860
88.0	57.24	28.601
88.5	76.35	38.162
89.0	114.56	57.272
90.0	∞	∞

Atomic Weights[†]

For the sake of completeness all known elements are included in the list. Several of those more recently discovered are represented only by the unstable isotopes. The value in parentheses in the atomic-weight column is, in each case, the mass number of the most stable isotope.[‡]

Name	Symbol	Z	International Atomic Weight		Valence
			1961	1959	
Actinium	Ac	89		(227)	
Aluminum	Al	13	26.9815	26.98	3
Americium	Am	95		(243)	3, 4, 5, 6
Antimony, stibium	Sb	51	121.75	121.76	3, 5
Argon	A	18	39.948	39.944	0
Arsenic	As	33	74.9216	74.92	3, 5
Astatine	At	85		(210)	1, 3, 5, 7
Barium	Ba	56	137.34	137.36	2
Berkelium	Bk	97		(249)	3, 4
Beryllium	Be	4	9.0122	9.013	2
Bismuth	Bi	83	208.980	208.99	3, 5
Boron	B	5	10.811	10.82	3
Bromine	Br	35	79.909	79.916	1, 3, 5, 7
Cadmium	Cd	48	112.40	112.41	2
Calcium	Ca	20	40.08	40.08	2
Californium	Cf	98		(251)	

[†]Reprinted from *Handbook of Chemistry and Physics*, 47th ed., The Chemical Rubber Company, Cleveland, 1966–1967, p. B–1.
[‡]The 1959 atomic weights are based on O = 16.000, whereas those of 1961 are based on the isotope ^{12}C.

Name	Symbol	Z	International Atomic Weight		Valence
			1961	1959	
Carbon	C	6	12.01115	12.011	2, 4
Cerium	Ce	58	140.12	140.13	3, 4
Cesium	Cs	55	132.905	132.91	1
Chlorine	Cl	17	35.453	35.457	1, 3, 5, 7
Chromium	Cr	24	51.996	52.01	2, 3, 6
Cobalt	Co	27	58.9332	58.94	2, 3
Copper	Cu	29	63.54	63.54	1, 2
Curium	Cm	96		(247)	3
Dysprosium	Dy	66	162.50	162.51	3
Einsteinium	Es	99		(254)	
Erbium	Er	68	167.26	167.27	3
Europium	Eu	63	151.96	152.0	2, 3
Fermium	Fm	100		(253)	
Fluorine	F	9	18.9984	19.00	1
Francium	Fr	87		(223)	1
Gadolinium	Gd	64	157.25	157.26	3
Gallium	Ga	31	69.72	69.72	2, 3
Germanium	Ge	32	72.59	72.60	4
Gold, aurum	Au	79	196.967	197.0	1,3
Hafnium	Hf	72	178.49	178.50	4
Helium	He	2	4.0026	4.003	0
Holmium	Ho	67	164.930	164.94	3
Hydrogen	H	1	1.00797	1.0080	1
Indium	In	49	114.82	114.82	3
Iodine	I	53	126.9044	126.91	1, 3, 5, 7
Iridium	Ir	77	192.2	192.2	3, 4
Iron, ferrum	Fe	26	55.847	55.85	2, 3
Krypton	Kr	36	83.80	83.80	0
Lanthanum	La	57	138.91	138.92	3
Lead, plumbum	Pb	82	207.19	207.21	2, 4
Lithium	Li	3	6.939	6.940	1
Lutetium	Lu	71	174.97	174.99	3
Magnesium	Mg	12	24.312	24.32	2
Manganese	Mn	25	54.9380	54.94	2, 3, 4, 6, 7
Mendelevium	Md	101		(256)	
Mercury, hydrargyrum	Hg	80	200.59	200.61	1, 2
Molybdenum	Mo	42	95.94	95.95	3, 4, 6
Neodymium	Nd	60	144.24	144.27	3
Neon	Ne	10	20.183	20.183	0
Neptunium	Np	93		(237)	4, 5, 6
Nickel	Ni	28	58.71	58.71	2, 3

Name	Symbol	Z	International Atomic Weight		Valence
			1961	1959	
Niobium (columbium)	Nb	41	92.906	92.91	3, 5
Nitrogen	N	7	14.0067	14.008	3, 5
Nobelium	No	102		(254)	
Osmium	Os	76	190.2	190.2	2, 3, 4, 8
Oxygen	O	8	15.9994	16.000	2
Palladium	Pd	46	106.4	106.4	2, 4, 6
Phosphorus	P	15	30.9738	30.975	3, 5
Platinum	Pt	78	195.09	195.09	2, 4
Plutonium	Pu	94		(242)	3, 4, 5, 6
Polonium	Po	84		(210)	
Potassium, kalium	K	19	39.102	39.100	1
Praeseodymium	Pr	59	140.907	140.92	3
Promethium	Pm	61		(145)	3
Protactinium	Pa	91		(231)	
Radium	Ra	88		(226)	2
Radon	Rn	86		(222)	0
Rhenium	Re	75	186.2	186.22	
Rhodium	Rh	45	102.905	102.91	3
Rubidium	Rb	37	85.47	85.48	1
Ruthenium	Ru	44	101.07	101.1	3, 4, 6, 8
Samarium	Sm	62	150.35	150.35	2, 3
Scandium	Sc	21	44.956	44.96	3
Selenium	Se	34	78.96	78.96	2, 4, 6
Silicon	Si	14	28.086	28.09	4
Silver, argentum		47	107.870	107.873	
Sodium, natrium	Na	11	22.9898	22.991	1
Strontium	Sr	38	87.62	87.63	2
Sulfur	S	16	32.064	32.066[a]	2, 4, 6
Tantalum	Ta	73	180.948	180.95	5
Technetium	Tc	43		(99)	6, 7
Tellurium	Te	52	127.60	127.61	2, 4, 6
Terbium	Tb	65	158.924	158.93	3
Thallium	Tl	81	204.37	204.39	1, 3
Thorium	Th	90	232.038	(232)	4
Thullium	Tm	69	168.934	168.94	3
Tin, stannum	Sn	50	118.69	118.70	2, 4

[a]Because of natural variations in the relative abundances of the isotopes of sulfur, the atomic weight of this element has a range of ± 0.003.

Name	Symbol	Z	International Atomic Weight		Valence
			1961	1959	
Titanium	Ti	22	47.90	47.90	3, 4
Tungsten (wolfram)	W	74	183.85	183.86	6
Uranium	U	92	238.03	238.07	4, 6
Vanadium	V	23	50.942	50.95	3, 5
Xenon	Xe	54	131.30	131.30	0
Ytterbium	Yb	70	173.04	173.04	2, 3
Yttrium	Y	39	88.905	88.91	3
Zinc	Zn	30	65.37	65.38	2
Zirconium	Zr	40	91.22	91.22	4

Miscellaneous Physical and Numerical Constants[†]

Constant	Symbol	Value
Avogadro's number	N	$= 6.02257 \times 10^{23}$ mole^{-1}
Loschmidt's number	n_0	$= 2.68702 \times 10^{19}$ cm^{-3}
Velocity of light	c	$= 2.997925 \times 10^{10}$ cm/sec
Electronic charge	e	$= 4.80296 \times 10^{-10}$ esu
Electronic rest mass	m_0	$= 9.10904 \times 10^{-28}$ g
Planck's constant	h	$= 6.62554 \times 10^{-27}$ erg sec
Atomic mass of electron	Nm_0	$= 5.485981 \times 10^{-4}$ amu
Atomic mass of proton	Nm_p	$= 1.00727663$ amu
Atomic mass of hydrogen	H	$= 1.00782522$ amu
Boltzmann constant	$k = R_0/N$	$= 1.38053 \times 10^{-16}$ erg/degree
Conversion unit from Siegbahn X units to milliangstroms	Λ	$= 1.002063$
Wavelength associated with 1 eV	λ_0	$= 1.239805 \times 10^{-4}$ cm eV
Energy associated with 1 eV	$e \times 10^8$	$= 1.602095 \times 10^{-12}$ ergs/eV
Gas constant	R_0	$= 8.31432 \times 10^7$ erg/(g-mole)(deg)
0°C		$= 273.18°$K
Base of natural logarithms	e	$= 2.7182818$

1 radian $= 57.29578$ degrees
π $\quad = 3.14159265$
1 cm $\quad = 10^8$ Å $= 10^4$ microns $= 0.39370$ in.
1 in. $\quad = 2.540005$ cm
$\log_e x$ $\quad = 2.302585 \log_{10} x$

[†]Physical constants are from *Handbook of Physical Constants*, revised edition, S. P. Clark, Jr. (editor), Geol. Soc. Amer., Memoir 97, 1966.

Name Index

Alexander, L. E., 6, 10, 11, 32, 35, 47–49, 155, 156, 161–166, 386, 529
Anderson, H. R., 296
Arlman, J. J., 151, 152, 160, 161
Arndt, U. W., 369, 372–375, 377
Arnett, L. M., 338
Arnott, S., 407, 408
Asp, E. T., 215, 221
Astbury, W. T., 53
Averbach, B. L., 437, 438, 440, 444, 450

Babinet, A., 282, 328
Bamford, C. H., 53
Banerjee, B. R., 124, 125
Baro, R., 110, 111, 126, 127
Barrett, C. S., 205–208
Bassett, D. C., 21, 22
Bassi, I. W., 364, 419
Bear, R. S., 102, 332, 334
Beeman, W. W., 107, 317, 318
Belbéoch, B., 338
Benoit, H., 313
Bernal, J. D., 59, 60
Bertaut, E. F., 437
Beu, K. E., 160
Bjørnhaug, A., 370–372
Blow, D. M., 392, 402, 403
Bolduan, O. E. A., 102, 334
Bonart, R., 24, 25, 130, 336–341, 425–432
Bowles, B. B., 266, 268
Brader, J. J., 193
Bragg, W. L., 36, 37, 39, 284, 373
Branson, H. R., 393
Bravais, M., 5
Breit, G., 31, 144, 145
Brown, C. J., 419

Brown, L., 399, 400
Brown, W., 113
Brumberger, H., 296
Buchanan, D. R., 423, 437, 438, 441, 444–451
Buerger, M. J., 57, 58, 60, 87, 93
Bunn, C. W., 7, 51, 55, 87, 272, 364, 366, 367, 393, 408, 413–419
Burgers, J. M., 451
Burton, R. L., 206, 209–211, 227, 228, 252

Čacković, H., 435
Caspar, D. L. D., 404, 405
Challa, G., 166–173
Chesley, F. G., 95–97
Chirer, E. G., 230
Clark, E. S., 76, 88, 91, 201, 361, 362, 369, 419
Clark, G. L., 262–265
Clark, S. P., Jr., 552
Coates, W. A., 372
Cochran, W., 393
Cohen, I., 233
Compton, A. H., 30
Conrad, C. M., 264
Corey, R. B., 393
Corradini, P., 8, 55, 149, 358, 364, 366, 419
Crawford, S. M., 274
Creely, J. J., 263, 264
Crick, F. H. C., 389, 390, 393, 401, 405
Crystal, E., 176–183
Cullity, B. D., 228

Daly, M. C., 385–388
Dana, E. S., 203, 204

Daubeny, R. de P., 272, 419
Davey, W. P., 55
Davies, D. R., 395, 399, 400
Debye, P., 43, 47, 280, 281, 296, 380, 458, 461
Decker, B. F., 215, 221
DeLuca, L. B., 265, 266
Desper, C. R., 140, 212, 213, 223, 252, 253, 270, 275
Dewaelheyns, A., 140
Dirac, P. A. M., 31, 144, 145
Dismore, P. F., 267, 347, 349
Dobson, G. M. B., 83
Dowling, P. H., 116, 118, 121
Dumbleton, J. H., 266, 268
Dunitz, J. D., 358

Ellefsen, O., 370–372
Elliott, A., 112
Elliott, G. F., 112, 113
Emde, F., 463
Ergun, S., 31, 472

Fankuchen, I., 105
Farina, M., 360
Farrow, G., 190–195
Field, J. E., 151
Finback, C., 370
Fischer, E. W., 343
Flory, P. J., 158
Fock, V., 31
Ford, W. E., 203, 204
Fournet, G., 106, 288, 299
Franklin, R. E., 402–405
Franks, A., 110–113
Freda, A., 228
Frey-Wyssling, A., 327
Frosch, C. J., 4, 393
Fuller, C. S., 4, 393
Fuller, W., 402, 403

Ganis, P., 366
Garner, E. V., 7
Geiger, H., 115, 117
Geil, P. H., 17, 23, 199, 342-349
Geisler, A. H., 230
Godard, G. M., 26, 332, 338
Goppel, J. M., 151, 152, 160, 161
Grün, F., 274
Guinier, A., 100, 101, 106, 280, 281, 288, 298-300, 311, 312, 315, 324, 325, 430-434

Harker, D., 215, 221
Hartree, D. R., 31
Hayes, J. E., Jr., 442
Heffelfinger, C. J., 206, 209–211, 227, 228, 252
Heikens, D., 77, 328, 330
Heine, S., 313
Hendee, C. F., 116
Hendus, H., 193
Hermans, J. J., 259, 262, 313, 426
Hermans, P. H., 77, 166–173, 241, 262, 328, 330
Hess, K., 24, 25, 332–336
Heyn, A. N. J., 327, 376–380
Higashimura, T., 50
Hight, R., Jr., 288
Holden, A. N., 229
Holmes, D. R., 87, 267, 393, 419
Holmes, K. C., 392, 402–404
Hosemann, R., 24, 25, 130, 313, 338–341, 423–436
Howells, E. R., 367
Huggins, M. L., 349, 358
Hughes, R. E., 361, 475
Hull, A. W., 55
Huxley, H. E., 113

Jaccodine, R., 16
Jagodzinski, H., 433, 435
Jahnke, E., 463
Jellinek, M. H., 105
Johann, T. H., 123
Johansson, T., 100, 109, 110, 123, 435
Jones, J. W., 223
Judge, J. T., 276

Kaesberg, P., 107, 317, 318
Kahovec, L., 288
Keller, A., 21, 22
Kendrew, J. C., 373, 405
Kiessig, H., 24, 25, 104, 105, 332–336
King, M. V., 87
Kinoshita, Y., 87, 419
Kirste, R., 313
Klug, A., 389, 390, 402–405
Klug, H. P., 6, 10, 11, 32, 35, 47–49, 529
Kobayashi, M., 361, 475
Kohler, T. R., 116
Kolsky, H., 274

Kratky, O., 107–110, 280, 281, 286–291, 297, 304–313, 318–327, 331, 332, 426
Krigbaum, W. R., 259–262
Krimm, S., 180
Kuhn, W., 274

Lang, A. R., 123, 124
Laue, M. v., 36, 51, 66, 85
Lauer, J. L., 361, 475
Leeper, H. M., 16
Lindenmeyer, P. H., 223
Lipson, H., 128–133
Liquori, A. M., 368, 393
Lorentz, H. A., 40, 41, 283, 284
Lustig, S., 223
Luzzati, V., 110, 111, 126, 127, 280, 281, 285, 290, 294, 295, 300, 303–305, 313–318

McCall, D. W., 194
McCullough, R. L., 426–428, 438, 447
Magill, J. H., 18–20, 23, 98, 338, 347
Matsubara, I., 53
Meibohm, E. P. H., 338
Meyer, K. H., 384
Michalik, E. R., 386
Miholic, G., 289, 291, 320–326
Milberg, M. E., 262, 376, 384–388
Miller, D. C., 120
Miller, R. G. J., 193, 194
Miller, R. L., 360–363, 423, 437, 438, 441, 444–451, 473
Misch, L., 384
Mitsuhashi, S., 21, 22
Müller, W., 115, 117
Muus, L. T., 369

Nagai, K., 361, 475
Natta, G., 8, 15, 55, 149, 161, 168, 360, 364, 366, 419
Neff, H., 232
Newman, S., 360
Nicolaieff, A., 295, 315
Nielsen, L. E., 360, 363
Nies, N. P., 48, 59
Noble, F. W., 442
Norman, N., 376, 382–384
Norris, F. H., 272–274
Nyburg, S. C., 160, 161

Ogilvie, R. E., 124
Ohlberg, S. M., 155, 156, 161–166
Okamura, S., 50
Orr, R. S., 265, 266
Oster, G., 461

Palmer, R. P., 267
Pape, N. R., 4, 393
Parrish, W., 116, 118
Patterson, A. L., 372, 381, 407, 458, 460
Pauling, L., 393, 418
Peraldo, M., 360
Perutz, M. F., 373
Peterlin, A., 348
Planck, M., 31
Polanyi, M., 263
Pollack, S. S., 45, 52, 53, 95, 98, 200, 338, 347
Porod, G., 280, 288–296, 306, 313, 327, 426
Predecki, P., 132, 339
Prelog, V., 358
Prins, J. A., 426

Rabjohn, N., 158
Ramachandran, G. N., 405
Reneker, D. H., 342, 345
Rich, A., 395, 399, 400, 405
Riley, D. P., 369–377, 461
Ritland, H. N., 107, 317, 318
Roe, R.-J., 259–262
Roess, L. C., 313
Ross, P. A., 110, 120, 144
Rugo, H. J., 334
Ruland, W., 31, 122, 126, 138–149, 166, 167

Sack, R. A., 259
Salo, T. P., 334
Samuels, R. J., 277
Scherrer, P., 47, 335, 338, 423, 424
Schlesinger, W., 16
Schmidt, G., 24, 343
Schmidt, P. W., 288
Schnell, G., 193
Schoening, F. R. L., 447–450
Schultz, L. G., 223, 226
Segal, L., 264
Sekora, A., 331, 332

Shaffer, M. C., 158
Shull, C. G., 313
Sisson, W. A., 206, 262–265
Skala, Z., 288, 289
Slichter, W. P., 194, 196
Smith, A. F., 338
Smith, D. J., 87, 419
Statton, W. O., 26, 77, 78, 102, 103, 132, 184–187, 191, 267, 282, 327, 328, 331–339, 343, 347, 349
Stein, R. S., 140, 180, 212, 213, 223, 241–245, 252–254, 270–276
Stokes, A. R., 442, 443, 467, 468, 471
Storks, K. H., 16
Straumanis, M. E., 91
Stuart, H. A., 24

Tadokoro, H., 365
Taylor, C. A., 128–133, 392
Taylor, G. R., 155, 156, 161–166
Taylor, J., 118
Tønnesen, B. A., 370–372
Treloar, L. R. G., 274
Trotter, I. F., 399, 400
Turner-Jones, A., 408, 410, 413–418

Ulmann, R., 313
Ulrey, C. T., 27

Vand, V., 393
Virgin, H. S., 176–183
Visser, J., 102

Waals, J. D. van der, 144
Wakelin, J. H., 176–183
Ward, I. M., 190–195
Warren, B. E., 437–440, 444, 450
Watson, J. D., 401
Weidinger, A., 77, 166–173, 328, 330
Weissenberg, K., 262
Wilchinsky, Z. W., 223, 228, 234–240, 245–259, 275
Wilke, W., 433–435
Williams, R. O., 259–262
Willis, H. A., 193, 194
Wilson, A. J. C., 91–95, 438
Witz, J., 126, 127, 295, 315
Wolff, P. M. de, 41, 100, 123
Wonacott, A. J., 407, 408
Wooster, W. A., 124
Worthington, C. R., 105, 106, 112
Wulff, G. V., 204
Wunderlich, W., 313
Wyckoff, H. W., 389, 390

Zernike, F., 426
Ziegler, K., 168

Subject Index

Absolute intensity measurements, 123, 126, 285, 286, 328–331
  by means of filters, 285
  by means of a perforated rotor, 285, 286
  by reference to a standard sample, 286, 328, 329
Absorption coefficient, calculation of, 28, 69
  linear, 27, 68, 69
  mass, 27–29, 69
    table of, 524–528
  significance of, in crystallinity determination, 152–154
  of a typical polymer, 68
Absorption corrections, for normal-beam-transmission technique, 69, 70
  in crystallinity determination, 151–154
  in pole-figure diffractometry, 235–237
  by reflection technique, 226, 227
  by transmission technique, 221, 222
  for symmetrical-reflection technique, 81
  for symmetrical-transmission technique, 70, 71
Absorption, edge, K-, 28
  L-, 28
  exponent, μt, measurement of, 222
  factor, 42
  of x-rays, 26–29
    by β-filters, 29, 540
    correction for, 70–72, 81
    discontinuities in, 28
    linear coefficient of, 27, 68, 69
    mass coefficient of, 27–29, 69
    by Ross balanced filters, 120–122
    by specimen, diffractometric measurement of, 155, 156

variation of, with wavelength, 28
  by zirconium, 28
Accuracy, statistical, of counter measurements, 67
Adhesives for powder specimens, 73
Affine extension of elastomers, 156
Air scattering of x-rays, 85, 159
  correction for, in crystallinity determination, 159, 162, 168
  darkening of film by, 85
Alignment, of Franks camera, 112
  of Kratky camera, 107, 110
  of microcameras, 96, 97
Aluminum foil, as reference specimen in diffractometry, 156
Ambiguities, in interpretation of small-angle scattering, 281, 282, 298
  in radial-distribution analysis, 372
American Society for Testing and Materials (ASTM), 48, 50
Amino-acid residue, 372, 373
Amorphous birefringence, 269, 270
  determination of, 270
    of axially oriented polyethylene, 272–274
    of biaxially oriented polyethylene, 274–277
Amorphous fraction, 151–154
Amorphous halos, 1, 43, 45, 380, 381
  anisotropic, 385
  of polybutadiene
    constancy of integrated intensity of, 163
    integral breadth of, 163, 164
    variation of, with extension, 162–164
  use of in crystallinity determinations, 151–165

Amorphous intensity, use of in crystal-
        linity determinations, 151–165
Amorphous regions, 18, 21, 25, 137, 190,
        337, 378, 409
    orientation of, 190, 191, 198, 385, 386
Amorphous specimens, 1, 2
    effect of orientation on density of,
        190–192
    scattering of x-rays by, 43–45
Anisotropy, optical, of monomeric units,
        269
    (shape) factor, 301
    of lysozyme, 319
Annealing of polymers, 25, 440
    effect of, on crystallite size, 447, 448
        on degree of crystallinity, 192
        on long spacings, 342, 343
    morphology of, 342–350
Asbestos, fiber pattern of, 58
ASTM x-ray powder diffraction data
        cards, 48, 50
Asymmetric film position, of Straumanis,
        91, 92
    of Wilson, 91, 92, 95
Asymmetric unit of structure, 415
    helical, 388–390
Atactic polymers, 16, 358
Atomic coordinates, 457
    of poly(ethylene adipate), 415
Atomic-density function, 459
Atomic radii, 418, 419
Atomic scattering factors, 30, 40, 44,
        457–459, 463
    mean-square, 139, 145
    table of, 533–539
Atomic weights table, 548–551
Attenuation of direct beam, with a per-
        forated rotor, 285, 286
    with filters, 285
Avogadro's number, 552
Axes, crystallographic, 5, 9
Axial (fiber) orientation, 50–62, 73, 87,
        199, 208–211, 241–252, 376–
        386, 406, 409
    analysis of, in nonorthogonal crystal
        systems, 246–249
    in orthorhombic polymers, 241–243
    analytical description of, 241–252
    azimuthal-reflection breadth as a mea-
        sure of, 262–268
    as component of complex textures, 245
    cylindrical symmetry of, 241
    limiting values of $< \cos^2 \phi >/f$ for,
        241, 242
    overall, birefringence as a measure of,
        268
    representation of, by inverse pole
        figures, 259–261
        on a ternary diagram, 241–243
        in sodium metaphosphate, 385, 386
    Stein's model for, 241–243
    stereographic representation of, 261
Axialites, 20–22, 344
Axially oriented specimens, structural
        analysis of, 376–386
Axial translations, 5
Azimuth, 73, 74
    specimen holders for varying, 73, 75,
        76, 154, 157
Azimuthal reflection breadth, as a mea-
        sure of fiber orientation, 262–
        268
    as an orientation index, 262–268

Babinet compensator, 271
Babinet reciprocity principle, 282, 327,
        328
Background, correction for, 214, 215, 229,
        235, 266, 441
    from continuous spectrum, 29
    from fluorescent x-rays, 32
    from incoherent scattering, 30
    from paracrystalline distortions, 424,
        425, 429
Back-reflection technique, 86
Balanced filters (Ross), 110, 120–122
    of cobalt and nickel, 120, 122
    of cobalt and nickel oxides, 144
    pass band of, 122
    table of, 541
    use of, with pulse-height discrimina-
        tion, 122
Base unit of a polymer chain, 358–360,
        363
Beam divergence, axial, 219, 220, 224
    restriction of, by Soller slits, 219, 220
    effect of, on reflection profiles, 217, 224,
        225

equatorial, 216–218, 225
Benzene ring, optical transform of, 132,
    133
Bernal charts, 59, 60
Bessel functions, 305, 378, 460–464
    in analysis of helical molecules,
        388–400
    spherical, 382
Beta (β) filters, 29, 85, 372, 384
    for common targets, table of, 540
    for copper radiation, 29
    for molybdenum radiation, 29
Biaxial orientation, analysis of, in orthor-
        hombic polymers, 254, 255
    definitions of, 252, 254
    determination of, from birefringence
        measurements, 270, 271
    Stein's analysis of, 254, 255
Binders for powder specimens, 73
Biological molecules, helical structures of,
        372–376, 392, 393, 399–405
    small-angle cameras for the study of,
        105–113
    small-angle diffractometer for the study
        of, 126, 127
Birefringence, 268–277
    amorphous, 268–270, 272–274, 276
    of axially oriented specimens, 269, 270
    of biaxially oriented specimens, 270,
        271
    crystalline, 269, 270
    definition of, 268
    experimental measurement of, 271, 272
    form, 269, 270
    intrinsic, 270
    as measure of overall orientation, 198,
        268
    in-plane, 270
    out-of-plane, 270
    of polyethylene, 272–277
    of a single crystal, 270
Boltzmann constant, 552
Bragg angle, Θ, 40, 79–82, 168, 169, 216,
        219–226, 263, 406
Bragg equation (law), 36, 39, 42, 48, 62
    in interpretation of small-angle reflec-
        tions, 332, 334, 335
Bragg spacings, see Interplanar (d)
    spacings

Bravais lattice, 5
Breit-Dirac recoil factor, 31, 144, 145
Buerger precession camera, 87
Burgers vector, 451
Bunn chart, 55, 56

Calcite crystals, use of, in double-crystal
        spectrometer, 106
Calibrating substances for cameras, 87,
        93, 102
Calibration of Debye-Scherrer cameras,
        93
Cameras, cylindrical, 66, 87–89
    Debye-Scherrer, 47, 48, 89–94;
        see also Debye-Scherrer cameras
    evacuation of, 85, 87, 94, 95, 103, 104,
        109, 110, 382
    flat-film, 85–87
    general-purpose, 87–90, 102–104
    Goppel-Arlman, 151, 152
    Guinier, 100, 433, 435
    Guinier-de Wolff, 100, 101
    Laue, 85
    micro-, 95–97
    monochromators for, 97–102
    parafocusing, 99–102
        with curved-crystal monochromator,
            99, 100
    small-angle, 102–113
        for biological specimens, 112, 113
        of A. Elliott, 112, 113
        of G. F. Elliott and Worthington, 112
        of Franks, 110–113
        of Kiessig, 104, 105
        of Kratky, 107–110
        of Luzzati-Baro, 110
        of Statton, 87, 102, 103
        of Worthington, 105
        resolution of, 104–107
    wide-angle, 85–102
    see also Cylindrical cameras
Carbon chains, conformations of, 364, 365
    identity periods of, 365
Carver laboratory press, 180
Cauchy reflection profiles, 439–443, 447,
        450
Cell, see Unit cell
Cellulose, air-swollen, Guinier plots of,
        324, 325

particulate parameters of, 323–326
small-angle intensity curves of, 320, 321, 324
small-angle scattering study of, 319–326
long spacings in, 335
microvoids in, 327–331
scattering power of, 328–331
small-angle scattering by, 327–331
Cellulose (cotton), determination of crystallinity index of, 180–183
orientation in fibers of, 263–266
preparation of specimens of, for crystallinity determination, 180
regression curve of, 178
scattering curves of, 177, 181–183
spiral angle in, 265–267
Cellulose I, orientation in fibers of, 264–266
unit cell of, 264
Cellulose fibers, *see* Fibers, cellulose
Cellulose nitrate, small-angle scattering study of, 310–313
Cellulose structure model of Meyer and Misch, 384
study of, by cylindrical distribution functions, 382–384
Centered cell, 5
Centered lattice, 5
Chain folding, 16, 21, 25, 342–350
Characteristic (correlation function), 296
Characteristic x-ray spectrum, 26
Chrysotile asbestos, 53
cis Configuration, 15
Cluster fraction, 322
of air-swollen cellulose, 323
Clustering, 292
of parallel fibrils, 432–434
Cobalt and nickel, as balanced filters for CuKα x-rays, 120, 122, 144
Coherent (unmodified) scattering of x-rays, 29–32, 138, 139, 144
calculation of, for polypropene, 145–147
Coiled-coil model of α-keratin, 405
Cold work, induction of preferred orientation by, 198
Collagen, small-angle scattering by, 334
small-angle study of, 105, 112, 113, 332, 334

triple-helix conformation of, 405
College of Science and Technology, Manchester, England, 128
Collimation errors, correction of, 287–290, 319
for infinitely long and narrow slits, 287, 288
for narrow slits of arbitrary height, 288
for slits of finite breadth, 289, 290
use of computers in, 288
use of convolution analysis in, 289
Collimation of x-ray beam, by Soller slits, 200, 219
in Debye-Scherrer cameras, 47, 48
in flat-film cameras, 74, 85
in microcameras, 95, 96
in pole-figure diffractometry, 216–221, 225, 226
in small-angle scattering, 286–290
by pinholes, 286, 287
by slits, 286, 287
Collimator, geometry,
in diffractometry, 79, 80, 218–220
in small-angle scattering, 286–290
Soller-slit, 219, 220
Complex conjugate, 457
Compton scattering, *see* Incoherent (modified) scattering of x-rays
Computer methods, 115, 184, 233, 234, 260, 288, 395, 408, 441, 442, 464, 469
Computers, use of, in controlling pole-figure diffractometers, 233
in processing pole-figure data, 233, 234
Concentration, electron, 293
Configuration, 15, 358–360
cis, 15
definition of, 358
gauche, 364
sensitivity of infrared absorption to, 192, 193
trans, 15, 364
Conformation, analytical description of, 360–366
by internal rotation angles, 364, 368
definition of, 358
factors determining 366–368

gauche, 364
helical, 360–368, 376, 388–405
    nomenclature of , 360–366
planar zigzag, 14, 358, 359, 360,
        368–371, 410–413
    of poly(ethylene adipate), 410, 412,
        416–418
trans, 364
Conformational nomenclature, helical,
        360–366
    Bunn's, 364–366
    helical-point-net system of, 360–364
    screw-axis system of, 360–364
    Tadokoro's, 365, 366
Constants, table of physical and numerical,
        552
Continuous x-ray spectrum, 26
    as a cause of background, 29
    correction for, in crystallinity determi-
        nation, 159
    lower wavelength limit of, 98
Convolution, analysis, of collimation-error
        corrections, 289
    of reflection profiles, 217, 218, 224
    (fold), 391, 437, 455–458, 467
    relation of, to multiplication, 456
Coordinates, positional, of an atom, 39
Copper, analysis of reflection profile of,
        470
Copper-target x-ray tube, spectrum of, 121
Copper x-rays, balanced filters for, 120,
        122, 144
    β-filter for, 29
Correlation, distance (length), 296
    function (characteristic), 296
$<\cos^2 \theta>$ functions, experimental
        evaluation of, 243–247
    in description of orientation, 241, 254,
        255
    numerical examples of, 249–252
Cotton-amine complexes, orientation in
        fibers of, 265
Cotton cellulose, *see* Cellulose (cotton)
Counter diffractometers, 113–126, 372,
        see also Diffractometry
    automatic, pole-figure measurements
        with, 232, 233
    automation of, 115
    components of, 113–115

counters for, 115–119
crystallinity measurements with,
        151–165, 168, 180–183
four-circle, pole-figure measurements
        with, 233, 234
    of Krimm and Stein, 180
    of Luzzati, Witz and Baro, absolute
        intensity measurements with,
        126
    use of, in study of protein solutions,
        126
    orientation measurements with, 264,
        266
    use of, in measuring scattering curve of
        polypropene, 144
Counters for x-ray measurements, 115–119
    comparative wavelength sensitivity of,
        117–119
    dead time of, 117
    Geiger-Müller, 115–118, 127, 159, 180,
        181, 319
    nonlinearity of response of, 117
    quantum-counting efficiency of,
        117–119
    resolving time of, 117
    scintillation, 116–121, 440
Counter measurements, sensitivity of, 67
    statistical accuracy of, 67
Crystal-defect concept, 3, 137
Crystal, ideal, 5–12, 51, 424, 467
    axial translations in, 5
    crystallographic axes of, 5
    crystallographic planes in, 9–12
    indexing of planes in, 9–12
    space lattice of, 5
    unit cell of, 5
Crystalline birefringence, 269, 270
    calculation of, 270
    fraction, 25, 137–197
    regions, 21, 25, 137, 141, 327, 328, 337,
        409, 424
    orientation of, by cold work, 198
    specimens, 1, 2
    structures, intermediate, 20, 21
Crystallinity, comparison of x-ray, with
        density crystallinity, 191, 192
    with infrared crystallinity, 191, 192
    with NMR crystallinity, 194–196
    concept of, 137, 138.

degree of, 25, 137–197
determination of, from density
    measurements, 189–191
from infrared measurements, 192, 193
from NMR measurements, 194–196
diffractometric measurement of, 144,
    159, 160, 168, 180, 181
relation between weight and volume
    fractions of, 190
volume fraction of, 269
x-ray determination of degree of,
    138–185
    with allowance for lattice imperfec-
        tions, 144–151
    from amorphous intensity, 151–165
    correction for scattering mass in, 151,
        169
    in cotton cellulose, 180–183
    from crystalline and amorphous
        intensities, 162–176
    as a crystallinity index, 176–185
    from differential intensities, 176–185
    diffractometric method for, 152–160
    effect of preferred orientation on
        accuracy of, 174–176
    in elastomers, 151–165
    as a function of temperature, 160–165
    Goppel-Arlman method for, 151–154
    Hermans-Weidinger method for,
        166–176
    from integrated crystalline and
        amorphous scatter, 143–151
    in isotactic polystyrene, 168–176
    in natural rubber, 158–160
    normalization of intensities in, 169,
        170, 179, 181
    in nylons 6 and 7, 149–151
    in polybutadiene, 160–166
    in poly(ethylene terephthalate),
        184–187
    in polypropene, 144–149
    precision of, 154
    Ruland's method for, 143–151
    use of regression curve in, 165–175
    Wakelin, Virgin and Crystal's method
        for, 176–183
Crystallinity index, as an expression of
    degree of ordering, 176
    correlation, 177, 178

determination of, corrections required
    in, 179, 181
    of cotton cellulose, 180–183
    of poly(ethylene terephthalate),
        184–187
    practical requirements for, 178, 179
    simplifications of procedure in, 179
    use of regression curve in, 177–180
    integral, 177, 178
    photographic method for determination
        of, 184
    reference specimens required for, 176,
        180
    theory of, 176–178
Crystallites, 327;
    size, analysis of, 423–452
    determination of, with allowance for
        lattice distortions, 429–452
    distribution, 450
    effect of annealing on, 447, 448
    number-average, 438, 444, 447–451
    weight-average, 447, 451
    from Scherrer equation, 335–338,
        423, 424
    see also Crystalline regions
Crystallization, 1-3
    from the melt, 17-21, 343, 344
    morphology of, 16-26, 342-350
    of polymers, 12-26
        as single crystals, 16
        axialitic, 20, 344
        effect of, on density, 189
        hedritic, 20, 344
        importance of stereoregularity in, 12,
            14, 15
        lamellar, 16-21, 25, 343-348
    primary, 18
    in relation to texture, 21-26
    secondary, 18
    from solution, 16-21
    spherulitic, 17-19, 344-346
Crystallographic axes, direct, 5, 9-12, 457
    data for polymers, Miller and
        Nielsen's tables of, 363
    Miller's tables of 473-523
    reciprocal, 457
Crystal-monochromatized x-rays, polariza-
    tion factor for, 40
Crystal monochromators, 29, 40, 85,

97–102, 234, 377, 435
beam intensity from, 99
curved, 99–102, 113, 123, 124, 127
in diffracted beam, 123, 124, 125, 168
in direct beam, 124–127, 180
exposure times required with, 99, 100
flat, 97–99
holder for, 99
improvement of patterns with, 91, 99
Johann, 123
Johansson, 100, 109, 110, 123
pentaerythritol, 97, 124
quartz, 110, 113, 123, 126, 285
rock-salt, 124, 370
RR and TR arrangements of, 123–125
    comparative merits of, 123
separation of $K\alpha$ doublet with, 123
Crystal structure, models, testing of, with
        optical diffractometer, 126–133
    of poly(ethylene adipate), 408–419
    of poly(ethylene adipate) and poly-
        ethylene, comparison of, 412,
        415–419
    of polymers, determination of, by trial
        and error, 407, 408
        limitations affecting determination of,
            406–408
        refinement of, 407, 408, 412–414
    from wide-angle-diffraction data,
        405–419
Crystal systems, 5
    unique directions in, 245
Cubic system, 5
310 Curve Resolver, du Pont, 442
Cylindrical cameras, 66, 87–89
    measurement of reflection intensities,
        with, 87
    preparation of fiber patterns with, 56,
        57, 87–89
    rotation of specimen in, 89
Cylindrical-distribution functions,
        381–384
    of cellulose fibers, 382–384
    difference, 262, 384–386
    helical, 376
Cylindrically symmetrical systems, 460,
        461
    radial analysis of, 377–379, 461
    radial-distribution functions of, 377–379

transform analysis of, 376–386
x-ray scattering by, 376–386
Cylindrical transforms, 376–386, 460, 461

Dead time, of an x-ray counter, 117
Debye intensity equation, 44, 380, 458
Debye-Scherrer, cameras, 47, 48, 89–94
    calibration of, 93
    film radius of, 91, 92
    four film positions in, 91, 92
    preparation of specimens for, 72, 91
    diffraction patterns, 91–95, 99, 370
    d-value scales for measuring, 48, 49
    of polystyrene, 50
    of representative polymers, 94, 95
    use of, in analysis, 47–50
    diffraction technique, 47, 48, 72, 91–94,
        99, 205, 370
Decamethylene dis-methylazodicarboxy-
        late, use of, as cross-linking
        agent, 158
Deconvolution, method of Ergun for, 472
    method of Stokes for, 442, 443, 467–472
Defects, propagation of, 342–345
Deformation of polymers, morphology of,
        349
Delta function, 457
Dense systems, 282
    analysis of, from small-angle scattering,
        327–350
Densitometers, see Microdensitometers
Density, amorphous, 189, 409
    crystalline, 189, 409
    determination of crystallinity from,
        189–191
    effect of crystallization on, 189
    optical, 83, 169
    x-ray determination of, 409
Deoxyribonucleic acid (DNA), A and B
        forms of, helical diffraction by,
        401, 402
    helical conformations of, 401, 403
Diatropic reflections, see Meridional
        (diatropic) reflections
Difference-distribution function, cylindri-
        cal, 384–386
    electronic, 385
Diffraction of light, Fraunhofer, 127

Diffraction patterns, x-ray,
  optical, as analogs of x-ray diffraction
    patterns, 126–128
  of paracrystalline models, 425, 426
  see also Diffraction of x-rays
Diffraction profiles, see Reflection profiles
Diffraction tube, microfocus, 96, 97, 110
  rotating-anode, 105, 328, 372
  target of, 26, 40
Diffraction of x-rays, 33–63,
  amplitude of, 455
  by cylindrically symmetrical systems,
    376–386, 460–461
  by fibers, 50–62, 199–202; see also
    Fiber diffraction patterns
  geometry of, 35–39
  by helical molecules, 388–405, 461–464
    effect of crystalline ordering on, 464
    intensity of, 390, 391
    meridional reflections in, 391
  intensity of, factors affecting, 39–43
  by a linear lattice, 35
  molecular structure from, 368–405
  by paracrystalline structures, 425–436
    Hosemann's theory of, 426–429
    small-angle, 429–433
    wide-angle, 433–436
  by preferentially oriented specimens,
    50–62, 199–202
  by randomly oriented specimens, 45–50,
    94, 95, 101, 102, 369–376,
    458–460
  reciprocal-lattice model of, 38, 46
  reflection analogy of, 36, 37
  small-angle, 33, 62, 63, 280–356;
    see also Small-angle scattering
  by three-dimensionally ordered systems,
    456, 457
  wide-angle, 33–62
    cameras for, 85–102
    crystal structure from, 405–419
    microstructure from, 357–422
  see also Scattering of x-rays
Diffractometer, optical, see Optical
  diffractometer
Diffractometers, see Counter diffractometers
Diffractometric techniques, advantages
  and disadvantages of, 67

comparison of, with photographic
  techniques, 67
Diffractometry, 66–82, 113–126,
  absorption corrections in, 69–72, 81
  counters for, 115–119
  geometrical arrangements for, 77–82
  monochromatization of x-rays in,
    119–126, 372
    by crystal monochromators, 122–126
    by pulse-height discrimination, 110,
      119, 122
    by Ross balanced filters, 110,
      120–122
  normal-beam-transmission technique
    in, 68–70
  pole-figure, reflection geometry for, 214,
    223–227
  transmission geometry for, 214–223
  preparation and mounting of specimens
    for, 77–82
  step-scanning procedures in, 114
  symmetrical-reflection technique in,
    77–82, 123
  symmetrical-transmission technique in,
    71, 72, 113, 123, 168, 372
  use of, in small-angle-scattering studies,
    113, 126
  use of balanced filters in, 113
  use of focusing geometry in, 79–82
  use of reference specimen in, 155, 156,
    169, 180
  use of strip-chart recorder in, 73, 227,
    229
  see also Counter diffractometers
Di-isotactic polymers, 359, 360
Dilute systems, 281, 282
Dislocations, lattice, 349, 450, 451
Disordered state, 1, 2, 142, 143
  "homogeneous", 143
Dispersant phase, 293, 295
Disperse phase, 293–295
Distance fluctuations in paracrystalline
  lattices, 429, 430, 433, 436
Distortion and size parameters, separation
  of, 427–452
  Buchanan and Miller's analysis of,
    437–452
  Fourier method for, 437, 438,
    447–452

integral-breadth method for,
438–440, 447–452
method of Hosemann for, 427–436
paracrystalline model for, 427–436,
447–451
Distortions, lattices, see Lattice distortions
Distortions (imperfections), of the first
kind, 141, 425–427
of the second kind (paracrystalline),
141, 425–427; see also Paracrystalline lattice distortions
see also Lattice distortions
Distribution functions, cylindrical-, 376,
381–384
of cellulose fibers, 382–384
electronic-, 370
orientation-, 259–262
Ditactic polymers, 358
Divergence of beam, see Beam divergence
Double-crystal spectrometer, 105–107
Double helix, of DNA, 401, 403
Drawing of polymers, 21, 26, 440
morphology of, 346–350
d-spacings, see Interplanar (d) spacings

Elastomers, diffractometric determination
of crystallinity in, 152–165
Electron concentration, 293
Electron density, distribution of, 455–458
in a paracrystal, 424
mean-square fluctuation of, 290, 291
of solute, from small-angle scattering,
304
Electronic charge, 552
Electronic rest mass, 552
Electrons, scattering of x-rays by, 29, 30
Electron units (of intensity), 31
Ellipsoid of revolution, as lysozyme model,
318
Elongation of elastomers, degree of, 156,
157
Enantiomorphism of α-helix, 373
Enstatite, 52
Equatorial (paratropic) reflections, 41
Equator of a fiber pattern, 73, 74
Equivalence postulate, 366, 367
Erythro-di-isotactic polymers, 359, 360
Evacuation of cameras, 85, 87, 94, 95, 103,
104, 109, 110, 382

Exposure time, in small-angle scattering,
112, 113, 285
with pinhole collimators, 102
with slit collimators, 104
of Franks cameras, 112
of Guinier cameras, 100
of microcameras, 96
with crystal monochromators, 99, 100
Extension of elastomers, affine, 156
dependence of crystallinity on, 158–160
device for, 156, 157
measurement of, with fiducial marks,
157, 158
Extension ratio, 156
Extinction bands, 19
Extrusion of polymer films, 26

Face, crystal, 10, 12
Faltung, see Convolution
Fiber diffraction patterns, 50–62,
cameras for, 85–89, 95–97
crystal-structure determination from,
406
cylindrical symmetry of, 54
effect of tilt of specimen on, 199–201
equator of, 51, 73
indexing of, 54–62
limitations on intensity measurements
from 406, 407
Lorentz factor for, 41
meridian of, 61, 73, 74
preparation of, 85–89, 95–99
radial and azimuthal coordinates of, 73,
205, 206
spherical averaging of, 140
of various polymers, 4, 53, 88, 91, 200,
201, 400, 410
Fiber orientation, see Axial (fiber)
orientation
Fibers, cellulose, 377–379
analysis of equatorial scattering of,
377–379
drawn, 25
microstructural distortions in, 433,
434
Fortisan rayon, 378, 379, 382–384
geometry of diffraction by, 74
jute, 378, 379
mounting of, 76–78, 87, 89

for orientation measurements, 264
radial analysis of, 377–379
ramie, 382–384
stiffness of, 138
tensile strength of, 138, 199
yield stress of, 138
Fibrils, 24
cellulose, diameter of, 377–379
distance between, 378, 379
clustering of, 432–434
diameter of, from small-angle scattering, 335, 430–432
ultra-, diameter of, 433, 434
Fiducial marks, use of, in measuring extension of elastomers, 157, 158
in calibrating Debye-Scherrer cameras, 93
Film, photographic, 82
Polaroid, 104
Films, polymer, 440
preparation of diffraction patterns of, 87
Filters, for beam attenuation, 285
for monochromatization, beta (β), 29
table of, 540
Ross balanced, 110, 120–122, 144
table of, 541
Flat-film, cameras, 85–87
diffraction technique, 47, 66, 85–87
geometry of, 73, 74
importance of, 67, 68
preparation of specimens for, 72, 73
use of, in recording fiber patterns, 58–60, 85–87
Fluorescent x-rays, 32, 33, 109
as a cause of background, 32
Fluorite crystals as monochromators, 98
Focal spot of x-ray tube, linear, 105, 218
Focusing cameras, 99–102
Focusing geometry for diffractometry, 79–82
Fold, see Convolution
Folding of molecular chains, 16, 21, 25, 342–350
Form birefringence, 269, 270
Form factor, atomic, 30
Fortisan rayon, 378–384
Fourier-Bessel, inversion theorem, 461
transforms, 378, 454–465

Fourier, coefficients, 442, 470
deconvolution, 442, 443, 467–472
integral theorem, 454, 456, 467
series, 469, 470
synthesis, 412–414
structural refinement by, 407
transforms, 87, 126, 127, 369, 437, 438, 454–467
one-dimensional, 454, 455
in separation of size and distortion parameters, 437, 438, 443, 444, 447–452
spherical, 458–460
three-dimensional, 456, 457
Four-point diagrams, small-angle, 24, 25, 338–340
Franks small-angle camera, 110–113
as modified, by A. Elliott, 112, 113
by G. F. Elliott and C. R. Worthington, 112, 113
Fraunhofer diffraction, 127
Frozen-in thermal motion, 149

Gauche configuration, 364
Gauche conformation, 364
Gaussian reflection profiles, 429, 434, 435, 439–443, 447, 450
Geiger-Müller counters, 115–118, 127, 159, 180, 181, 319
General radiation, see Continuous x-ray spectrum
Globular particles, small-angle scattering by, 300–304, 314–319
Goniometer head, 87, 88
Goniometers, four-circle, 233, 234
pole-figure (texture), 229–234
three-circle, 227
Goppel-Arlman camera, 151, 152
Guinier, law of, 298, 299
Guinier camera, 100, 433, 435
Guinier-de Wolff camera, 100, 101
Guinier plots, 299, 301, 302, 311, 312, 315, 324, 325, 430–434
Gutta percha, 14, 16
Gyration, radius of, see Radius of gyration

Halos, amorphous, see Amorphous halos
Hankel inversion theorem, 461

Harmonics, 29, 98
  spherical, in texture analysis, 259–262
  x-ray, exclusion of, by pulse-height
    discrimination, 119
  by reduction of x-ray voltage, 285
Heat treatment of polymers, see Annealing
  of polymers
Hedrites, 20, 23, 344
Helical conformation; 360–368, 376,
  388–405
Helical models, of polypeptide chains,
  373–375, 395–400
  optical transforms of, 391, 392
Helical molecules, conformational analysis
  of, by helical structure-factor
  calculations, 395–400, 463, 464
  from diffraction patterns, 388–400,
    461–464
  selection rule in, 388–394, 397, 462
  diffraction of x-rays by, 388–405,
    461–464
  effect of crystalline ordering on, 464
  layer-line intensities in, 390, 391
  meridional reflections in, 391
Helical net, 389
Helical point net, 360–364
Helical structure factors, 463, 464
  calculation of, for polypeptide models,
    395–400
Helical transforms, 388–391, 461–464
  ℓ, n net of, 389, 390
α-helix, 372–376, 393–400, 405
  ℓ, n net of, 394
  theoretical scattering curve of, 377
γ-helix, 395–400
π-helix, 395–400
Helix, 360–366
  axis of, 360
  continuous, 461, 462
  discontinuous, 391, 462
  double, of DNA, 401, 403
  identity period of, 361, 393, 401, 462
  motif of, 361, 388–391
  motif periodicity of, 391, 462
  pitch of, 360
  radial projection of, 389
  rotational and translational operations
    of, 360–363
Hemoglobin, 373

Hermans-Weidinger method for crystal-
  linity determination, 166–176
Hess-Kiessig model, 24, 25, 332
Hexagonal division, of hexagonal system,
  5
Hexagonal system, 5
  hexagonal division of, 5
  rhombohedral division of, 5
Hexamethylenediisocyanate, 409
Hexamethylenetetramine, 442
($hk\ell$) (Miller indices), 11
Holotactic polymers, 358
"Homogeneous" disorder, 143
Homometric structures, 372
Hosemann's theory of paracrystalline
  diffraction, 426–429
Hull-Davey charts, 55
Hydrogen bonds, 144, 367

Identity period, fiber, 409, 412, 417
  helical, 361, 393, 401, 462
  molecular, 361, 362, 365
Imperfection factor, determination of, for
  nylons 6 and 7, 149–151
  for polypropene, 144–149
  lattice-, 139–143
Imperfections (distortions), 4, 18, 406,
  408, 414,
  of the first kind, 141, 144–151, 425–427
  of the second kind (paracrystalline),
    141, 143, 149–151, 425–427,
    447–450; see also Paracrystalline
    lattice distortions
  see also Lattice distortions
Incoherent (modified) scattering of x-rays,
  29–32
  calculation of, for polypropene, 144–146
  as a cause of background, 30
  intensity of, 31
  phase of, 30
  wavelength of, 30
Independent scattering of x-rays, 31
Indexing, of fiber patterns, 54–62
  of low symmetry, 61, 62
  use of reciprocal lattice in, 56–62
  of orthorhombic fiber patterns, 54–62
  chart for, 54, 55
Index of refraction, 268–271
Indices, Miller, 10–12

Infrared absorption, relation of, to crystal-
　　line and amorphous fractions,
　　192, 193
　　to tacticity, 192, 193
　　of poly(ethylene terephthalate), 194
Inhomogeneity length, 296
　　reduced, 296, 297
Instrumental broadening, correction for,
　　429, 435, 437, 442, 443, 447,
　　448, 467–472
Instrumentation useful in polymer studies,
　　66–135
Insulin, 373
Intensity, in electron units, 31
　　of coherent scattering, 138, 139
　　of incoherent scattering, 31
　　of small-angle scattering, 63, 283–290,
　　342, 347, 348
　　　　absolute, 126, 285, 286, 328–331
　　　　relative, 285
Intensity, distribution, difference-, 384,
　　385
　　of primary beam, meaning of, 290
　　measurement of, 123, 126, 285, 286,
　　290, 328, 329
　　relation of, to absolute intensity
　　　　measurements, 123, 126, 285,
　　　　290
　　of a reflection ($hk\ell$), correction of, for
　　　　background, 214, 215, 229
　　　　ideal, 40
　　　　factors modifying, 39–43
　　　　integrated, 214
　　photographic measurement of, 66, 67,
　　　　82–84, 409, 410
　　transform, 381
　　of x-ray diffraction, diffractometric
　　　　measurement of, 66–82, 113–126
　　　　for pole figures, 209–234
　　normalization of, 169, 170, 179, 181,
　　　　222, 223, 227, 235–240
　　photographic measurement of, 66, 67,
　　　　82–84, 329, 409, 410
　　relation of, to thickness of specimen,
　　　　68–72, 77, 79
　　separation of crystalline and amor-
　　　　phous, 166, 170–174
Intercepts of a plane, 9–13
Interchain distance, determination of,

　　from amorphous halo, 379–381
　　in polybutadiene, 163
Interference patterns, optical, 18, 19
Internal-rotation angles, 364, 368
　　in poly(ethylene adipate), 416, 417
Internal solvation, 300
　　of lysozyme, 316, 317
　　ratio, 300
　　from small-angle scattering, 304
Interplanar (d) spacings, 11, 12, 35, 47,
　　104, 110, 112, 126, 319, 335,
　　380, 451
　　formulas for calculating, 542, 543
　　mean fluctuation of, 429, 433, 436
　　scales for measuring, 48, 49
Invariant, 291–293
Inverse pole figures, 259–261
Isotactic polymers, 15, 358–360
　　definition of, 358, 359

Johann-type monochromator, 123
Johansson-type monochromator, 100, 109,
　　110, 435
Jute, 378–380

K-absorption edge, 28, 120, 122
$K\alpha$ doublet, 26
　　correction for, 470
　　separation of, with crystal monochro-
　　　　mator, 123
Keratin, small-angle scattering by, 332,
　　334
α-keratin,
　　coiled-coil model of, 405
Kiessig small-angle camera, 104, 105
Kratky small-angle (U-bar) camera,
　　107–110
$K\alpha$ x-rays, 26, 29
$K\beta$ x-rays, 26, 29
kX unit, 529

L-absorption edges, 28
Lamellae, 16–21
Lamellar crystallization, 16–21, 25,
　　343–348
Lamellar particles, thickness of, from
　　　　small-angle scattering, 306, 307
Laminated specimens, 73–76

Lattice, Bravais, 5
centered, 5
direct, 33
primitive, 5
reciprocal, 33–35, 457
in relation to diffraction geometry, 38, 46, 52
space, 5, 6, 10, 11, 33
Lattice distortions, 4, 18, 406, 408, 414
analysis of, 423–452
with allowance for crystallite size, 429–452
dislocation theory of, 450, 451
in drawn fibers of polyethylene, 433, 434
of the first kind, 141, 425–427
in nylons 6 and 7, 149-151
in polypropene, 144-149
local, 437, 450
"maximum", 438–440, 447, 448
mean-square, 438
of the microstrain type, 437, 447, 450, 451
of the second kind (paracrystalline), 141, 143, 425-427, 447-450
as a cause of background, 424, 425, 429
distance fluctuations in, 429, 430, 433, 436
in nylons 6 and 7, 149-151
in polyethylene, 429-436
in single crystals of polyethylene, 435, 436
Lattice-imperfection factor, 139, 140, 143
Lattice imperfections, 4, 18, 406, 408, 414
Laue equations, 30, 51
Layer lines, intensity of, in helical diffraction, 390–393, 462
Least-squares structural refinement, 407, 408
Arnott and Wonacott's method of, 407, 408
Legendre polynomials, in analysis of cylindrically symmetrical systems, 381–384
in texture analysis, 259–262
Light, velocity of, 552
Lignin, 378
Linear absorption coefficient, 27, 68, 69

Linear polymers, randomly coiled, 307–310
mean-square end-to-end distance of, 307, 308
persistence length of, 307-313
small-angle-scattering by, 308–313
wormlike-chain model of, 307, 312
tacticity of, 15, 16
Liquids, small-angle scattering by, 110
Long-range order, 425, 426
Long spacings, effect of annealing on, 342, 343
in drawn fibers, 335
in natural fibers, 332, 334
paracrystalline fluctuations of, 430
in polyethylene, 335, 338
in single-crystal mats, 337
in small-angle scattering, 332–342
in synthetic fibers, 334–342
Lorentz factor, for crystalline powder, 41
for rotating crystal, 40, 41
in small-angle scattering, 283, 284
tables of, 545, 547
Lorenz-Lorentz equation, 272
Loschmidt's number, 552
Luzzati-Baro small-angle camera, 110, 111
Lysozyme, ellipsoidal model of, 318
Guinier plots of, 315
particulate parameters of, 317, 318
small-angle scattering study of, 314–319
validity of Porod's rule for, 295

Machine direction in oriented specimens, 74, 75, 208
Maltese cross, 19
Marlex 50, 95
Masks for optical diffractometer, preparation of, 130
Mass absorption coefficients, 27–29, 69
of a particle, from small-angle scattering, 302, 303
per unit length, from small-angle scattering, 305, 306
of cellulose fibrils, 326
table of, 524–528
Mellon Institute, 154
Meridian, of a fiber pattern, 61, 73, 154

Meridional (diatropic) reflections, 41
    conditions for production of, 61,
        199–202
    enhancement of, by tilting fiber axis,
        199–202
    in helical diffraction patterns, 391
    small-angle, 25, 332–342
    study of, with optical diffractometer,
        340, 341
Mica, as x-ray window material, 110
Microbeam techniques, 95
Microcameras, 95-97
    alignment of, 96, 97
    applications of, 95
    of Chesley design, 95–97
    collimators for, 95, 96
    preparation of fiber patterns with, 95, 96
Microcrystalline specimens, preferentially
        oriented, diffraction of x-rays by,
        50–62
    randomly oriented, diffraction of x-rays
        by, 45–50, 369–376
Microdensitomers, 82-85, 329, 377, 378,
        382
    components of, 83
    Dobson design, 83, 84
    Knorr-Albers, 184
    recording, 83, 85
    required features of, 83, 84
Microfocus x-ray tube, 96, 97, 110
Microphotometers, see Microdensito-
        meters
Microscopy, light-optical, 18
Microstrains, 437, 447, 450, 451
Microvoids in polymers, 21, 63
    as cause of diffuse small-angle scatter-
        ing, 63, 281, 327–332
    elongation of, by extension of fibers, 332
    evidence for, in synthetic fibers, 331
    presence of, demonstrated in cellulose,
        327–331
Miller indices, 10-12
Miller's tables of crystallographic data for
        polymers, 473–523
Molecular chains, arrangement of, in
        poly(ethylene adipate), 412, 415
    in poly(hexamethylene adipamide),
        7, 8
    orientation of, 246, 263, 264, 385, 386

    by cold work, 198,199
    structure of, 357–405
Molecular motion, sensitivity of NMR to,
        194
Molecular structure, analytical description
        of, 358–366
    determination of, by radial distribution
        analysis, 368–376
    from wide-angle diffraction, 368–405
    in randomly oriented systems,
        368–376
Molecular weight, from small-angle
        scattering, 302, 303
    of lysozyme, 316, 317
Molybdenum-target x-ray tube,
        spectrum of, 27
Molybdenum x-rays, β-filters for, 29
Monitoring of direct beam, 124, 126, 156,
        180, 181
Monochromatization, in optical diffracto-
        meter, 128
    of x-rays, by β-filters, 29, 85, 372, 384
    in diffractometry, 119-126
    by pulse-height discrimination, 110,
        119, 372, 440
    by reflection from crystals, 29, 40, 85,
        97–102, 109, 110, 113, 122–126,
        382, 433, 435, 440
    by Ross balanced filters, 110,
        120–122
    by total reflection, 112
Monochromators, see Crystal mono-
        chromators and Monochromati-
        zation, of x-rays
Monoclinic system, 5, 8, 9, 409
    reciprocal lattice of, 33, 34
Monodisperse systems, 298, 303
Monomer units (residues), 7, 15, 358,
        363, 366, 393, 400, 407–410,
        415, 425, 459
Monotactic polymers, 358, 359
Morphology, of crystallization of polymers,
        16–26, 342–350
    of deformation of polymers, 349
Mosaic single crystals, 436
Motif, 8, 366, 425, 457
    of a helix, 361–364, 388–391
Mounting of specimens, 68–81
Multiplicity factor, 41, 42

Muscle, 405
  small-angle study of, 113
Myoglobin, 373

Nickel, as a β-filter for copper x-rays, 29
  cobalt, as balanced filters for $CuK\alpha$
    x-rays, 120, 122, 144
Nicol prism, 19, 271
Niobium, as a β-filter for molybdenum
    x-rays, 29
NMR curve of poly(ethylene terephtha-
    late), 194
NMR determination of crystallinity,
    194–196
Noise, electronic, exclusion of, 119
Noncrystalline specimens, 1, 2
  radial-distribution analysis of, 43–45
  scattering of x-rays by, 43–45, 137, 138,
    143, 429
Normal-beam transmission technique,
    68–70
  absorption corrections for, 69, 70
Normal direction, in oriented specimens,
    75, 208
Normalization of intensities, in crystal-
    linity determination, 169, 170,
    179, 181
  in pole-figure diffractometry, 222, 223
    227, 235–240
Nuclear magnetic resonance (NMR),
    194–196
Nucleation, 17, 19
Nucleic acids, 400–403
  nucleotides in, 401, 403
Nucleotides, 401, 403
Numerical constants, table of, 552
Nylon 56, 20, 21, 95, 96
  microdiffraction pattern of, 98
Nylon 6, 144, 149, 335
Nylon 66, see Poly(hexamethylene
    adipamide)
Nylon 6·10, 338
Nylons 6 and 7, determination of degree
    of crystallinity in, 149–151
  determination of lattice-imperfection
    factor in, 149–151
Nylon 7, 144, 149–151
Nylon 77, 18, 19

Optical density, 83, 169
Optical diffraction, as analog of x-ray
    diffraction, 126–128
Optical diffractometer, 126–133, 464
  of Bonart and Hosemann, 130
  calibration of, 131, 132
  principles of, 127, 128
  of Taylor and Lipson, 128–133
    components of, 128, 129
    masks for, 128, 130
    optical system of, 128, 129
    resolution attainable with, 128
Optical transforms, 405, 464, 465
  as analogs of x-ray transforms, 126–128,
    464
  of benzene ring, 132, 133
  of helical molecular models, 391, 392
  of paracrystalline models, 425, 426
  two-dimensional limitation of, 465
Order, degree of, 142, 143
  long-range, in relation to lattice distor-
    tions, 425, 426
  one-dimensional, 142, 143
  three-dimensional, 140–143
  two-dimensional, 142, 143
Ordered state, 1, 2, 142, 143
Ordering, degree of, expressed as a
    crystallinity index, 176
  of helices, crystalline, 401
Orientation,
  preferred, see Preferred orientation and
    Textures in polymers
  random, 19, 140, 368, 369
  randomness of, in bulk and molded
    specimens, 45
  in spherulites, 19
  see also Textures in polymers
Orientation functions, 241–243, 270
  azimuthal, 254, 255
  of biaxially-oriented polyethylene, 275,
    276
  description of axial orientation with,
    241–243
  description of biaxial orientation with,
    254–256
  experimental evaluation of, 243–247
  least-squares calculation of, 259
  relationship of, to $<\cos^2 \phi>$ values,
    241–243

statistical accuracy of, 259
Orientation index, 262, 264, 267
Oriented specimens, 73–77
   azimuthal holders for, 73, 75, 76, 154,
      157
   diffraction by, 50–62, 119–202
   machine, normal, and transverse direc-
      tions in, 74, 75, 208
Orthogonality relationship in texture
      analysis, 241, 247–250, 255, 256
Orthorhombic system, 5
   indexing fiber patterns in, 54–62

Pantograph punch for preparation of
      optical-diffractometer masks,
      130, 131
Paracrystalline lattice distortions, 141,
      143, 425–427, 447–450
   as a cause of background, 424, 425, 429
   in nylons 6 and 7, 149–151
   in polyethylene, 429–436
Paracrystalline lattices, distance fluctua-
      tions in, 429, 430, 433, 436
   two-dimensional, 425
Paracrystals, concept of, 424, 425
   diffraction by, small-angle, 429–433
     wide-angle, 433–436
   electron-density distribution in, 424
   Hosemann's theory of diffraction by,
      426–429
   unit-cell dimensions of, 424
Parafocusing cameras, 99–102
Paramyosin, 405
Parasitic scattering, 107
   exclusion of, 107–109
Paratropic (equatorial) reflections, 41
Partial specific volume, 303
   electronic, 293
Particulate scattering, 281, 282
   factors favoring, 281, 282
Particulate systems, dilute, small-angle
      scattering by, 297–307
Pass band, spectral, of balanced filters, 122
Patterson function, 458
   cylindrical, 460
Pentaerythritol, use of, as crystal mono-
      chromator, 97, 124
Periodicity, axial, of a helix, 361, 393, 401,
      462

Perlon L, 335
Persistence length, 307–313
   from small-angle scattering, 309–312
   of cellulose nitrate, 310–313
Perspex, use of, as reference specimen, 169
Phase, of structure factor, 40
Phenol-formaldehyde resins, 15
Phosphate chains, in sodium meta-phos-
      phate fibers, 385, 386
Photographic diffraction techniques,
      82–113
   advantages and disadvantages of, 67
   comparison of, with diffractometry, 67
   measurement of intensity with, 409, 410
   use of, in determination of crystallinity,
      184–187
Photographic film, characteristics of, 82
   choice of, 82, 184
   Polaroid, 104
Photographs, x-ray diffraction, densitome-
      try of, 82–84
Photomultiplier tube, 118, 119
Photons, x-ray, 67, 117
Physical constants, table of, 552
Pinhole collimation in small-angle scatter-
      ing, 102–104, 286, 287
Pitch, of a helix, 360
Planar orientation, 208–211
Planar zigzag conformation, 14, 358–360,
      368–371, 410–413
Plan-axial orientation, 208–211
Planck's constant, 31, 552
Planes, crystallographic, 9–13, 35
   axial intercepts of, 9–13
   interplanar spacing of, 11, 12
   Miller indices of, 10–12
   poles of, 202–215
Polanyi's equation, 262, 263
Polarization factor, 40
   for crystal-monochromatized x-rays, 40
   tables of, 544–547
Polaroid film cassette, 104, 105, 132, 133
Polaroid polarizers, 19, 271
Polar stereographic net, 204, 205
Pole figures, diffractometric measurement
      of intensities for, 209–234
   by reflection, 214, 223–227
   by transmission, 214, 215–223
   illustrative preparation of (polypro-

pene), 234–240
importance of integrated intensities for, 214
inverse, 259–261
latitude and longitude coordinates of, 203, 208
normalization of intensities for, 222, 223, 227
of polyethylene film, 209, 212, 213, 253
of polypropene film, 240
preparation of, 209–240
representation of orientation by, 202–209
Pole-figure (texture) goniometers, 229–234
automatic, 230–232
absorption corrections for, 232
use of X-Y recorder with, 232
Poles of crystallographic planes, 202–215
Polyacrylonitrile fibers,
small-angle scattering of, 281, 282
α-poly-L-alanine, 408
diffraction patterns of, 399, 400
helical analysis of, 395, 400
models of, 395–400
β-poly-L-alanine, 53
Polyallene, 101, 102
Polyamides, 144, 367
long spacings in fibers of, 335
Poly-γ-benzyl-L-glutamate, 375, 377
Polybutadiene, amorphous, intermolecular distance in, 163
amorphous orientation of, 161, 386
cis 1, 4-, 8
crystallization of, 165
determination of crystallinity in, 160–166
diffraction patterns of, 162
isotactic, 366
molecular orientation in, 161, 386
pseudohexagonal phase of, 161
variation of amorphous halo of, with extension, 162–164
Poly-l-butene, isotactic, 366
Poly(butyl vinyl ether), isotactic, 366
Polycaprolactam (nylon 6), 144, 149, 335
Polydispersity, 282, 287, 292, 313, 314
ambiguities in analysis of, 313, 314
Polyesters, 408, 412

long spacings in fibers of, 335
Polyethylene, 14, 166, 338
axial orientation in films of, 267
banded spherulites of, 346–348
biaxially oriented, ternary diagram of, 275
calculation of crystalline birefringence of, 272, 275, 276
comparison of x-ray, infrared, and NMR crystallinities of, 193–195
crystallite size in, 433–436
determination of amorphous birefringence of, 272–277
effect of annealing on single crystals of, 342–344
fiber patterns of, 201
four-point diagrams of, 339
Guinier plots of, 431–434
lattice distortions in, 429–436
low-molecular-weight linear, 95
microstructural distortions in, 434–436
orientation functions of, 273–276
paracrystalline lattice distortions in, 429–436
pole figures of, 209, 212, 213, 253
precession patterns of, 87, 88
single-crystal mats of, 337, 342
single crystals of, 16, 17, 344
mosaic, 436
small-angle scattering by, 24, 336–339, 430
strained, axial textures in, 262
structural comparison of, with poly-(ethylene adipate), 412, 415, 417, 419
ternary orientation diagram of, 275
variation of birefringence of, with elongation, 273, 274
Poly(ethylene adipate), arrangement of molecular chains in, 412, 415, 417, 418
atomic coordinates of, 415
determination of crystal structure of, 408–419
fiber patterns of, 409, 410
molecular configuration of, 416
molecular conformation of, 410, 412, 416–418

observed and calculated structure
factors of, 410, 411
β-phase of, 409
structural comparison of, with polyethylene, 412, 415, 417, 419
unit cell of, 412, 413
Polyethylene films, measurement of, with
texture goniometer, 230
unidirectionally crystallized, pole figures
of, 252–253
Poly(ethylene oxide), 20, 23, 400
Poly(ethylene terephthalate), 26, 53, 408,
414
comparison of x-ray and NMR crystallinities of, 195, 196
comparison of x-ray density, and infrared crystallinities of, 190–192
crystalline and amorphous reference
specimens of, 185
densities of unoriented and oriented,
192, 193
determination of crystallinity index of,
184–188
infrared absorption of, 194
intensity curves of, 184, 185
NMR curve of, 195
regression curve of, 184–186
small-angle scattering by, 24
study of texture in, 266–269
unit cell of, 266
Polyethylidene, 359
Poly-ω-heptanolactam (nylon 7), 144,
149–151
Poly(hexamethylene adipamide) (nylon
66), 7, 69, 338, 347
axial orientation in, 267, 268
calculation of absorption coefficient of,
68, 69
radial distribution analysis of, 370–372
small-angle scattering by, 333
Polyisobutene, 4, 14, 368
helical analysis of, 393–397
Polyisoprene, 31
scattering curve of, 32
Polymerization, stereospecific, 14, 15
Poly-γ-methyl-L-glutamate, helical
analysis of, 392, 393, 395
Poly-4-methylpentene-1 fiber, analysis of
texture in, 250–252

Poly-m-methylstyrene, 400
Poly(N, N-dibutyl acrylamide), isotactic,
366
Polyoxymethylene, helical conformation
of, 362, 363
Poly-1-pentene, isotactic, 366
Poly-2-pentene, 360
Polypeptide chains, helical models of,
373–375, 395–400
Polypeptides, 372, 373
synthetic, 375–377, 392–400
Polypropene, analysis of texture in film of,
247–249
atactic, 144
calculation of incoherent and coherent
scattering of, 144–147
c-axis orientation in film of, 256–259
determination of degree of crystallinity
in, 144–149
determination of imperfection factor of,
144–149
illustrative preparation of (040) pole
figure of, 234–240
isotactic, 144, 166, 366
scattering curve of, 146, 147
Poly(propylene oxide), 20, 95
Polystyrene (isotactic), 166, 366, 367
analysis of crystallite size and lattice
distortions in, 440–451
determination of crystallinity in,
168–176
intensity curves of, 170, 171, 440, 441
powder-diffraction pattern of, 45, 50
preparation of amorphous specimen of,
168
preparation of specimens of, 440
reflection breadths of, 443
Polytetrafluoroethylene, 345, 346
crystal-monochromatized pattern of, 99
fiber patterns of, 89, 91
helical conformation of, 368, 369
Polytetramethyl-p-silphenylenesiloxane,
347
Poly(vinyl alcohol), 14, 15, 95
Poly(vinyl chloride), 12, 14, 55, 366
indexing of fiber pattern of, 55, 56, 59,
61
Poly(vinylidine chloride), 14
Porod's rule, 293–295

Positional coordinates of an atom, 39
Potential energy, relation of, to conformation, 367
Powder-diffraction, data cards (ASTM), x-ray, 48, 50
    patterns, 45–50
    d-value scales for measuring, 48, 49
    of polystyrene, 45, 50
    of representative polymers, 94, 95
    use of, in analysis, 47–50
Precession camera (Buerger), 87
Precession photographs, 88
Precision film position, 91, 92
Preferred orientation, 21, 26, 50–62, 198–279,
    of amorphous poly(ethylene terephthalate), 190–192
    of amorphous regions, 190, 191, 198, 385, 386
    analysis of, 198–277
        Milberg's method for, 262, 384–386
        number of crystallographic planes required for, 247, 248
        orthogonality relationships in, 241, 247–250, 255, 256
        by pole figures, 202
        Roe and Krigbaum's method for, 259–262
        Sack's method for, 259
        Stein's method for, 241–243, 254, 255
        using difference-distribution functions, 384–386
        Wilchinsky's method for, 245–252, 255–259
    analytical description of, 241–268
    axial, see Axial (fiber) orientation
    biaxial, see Biaxial orientation
    classification of modes of, 206–211
    complex modes of, 244, 252–259, 274
    axial component of, 244
    of crystalline regions, 198, 199
    definition of, 198
    degree of, 199–202
    effect of, on crystallinity determination, 174–176
    induction of, by cold work, 198
    measurement of, by optical birefringence, 198
    of molecular chains, 198, 199, 246

    planar, 208–211
    plan-axial, 208–211
    randomization of, 140
    relation of, to azimuthal reflection breadth, 262–268
    representation of, 202–209
        by ternary diagrams, 241, 243, 256–258
    semiquantitative measurement of, 262–268
    spiral, in cellulose fibers, 264–267
    uniplanar, 208–211
    uniplanar-axial, 208–211
    see also Textures in polymers
Primitive cell, 5
Primitive lattice, 5
Profiles of reflections, see Reflection profiles
Proportional counters, 116–120, 144
Proteins, helical characterization of, 372–376
    radial-distribution analysis of, 372–376
    use of Luzzati-Baro camera in study of, 110
Pulse-height discrimination, 110, 119, 372, 440
    window width in, 119

Quadrant reflections, small-angle, 24, 25, 338–340
Quantum-counting efficiency of an x-ray counter, 117–119
Quartz, use of, as crystal monochromator, 110, 113, 123, 126, 285
    in double-crystal spectrometer, 106

Radial-distribution, analysis, 43–45, 458–460
    of noncrystalline specimens, 43–45
    of poly(hexamethylene adipamide), 370–372
    of proteins, 372–377
    function, electronic, 370
Radial projection, of a helix, 389
Radius of gyration,
    apparent, 302
    for bodies of particular shapes, 300, 301
    of cellulose nitrate, 310–313

cross-sectional, of air-swollen cellulose, 324
of cross section of a rod, 305, 306, 324
electronic, 298
of lysozyme, 315, 317
of randomly coiled linear molecules, 308
from small-angle scattering, 299–302, 305, 306
of a sphere, 301
Ramie, 382–384
orientation in fibers, of, 264, 265
small-angle scattering by, 328
Random-flight model, 307
Randomly oriented systems,
radial-distribution analysis of, 43–45, 368–376, 458–460
x-ray diffraction by, 45–50, 94, 95, 101, 102, 369–376, 458–460
Rat-tail tendon, small-angle study of, 112
Rayon, Fortisan, 377–384
scattering power of, 329–331
small-angle scattering by, 329–331
Receiving-aperture, height, choice of, 221, 224
width, choice of, 218–221, 225, 226
effect of, on reflection profile, 217, 224, 225
Reciprocal lattice, 33–35, 457
coordinates of, 56–62
axial and radial, 455
Bernal charts for determining, 59–60
levels of, 51, 55, 56
of monoclinic space lattice, 33, 34
nets of, 51
point ($hk\ell$) of, 35, 37, 52
in relation to diffraction geometry, 37, 38, 46, 52
unit cell of, 33, 34
use of, in indexing fiber patterns, 56–62
vector ρ_{hkl} of, 35, 37, 38, 138, 455
Reciprocal space, 33, 454
Reciprocal unit cell, 33, 34
Reciprocity principle of Babinet, 282, 327, 328
Recoil factor, Breit-Dirac, 31, 144, 145
Recorder, strip-chart, 73, 227, 229
X-Y, in pole-figure diffractometry, 232, 233

Reference specimen, for absolute intensity measurements, 286, 328, 329
aluminum foil as, 156
gold or silver sol as, 286, 329
nickel foil as, 180
Perspex plate as, 169
polyethylene as, 286
poly(methyl methacrylate) as, 329
scattering power of, in small-angle scattering, 286
use of, in correction for instrumental broadening, 442, 470, 471
in crystallinity determinations, 156, 169, 180
see also Calibrating substances
Refinement of crystal structures of polymers, 407, 408, 412–414
Reflection, sphere of, 37, 39, 46, 51
Reflection (hkl), experimental factors modifying intensity of, 40–43
Reflection analogy of x-ray diffraction, 36, 37
Reflection profiles, analog computer for analysis of, 442
azimuthal breadths of, 263–268
Cauchy, 218, 439–443, 447, 450
components of, 217–224
convolution synthesis of, 217, 218, 224
correction of, for instrumental broadening, 429, 435, 437, 442, 443, 447, 448, 467–472
Gaussian, 217, 218, 429, 434, 435, 439–443, 447, 450
paracrystalline broadening of, 429
pure, 423, 424, 435, 439, 442, 467–470
radial breadths of, 217–221, 224, 423, 424, 429, 435, 436
integral, 435, 438–447, 450
size and distortion broadening of, 423, 424, 427, 429, 437, 439, 440, 467, 470, 471
Reflections, diffractometric measurement of, 209, 214
equatorial (paratropic), 41
indexing of, 54–62
meridional (diatropic), 41
photographic measurement of, 66, 67, 82–84, 409, 410
resolution of, 216–221, 224, 225, 442

see also 'Reflection profiles
Reflection technique, in pole-figure
    diffractometry, 214, 223–227
symmetrical-, *see* Symmetrical-reflec-
    tion technique
Refractive indices, 268–271
of the *n*-paraffin $C_{36}H_{74}$, 272
principal, 271
Registration plane, 283–285
Regression curve, in crystallinity determi-
    nation, 165–176
of cotton cellulose, 178
in crystallinity-index determination,
    177–180
of isotactic polystyrene, 172, 173
least-squares, 180
    correlation coefficient of, 180
of poly(ethylene terephthalate), 186
Regular film position, 91, 92
Reliability index, 408, 414, 415
Resolution, of a crystal-structure solution,
    406
of overlapping reflections, 442
of reflections, factors affecting, 216–221,
    224, 225
of small-angle cameras, 104–107
Resolving time of an x-ray counter, 117
Resolving-time losses, correction for, 159
Rhombohedral division of hexagonal
    system, 5
Ribonuclease, 373
Ribonucleic acid (RNA), 401, 403
Rock salt, use of, as crystal monochro-
    mator, 98, 124, 370
Rodlike particles, cross-sectional area of,
    306
cross-sectional radius of gyration of,
    305, 306
mass per unit length of, 305, 306
small-angle scattering by, 319, 326
Rolling of polymers, 21, 26
Ross filters, *see* Balanced filters (Ross)
Rotating-anode x-ray tubes, 105, 328, 372
Rotating-crystal diffraction technique, 51,
    52
Rotation photographs, 51, 52
Rubber, natural, 14
determination of crystallinity in, 151,
    158–161

Sample, *see* Specimens and Standard
    sample
Scattering curve, of cellulose (cotton),
    177, 181–183
of polyisoprene, 32
of polypropene, 146, 147
Scattering-equivalent systems, 281, 282,
    296, 372
Scattering factors, atomic, 30, 40, 44,
    457–459, 463
mean-square, 139, 145
table of, 533–539
electronic, 31, 298, 459
radial, of a cylinder, 461
Scattering mass, correction for, in crystal-
    linity determination, 151, 169
Scattering power, 290, 291, 348
concept of, 290, 291
of cellulose, quantitative study of,
    328–331
of generalized systems, 290, 291
of reference specimens in small-angle
    scattering, 286
Scattering of x-rays, 29–33, by air, 85, 159
by amorphous specimens, 43–45, 137,
    138, 143, 429
coherent (unmodified), 29–32, 138, 139
    by polypropene, 144–146
Compton, 30
crystalline, 137–139, 143
by cylindrically symmetrical systems,
    376–386, 460, 461
diffuse, 137, 138
    due to atomic thermal vibrations, 139
    due to lattice imperfections, 139
by electrons, 29, 30
by helical molecules, 388–405, 461–464
incoherent (modified), 29–32
    by polypropene, 144, 145
independent, 31
by noncrystalline specimens, 43–45,
    137, 138, 143, 429
parasitic, 107, 109
by randomly oriented systems, 369–376
separation of crystalline and amorphous,
    166, 170–172
small-angle, 33, 62, 63, 102–113,
    280–356; *see also* Small-angle
    scattering

see also Diffraction of x-rays
Scherrer equation, 335, 338, 423, 424
Scintillation counters, 116–121, 440
Screw axis, 360–364
    nomenclature, 360–364
Selection rule in helical analysis, 388–394,
        397, 462
Selective uniaxial orientation, 210, 211
Serum albumin, 373
Shape (anisotropy) factor, 301
    of lysozyme, 319
Silicone elastomer, molecular orientation
        in, 386
Silk, 53
    long spacings in, 335
Single-crystal orienter, 227
Single crystals of polymers, 16
    effect of annealing on, 342–344
    mats of, 337, 342, 435
    microstructural distortions in, 435, 436
    mosaic, 436
    of polyethylene, 16, 17, 342–344
Size and distortion parameters, *see* Dis-
        tortion and size parameters
Slit collimation in small-angle scattering,
        104–110, 286–290
    corrections for, 287–290
    smearing of intensity curve by, 287
Small-angle cameras, *see* Cameras, small-
        angle
Small-angle diffraction of x-rays, *see*
        Small-angle scattering
Small-angle reflections, meridional,
        332–342
Small-angle scattering, 33, 62, 63,
        102–113, 280–356
    by air-swollen cellulose, 319–326
    alternation of crystalline and amorphous
        regions as cause of, 63
    angular nomenclature of, 283
    by cellulose, 327–331
    by cellulose nitrate, 310–312
    by collagen, 332, 334
    comparison of, by dilute and dense
        systems, 282
    correction of collimation errors in,
        287–290
    cross-sectional area of rods from, 306
    by dense systems, 327–350

diffractometric measurement of, 113,
        126
diffuse, 63, 281–332, 429–433
    ambiguities in interpretation of, 281,
        282, 298
    clusters of fibrils as cause of, 432, 433
    microvoids as cause of, 63, 281,
        327–332, 433
    by dilute particulate systems, 297–307
    discrete, 63, 281, 332–348
    disperse and dispersant phases in, 293,
        295
    electron density of solute from, 304
    experimental requirements for, 63
    by fibers, 24
    by generalized systems, 290, 291
    geometry of, 284
    by globular particles, 300–304, 314–319
    inhomogeneities of colloidal dimensions
        as cause of, 62, 281
    instrumentation for, 102–113
    intensity curve of, smearing of, by slits,
        287, 288
        unsmearing of, 287, 288
    intensity of, 63, 283–290
        absolute, 285
        factors affecting, 283, 284
    internal solvation from, 300, 304
    interparticle interferences in, 281, 282,
        298
    by keratin, 332, 334
    by lamellar particles, 306, 307
    by linear polymers in solution, 308–313
    Lorentz factor in, 283, 284
    by lysozyme, 314–319
    mass of particle from, 302, 303
    mass per unit length from, 305, 306
    molecular weight from, 302, 303
    by paracrystalline structures, 425–436
    particulate, 281, 282
        factors favoring, 281, 282
    pinhole collimation in, 102–104,
        286–290
    radius of gyration from, 299–302, 305,
        306
    by ramie, 328
    by randomly coiled linear molecules,
        308–313
    by rodlike particles, 305, 306, 319–326

scattering equivalence in, 281, 282
slit collimation in, 104–110, 286–290
symbols used in, 350–352
thickness of lamellar particles from, 306, 307
by two-phase systems, 293–312
use of, in studying biological specimens, 112, 113, 315–319
Smearing of small-angle scattering curve, 287, 288
Sodium iodide, use of, in scintillation counters, 118
Sodium metaphosphate fibers, difference-distribution function of, 388
molecular conformation in, 385, 386
molecular orientation in, 385, 386
phosphate chains in, 385, 386
Sodium metasilicate, 386
Soller slits, 219, 220
Solvation, internal, 300
Solvation ratio, internal-, 300
Space group, 409
Space lattice, 5, 6, 10, 11, 33
Specific surface, of air-swollen cellulose, 325
inner, 294–296
of lysozyme, 316, 317
Specific volume, effect of crystallization on, 189
partial, 303
partial electronic, 293
Specimen holders, azimuthal, 73, 75, 76, 154, 157
Specimens, azimuthal holders for, 73, 75, 76, 154, 157
fiber, preparation and mounting of, 76–78, 87, 89, 264, 328
film, preparation of, 440
laminated, preparation and mounting of, 73–75, 227, 228
liquid, handling of, 110
mounting of, 68–81, 87, 88
oriented, reference directions in, 208, 214
preparation and mounting of, 68–82
for crystallinity determination, 168, 180
for Debye-Scherrer cameras, 72
for diffractometry, 77–82

for flat-film cameras, 72, 73
for Guinier-de Wolff camera, 101, 102
for pole-figure diffractometry, 227, 228, 234, 244
randomly oriented, for normalization of intensities, 222, 223, 227
preparation and mounting of, 72, 73, 223
reference, 155, 156, 169, 180, 286
relation of thickness of, to intensity, 68–72, 77, 79
rod-shaped, 228, 244
rolled-film, 228, 244
temperature control of, 103, 104, 110, 161, 162
thickness of, in relation to geometrical broadening, 69, 79, 80, 216, 224
in relation to intensity, 216, 224
Specimen-to-film distance, determination of, 87
maintenance of, with a gauge block, 87
Spectral dispersion, effect of, on reflection profile, 217, 225
Spectrometer, double-crystal, 105–107
Spectrum (x-ray), 26–29
characteristics, 26
continuous, 26
as cause of background, 29
of copper-target tube, 121
of molybdenum-target tube, 27
Sphere of reflection, 37, 39, 46, 51
Spherical harmonics, in texture analysis, 259–262
Spherical transforms, 369–376, 458–460
Spherically symmetrical systems, transform analysis of, 458–460
Spherulites, 17–20, 344–346
banded, 19, 346–348
Spherulitic crystallization, 17–19
Staggered bonds, principle of, 365–367
Standard sample, use of, in absolute intensity measurements, 286, 328, 329
Statton camera, 87, 102, 103
Stereographic projection, 204–208
meridional (Wulff), 204, 207
of polyethylene, 261
polar, 204–206

Stereoregularity of polymer chains, 12–16, 357–360
  atactic, 16, 358
  di-isotactic, 359, 360
  ditactic, 358
  erythro-di-isotactic, 359, 360
  holotactic, 358
  isotactic, 15, 358–360
  monotactic, 358
  nomenclature of, 358–360
  syndiotactic, 15, 359, 365, 366
  threo-di-isotactic, 359, 360
Stereospecific polymerization, 14, 15
Stiffness of fibers, in relation to crystal-linity, 138
Stokes' method of analyzing reflection profiles, 442, 443, 467–472
Strip-chart recorder, 73, 227, 229
Structure factors, 39, 40, 457
  calculated, 407
  helical, 395–400, 463, 464
  modulus of, 40
  observed, 407
  of poly(ethylene adipate), 410, 411
  real and imaginary terms of, 40
Structure models, testing of, with optical transforms, 464, 465
  trial, 407
Surface, specific inner, 294–296, 316, 317, 325
Swelling factor, of air-swollen cellulose, 323
  of a solute particle, 300, 319, 322
Symbols used in small-angle scattering, 350–352
Symmetrical-reflection technique, 72–82, 144, 440
  absorption corrections for, 81
  advantages and disadvantages of, 81, 82
  broadening of focus in, 79, 80
  optimum specimen thickness for, 80
  optimum width of receiving slit for, 80
  suitable 2θ-range for, 81
Symmetrical-transmission technique, 70–72, 168, 372, 440
  absorption corrections for, 71, 72
  advantages and disadvantages of, 82
  suitable 2θ-range for, 82
  usefulness of, in diffractometry, 70, 71

Symmetry, crystallographic, in relation to texture analysis, 247
Syndiotactic polymers, 15, 359, 365, 366
  definition of, 359
Systems, crystal, 5
  unique directions in, 245

Tacticity of linear polymers, 12–16, 358–360
  definition of, 358
Take-off angle of x-ray tube target, 105
Target of x-ray tube, 26, 40
Taylor-Lipson optical diffractometer, 128–133
Temperature factor, 42, 43
  anisotropic, 43
  anomalously large, of polymers, 406
  isotropic, 42, 43, 373, 414
Temperature of specimen, control and measurement of, 103, 104, 110, 161, 162
Tensile strength, relation of, to crystal-linity, 138
  to molecular orientation, 199
Ternary diagrams, for representation of orientation, 241, 243, 256–258
  of polyethylene, 275
Tetragonal system, 5
Textile Research Institute (Princeton), 124
Texture goniometers, 229-234
Textures in polymers, 21, 24-26, 204
  analysis of,
    in a monoclinic polymer, 247-249
    in poly(ethylene terephthalate), 227, 228
    in poly-4-methylpentene-1, 250-252
    in polypropene film, 247-249, 277
    relation of crystallographic sym-metry to, 247
    in a tetragonal polymer, 250-252
    by x-rays combined with other tech-niques, 277
  biaxial, 252, 245, 255, 270, 271
  complex, 244, 252–259, 274
  fiber, see Axial (fiber) orientation
  see also Preferred orientation
Thermal diffuse scatter, 139
Thermal motion, frozen-in, 149

Thermal vibrations, 139, 141, 414, 425
  in polypropene, 149
Thickness of lamellar particles from small-
      angle scattering, 306, 307
Thickness of specimen, effect of, on re-
      flection profiles, 217, 224, 225
  optimum, 68, 72
    in pole-figure diffractometry, 216, 224
  relation of, to geometrical broadening,
      69, 79, 80, 216, 224
    to intensity, 68–72, 77, 79, 216, 224
Threo-di-isotactic polymers, 359, 360
Tobacco mosaic virus, diffraction pattern
      of, 402, 404
  helical structure of, 401–404
Toroidal mirror, use of, in collimation of
      x-rays, 112
trans configuration, 15
trans conformation, 364
Transform, of difference-intensity distri-
      bution, 384, 385
Transforms, continuous, 457
  cylindrical, 376–386, 460, 461
  discontinuous, 457
  Fourier, 87, 126, 127, 369, 437, 438,
      447–467
    in separation of size and distortion
        parameters, 437, 438, 447–452
  Fourier-Bessel, 378, 454–465
  helical, 388–391, 461–464
    ℓ, n net of, 389, 390
  intensity, 381, 384
  optical, 405, 464, 465
    as analogs of x-ray transforms,
        126–128, 464
    of helical models, 391, 392
    of paracrystalline models, 425, 426
    two-dimensional limitation of, 465
  spherical, 369–376, 458–460
Transmission factor, 42
Transmission technique, geometry of,
      68–72
  normal-beam-, 68–70
    absorption corrections for, 69, 70
    in pole-figure diffractometry, 214,
        215–223
  symmetrical-, 70–72, 168, 372, 440
    absorption corrections for, 71, 72
    advantages and disadvantages of, 82

suitable 2Θ range for, 82
  usefulness of, in diffractometry, 70,
      71
Transversal length, 296, 297
Transverse direction in oriented speci-
      mens, 74, 208
Trial-and-error refinement of crystal
      structures, 407, 408, 412, 414
Triclinic system, 5, 7
Tube, diffraction, see X-ray tube
Tube, x-ray, see X-ray tube
Two-phase concept, 3, 137, 149, 269
Two-phase systems, scattering power of,
      293
  small-angle scattering by, 293–312

Uniaxial orientation, 210, 211, see also
      Axial (fiber) orientation
Uniplanar-axial orientation, 208, 211
Uniplanar orientation, 208–211
Unique crystallographic directions, 245
Unit cell, 1, 5–8, 39, 142, 437, 457
  centered, 5, 6
  determination of, from a fiber pattern,
      54–62
  monoclinic, of poly(ethylene adipate),
      409
  of a paracrystal, 424
  primitive, 5
  reciprocal, 33, 34
Unit-cell constants, 5
Unsmearing of small-angle-scattering
      curve, 287–290
Urea nitrate, use of, as crystal monochro-
      mator, 97

Van der Waals forces, 144, 367
Van der Waals radii, atomic, 367, 418, 419
Vector space, 454, 458
Vinyl polymers, 365, 366
Voids in polymers, see Microvoids, in
      polymers
Volume of a particle, from small-angle
      scattering, 304
Volume and weight fractions of crystal-
      line polymer, relation between,
      190

Wavelength limit of continuous spectrum, 98
Wavelengths, x-ray, table of, 529–532
Weight and volume fractions of crystalline polymer, relation between, 190
Wide-angle diffraction of x-rays, 33–62
cameras for, 85–102
crystal structure from, 405–419
microstructure from, 357–422
molecular structure from, 368–405
see also Scattering of x-rays
Window width, of pulse-height discriminator, 119
Wormlike model of polymer chain, 307, 312
Wulff stereographic net, 204, 207

X-ray beam, primary, measurement of intensity of, 123, 126, 285, 286, 290, 328, 329
monitoring of, 124, 126, 156, 180, 181
X-ray powder-diffraction data cards (ASTM), 48, 50
X-rays, absorption of, 26–29
characteristic, 26
continuous, 26
diffraction of, 33–63; see also Diffraction of x-rays
fluorescent, 32, 33, 109
as a cause of background, 32
interaction of, with matter, 26–33
$K\alpha$, 26, 29

$K\beta$, 26, 29
properties of, 26–29
see also Absorption of x-rays
scattering of, 29–33; see also Scattering of x-rays
small-angle scattering of, 33, 62, 63, 102–113, 280–356; see also Small-angle scattering
white, 26
X-ray source, effect of, on reflection profiles, 217, 225
linear, 218
X-ray spectrum, 26–29
characteristic, 26
continuous, 26
as cause of background, 29
of copper-target tube, 121
X-ray tube, microfocus, 96, 97, 110
rotating-anode, 105, 328, 372
target of, 26, 40
X-ray wavelengths, table of, 529–532
X-Y recorder, in pole-figure diffractometry, 232, 233

Yield stress of fibers, relation of, to crystallinity, 138

Zigzag conformation, planar, 14, 358–360, 368–371, 410–413
Zirconium, as a β-filter for molybdenum radiation, 29
x-ray absorption of, 28